Interaction of Mechanics and Mathematics

For further volumes:
www.springer.com/series/5395

Victor L. Berdichevsky

Variational Principles of Continuum Mechanics

I. Fundamentals

With 79 Figures

Author

V.L. Berdichevsky
Professor of Mechanics
Department of Mechanical Engineering
Wayne State University
Detroit, MI 48202
USA
vberd@eng.wayne.edu

ISSN 1860-6245 e-ISSN 1860-6253
ISBN 978-3-540-88466-8 e-ISBN 978-3-540-88467-5
DOI 10.1007/978-3-540-88467-5
Springer Heidelberg Dordrecht London New York

Library of Congress Control Number: 2008942378

Cover design: WMXDesign GmbH, Heidelberg

Printed on acid-free paper

Springer is part of Springer Science+Business Media (www.springer.com)

Our mind is frail as our senses are; it would lose itself in the complexity of the world if that complexity were not harmonious; like the short-sighted, it would only see details, and would be obliged to forget each of these details before examining the next because it would be incapable of taking in the whole. The only facts worthy of our attention are those which introduce order into this complexity and so make it accessible to us.

H. Poincaré, Science and Method

Preface

There are about 500 books on variational principles. They are concerned mostly with the mathematical aspects of the topic. The major goal of this book is to discuss the physical origin of the variational principles and the intrinsic interrelations between them. For example, the Gibbs principles appear not as the first principles of the theory of thermodynamic equilibrium but as a consequence of the Einstein formula for thermodynamic fluctuations. The mathematical issues are considered as long as they shed light on the physical outcomes and/or provide a useful technique for direct study of variational problems.

The book is a completely rewritten version of the author's monograph *Variational Principles of Continuum Mechanics* which appeared in Russian in 1983. I have been postponing the English translation because I wished to include the variational principles of irreversible processes in the new edition. Reaching an understanding of this subject took longer than I expected. In its final form, this book covers all aspects of the story. The part concerned with irreversible processes is tiny, but it determines the accents put on all the results presented. The other new issues included in the book are: entropy of microstructure, variational principles of vortex line dynamics, variational principles and integration in functional spaces, some stochastic variational problems, variational principle for probability densities of local fields in composites with random structure, variational theory of turbulence; these topics have not been covered previously in monographic literature. Other than that, the scope of the book is the same though the text differs considerably due to many detailed explanations added to make the level of the book suitable for graduate students.

Grass Lake, Michigan *V.L. Berdichevsky*

Acknowledgements

This book has been affected by many influences. The most essential one was that of my teacher, L.I. Sedov, who set up the standards of scholarly work. My late friend P.P. Mosolov, the most knowledgeable expert in variational calculus I knew, explained to me the notion of "feeling the constraints" (Sect. 5.5). He insisted that the first 30 pages of any good scientific book must be written at a level undergraduate students can comprehend. His point influenced the way the beginning of the book is written; it seems that the limit suggested is exceeded. One short chat with V.I. Arnold in the 1980s advanced considerably my understanding of thermodynamics for ergodic Hamiltonian systems. Further work on this topic has resulted in a series of my papers on the subject; an overview of the "derivation of thermodynamics from mechanics" is given in Chap. 2. I learned the asymptotics in homogenization of periodic structures from N.S. Bakhvalov (Sect. 17.2), its generalization to random structures from S. Kozlov (Sect. 17.4), and conditions of elastic phase equilibrium from M. Grinfeld (Sect. 7.4).

The last but not the least: two decisive contributions to the book were made by my daughter Jenia and my former student Partha Kempanna. Jenia translated the Russian edition into English, and Partha prepared the Tex file and shaped the outlook of the book. At the final phase of the work I received help from Maria Lipkovich. The advice of Vlad Soutyrine at each stage of this work was, as always, very helpful. Dewey Hodges, Le Khanh Chau and Lev Truskinovsky read the manuscript and made valuable comments. I had many discussions of the issues considered here with Boris Shoykhet.

The usual words of gratitude can hardly convey my special feelings toward everyone mentioned.

Contents - I. Fundamentals

Contents - II. Applications

Part III Some Applications of Variational Methods to Development of Continuum Mechanics Models

Introduction

A variational principle is an assertion stating that some quantity defined for all possible processes reaches its minimum (or maximum, or stationary) value for the real process. Variational principles yield the equations governing the real processes. The equations following from a variational principle possess a very special structure. The major feature of this structure is the reciprocity of physical interactions: action of one field on another creates an opposite and, in some sense, symmetric reaction. All equations of microphysics possess such a structure. Perhaps this is the most fundamental law of Nature revealed up to now.

Macrophysics operates with the averaged characteristics of microfields. The variational structure of microequations affects the structure of macroequations. In particular, for equilibrium processes, the variational structure of microequations brings up the classical equilibrium thermodynamics. In the case of non-equilibrium reversible processes the variational structure of microequations yields a variational structure of macroequations. The governing equations of irreversible processes also possess a special structure. This structure, however, is not purely variational.

The above-mentioned explains the fundamental role of the variational principles in modeling physical phenomena. If the interactions between various fields are absent or simple enough, then one does not need the variational approach to construct the governing equations. However, if the interactions in the system are not trivial (e.g. nonlinear and/or involving high derivatives, kinematical constraints, etc.) the variational approach becomes the only method to obtain physically sensible governing equations.

Another important use of the variational principles is the direct qualitative and quantitative analysis of real processes which is based solely on the variational formulation and does not employ the governing equations. Such analysis is very advanced for solids while for fluids the major developments are still ahead.

The book aims to review the two above-mentioned sides of the variational approach: the variational approach both as a universal tool to describe physical phenomena and as a source for qualitative and quantitative methods of studying particular problems. In addition, a thorough account of the variational principles discovered in various branches of continuum mechanics is given, and some gaps are filled in.[1]

The book consists of three parts. Part I presents basic knowledge in the area, including variational principles for systems with a finite number of degrees of freedom, "the derivation of thermodynamics from mechanics," a review of basic concepts of continuum mechanics and general setting of variational principles of continuum mechanics. Part I also contains an exposition of the direct methods of calculus of variations. The major goal here is to prepare the reader to understand and to speak the "energy language," i.e. to be able to withdraw the necessary information directly from energy without using the corresponding differential equations. An important component of the energy language is the ability to work with energy depending on a small parameter. A way to do that (variational-asymptotic method) is discussed in detail. Another important component, duality theory, is also covered in detail. The variational-asymptotic method and duality theory are widely used throughout the book. Part II gives an account of variational principles for solids and fluids. Part III is concerned with applications of variational methods to shell and plate theory, beam theory, homogenization of periodic and random structures, shallow water theory, granular media theory and turbulence theory. The consideration of random structures is preceded by a review of stochastic variational problems. Some interesting variational principles that are beyond the main scope of the book are placed in Appendices. The details of some derivations that can be skipped without detriment for understanding of further material are also put in the appendices. By publisher's suggestion, the book is published in two volumes with volume 1 containing the first two parts of the book.

It is assumed that the reader knows the basics of calculus and tensor analysis. The latter, though, is not absolutely necessary as all tensor notations used are briefly outlined. Part I was used by the author as notes for the course Fundamentals of Mechanics, some chapters of Parts II and III were used in courses on elasticity theory and advanced fluid mechanics. Every effort was made to unify the notation for the broad range of the subjects considered. The notation is summarized at the end of each volume.

---

[1] Following the tradition, variational principles are named after their author; the references are given in Bibliographical Comments at the end of the book. Most of the variational principles with no name attached appeared first in the previous edition of this book.

Part I
Fundamentals

Chapter 1
Variational Principles

1.1 Prehistory

Mechanics is a branch of physics studying motion. The history of mechanics, as well as the history of other branches of science, is a history of attempts to explain the world by means of the smallest possible number of universal laws and general principles. The most successful and fruitful attempts stem from the idea that the observable events are extreme in their character and that the general principles sought are variational, i.e. they assert that certain parameters obtain their maximum or minimum values in realizable physical processes.

This idea seems to endow Nature with some goal and appeared a long time ago. Aristotle (384–322 B.C.) claimed in his *Physics*, which served as the major source for natural philosophers for over 2000 years, that in all its manifestations, Nature follows the easiest path that requires the least amount of effort. However, this idea is, perhaps, even older. As Euler mentioned [99], "...It seems, Aristotle borrowed this dogma from his predecessors rather than invented it independently." It was a long way from Aristotle's vague assertion to a precise quantitative formulation. The major breakthrough occurred in the seventeenth century along with other key advances in mathematics and physics.[1]

The figures whose contributions are most closely related to our consideration have been Galileo (1564–1642), Descartes (1596–1650), Fermat (1601–1665), Newton (1643–1727), Leibnitz (1646–1716), R. Hooke (1635–1703) and J. Bernoulli (1667–1748).

Galileo discovered the universal features of motion which formed the experimental basis of Newtonian mechanics: acceleration of a falling body does not depend on its mass; pendulum vibration frequency does not depend on the mass of the pendulum; a falling body passes distances proportional to the second power of time. Galileo also introduced the two principles which later became the cornerstones of Newtonian mechanics: the invariance of the laws of mechanics with respect to change of inertial frames, and the inertia principle – motion of a body

[1] As B. Russell put it, we would live in a quite different world if in the seventeenth century 100 scientists were killed in their childhood.

V.L. Berdichevsky, *Variational Principles of Continuum Mechanics*,
Interaction of Mechanics and Mathematics, DOI 10.1007/978-3-540-88467-5_1,

which does not interact with other bodies will remain uniform indefinitely in an unbounded space. Note that both principles, being in complete accord with the spirit of Euclidian geometry, were purely a mind game: they cannot be checked experimentally because there are neither isolated bodies nor inertial frames, not to mention that one can hardly justify the unboundedness of our space. Galileo was also known for his experiments with telescopes and astronomical observations. Here is Lagrange's appreciation [168] of Galileo's achievements:

> ...To discover the satellites of Jupiter, the phases of Venus, the spots on the Sun, etc., one needs only a telescope and a power of observation, but an exceptional genius is needed to establish the laws of Nature for phenomena which were in everyone's plain sight, but, nevertheless, escaped the attention of philosophers.

Another giant of the seventeenth century, Descartes, introduced the analytical approach to geometry and emphasized the method of orthogonal coordinates which allows one to study geometrical objects in terms of equations. Thus, for example, an ellipse, considered before Descartes as a cross-section of a cone, becomes a solution of an algebraic equation of second order.

Newton formulated the basic laws of dynamics and created the theory of gravitation. In particular, he introduced the concept of mass, a characteristic of bodies which is different from weight; discovered the key dynamic law: *mass* $\times$ *acceleration* $=$ *force*; gave the general formulation of the principle of the parallelogram of forces, and formulated the law of action and reaction.[2]

Discovery of the laws of mechanics is inseparable from development of the differential calculus by Leibnitz and Newton.

And, finally, it was Fermat who set up the beginning of the story which is the subject of this book.

One of the topics widely discussed at the time was reflection and refraction of light. It was known for centuries that the beam of light hitting a mirror at some angle α reflects from the mirror at the same angle (Fig. 1.1).

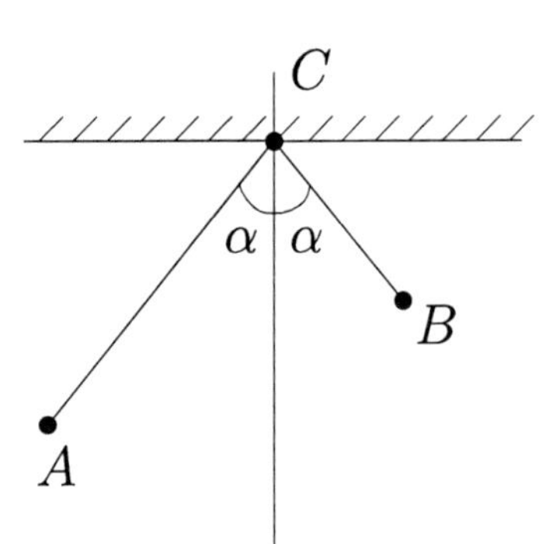

Fig. 1.1 The law of light reflection

[2] Interestingly, all the laws of statics were known to Archimedes (287–212 B.C.). It took about 19 centuries before the next step was made.

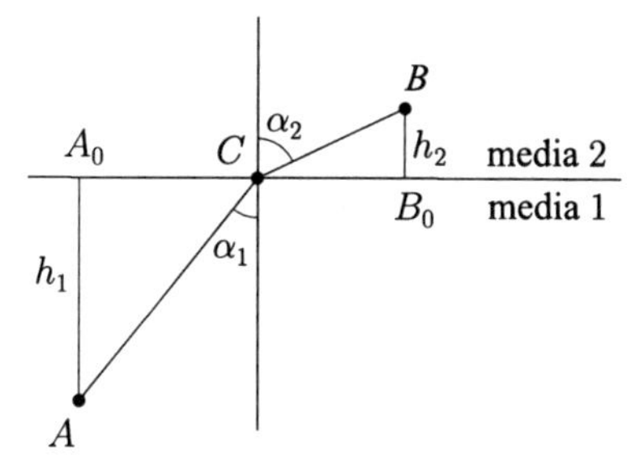

Fig. 1.2 The law of light refraction

It was also established experimentally that if a beam of light falls on the interface of two transparent media at some incidence angle α_1 from the side of medium 1, it penetrates into medium 2 at a different angle, α_2 (Fig. 1.2).

It is remarkable that, for any two media 1 and 2, the ratio $\frac{\sin\alpha_1}{\sin\alpha_2}$ remains the same for all incidence angles α_1:

$$\frac{\sin\alpha_1}{\sin\alpha_2} = n = \text{constant}. \tag{1.1}$$

This ratio, n, depends only on the materials (for example water and air, or glass and air). Moreover, the ratio remains the same if the direction of light is reversed: the beam of light falling on the interface from the side of the medium 2 at angle α_2 penetrates the medium 1 at the angle α_1 determined by the formula (1.1). This law was established by working up the experimental data by Snell in 1621.

The challenge for theoreticians was to find some underlying reasons for the peculiar behavior of light. The first attempt was made by Descartes. Descartes envisioned light as a set of small elastic balls and attempted to derive the diffraction law from the laws of elastic collisions. He obtained the correct answer. Descartes' derivation assumed, however, that light propagates in a denser medium, say, water, faster than in a less dense one, like air. Experimental verification of such features was beyond the technical possibilities at the time: methods for determining the speed of light appeared much later.

The proposition about faster light propagation in a denser medium seemed quite questionable for Fermat, and he attempted to consider diffraction from another perspective. As the basis for his derivation, he used the following postulate: Nature takes the easiest and most accessible paths. However, what is the measure of "easiness of path"? The simplest candidate is the length of the path. Consider whether it works in the reflection phenomenon.

Let light go from point A to point B reflecting from the mirror at some point (Fig. 1.3). Let C be the point for which the incidence angle is equal to the reflection angle. Consider two light trajectories, one reflects from the mirror at the point C, another at some point C'. The length of the path $AC'B$ is greater than the length

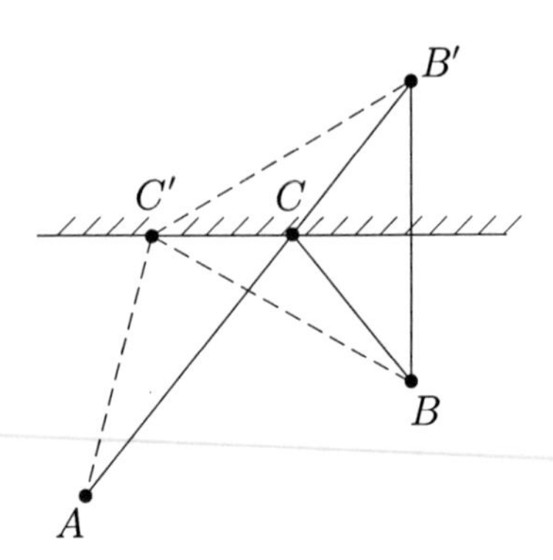

Fig. 1.3 The principle of minimum distance

of the path ACB. To see that, we[3] introduce the mirror image of point B, point B'. Obviously, the lengths of lines ACB' and $AC'B'$ are equal to the lengths of lines ACB and $AC'B$, respectively. Since the line ACB' is straight, the length of $AC'B'$ is greater than the length of ACB', and, correspondingly, the length of $AC'B$ is greater than that of ACB. We arrived at the simplest example of a variational principle: light moving from point A to a mirror and then to point B chooses a trajectory such as to minimize the distance traveled.

This example contains the two major "entries" of any variational principle: the set of admissible trajectories, in this case all lines connecting the starting point A with the reflection point C and the destination point B, and the quantity to be minimized, in this case the length of the trajectory. Note that the minimizing trajectory is highly sensitive to the choice of admissible paths. If, for example, we choose as admissible all the paths connecting the point A and the point B, then the minimizing trajectory is the straight line connecting A and B which is not the correct answer for the initial physical problem.

The principle of minimum distance works for reflection, but does not work for refraction: if point B is on the other side of the interface plane, then the minimum distance corresponds to the straight line, i.e. $\alpha_1 = \alpha_2$, in contradiction to the experimental observations.

Fermat suggested that the experimentally observed refraction law corresponds to the principle of minimum time: light moving from point A to point B chooses the trajectory for which the travel time is minimum.

Let us derive the Snell law (1.1) from the Fermat principle. To proceed, we have to introduce the speeds of light in both media, c_1 and c_2, and distances from points A and B to the interface, h_1 and h_2 (Fig. 1.2).

We are looking for point C such that the travel time from A to B is minimum. In each medium, 1 and 2, the trajectory of light is straight, because, for constant velocity, minimum travel time between two points, A and C or C and B, corresponds to the paths with minimum lengths, i.e. to the straight segments AC and CB. Let A_0 and B_0 be the orthogonal projections of points A and B onto the interface. Denote

[3] Here and in what follows "we" means "the reader and the author."

the distances A_0C and A_0B_0 by x and a, respectively. The distance x is unknown while the distance a is given. The travel time from A to B is a function of x:

$$f(x) = \frac{\sqrt{h_1^2 + x^2}}{c_1} + \frac{\sqrt{h_2^2 + (a-x)^2}}{c_2}.$$

We consider this function on the segment $[0, a]$ and seek the value of x which delivers the minimum value to this function. Note that the second derivative of $f(x)$,

$$\frac{d^2 f(x)}{dx^2} = \frac{h_1^2}{c_1 \left(h_1^2 + x^2\right)^{\frac{3}{2}}} + \frac{h_2^2}{c_2 \left(h_1^2 + (a-x)^2\right)^{\frac{3}{2}}},$$

is positive. Therefore, $f(x)$ is a convex function[4] and has the form shown in Fig. 1.4.

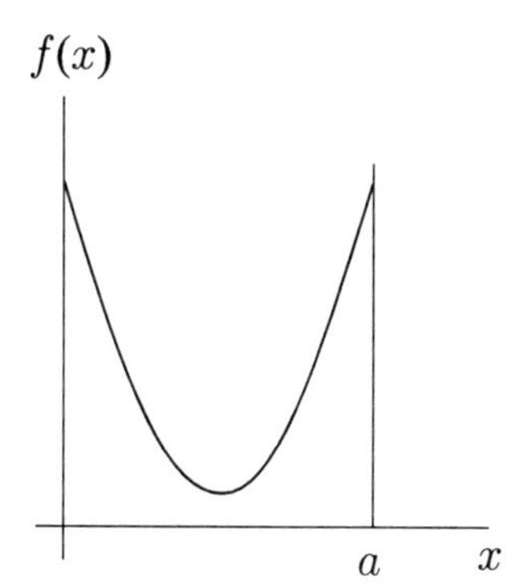

Fig. 1.4 Qualitative dependence of the travel time on the position of point C

It is seen that $f(x)$ has only one minimum. This minimum is achieved at the point x where $\frac{df(x)}{dx} = 0$, i.e.

$$\frac{x}{c_1\sqrt{h_1^2 + x^2}} - \frac{a-x}{c_2\sqrt{h_2^2 + (a-x)^2}} = 0. \tag{1.2}$$

The ratio $\frac{x}{\sqrt{h_1^2 + x^2}}$ is equal to $\sin \alpha_1$. Similarly, $\frac{a-x}{c_2\sqrt{h_2^2 + (a-x)^2}} = \sin \alpha_2$. Equation (1.2) becomes

$$\frac{\sin \alpha_1}{c_1} = \frac{\sin \alpha_2}{c_2}. \tag{1.3}$$

---

[4] Here we rely on the reader's knowledge of calculus. The notion of a convex function and its relations to variational problems will be discussed in detail later in Chap. 5.

Equation (1.3) yields the Snell law (1.1) because, as expected, c_1 and c_2 depend only on the material properties. Equation (1.3) also contains additional information: the material constant n can be expressed in terms of the speeds of light in both media:

$$n = \frac{c_1}{c_2}. \tag{1.4}$$

It was established experimentally that $n \approx 0.75$ if the first medium is water and the second medium is air. This is why we see stones at the bottom of the river closer than they are in reality (Fig. 1.5).

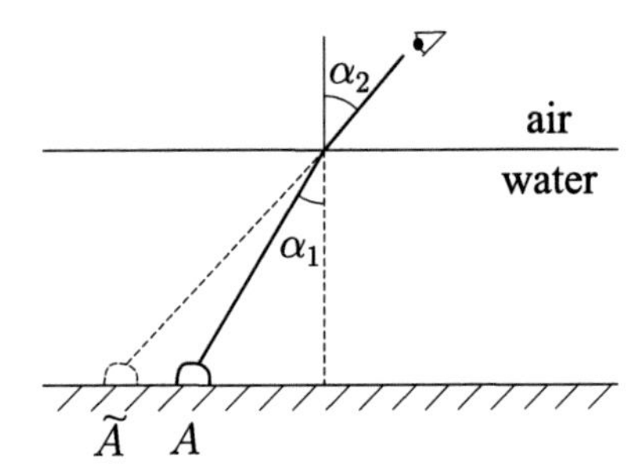

Fig. 1.5 Refraction in water and air

According to (1.4), the speed of light in water is about 3/4 of that in air. This contradicts the Descartes claim that the speed of light is greater in denser media.

Descartes' conclusion was supported by Leibnitz. Like Fermat, he based his consideration on a variational principle. But he suggested that Nature chooses the easiest path defined as a path of least resistance. The resistance to light propagation is different in different media. Leibnitz introduced the notion of "difficulty" which is equal to the product of time and resistance, and postulated that light chooses the trajectory for which the sum of all difficulties is minimum. Eventually, he arrived at the Snell law. According to Leibnitz, the denser the medium, the greater its resistance. This fact, he argued, yields a greater velocity of light in denser media: the dispersion of light is smaller, hence the beam of light is more compressed along its way, and, similar to water in a narrow river-bed, moves faster. Apparently, he envisioned the light flux as a flow of some medium.

It became clear much later that in this controversy Fermat was right.

Successful applications of the variational ideas to optics encouraged the search for analogies in mechanics. In 1696, in Leipzig journal *Acta Eruditorum*, Johann Bernoulli published the following note:

> Two points, A and B, are given in a vertical plane (see Fig. 1.6). Find the trough AMB, which minimizes the travel time of a body moving from point A to point B by the force of its own weight. To arouse the interest of amateurs in such matters and to encourage their enthusiasm in attempting to obtain the solution, I will say that solving this problem is not a pure intellectual speculation deprived of any practical application whatsoever, as it may seem. Actually, this problem is of great practical interest and not just in the subject

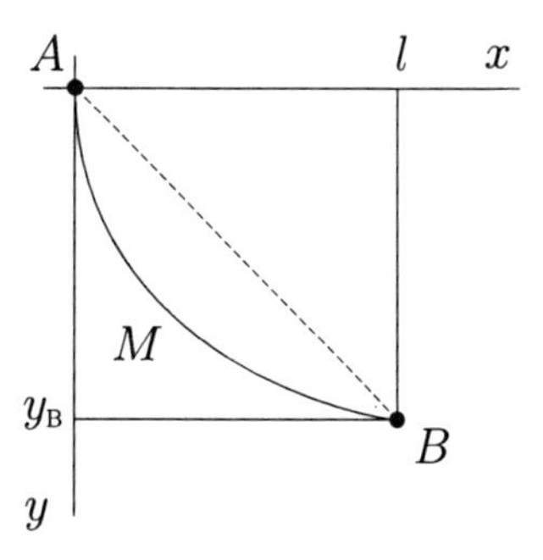

Fig. 1.6 Bernoulli's problem

of mechanics, but also in other disciplines, which may seem unbelievable. By the way (I am mentioning the following fact in order to avert a possible misjudgment), although the segment AB is the shortest distance between points A and B, the time it takes for the body to travel this distance is not the shortest possible, and there exists the (minimum time) curve *AMB* well known to geometers. I will disclose what this curve is if, during the course of the year, nobody else announces the answer.

Bernoulli sent his solution to Leibniz, in order for Leibniz to publish it in a year. This, as Leibnitz put it, "such wonderful and hitherto unheard problem" which "entices by its beauty just as the apple enticed Eve," attracted the attention of many scholars. In particular, it is believed that some anonymous solutions were given by Jacob Bernoulli and Isaac Newton. The notorious curve turned out to be a cycloid, a curve "well known to geometers."

Bernoulli's problem is much more difficult than Fermat's problem since the function to be minimized depends on the curve connecting the points A and B, i.e. on an infinite number of variables. Such functions are now called functionals.

If $y = y(x)$ is the equation of the curve connecting points A and B, then the time of motion along this curve, I, depends only on the function $y(x)$. One writes

$$I = I(y(x))$$

and says that I is a functional of $y(x)$. In Bernoulli's problem,

$$I(y(x)) = \int_0^l \sqrt{\frac{1 + \left(\frac{dy}{dx}\right)^2}{2gy(x)}} dx. \tag{1.5}$$

where g is the acceleration of gravity.

To see that, we note that motion of the body is frictionless; therefore, the total energy of the particle is conserved. Energy is the sum of kinetic energy K and potential energy U. Kinetic energy is the mass m times half of the squared velocity, $\left(\frac{dx}{dt}\right)^2 + \left(\frac{dy}{dt}\right)^2$. Since the point moves along the curve $y = y(x)$,

$$K = \frac{m}{2}\left(\left(\frac{dx}{dt}\right)^2 + \left(\frac{dy}{dx}\right)^2\left(\frac{dx}{dt}\right)^2\right) = \frac{m}{2}\left(1 + \left(\frac{dy}{dx}\right)^2\right)\left(\frac{dx}{dt}\right)^2.$$

The potential energy is (remember that the positive direction of the y-axis is down, therefore the smaller y the bigger U)

$$U = -mgy.$$

Initial value of energy, $K + U$, is zero; therefore at all times

$$\frac{m}{2}\left(1 + \left(\frac{dy}{dx}\right)^2\right)\left(\frac{dx}{dt}\right)^2 - mgy = 0.$$

Hence,

$$\frac{dx}{dt} = \sqrt{\frac{2gy}{1 + \left(\frac{dy}{dx}\right)^2}}. \tag{1.6}$$

The travel time is

$$I = \int_0^l \frac{dx}{\frac{dx}{dt}}. \tag{1.7}$$

Formula (1.5) follows from (1.7) and (1.6).

The functional (1.5) must be minimized on the set of curves $y(x)$ which satisfy the following conditions:

$$y(0) = 0, \qquad y(l) = y_B. \tag{1.8}$$

The first condition fixes the choice of the origin, the second prescribes the y-coordinate of the given point, B.

The solutions obtained at the time did not provide a general method of investigating this type of problem. It was found 50 years later by a student of Johann Bernoulli, Leonard Euler, who laid down the cornerstones of modern calculus of variations.

In addition to the constraints (1.8) the admissible functions $y(x)$ should obey some other restrictions which we have not mentioned explicitly yet. For example, the integrand in (1.5) contains the derivative dy/dx, and therefore the admissible function must be differentiable at least.

If we accept for consideration all non-negative functions $y(x)$ with piecewise continuous derivatives, then the integral (1.5) makes sense. Note that for some functions, e.g., quadratic functions, $y(x) = const\, x^2$, the integral is equal to $+\infty$. Such functions are automatically sorted out because we seek the minimum value of the functional (1.5).

The insights of Fermat and Bernoulli imparted a new importance to the question of whether some goal parameter obtains its minimum in realizable motion of bodies or, in the language of the time, whether the body, moving from one point to another chooses such a way that the payment, if the body is to pay for its motion, is minimized.

1.2 Mopertuis Variational Principle

The first formulation of the variational principle in mechanics was made by Pierre Mopertuis in 1744. According to Mopertuis' principle, in real motion, the product of the mass of the body, its speed and the distance it has traveled, is minimum. This quantity,

$$I = mvs, \tag{1.9}$$

Mopertuis, following Leibniz, called action.

An important testing ground for Mopertuis was the law of collision of elastic balls. Let us first show that Mopertuis' principle yields the correct law of reflection of a rigid ball from the wall when the ball moves along the normal line to the wall. To emphasize an analogy with the Fermat principle, we consider a trajectory of the ball in the two-dimensional space-time plane (see Fig. 1.7).

If at the instants $t = 0$ and $t = \theta$ the ball was at the points A and B, and correspondingly, h_1, h_2 are the distances from these points to the wall, and t_c is the time of the collision, then the action is

$$I = m\frac{h_1}{t_c}h_1 + m\frac{h_2}{\theta - t_c}h_2. \tag{1.10}$$

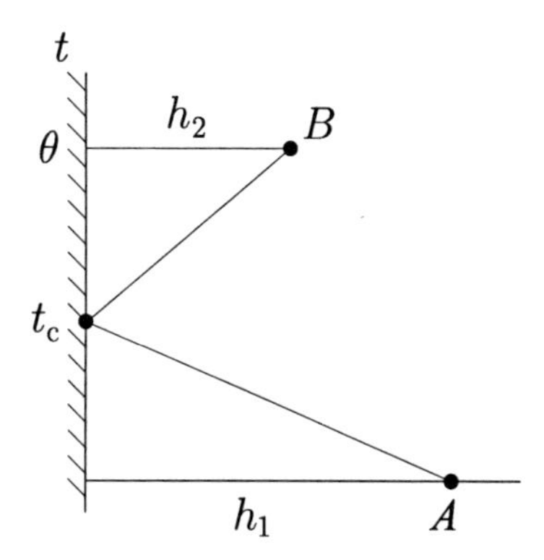

Fig. 1.7 Reflection of the ball from the wall

Note that the initial and final positions of the ball and the total time of motion θ are assumed to be given, so action I is a function of one variable, t_c, $I = I(t_c)$. Function $I(t_c)$ is defined on the segment $[0, \theta]$. The second derivative of this function,

$$\frac{d^2 I}{dt_c^2} = 2m\frac{h_1^2}{t_c^3} + 2m\frac{h_2^2}{(\theta - t_c)^3},$$

is positive. Thus, the function, $I(t_c)$, is convex and has only one minimum. The minimum is achieved at the point where the first derivative of $I(t_c)$ vanishes, i.e.

$$-m\frac{h_1^2}{t_c^2} + m\frac{h_2^2}{(\theta - t_c)^2} = 0.$$

This equation means that the velocity of the ball after the collision, $\frac{h_2}{\theta - t_c}$, is equal to the velocity of the ball before the collision, $\frac{h_1}{t_c}$, in full compliance with the experimentally established law of elastic collisions. The reader may check that the correct law of elastic collisions follows from Mopertuis' principle for an inclined collision with a flat wall (Fig. 1.8a) or a collision with a curved wall (Fig. 1.8b).

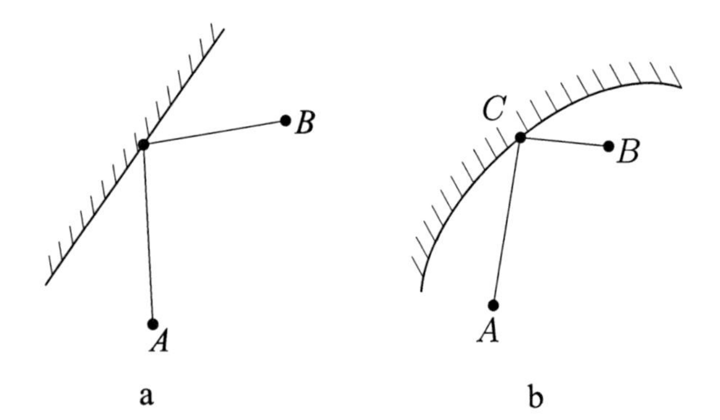

Fig. 1.8 Elastic collision of the ball with a flat and a curved wall

Collision with a curved wall is actually a more delicate problem. It turns out that multiple collision points are possible, and some collision points could correspond to the maximum value of action.

To clarify this issue, consider the following model problem. Let points A and B be inside a circular billiard table of radius R (see Fig. 1.9). The coordinates of points A and B are $(-a, b)$ and (a, b), respectively. The point of possible reflection from the wall is denoted by C, as before. Point C has coordinates $(R\cos\varphi, R\sin\varphi)$. The action is a function of two variables: collision time t_c and angle φ:

$$I(t_c, \varphi) = m\frac{(R\cos\varphi + a)^2 + (R\sin\varphi - b)^2}{t_c} + m\frac{(R\cos\varphi - a)^2 + (R\sin\varphi - b)^2}{\theta - t_c}.$$

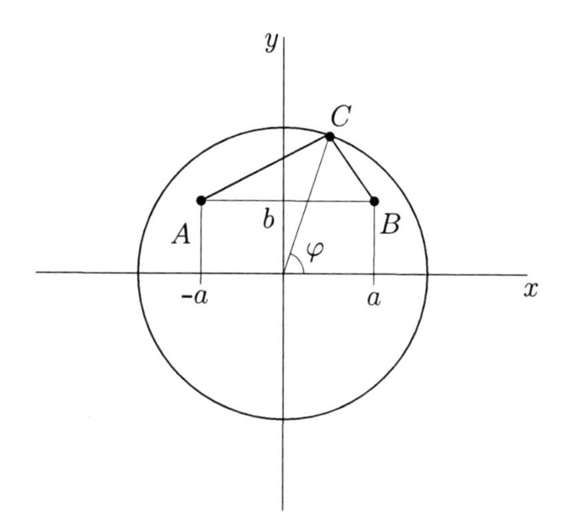

Fig. 1.9 Circular billiard table

All other parameters of the action, m, R, a, b and the motion time θ, are assumed to be given. Function $I(t_c, \varphi)$ is defined for $0 < t_c < \theta$ and $0 \leq \varphi \leq 2\pi$. For each given φ, the function $I(t_c, \varphi)$ has a unique minimum with respect to t_c (check). For each given t_c, $I(t_c, \varphi)$ is periodic and smooth with respect to φ. Therefore, if it has a minimum with respect to φ, it must have a maximum with respect to φ as well. The graphs of the function of φ $\min_{t_c} I(t_c, \varphi)$, are shown in Fig. 1.10 for two choices of (a, b): $a/R = 0.3$, $b/R = 0.3$ (upper curve) and $a/R = 0.8$, $b/R = 0.3$ (bottom curve). It is seen that the point $\varphi = \pi/2$, which apparently fits the law of elastic collisions, can be a point of minimum (left graph) and a point of maximum (right graph).

Mopertuis proclaimed the variational principle the general law of Nature and the most fundamental proof of the existence of God.

The mathematical context to the Mopertuis principle was imparted by Euler. In particular, Euler realized that the Mopertuis principle is applicable only to infinitely small segments of the path, ds, and in order to obtain the action for the whole path, one needs to sum the actions of all segments. Therefore, the action should be written as

$$I = \int mv ds$$

or, since $v = \dfrac{ds}{dt}$, as

$$I = \int_{t_0}^{t_1} mv^2 dt.$$

The quantity being integrated is equal to the kinetic energy up to factor 2 (Leibniz called the kinetic energy the living force, as the opposite of pressure – the dead force).

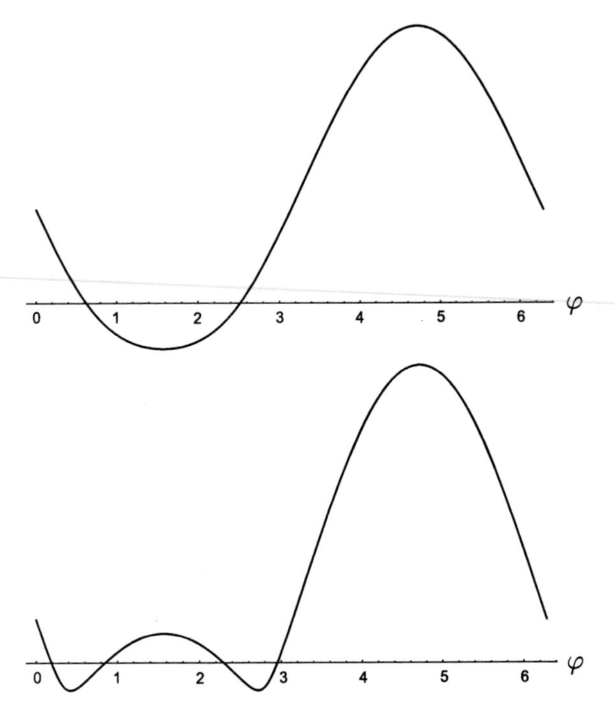

Fig. 1.10 Function $\min_{t_c} I(t_c, \varphi)$ of φ for two values of $(a, b) : a/R = 0.3$, $b/R = 0.3$ (upper curve), $a/R = 0.8, b/R = 0.3$ (bottom curve). The point $\varphi = \pi/2$ can be a point of minium (left) and a point of maximum (right)

In the 1750s the letters claimed to have been written by the late Leibniz were published, from which one may conclude that Leibniz knew about the extreme properties of action. Besides, it was asserted that the action can achieve in a real process not only some minimum, but also maximum value. This publication provoked great controversy which, in turn, raised some philosophical, moral and priority issues. It was a rare event when a purely scientific issue became a matter of public interest. Philosophers, writers, kings and their courts, took part in heated debates. One of the echoes of these debates reached us in the form of Voltaire's pamphlet "Histoire du docteur Akakia et du natif de Saint-Malo" (1752).

Finally, the following point of view crystallized: for some motions, the action reaches its minimum value, while for the others it may have the maximum value; or, in the language of the time, "Nature is a thrifty mother, who manages with the least possible, if she can do so; but, if not, she pays honestly and as much as possible, so as not to be reputed a miser".[5]

Although it was understood from the very beginning that action may have either minimum, maximum or stationary value, historically the term "principle of least

[5] From a letter by Kraft to Euler (1753).

action" gained a foothold. We will follow the tradition and use this term, though one should bear in mind that in some cases the term "principle of stationary action" might be more appropriate.

1.3 Euler's Calculus of Variations

The discovery of the least action principle made obvious the necessity of a technique to deal with the so-called integral functionals,

$$I(x(t)) = \int_{t_0}^{t_1} L\left(x(t), \frac{dx}{dt}, t\right) dt. \tag{1.11}$$

This technique was developed by L. Euler. The first question to answer was: if $\check{x}(t)$ is the minimizing function of the functional (1.11), which equation must it obey? The key role in establishing such an equation is played by the variation, δI, of the functional I. To define δI, consider infinitesimally small variations, $\delta x(t)$, of some function, $x(t)$, and the difference,

$$I(x(t) + \delta x(t)) - I(x(t)). \tag{1.12}$$

The difference (1.12) in which one keeps only terms of the first order with respect to $\delta x(t)$ and neglects all terms of higher orders is called the variation δI of the functional $I(x(t))$. The variation δI is a functional of two arguments, $x(t)$ and $\delta x(t)$. To underline that δI is a functional of two functions, $x(t)$ and $\delta x(t)$, one also uses the notation

$$\delta I = I'(x(t), \delta x). \tag{1.13}$$

Here the prime emphasizes the similarity with usual derivative.

Since all the terms of higher orders are omitted in the difference (1.12), δI depends on δx linearly, i.e. for any two functions δx_1 and δx_2,

$$I'(x(t), \delta x_1 + \delta x_2) = I'(x(t), \delta x_1) + I'(x(t), \delta x_2), \tag{1.14}$$

and for any number λ,

$$I'(x(t), \lambda\delta x) = \lambda I'(x(t), \delta x). \tag{1.15}$$

Functions, $x(t)$, may be subject to some constraints. Then $\delta x(t)$ are not arbitrary because $\check{x}(t) + \delta x$ must obey the constraints. Functions $x(t)$ and δx obeying the constraints are called admissible.

If $\check{x}(t)$ is the minimizing element of the functional, $I(x(t))$, then the variation of the functional, $I(x(t))$, computed at the function, $\check{x}(t)$, must vanish for any admissible variations, $\delta x(t)$,

$$\delta I = I'(\check{x}(t), \delta x) = 0. \tag{1.16}$$

Indeed, assume the opposite: $\delta I \neq 0$ for some $\delta x_0 \neq 0$. Then, there is δx_1, for which $\delta I < 0$: if $I'(\check{x}(t), \delta x_0) < 0$, we put $\delta x_1 = \delta x_0$; if $I'(\check{x}(t), \delta x_0) > 0$, we put $\delta x_1 = -\delta x_0$ and, due to linearity of δI with respect to δx, $I'(\check{x}(t), \delta x_1) = I'(\check{x}(t), -\delta x_0) = -I'(\check{x}(t), \delta x_0) < 0$. For sufficiently small δx_1, $I(\check{x}(t) + \delta x_1) - I(\check{x}(t)) \approx \delta I$, and δI being negative means that the value of the functional $I(x(t))$ on the function $\check{x}(t) + \delta x_1$ is smaller than that on the function $\check{x}(t)$. We arrive at a contradiction. Hence, (1.16) holds true. This equation is sometimes called the Euler equation for the functional I.

Let us find the variation of the functional (1.11). Since

$$I(x(t) + \delta x(t)) - I(x(t)) =$$
$$= \int_{t_0}^{t_1} L\left(x(t) + \delta x, \frac{d}{dt}(x(t) + \delta x), t\right) dt - \int_{t_0}^{t_1} L\left(x(t), \frac{dx}{dt}, t\right) dt$$
$$= \int_{t_0}^{t_1} \left(L\left(x(t) + \delta x, \frac{d}{dt}(x(t) + \delta x), t\right) - L\left(x(t), \frac{dx}{dt}, t\right)\right) dt,$$

computing of δI is reduced to keeping the linear terms with respect to δx in the integrand. We have

$$\delta I = \int_{t_0}^{t_1} \left(\frac{\partial L}{\partial x}\delta x + \frac{\partial L}{\partial \dot{x}}\frac{d\delta x}{dt}\right) dt. \tag{1.17}$$

Here and in what follows dot denotes the time derivative.

Now we have to draw the consequences from (1.11),

$$\delta I = \int_{t_0}^{t_1} \left(\frac{\partial L}{\partial x}\delta x + \frac{\partial L}{\partial \dot{x}}\frac{d\delta x}{dt}\right) dt = 0. \tag{1.18}$$

Consider first a simpler issue: let for some function $A(t)$ and for any continuous function $\delta x(t)$

$$\int_{t_0}^{t_1} A(t)\,\delta x(t)\,dt = 0. \tag{1.19}$$

What can one say about the function $A(t)$? If the function $A(t)$ is continuous then

$$A(t) = 0. \tag{1.20}$$

This is obvious: if at some point t^*, $A(t^*) \neq 0$, then $A(t)$ is not zero and does not change its sign in some small vicinity, V, of the point t^*. Choosing $\delta x(t)$ zero outside V and positive inside V we obtain

$$\int_{t_0}^{t_1} A(t)\,\delta x(t)\,dt = \int_V A(t)\,\delta x(t)\,dt.$$

This integral is not equal to zero because $A(t)\,\delta x(t)$ does not change the sign inside V, and we arrive at a contradiction with (1.19).

The statement made is called the main lemma of calculus of variation.

We may strengthen the main lemma: Equation (1.20) remains valid if we narrow the admissible functions $\delta x(t)$ in (1.19) by the functions δx vanishing at any finite number of points of the segment $[t_0, t_1]$ including the end points. Indeed, from the previous reasoning $A(t) = 0$ at all points excluding the points where $\delta x(t) = 0$. By continuity, $A(t) = 0$ on the entire segment $[t_0, t_1]$.

This extension of the main lemma yields an important consequence: if

$$\int_{t_0}^{t_1} A(t)\,\delta x(t)\,dt + B_0\delta x(t_0) + B_1\delta x(t_1) = 0, \tag{1.21}$$

for any function $\delta x(t)$, then

$$A(t) = 0, \quad B_0 = 0, \quad B_1 = 0. \tag{1.22}$$

Indeed, considering (1.21) for $\delta x(t_0) = \delta x(t_1) = 0$, we obtain from the extension of the main lemma, that $A(t) = 0$. Then, from (1.21) for arbitrary $\delta x(t_0)$ and $\delta x(t_1)$,

$$B_0\delta x(t_0) + B_1\delta x(t_1) = 0. \tag{1.23}$$

Setting first $\delta x(t_1) = 0$ and $\delta x(t_0)$ arbitrary, we find from (1.23) that $B_0 = 0$. This equation along with (1.23) yield, for arbitrary $\delta x(t_1)$, $B_1 = 0$.

If we have a number of functions, $A_1(t), \ldots, A_n(t)$ and a number of variations, $\delta x_1(t), \ldots, \delta x_n(t)$, and

$$\int_{t_0}^{t_1} (A_1(t)\,\delta x_1(t) + \ldots + A_n(t)\,\delta x_n(t))\,dt = 0, \tag{1.24}$$

for any choice of functions $\delta x_1(t), \ldots, \delta x_n(t)$, then

$$A_1(t) = 0, \ldots, A_n(t) = 0. \tag{1.25}$$

This statement can be derived from the main lemma: first we set $\delta x_2 = \ldots = \delta x_n = 0$. Then (1.24) transforms to (1.19), and we obtain $A_1 = 0$. Putting $\delta x_3 = \ldots = \delta x_n = 0$ we transform (1.24) to the equation

$$\int_{t_0}^{t_1} A_2(t)\,\delta x_2(t)\,dt = 0.$$

Hence, $A_2(t) = 0$. Continuing this procedure, we obtain (1.25).

Formally, (1.18) has the form (1.24) with

$$A_1 = \frac{\partial L}{\partial x}, \quad \delta x_1 = \delta x, \quad A_2 = \frac{\partial L}{\partial \dot{x}}, \quad \delta x_2 = \frac{d\delta x}{dt}.$$

We cannot conclude, however, that $\partial L/\partial x$ and $\partial L/\partial \dot{x}$ are zero, because the variations δx and $d\delta x/dt$ are not independent: for the prescribed $\delta x_1 = \delta x$, the variation $\delta x_2 = d\delta x/dt$ is determined completely.

To put (1.18) into the form suitable for the application of the main lemma we integrate the second term by parts

$$\begin{aligned}
&\int_{t_0}^{t_1} \left(\frac{\partial L}{\partial x}\delta x + \frac{\partial L}{\partial \dot{x}}\frac{d\delta x}{dt} \right) dt = \\
&= \int_{t_0}^{t_1} \left(\frac{\partial L}{\partial x}\delta x + \frac{d}{dt}\left(\frac{\partial L}{\partial \dot{x}}\delta x\right) - \delta x \frac{d}{dt}\frac{\partial L}{\partial \dot{x}} \right) dt = \\
&= \int_{t_0}^{t_1} \left(\frac{\partial L}{\partial x} - \frac{d}{dt}\frac{\partial L}{\partial \dot{x}} \right) \delta x\, dt + \left.\frac{\partial L}{\partial \dot{x}}\delta x\right|_{t=t_1} - \left.\frac{\partial L}{\partial \dot{x}}\delta x\right|_{t=t_0} = 0.
\end{aligned} \tag{1.26}$$

Since the function $\delta x(t)$ is arbitrary, from (1.26) we obtain for the minimizing function the ordinary differential equation

$$\frac{\partial L}{\partial x} - \frac{d}{dt}\frac{\partial L}{\partial \dot{x}} = 0, \tag{1.27}$$

and the boundary conditions

$$\left.\frac{\partial L}{\partial \dot{x}}\right|_{t=t_1} = 0, \quad \left.\frac{\partial L}{\partial \dot{x}}\right|_{t=t_0} = 0. \tag{1.28}$$

If $\partial L/\partial \dot{x}$ does depend on $\dot{x}$, (1.27) is an equation of second order. Supplementing this equation with the two boundary conditions (1.28), we obtain a sensible boundary value problem.

Equations (1.27) and (1.28) are called Euler equations of the minimization problem. Their equivalent form is the equation $I'(\check{x}(t), \delta x) = 0$.

The admissible functions in variational principles may obey some kinematic constraints. For example, the values of the admissible functions at the ends, t_0 and t_1, can be given:

$$x(t_0) = x_0, \quad x(t_1) = x_1. \tag{1.29}$$

Then, since the varied functions, $x(t) + \delta x(t)$, must obey the same conditions,

$$x(t_0) + \delta x(t_0) = x_0, \quad x(t_1) + \delta x(t_1) = x_1,$$

the variations must vanish at the ends:

$$\delta x(t_0) = 0, \quad \delta x(t_1) = 0.$$

The last two terms in (1.26) become zero, and the only consequence of (1.26) is (1.27). It must be accomplished with the two boundary conditions (1.29). We have again a sensible boundary value problem. We observe here a general feature of the variational approach: in a generic case, the number of equations it provides is as many as necessary to solve the problem.

The integral functional of many functions, $x_1(t), \ldots, x_n(t)$,

$$I(x_1(t), \ldots, x_n(t)) = \int_{t_0}^{t_1} L(x_1, \ldots, x_n, \dot{x}_1, \ldots, \dot{x}_n, t)\, dt,$$

is considered similarly. We have

$$\begin{aligned}\delta I &= \sum_{i=1}^{n} \int_{t_0}^{t_1} \left(\frac{\partial L}{\partial x_i} \delta x_i + \frac{\partial L}{\partial \dot{x}_i} \frac{d\delta x_i}{dt} \right) dt = \\ &= \sum_{i=1}^{n} \left(\int_{t_0}^{t_1} \left(\frac{\partial L}{\partial x_i} - \frac{d}{dt} \frac{\partial L}{\partial \dot{x}_i} \right) \delta x_i dt + \left[\frac{\partial L}{\partial \dot{x}_i} \delta x_i \right]_{t_0}^{t_1} \right).\end{aligned}$$

If $x_i(t)$ are the minimizing functions, then they satisfy the equation

$$\frac{\partial L}{\partial x_i} - \frac{d}{dt} \frac{\partial L}{\partial \dot{x}_i} = 0. \tag{1.30}$$

A few words about terminology. When dealing with functions of a finite number of variables, $f(x_1, \ldots, x_n)$, it is convenient to use a geometrical language considering $x_1, \ldots, x_n$ as coordinates of a point x in some n-dimensional space. Accordingly, every time when it cannot cause confusion, we suppress indices and write $f(x)$ instead of $f(x_1, \ldots, x_n)$ or $f(x_i)$. We keep the notation $f(x_1, \ldots, x_n)$ or $f(x_i)$ only if it is worth emphasizing that f is a function of many variables. The major feature of the x-space used is the linearity of the space: for any number λ and the point x, the space contains the point λx (its coordinates are $\lambda x_1, \ldots, \lambda x_n$), and for any two points x' and x'' the space contains their sum $x' + x''$ (if $x_1', \ldots, x_n'$ and $x_1'', \ldots, x_n''$ are the coordinates of the points x' and x'', respectively, then, by definition, the coordinates of the point $x' + x''$ are $x_1' + x_1'', \ldots, x_n' + x_n''$).

Similarly, considering a functional $I(x(t))$, it is convenient to think of its argument as a point in some space of functions. The space of functions (or functional space) is linear because for each number λ and each function $x(t)$ the function, $\lambda x(t)$, is defined, and for each two functions, $x_1(t)$ and $x_2(t)$, the functions $x_1(t) + x_2(t)$ is defined. In accordance with such a view, the point of the functional space at which $\delta I = 0$ is called the stationary point of the functional, and the value of the functional at this point the stationary value. We will also use the following notation: the point of maximum of the functional $I(x(t))$ is marked by "hat", $\hat{x}(t)$, the point of minimum by the "check" sign, $\check{x}(t)$, and the stationary point by the cross sign, $\overset{\times}{x}(t)$. Of course, the points of minimum and maximum are also the stationary points, and the notation $\overset{\times}{x}$ is used if we seek a stationary point or if we are not certain about the type of the stationary point.

1.4 Lagrange Variational Principle

The development of the least action principle was completed by Lagrange. To present his final version of the principle in modern terms we first consider the notion of generalized coordinates.

The construction of any mathematical model of a mechanical system begins with a description of its kinematics, i.e. a description of all possible configurations. To do this, one has to specify a set of numbers, $q_1, \ldots, q_n$, which determine the configuration of the system. Then motion of the system is described by functions of time t, $q_1(t), \ldots, q_n(t)$. "To know motion" means "to know the functions $q_1(t), \ldots, q_n(t)$." The n-dimensional space Q of points q with coordinates $q_1, \ldots, q_n$ is called the configuration space , or Q-space. The coordinates, $q_1, \ldots, q_n$, are called the generalized coordinates if they are independent in the sense that any curve in Q-space represents some motion. The number n is called the number of degrees of freedom.

Example 1[6] The position of a particle on a line is given by one number q – the coordinate of the point on the line. The motion of a particle is described by the

---

[6] Examples are numbered separately within each section.

function $q(t)$. The coordinate space is the line. If the particle moves between two walls with coordinates 0 and a, the admissible values of q are the numbers between 0 and a (Fig. 1.11).

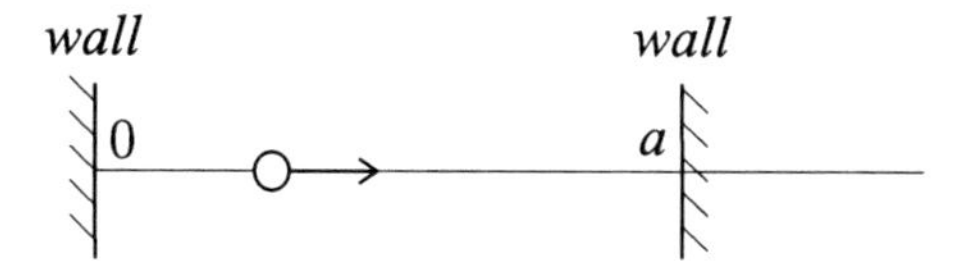

Fig. 1.11 Particle moving along the line between the two walls

Example 2. The position of a pendulum (Fig. 1.12a) in a plane is given by one number, q, the angle between the rod and the vertical axis. So, the pendulum is a system with one degree of freedom. The angle q can take any value between 0 and 2π. Two values of q, 0 and 2π, correspond to the same position of the pendulum. Combining these two points of the segment, one gets a circle, so one can say that the Q-space is a circle. The Cartesian coordinates of the position of the pendulum x, y are not generalized coordinates because they are dependent: they obey the equation $x^2 + y^2 = \ell^2$, where ℓ is the length of the rod.

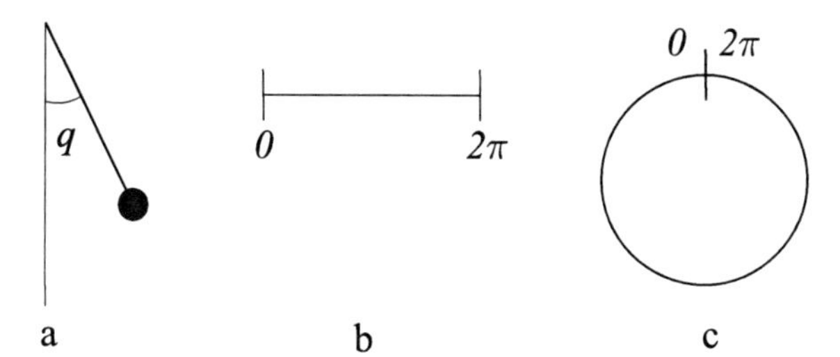

Fig. 1.12 Generalized coordinate and Q-space for a pendulum

Example 3. Each position of a double pendulum (Fig. 1.13) in a plane is specified by two angles, q_1 and q_2. This is an example of a system with two degrees of freedom. Each angle can have any value between 0 and 2π, and the Q-space is a square (Fig. 1.14a).

The points on the segments OA (by O we denote the origin) and BC with the same q_2 coordinate correspond to the same position of the pendulum. Combining these two segments, one gets a cylinder (Fig. 1.14b); the position of the pendulum corresponds to the points on the lateral surface of the cylinder. The points on the top and the bottom circles of the cylinder in Fig. 1.14b also correspond to the same q_2 positions of the pendulum. Combining these circles one gets a torus (Fig. 1.14c). There is a one-to-one correspondence between points of the torus and positions of the pendulum. One can say that the Q-space of the double pendulum is the torus. Any motion of the double pendulum is displayed by a curve on the torus.

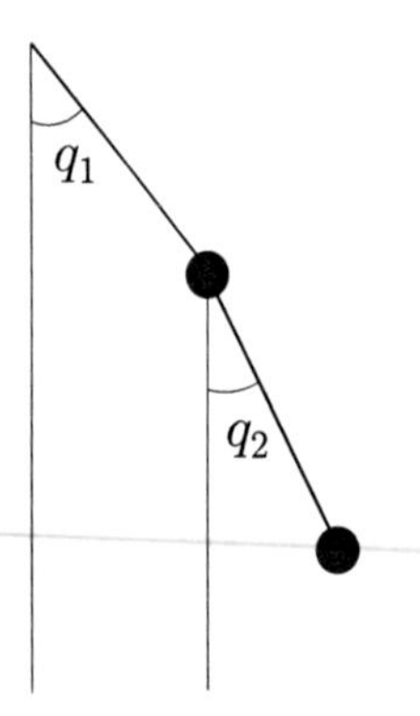

Fig. 1.13 Generalized coordinates for a double pendulum

Example 4. A particle in three-dimensional space is, obviously, a mechanical system with three degrees of freedom. The generalized coordinates q_1, q_2, q_3 coincide with the Cartesian coordinates of the particle in space. N particles in three-dimensional space form a system with $3N$ degrees of freedom.

Consider a mechanical system which is described by a finite number of generalized coordinates $q_1, \ldots, q_n$. The set of coordinates q_i $(i = 1, \ldots, n)$ will be denoted by q. Each motion of the system corresponds to a trajectory $q = q(t)$ in the configurational space Q.

Inertial properties of the system are characterized by a function of q and $\dot{q}$, $K(q, \dot{q})$, which is called kinetic energy. By definition, in an inertial observer's frame, the kinetic energy is one-half of the sum of the products of the masses and squared velocities of all parts of the mechanical system. Usually, velocities depend on $\dot{q}_i$ linearly, and kinetic energy is quadratic with respect to $\dot{q}_i$:

$$K = \frac{1}{2} \sum_{i,j} m_{ij} \dot{q}_i \dot{q}_j. \tag{1.31}$$

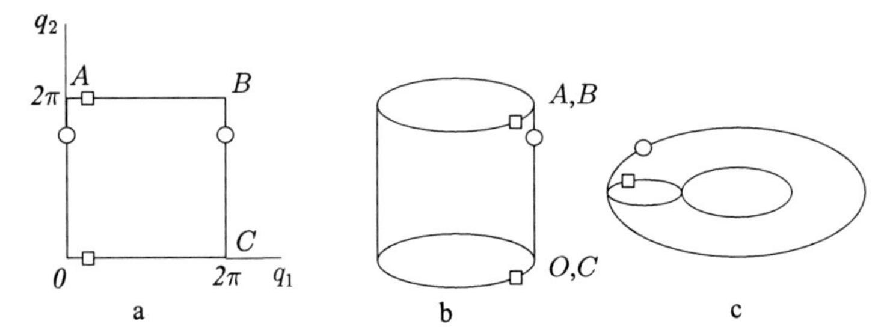

Fig. 1.14 Q-space of a double pendulum; *small circles* and *squares* mark the points in Q-space which correspond to the same position of the pendulum and should be identified

Kinetic energy may be considered as a primary characteristic of mechanical systems. Then masses can be defined, as was suggested by H. Poincaré, as coefficients of this quadratic form. Formula (1.31) is valid only in inertial frames. In all further considerations, the observer's frame is assumed to be inertial. The corresponding results in non-inertial frames are obtained by change of coordinates.

In the Lagrange variational principle, it is not essential that K is a quadratic form; kinetic energy can be any positive-valued function $K(q, \dot{q})$ homogeneous of second order with respect to $\dot{q}$. Homogeneity means that for any λ,

$$K(q, \lambda\dot{q}) = \lambda^2 K(q, \dot{q}). \tag{1.32}$$

Homogeneous functions of the second order obey the identity

$$\sum_i \dot{q}_i \frac{\partial K}{\partial \dot{q}_i} = 2K(q, \dot{q}). \tag{1.33}$$

This identity follows from (1.32) by differentiation of (1.32) with respect to λ and setting $\lambda = 1$. Obviously, quadratic forms obey both (1.32) and (1.33).

Formulas (1.31) and (1.33) can be written shorter if we employ notation from tensor analysis: repeated indices in a formula implies summation over these indices. This convention allows us to drop the summation sign in (1.31) and (1.33) to write

$$K = \frac{1}{2} m_{ij}\dot{q}_i\dot{q}_j, \quad \dot{q}_i \frac{\partial K}{\partial \dot{q}_i} = 2K(q, \dot{q}). \tag{1.34}$$

Summation over repeated indices is always assumed throughout the book.

Another notation from tensor analysis which we will employ is the use of low and upper indices. The reader who is not familiar with tensor analysis may assume that quantities with upper and lower indices coincide, for example, $q^1 = q_1$. This is always true in Cartesian coordinates. The differences appear only in curvilinear coordinates. The summation rule in invariant[7] form requires the summation to be taken over repeated one upper and one low indices. For example, an invariant form of the first Equation (1.34) is

$$K = \frac{1}{2} m_{ij}\dot{q}^i\dot{q}^j. \tag{1.35}$$

The reader may ignore these nuances and identify the quantities with upper and low indices until Chap. 3, where an adequate treatment of continuum mechanics requires the invariant tensor language. We also prefer to write all the equations for mechanical systems with a finite number of degrees of freedom in an invariant form.

Internal interactions in a mechanical system are characterized by its internal energy, a function $U(q)$. For any isolated system the law of conservation of energy holds:

[7] I.e. independent of the choice of the coordinate system.

$$K(q, \dot{q}) + U(q) = E = const. \tag{1.36}$$

Consider two points, q_0 and q_1, in the configurational space Q and the trajectories, $q = q(t)$, which begin at a point q_0 at time t_0 and end at a point q_1 at time t_1 (see Fig. 1.15):

$$q(t_0) = q_0, \quad q(t_1) = q_1. \tag{1.37}$$

We denote the set of such trajectories by $\mathcal{M}$.

The law of conservation of energy (1.36) does not put constraints on the possible trajectories: for any trajectory, $q(t)$, connecting the points q_0 and q_1, one can require that (1.37) holds. This equation will determine the rate of passing the trajectory, and, thus, the time, t_1, of arrival at the point q_1.

Indeed, according to (1.32), we can write

$$K(q, \dot{q}) = \frac{K(q, dq)}{dt^2}.$$

Therefore, for a given constant E,

$$dt = \frac{\sqrt{K(q, dq)}}{\sqrt{E - U(q)}}, \tag{1.38}$$

and the time of arrival at the point q is determined by the integral over the trajectory going from q_0 to q:

$$t = t_0 + \int_{q_0}^{q} \sqrt{\frac{K(q, dq)}{E - U(q)}}.$$

Accordingly, the arrival time at the point q_1 is

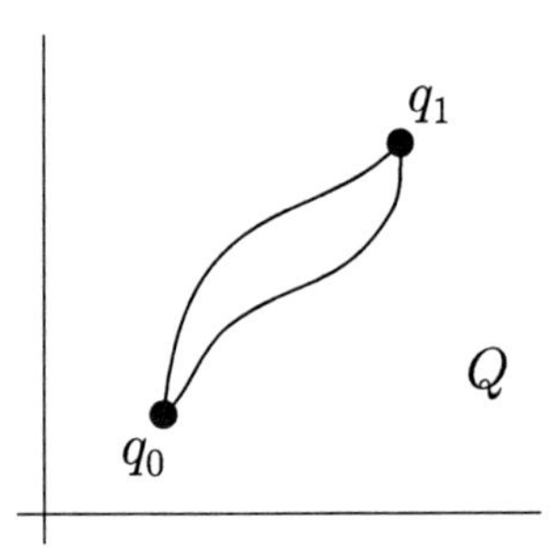

Fig. 1.15 Admissible trajectories in the configurational space Q

$$t_1 = t_0 + \int_{q_0}^{q_1} \sqrt{\frac{K(q, dq)}{E - U(q)}}.$$

The arrival time depends on the trajectory.

Let us introduce action as the time integral of kinetic energy:

$$I = \int_{t_0}^{t_1} K(q, \dot{q})\, dt. \tag{1.39}$$

Lagrange variational principle. *The true motion is a stationary point of the action functional (1.39) on the set $\mathcal{M}$ narrowed by an additional constraint: energy is conserved along each path.*

Note that in the Lagrange variational principle the time at which the system arrives at the point q_1 is not fixed. This time is determined by the trajectory. The arrival time is varied when one varies the trajectory.

We will see that the equations, determining the true trajectory, are

$$\frac{\partial L}{\partial q^i} - \frac{d}{dt}\frac{\partial L}{\partial \dot{q}^i} = 0, \tag{1.40}$$

where L is the difference of kinetic and potential energies,

$$L = K - U. \tag{1.41}$$

The left hand side of (1.40) is usually denoted by the symbol,

$$\frac{\delta L}{\delta q^i} \equiv \frac{\partial L}{\partial q^i} - \frac{d}{dt}\frac{\partial L}{\partial \dot{q}^i},$$

which is called the variational derivative of the function $L(q, \dot{q}, t)$, while $L(q, \dot{q}, t)$ is called Lagrange function or Lagrangian.

The equations of the type (1.40) are the basic equations of classical mechanics. They are called Lagrange equations.

The derivation of (1.40) from Lagrange variational principle is given in Appendix D. Equation (1.40) will be obtained further from the Hamilton variational principle which requires fewer technicalities.

The principle of least action was not explained very clearly in Lagrange's *Analytical Mechanics*. As Jacobi pointed out in his *Lectures on Dynamics* [140], "This principle is presented in almost all textbooks, even in the best ones, like Poisson's, Lagrange's and Laplace's, in such a way that, in my opinion, it is beyond comprehension." The derivation given in Appendix D was published 3 years after Lagrange's death by Rodrigues [257]. The obscurities instigated Hamilton,

Ostrogradsky, Jacobi, and Poincaré to examine the subject further. This brought up a number of modifications of the principle of least action.

1.5 Jacobi Variational Principle

Jacobi proposed to eliminate time from the variational principle by means of the energy equation. Indeed, using relation (1.38), one can write the action functional in the form

$$I = \int_{t_0}^{t_1} K dt = \int_{q_0}^{q_1} \sqrt{E - U(q)} \sqrt{K(q, dq)}. \tag{1.42}$$

Here integration is taken over a path connecting the points q_0 and q_1. We arrive at **Jacobi variational principle.** *The true motion is the stationary point of functional (1.42) on the set of all trajectories which begin at point q_0 and end at point q_1 and which satisfy the law of conservation of energy (1.36).*

The action functional taken in the form (1.42) does not depend on the rate with which the path is passed. The rate is controlled by the energy equation.

The Jacobi principle takes an especially beautiful form in the case when internal energy does not depend on q. Then, up to a constant factor,

$$I = \int_{q_0}^{q_1} \sqrt{2K(q, dq)},$$

(the factor 2 is put in for convenience). For a quadratic function, $2K(q, \dot{q}) = g_{ij}(q)\dot{q}^i\dot{q}^j$, and

$$I = \int_{q_0}^{q_1} \sqrt{g_{ij}(q) dq^i dq^j}.$$

If one measures distances in Q-space by means of kinetic energy form, i.e. the squared distance between the points q and $q+dq$ is set equal to $g_{ij}(q) dq^i dq^j$, then the principle of least action states that true motion corresponds to geodesic lines in Q-space, i.e. the shortest paths connecting the points q_0 and q_1.

1.6 Hamilton Variational Principle

Hamilton put the principle of least action in the form which is used nowadays most widely. He noticed that it is not necessary to take into account the law of conservation of energy if the action functional is taken in the form

$$I = \int_{t_0}^{t_1} L(q, \dot{q}, t)\, dt. \tag{1.43}$$

$$L(q, \dot{q}, t) = K(q, \dot{q}, t) - U(q, t).$$

It turns out that energy will be conserved automatically due to Euler equations for the functional (1.43) if K and U do not explicitly depend on time. An advantage of this modification of the least action principle is the possibility to consider time t_1 fixed and to avoid the energy constraint (1.36) on the trajectories.

Hamilton variational principle. *The true motion is the stationary point of the functional (1.43) on the set of all paths beginning at point q_0 and instant t_0 and ending at point q_1 at instant t_1.*

Somewhat later, and independently of Hamilton, Ostrogradsky arrived at the same statement.

Hamilton variation principle yields the dynamical equations of mechanics (1.40): this follows from (1.30). Consider these equations for the above-mentioned examples.

Example 1 of Sect. 1.5[8] *(continued)* Denote the mass of the particle by m. Then

$$K = \frac{1}{2} m \dot{q}^2.$$

Potential energy contains only the particle-wall interaction energy; denote it by $\Phi_\varepsilon(q)$. So,

$$L = \frac{1}{2} m \dot{q}^2 - \Phi_\varepsilon(q). \tag{1.44}$$

Substituting (1.44) into (1.40) we get the equation of particle motion:

$$m\ddot{q} + \frac{d\Phi_\varepsilon(q)}{dq} = 0. \tag{1.45}$$

If the function $\Phi_\varepsilon(q)$ is of the form shown in Fig. 1.16a, i.e. it is equal to zero on the segment $[\varepsilon, a - \varepsilon]$ and grows to infinity in the vicinities of the points $q = 0$ and $q = a$, then (1.45) describes elastic collisions of the particle and the walls. That means, by definition, that the particle, moving toward a wall with velocity v, has the velocity $-v$ after collision. Indeed, (1.45) admits the reduction of the order: multiplying it by $\dot{q}$ we have

$$\frac{d}{dt}\left(\frac{1}{2} m \dot{q}^2 + \Phi_\varepsilon(q)\right) = 0$$

or

$$\frac{1}{2} m \dot{q}^2 + \Phi_\varepsilon(q) = E \tag{1.46}$$

[8] The section number is mentioned only when we refer to the examples from a different section.

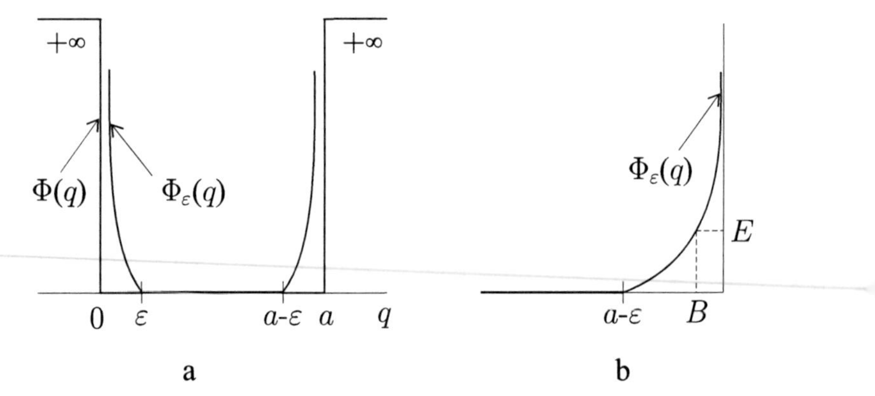

Fig. 1.16 Particle-wall interaction energy

where E is a constant, depending on the initial conditions. Consider a particle, moving toward the wall $q = a$ ($\dot{q} > 0$). When q is outside the particle-wall interaction region, $\Phi_\varepsilon(q) = 0$, and, as follows from (1.46), the velocity of the particle is constant. As soon as the particle comes into the interaction zone $[a - \varepsilon \leq q \leq a]$, the particle velocity begins to change. As the particle penetrates the interaction zone, the interaction energy grows and, as follows from (1.46), the velocity decreases. At point B on Fig. 1.16b, where the interaction energy Φ_ε is equal to the initial energy of the particle E, the velocity $\dot{q}$ is equal to zero in accordance with (1.46). However, this point is not an equilibrium point because, as follows from (1.45), the acceleration $\ddot{q}$ is not equal to zero at B. Since $\dfrac{d\Phi_\varepsilon(q)}{dq} < 0$ at B, the particle starts moving in the opposite direction. The particle's velocity grows because $\Phi_\varepsilon(q)$ decreases; and, when the particle escapes the interaction zone, the velocity becomes equal to the initial velocity with the opposite sign. So, the function $\Phi_\varepsilon(q)$ really describes an elastic collision. In the limit $\varepsilon \to 0$, function $\Phi_\varepsilon(q)$ becomes the function

$$\Phi(q) = \begin{cases} 0 & \text{if } 0 \leq q \leq a \\ +\infty & \text{if } q < 0 \text{ or } q > a \end{cases} \tag{1.47}$$

Every time we use this function we should bear in mind a smoothed function $\Phi_\varepsilon(q)$ with a parameter ε, which is much smaller than any characteristic length of the problem. Note that the functions $\Phi_\varepsilon(q)$ and $\Phi(q)$ depend on the position of the wall a. If one of the walls moves, a depends on time. Then, the potential energy and, thus, the Lagrange function explicitly depend on time.

If the particle is attached to a spring, and the spring is stress-free when $q = q_0$, then

$$U(q) = \frac{1}{2}k(q - q_0)^2,$$

k being the spring rigidity. The Lagrange function of the particle on a spring is

$$L = \frac{1}{2}m\dot{q}^2 - \frac{1}{2}k(q - q_0)^2,$$

and the Lagrange equations of motion (1.40) transform to the usual dynamical equation of an oscillator:

$$m\ddot{q} + k(q - q_0) = 0.$$

Now we proceed to a less elementary case.

Example 2 of Sect. 1.5 (continued) Let us find the Lagrange function of a pendulum. We refer the motion of the pendulum to Cartesian coordinates $(x, \ y)$, shown in Fig. 1.17. Assume that the rod is massless and the mass m is concentrated at the end of the rod. If $x(t)$, $y(t)$ are the coordinates of the mass point, then the kinetic energy is

$$K = \frac{1}{2}m\left(\dot{x}^2 + \dot{y}^2\right). \tag{1.48}$$

From Fig. 1.17 we have

$$x = \ell \sin q, \quad y = \ell \cos q. \tag{1.49}$$

Let us suppose that the length ℓ can also change with time; this corresponds, for instance, to the case of a mass on a rope, which could be pulled up through a hole (Fig. 1.18). The dependence of ℓ on time is supposed to be given. Differentiating (1.49), we get

$$\dot{x} = \dot{\ell} \sin q + \ell \cos q \ \dot{q}, \quad \dot{y} = \dot{\ell} \cos q - \ell \sin q \ \dot{q}. \tag{1.50}$$

Substituting (1.50) into (1.48), we find the kinetic energy:

$$K = \frac{1}{2}m\left(\dot{\ell}^2 + \ell^2\dot{q}^2\right). \tag{1.51}$$

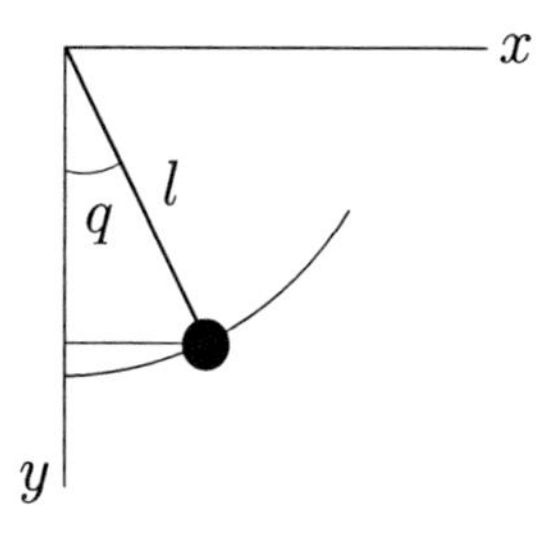

Fig. 1.17 To calculation of the kinetic energy of a pendulum

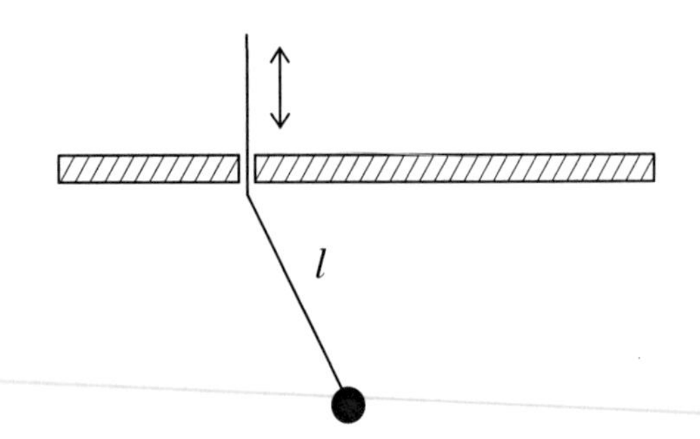

Fig. 1.18 The case when l depends on time

Suppose that gravity is the only force acting on the pendulum. The gravitational potential energy is equal to the product of the mass m, the gravity acceleration constant g, and the height of the mass above some fixed level. One can choose the bottom position of the pendulum as the level of zero potential energy. Then, it is seen from Fig. 1.17 that

$$U = mg(\ell - \ell \cos q).$$

So,

$$L = \frac{1}{2}m\dot{\ell}^2 + \frac{1}{2}m\ell^2\dot{q}^2 - mg\ell(1 - \cos q). \tag{1.52}$$

The Lagrange equation of motion takes the form

$$\frac{d}{dt}\left(m\ell^2\dot{q}\right) + mg\ell \sin q = 0. \tag{1.53}$$

The variation of the first term in (1.52) is zero because $\ell(t)$ is given.

If ℓ is constant, then (1.53) is reduced to the equation of a "mathematical pendulum":

$$\ddot{q} + \frac{g}{\ell}\sin q = 0. \tag{1.54}$$

Note that (1.54) does not contain the mass of the pendulum. This yields the remarkable feature of pendulum vibrations: the frequency of vibration does not depend on the pendulum mass. This property was first experimentally observed by Galileo and was a cornerstone of Newtonian mechanics.

If ℓ is not constant, then an additional "force" enters the governing equation:

$$\ddot{q} + \frac{g}{\ell}\sin q = -2\dot{q}\frac{\dot{\ell}}{\ell}.$$

It is quite cumbersome to obtain these equations from the "force concept."

Example 3 of Sect. 1.5 (continued). The reader can check that the double pendulum has the following Lagrange function:

$$L = \frac{m_1 + m_2}{2}\ell_1^2\dot{q}_1^2 + \frac{m_2}{2}\ell_2^2\dot{q}_2^2 + m_2\ell_1\ell_2\dot{q}_1\dot{q}_2\cos(q_1 - q_2)$$
$$+(m_1 + m_2)g\ell_1\cos q_1 + m_2 g\ell_2\cos q_2.$$

The notation is shown in Fig. 1.19; $\ell_1, \ \ell_2$ are assumed constant.

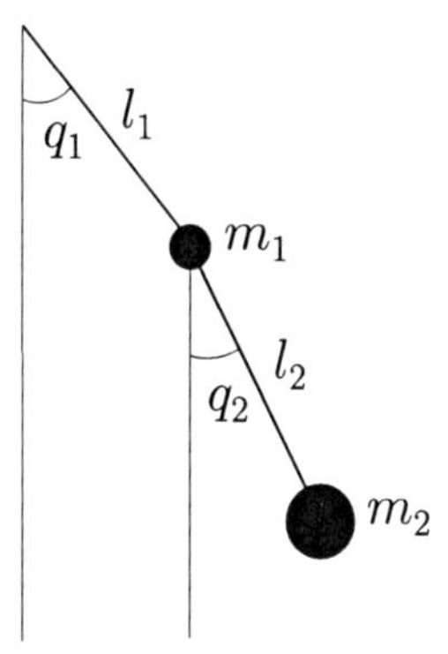

Fig. 1.19 Notation for the double pendulum

Example 4 of Sect. 1.5 continued. The Lagrange function of a particle in a box is

$$L = \frac{1}{2}m\left(\dot{x}^2 + \dot{y}^2 + \dot{z}^2\right) - \Phi(x) - \Phi(y) - \Phi(z),$$

where Φ is the function (1.47). The set of N particles with masses $m_1, \ldots, m_N$ has the Lagrange function

$$L = \frac{1}{2}m_1\left(\dot{x}_1^2 + \dot{y}_1^2 + \dot{z}_1^2\right) + \cdots + \frac{1}{2}m_N\left(\dot{x}_N^2 + \dot{y}_N^2 + \dot{z}_N^2\right) - \qquad (1.55)$$
$$-\Phi(x_1) - \Phi(y_1) - \Phi(z_1) - \cdots - \Phi(x_N) - \Phi(y_N) - \Phi(z_N).$$

Note that the particles interact in this case only with the walls, and two or more particles are allowed to be at the same space point. The model of spherical particles with some finite radius a, experiencing elastic collisions, is more realistic. In this case one has to include in (1.55) the particle-particle interaction energy:

$$\sum_{\alpha<\beta}\Psi(|r_\alpha - r_\beta|).$$

Here, Greek indices run through the values $1, 2, \ldots, N$; r_α is the position vector of the center of the αth sphere, $|r_\alpha|$ is the length of the vector, r_α, and the function Ψ

is shown in Fig. 1.20a. As before, in dealing with function Ψ, one should bear in mind a smoothed function Ψ_ε (Fig. 1.20b) and consider the limit $\varepsilon \to 0$. In some physical models an attraction of particles at long distances is added (Fig. 1.20c).

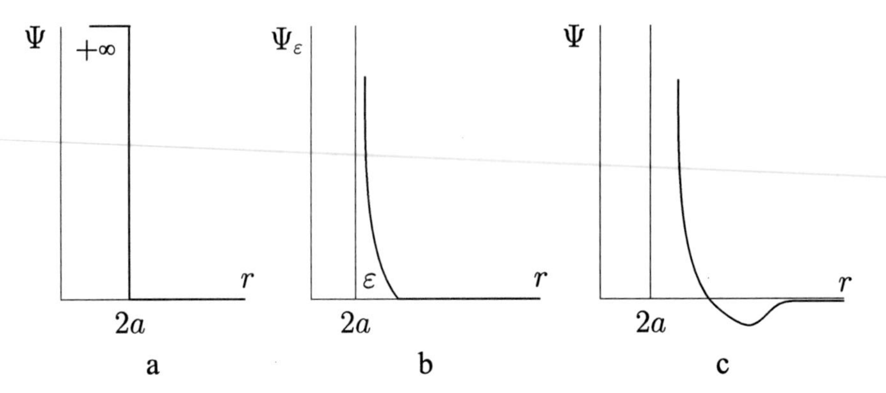

Fig. 1.20 Particle-particle interaction energy

For Hamilton, the formulation of the variational principle was a by-product of the study on integration of dynamical equations. He considered the action functional as a function of initial and final points of the trajectory, $I = I(q_1, q_2)$, and derived the partial differential equation for $I(q_1, q_2)$ which is equivalent to dynamical equations. This construction was used later by Schrödinger in motivations for the basic equation of quantum mechanics.

1.7 Hamiltonian Equations

The equations of mechanics (1.40) form a system of n ordinary differential equations of second order. Usually, it is more convenient to deal with a system of $2n$ ordinary differential equations of first order. Let us put the equations of motion (1.40) into this special form.

First, consider the systems with one degree of freedom and the Lagrange function

$$L = \frac{1}{2}m\dot{q}^2 - U(q).$$

The equation of motion (1.40) is the second-order equation

$$\frac{d}{dt}(m\dot{q}) + \frac{\partial U(q)}{\partial q} = 0. \tag{1.56}$$

To rewrite it as a system of two equations of first order, one has to introduce a new required function, the momentum $p = m\dot{q}$. Then (1.56) can be written as a system of two ordinary differential equations of the first order,

$$\dot{q} = \frac{1}{m}p, \quad \dot{p} = -\frac{\partial U}{\partial q}.$$

To make a similar transformation in the general case we introduce the new unknown functions, the momenta, by the relations

$$p_i = \frac{\partial L\left(q^k, \dot{q}^k, t\right)}{\partial \dot{q}^i}. \tag{1.57}$$

Suppose that we can solve (1.57) with respect to $\dot{q}_k$:

$$\dot{q}^k = A^k(q^i, p_i, t). \tag{1.58}$$

If L is a quadratic function of $\dot{q}$, then (1.57) is a system of linear equations with respect to $\dot{q}$, and velocities, $\dot{q}$, depend on momenta p linearly. This is the usual case. There are, however, mechanical systems for which L is not a quadratic function of velocities (see Sects. 9.6 and 9.7). Therefore, we proceed without the assumption that (1.57) is a linear system of equations.

Using (1.58) one can rewrite Lagrange equations (1.40) as a system of the first-order equations,

$$\dot{q}^k = A^k(q, p, t), \quad \dot{p}_k = \left.\frac{\partial L\left(q, \dot{q}, t\right)}{\partial q^k}\right|_{\dot{q}=A(q,p,t)}. \tag{1.59}$$

It is implied that the expression of $\dot{q}$ in terms of p and q is substituted in the second equation (1.59).

It turns out that equations (1.59) take an especially elegant form if they are written in terms of a function of the variables p and q, defined by the relation

$$H(p, q, t) = p_i\dot{q}^i - L(q^i, \dot{q}^i, t). \tag{1.60}$$

To calculate H one has to substitute in the right-hand side of (1.60) the expressions of $\dot{q}$ in terms of p and q (1.58), found from (1.57). The function H obtained in this way is the Legendre transformation of the Lagrange function with respect to the generalized velocities $\dot{q}_i$; the Legendre transformation is considered in detail in Chap. 5.

The functions A_k can be written in terms of function H in a simple form:

$$\dot{q}^k = A^k = \frac{\partial H(p, q, t)}{\partial p_k}. \tag{1.61}$$

Indeed, differentiating (1.60) with respect to p_k and taking into account that $\dot{q}_i$ in the right-hand side of (1.60) are functions of p and q found from (1.57), we have

$$\frac{\partial H}{\partial p_k} = \dot{q}^k + p_i \frac{\partial \dot{q}^i}{\partial p_k} - \frac{\partial L}{\partial \dot{q}^i} \frac{\partial \dot{q}^i}{\partial p_k}. \tag{1.62}$$

The two last terms in (1.62) are equal, due to (1.57), and we see that (1.61) holds true. Similarly, differentiating (1.60) with respect to q_k we obtain

$$\frac{\partial L}{\partial q^k} = -\frac{\partial H}{\partial q^k}.$$

So, the equation of motion can be rewritten in the form

$$\dot{q}^k = \frac{\partial H(p, q, t)}{\partial p_k}, \qquad \dot{p}_k = -\frac{\partial H(p, q, t)}{\partial q^k}. \tag{1.63}$$

These equations and function H were introduced by Hamilton [123]. They are called the Hamiltonian equations and Hamilton function (or Hamiltonian).

If the Lagrange function is the difference between the kinetic energy K and the potential energy U, and K is quadratic with respect to $\dot{q}_i$, then H can be interpreted as the total energy:

$$H = K + U. \tag{1.64}$$

The reader is invited to check (1.64).

The derivative of the Hamiltonian function with respect to time along a trajectory of the system can be found by using (1.63) and is equal to

$$\frac{dH}{dt} \equiv \frac{\partial H}{\partial q^k} \dot{q}^k + \frac{\partial H}{\partial p_k} \dot{p}_k + \frac{\partial H}{\partial t} = \frac{\partial H(p, q, t)}{\partial t}. \tag{1.65}$$

As follows from (1.60), the partial derivatives of H and L are linked as

$$\frac{\partial H(p, q, t)}{\partial t} = -\frac{\partial L(q, \dot{q}, t)}{\partial t}.$$

The explicit dependence of H and L on time appears if some external objects act on the system (such as in Example 2 where this dependence is due to the change of ℓ with time). If the system is isolated and does not interact with its surroundings,

H and L do not depend on time explicitly. Then (1.65) transforms into the law of conservation of energy:

$$\frac{dH\,(p(t),q(t))}{dt}=0 \quad \text{or} \quad H(p(t),q(t))=E=\text{const.} \tag{1.66}$$

Example 5 of Sect. 1.5 (continued). The Lagrange function of the pendulum is

$$L=\frac{1}{2}ml^2\dot{q}^2-mgl(1-\cos q).$$

Therefore, the generalized momentum, p, is given by

$$p=\frac{\partial L}{\partial \dot{q}}=ml^2\dot{q}. \tag{1.67}$$

From (1.60) we find the Hamilton function

$$H(p,q)=p\dot{q}-L=\frac{p^2}{2ml^2}+mgl(1-\cos q). \tag{1.68}$$

The Hamilton function is the total energy (the sum of the kinetic and potential energies). The equations of motion in Hamiltonian form are

$$\dot{p}=-\frac{\partial H}{\partial q}=-mgl\sin q, \quad \dot{q}=\frac{\partial H}{\partial p}=\frac{p}{ml^2}.$$

If l changes with time due to external action, the Hamilton function depends on time explicitly through l.

Poincaré constructed a different version of the least action principle, in which the action functional is defined on the trajectories in the phase space – the space of generalized coordinates $q^1,\dots,q^n$ and momenta $p_1,\dots,p_n$. To obtain the expression for the action functional one can plug in (1.43) function L determined from (1.60):

$$I=\int_{t_0}^{t_1}\left(p_i\dot{q}^i-H(p,q)\right)dt. \tag{1.69}$$

It is easy to check that the stationary points of functional (1.43) are also the stationary points of functional (1.69) on the set of all functions $q\,(t)$, $p\,(t)$ which satisfy the conditions $q\,(t_0)=q_0$, $q\,(t_1)=q_1$. Requiring the variation of functional (1.69) to vanish yields the Hamiltonian equations.

The functional

$$A = \int_{t_0}^{t_1} p_i \dot{q}^i dt,$$

is sometimes called the shortened action.

In physics, the methods of derivation of equations using functional (1.43) and functional (1.69) are called the Lagrangian and Hamiltonian formalisms, respectively.

Both versions of the least action principle enjoyed the most popularity, and in current literature, the versions of Lagrange and Jacobi are rarely encountered.

1.8 Physical Meaning of the Least Action Principle

Forces vs energy. Usually, the laws of mechanics are formulated in an alternative way using the notion of force. To discuss the differences, consider motion of a particle of mass m. In Cartesian coordinates x^i, $i = 1, 2, 3$, motion of the particle is described by the dependence of its coordinates on time: $x^i = x^i(t)$. According to Newton's second law, mass times acceleration of the particle is equal to the force, F^i, acting on the particle:

$$m \frac{d^2 x^i(t)}{dt^2} = F^i. \tag{1.70}$$

The right-hand-side of this equation, force, represents the external action on the particle. If there are no external actions, the right-hand-side of (1.70) is zero. This fact has important consequences. First, (1.70) incorporates Galileo's principle: an isolated particle moves with constant velocity. Second, the frame of reference used in writing (1.70) is inertial; an inertial frame is a frame of reference in which motion of a free mass is uniform.

From an experimentalist's point of view, (1.70) is nothing but a definition of force. To find the force one has to measure the mass of the particle and its acceleration, then the force is their product. Remarkably, in many cases, the force determined in this way turns out to be a universal function of the position of the particle, x^i, and its velocity, $\frac{dx^i}{dt}$. Then (1.70) becomes an ordinary differential equation of second order. It allows one to find motion if the initial position and initial velocity of the particle are given. The possibility to prescribe arbitrarily initial position and initial velocity is another remarkable feature of our world (consider, for example, throwing a stone: one can choose any starting point and any velocity, then the further path of the stone is determined uniquely). The force is not necessarily a function of x^i and $\dot{x}^i$: it may depend on the entire prehistory of motion (e.g. a body sinking in water; see Sect. 13.3).

Compare (1.70) with the Lagrange equation,

$$\frac{\partial L}{\partial x^i} - \frac{d}{dt}\frac{\partial L}{\partial \dot{x}^i} = 0. \tag{1.71}$$

We set Lagrange function to be the difference between kinetic and potential energy:

$$L = \frac{1}{2}m\frac{dx^i}{dt}\frac{dx_i}{dt} - U(x,t). \tag{1.72}$$

By x, as usual, we denote the set of three coordinates, x^1, x^2 and x^3. The governing dynamic equation (1.71) takes the form

$$m\frac{d^2x_i(t)}{dt^2} = -\frac{\partial U(x,t)}{\partial x^i}.$$

We see that the force corresponding to the Lagrange equation is always potential:

$$F_i = -\frac{\partial U(x,t)}{\partial x^i}. \tag{1.73}$$

Such a property may seem to be too strong a constraint on possible forces encountered in Nature. The striking feature of our world is that all physical laws do follow the potentiality rule at the microlevel. At least until now no experimental evidence has been found to contradict (1.73) (see also further comments in Sect. 2.6). Perhaps the principle of least action brings an adequate theory of micromotion. At the macrolevel some additional (dissipative) terms may appear in (1.73). Further discussion of this issue is given in Chap. 2.

Admitting that (1.73) holds we state that energy always exists. Then one can eliminate forces from consideration and discuss the physical properties of the system directly in terms of energy. This is a way which is accepted in modern physics.

Reciprocity of interactions. Existence of energy puts very strong constraints on possible interactions in the system. We discuss these constraints for two examples.

Consider a two-dimensional motion of a mass particle attached to a set of springs (Fig. 1.21). We choose the origin of the Cartesian coordinate system (x, y) at the equilibrium position of the particle. Let the deviations of the particle from the equilibrium position, x and y, be small. Then the potential energy of the system, U, may be expanded over x and y in the vicinity of the point $x = y = 0$. Let us ignore terms of the third and higher degrees in x and y:

$$U = U_0 + a_x x + a_y y + \frac{1}{2}k_x x^2 + k_{xy}xy + \frac{1}{2}k_y y^2. \tag{1.74}$$

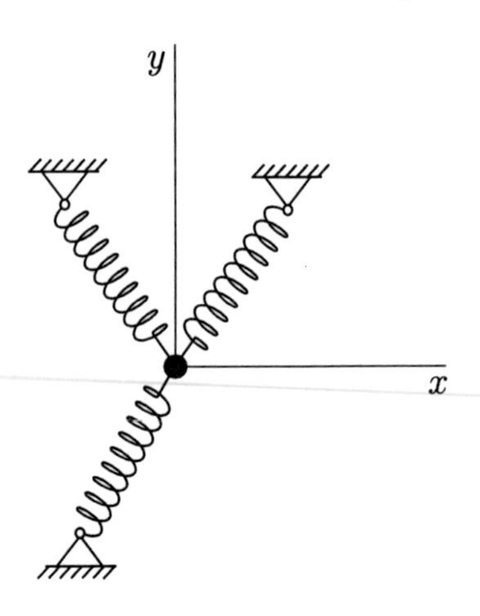

Fig. 1.21 Sketch for the first example

Then the forces acting on the particle in the axis directions, F_x and F_y, are

$$F_x = -\left(a_x + k_x x + k_{xy} y\right),$$
$$F_y = -\left(a_y + k_{xy} y + k_y y\right). \tag{1.75}$$

Note right away the feature which will be used further without mention: linear relations between forces and displacements correspond to the quadratic dependence of energy on displacements.

Since the particle position $x = y = 0$ is equilibrium, force must be zero at $x = y = 0$. Thus, $a_x = a_y = 0$, i.e. the linear terms in energy are zero. The additive constant U_0 does not affect the equations of motion. Due to this, one usually says that energy is defined up to a constant. This is true for the case under consideration. Later on we will encounter the situation where an additive constant may play an important role (see Sect. 7.4). So, forces are linked to displacements by the relations

$$F_x = -k_x x - k_{xy} y,$$
$$F_y = -k_{xy} y - k_y y. \tag{1.76}$$

Such types of relations are usually called constitutive equations. The coefficients in (1.76) are called rigidities of the system.

The term $-k_x x$ in (1.76) describes the force making the particle move to the equilibrium position: if $x > 0$, the force $-k_x x$ is negative; if $x < 0$, the force $-k_x x$ is positive; this, of course, assumes that the coefficient k_x is positive. The force $-k_y y$ has a similar meaning. The most important for our discussion are the interaction forces $-k_{xy} y$ and $-k_{xy} x$. The force $-k_{xy} y$ is the force in the x-direction caused by the vertical deflection of y. The force $-k_{xy} x$ is the force in the y-direction caused by the horizontal deflection of x. In general, these two forces may be expected to have different rigidities, say, $-k_1 y$ and $-k_2 x$. The existence of energy implies that k_1

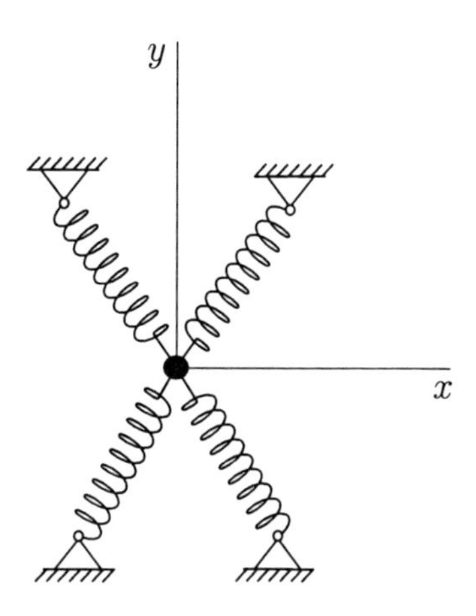

Fig. 1.22 Sketch for the second example

must be equal to k_2. Remarkably, this is a well-supported experimental fact provided that all springs can be considered as purely elastic.

The existence of energy brings more sophisticated constraints on the admissible constitutive equations in a nonlinear situation. Consider, for simplicity, the case of the system which is symmetric with respect to the reflection $x \rightarrow -x$, like the system shown in Fig. 1.22. Due to the symmetry, energy U must be an even function of x: the system gains the same energy as the particle is deflected in the x-direction for x or $-x$. The rigidity k_{xy} must be zero for all such systems. Let us write down energy keeping the terms of the third order. There are four terms of the third order: x^3, x^2y, xy^2 and y^3. The terms x^3 and xy^2 cannot appear due to the evenness of energy with respect to x. The term y^3 brings a small correction to the quadratic term y^2. Thus, the only interesting effect may be expected from the interaction term x^2y. So,

$$U = \frac{1}{2}k_x x^2 + \frac{1}{2}k_y y^2 + Ax^2 y. \tag{1.77}$$

For forces, we have

$$\begin{aligned} F_x &= -\left(k_x x + 2Axy\right), \\ F_y &= -\left(Ax^2 + k_y y\right). \end{aligned} \tag{1.78}$$

We see that the deflection in the x-direction causes a vertical force proportional to the squared deflection, while the deflection in the y-direction causes a horizontal force proportional to the product of the deflections in both directions. Moreover, the rigidity in the horizontal force is twice that in the vertical force. It would be hard to anticipate such a peculiar behavior of the system staying entirely within the force concept; it is caused only by the existence of energy. Remarkably, constitutive equations of the form (1.78) are supported experimentally in a similar phenomenon of the torsion-extension interaction in isotropic elastic beams, the so-called Poynting effect.

Inertial reciprocities. Accepting Lagrange equations as the basic governing equations of mechanics assumes more than just the existence of energy. We, in fact, claim the existence of the Lagrange function. The Lagrange function contains the complete information on the physical properties of the system. The special form of Lagrange equations (1.71) indicates that another kind of reciprocity appears if mass depends on generalized coordinates. Such a situation is encountered, for example, for rigid bodies moving in ideal fluids (see Sect. 13.3). In this case, Lagrange function coincides with the total kinetic energy of the system, which is the sum of kinetic energy of the rigid body and kinetic energy of fluid. The latter is proportional to squared velocity of the body; the coefficient, called the attached (or apparent) mass of the body, depends on the geometry of the flow region and, thus, on the position of the body. Consider, for example, a body moving toward (or away from) the wall (Fig. 1.23). Kinetic energy of the body is $\frac{1}{2}m_0(dy/dt)^2$. Kinetic energy of the fluid motion caused by the motion of the body is proportional to $(dy/dt)^2$. The coefficient, $\frac{1}{2}m_a$, depends on the distance between the body and the wall, y: $m_a = m_a(y)$, $m_a(y)$ being the attached mass.

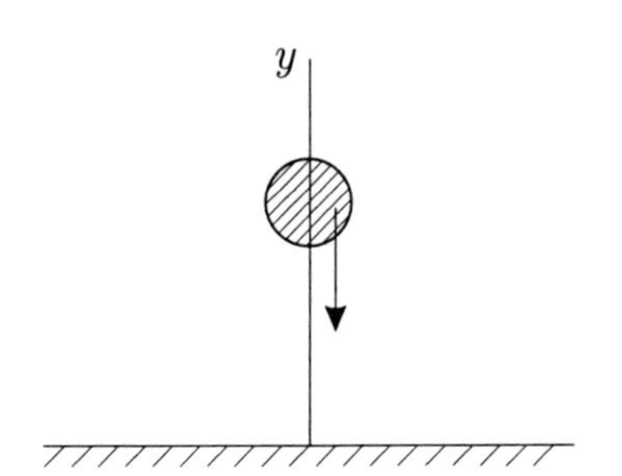

Fig. 1.23 Sketch for the third example

The Lagrange function of the system is

$$L = \frac{1}{2}\left(m_0 + m_a\left(y\right)\right)\left(\frac{dy}{dt}\right)^2,$$

The dynamical equation (1.71) takes the form

$$\frac{d}{dt}\left(m_0 + m_a\left(y\right)\right)\frac{dy}{dt} - \frac{1}{2}\frac{dm_a\left(y\right)}{dy}\left(\frac{dy}{dt}\right)^2 = 0.$$

We see that the force acting on the body is quite peculiar:

$$F = -\frac{d}{dt}\left(m_a(y)\frac{dy}{dt}\right) + \frac{1}{2}\frac{dm_a(y)}{dy}\left(\frac{dy}{dt}\right)^2 =$$

$$= -m_a\frac{d^2y}{dt^2} - \frac{1}{2}\frac{dm_a(y)}{dy}\left(\frac{dy}{dt}\right)^2$$

It requires quite an effort to obtain this relation without the variational approach.

Action and reaction. Consider two bodies, A and B, with masses m_A and m_B. The bodies move in unbounded space. We model the bodies by point masses and denote their coordinates by x^i and y^i, respectively. The bodies interact, and therefore the potential energy of the system is a function of x^i and y^i : $U = U(x^i, y^i)$, while the Lagrange function is

$$L = \frac{1}{2}m_A\frac{dx^i}{dt}\frac{dx_i}{dt} + \frac{1}{2}m_B\frac{dy^i}{dt}\frac{dy_i}{dt} - U(x^i, y^i)$$

To determine the interaction forces between the bodies, we write down the dynamical equations of the system:

$$m_A\frac{d^2x^i}{dt^2} = F^i_{AB} = -\frac{\partial U(x^i, y^i)}{\partial x^i}, \qquad m_B\frac{d^2y^i}{dt^2} = F^i_{BA} = -\frac{\partial U(x^i, y^i)}{\partial y^i}$$

Here F^i_{AB} is the force acting on the body A from the body B, and F^i_{BA} the force acting on B from A. In general, these two forces are different.

Assume that the translation of both bodies for an arbitrary vector, c^i, does not change the potential energy:

$$U(x^i + c^i, y^i + c^i) = U(x^i, y^i) \tag{1.79}$$

Then the potential energy is a function of only the difference, $x^i - y^i$ (it is enough to set $c^i = -y^i$ in (1.79), then $U(x^i, y^i) = U(x^i - y^i, 0)$). Denote this function by $\Phi(z^i)$:

$$U(x^i, y^i) = \Phi(x^i - y^i)$$

Plugging this relation into the formulas for the forces we obtain

$$F^i_{BA} = -F^i_{AB},$$

i.e. the force acting on the body B from the body A is equal in magnitude and opposite in direction to the force acting from the body B on the body A. This is the interaction law suggested by Newton. We see that this law has a simple underlying cause in the energy terms.

Meaning of the term "Hamiltonian structure". Hamiltonian form of the governing equations implies quite essential features of physical interaction. We need, however, make one important addition to the definition of the terms "Hamiltonian form" and "Hamiltonian structure." If these terms are understood "naively" as a way to write a system of ordinary differential equations in the form of Hamiltonian equations,[9] then a system possessing a Hamiltonian structure is not a special one because any system of ordinary differential equations can be written in Hamiltonian form. Indeed, consider a system of ordinary differential equations:

$$\frac{dq^i}{dt} = Q^i(q), \quad i = 1, \ldots, n. \tag{1.80}$$

Let us introduce the additional required functions, $p_i(t)$, and determine $p_i(t)$ from the equations,

$$\frac{dp_i}{dt} = -p_k \frac{\partial Q^k(q)}{\partial q^i}. \tag{1.81}$$

Equations (1.80) and (1.81) form a Hamiltonian system of equations with the Hamiltonian,

$$H(p, q) = p_k Q^k(q).$$

This emphasizes that the term "Hamiltonian structure" must also include fixing the phase space of the system. By Hamiltonian structure of a mechanical system we mean that the mechanical system is characterized by certain generalized coordinates, q, and certain generalized momenta, p, and their dynamics is governed by Hamiltonian equations. Stating that, we imply that the characteristics chosen, p and q, are linked by the "Hamiltonian reciprocities."

Lagrange function vs energy. Lagrange function is the difference of kinetic energy and potential energy,

$$L = K - U.$$

Such a difference appears only in the least action principle and is not encountered in other relations of classical mechanics. This causes some perplexity: what meaning could one associate with minimization of the time integral of L? Here is an excerpt from Principles of Mechanics by H. Hertz [130]:

[9] Hamiltonian form of the equations is also often understood as a formal structure based on Poisson's brackets. Such a structure captures the features of differential equations related to integrability. We do not dwell on these issues here because they are not used further in the book. The corresponding theory can be found in [5, 328].

> ... The Hamilton principle ... does not only make the present events dependent on the future outcome, supposing the existence of intentions in Nature, but, which is worse, it supposes the existence of meaningless intentions. The integral, the minimum of which the Hamilton principle requires, does not have any simple physical meaning; what is more, Nature's purpose in achieving the minimum in a mathematical expression or making the variation of this expression equal to zero seems unintelligible....

An answer to the question on "Nature's intention" was found after the creation of the relativity theory. If a particle with the rest mass m_0 is considered in a force field creating the potential energy $U(q)$, then the total energy of the particle is $m_0c^2 + U(q)$, c being the speed of light. Let t be the observer's time. Then the proper particle time t^* is linked to the observer's time by the relation

$$dt^* = \sqrt{1-\frac{v^2}{c^2}}dt.$$

Consider the integral of total energy over the proper time,

$$I = \int (m_0c^2 + U(q))dt^* = \int_{t_0}^{t_1} \left(m_0c^2 + U(q)\right)\sqrt{1-\frac{v^2}{c^2}}dt.$$

If we take into account that

$$\frac{v}{c} << 1, \quad \frac{U(q)}{m_0c^2} << 1,$$

and keep only the leading terms in the integrand,

$$\left(m_0c^2 + U(q)\right)\sqrt{1-\frac{v^2}{c^2}} \approx m_0c^2\left(1+\frac{U(q)}{m_0c^2}\right)\left(1-\frac{v^2}{2c^2}\right) \approx$$
$$\approx m_0c^2 + U(q) - \frac{m_0v^2}{2},$$

then, after neglecting the additive constant m_0c^2, we get the Lagrange function (with an opposite sign).

The quantity $m_0c^2 + U(q)$ is the rest energy of the particle. Therefore, the action has a clear physical meaning: it is a sum over the particle's proper time of the quantity which is the total rest energy of the particle. The "strange" expression for the Lagrange function as the difference of the kinetic and the potential energies is the Newtonian limit in the observer's frame.

Nonlocal nature of the least action principle. Another perplexity is caused by the nonlocal character of the least action principle: the particle trajectory is selected by prescribing the initial and final positions of the particle instead of initial position and initial velocity, as is done in the usual form of Newtonian mechanics. Here is an excerpt from H. Poincaré [242]:

...There is something unacceptable to the intellect in the way the least action principle is phrased. To reach one point from another, a molecule not acted upon by any forces but constrained to a surfacc will move along the geodesic line, i.e. along the shortest possible path. This molecule is moving as though it knows the point where it is supposed to go, foresees the time it will take to reach that point taking one or another path, and then chooses the most appropriate one. The way the principle is stated presents the molecule as an animated being exercising free will. It is clear that it would be better to rephrase the formulation of the least action principle by less striking on, in which, as philosophers say, the final goals do not seem to be replacing the actual causes.

Perhaps, nonlocality of the least action principle has its roots in quantum mechanics. In quantum mechanics all particle trajectories are possible. The key characteristics of the particle motion is the probability of transition from one point to another. A link between this probability and Lagrange function was found by R. Feynman. He considered the quantity

$$\Phi\left(t_1, q_1; t_0, q_0\right) = \sum_{\text{all trajectories}} e^{\frac{i}{h}\int_{t_0}^{t_1} L\left(q(t), \frac{dq(t)}{dt}\right)dt} \tag{1.82}$$

where h is Plank's constant,[10] the sum is taken over all possible trajectories connecting the initial position of the system q_0 and the final position q_1. Of course, some technical difficulties must be overcome in order to provide a mathematical meaning of such a sum. It turns out that Φ, as a function of t_1 and q_1, obeys the Schrödinger equation , while $|\Phi|^2$ is the probability of transition of the system from the state q_0 at instant t_0 to state q_1 at instant t_1. If the magnitude of the action is much larger than h, then the major contribution to the sum (1.82) is provided by the trajectory on which action is stationary (one can apply the method of stationary phase to the sum). Such a case corresponds to classical (non-quantum) motion of the system. Quantum laws become essential if the action is of the order of h. Plank's constant is very small; therefore, motion of all systems with a large mass (in particular macroscopic bodies) is governed by the least action principle.

Minimum action vs stationary action. The most important outcome of the least action principle is that the governing equations possess a very special form which we will call the Hamiltonian structure. From this perspective, it is not essential whether the action functional has a minimum value or a stationary value; only the structure of the governing equations matters. In the formulation of the variational principle, one can always choose the final time, t_1, close to the initial time, t_0. Then, as will be seen in Sect. 8.1, in typical cases the action functional is minimum on the true trajectory for systems with a finite number of degrees of freedom. Therefore, the term "least action principle" is precise for sufficiently small $t_1 - t_0$.

---

[10] Note that Plank's constant has the dimension of action = dimension of energy $\times$ time; thus, the expression in the exponent is dimensionless.

Chapter 2
Thermodynamics

2.1 Thermodynamic Description

In some cases, mechanical systems with many degrees of freedom admit a simpler description using a small number of parameters. Consider, for example, modeling of a gas in a vessel. The vessel is closed with a piston to which some force P is applied. The force compresses the gas (Fig. 2.1). The gas is envisioned as a system of a large number of rigid balls representing its molecules. The balls move inside the vessel colliding elastically with the walls and the other balls. Usually one is not interested in knowing the molecule motion. It is of interest to determine how the volume occupied by the gas depends on the applied force. This is a typical "thermodynamic" question: one is concerned with some integral characteristics of the system and the relations between them. The characteristics used for the reduced description of the system are called thermodynamic parameters. Traditionally, thermodynamics is presented as a field which is logically and conceptually independent of mechanics. In such treatments, the central notion of thermodynamics, entropy, remains vague; and achieving an understanding of thermodynamics is a similar process to that in quantum mechanics, where, by Feynman's words, "to understand" means "to get used to and learn how to apply." In fact, thermodynamics may be derived from mechanics. Such a derivation makes clear the notions used and provides the conditions which are necessary for the basic thermodynamical laws to be true. "The mechanical view" on thermodynamics is outlined in this chapter. We focus only on the basic ideas and skip a derivation if it is lengthy. For more details the reader is referred to [46, 50].

The reason why some universal thermodynamic relations may exist was uncovered by Boltzmann and Helmholtz: the rate of change of the thermodynamic parameters is much smaller than the rate of change of the generalized coordinates and momenta of the system. In the system of enclosed gas forced by the piston, a thermodynamic description is possible if the piston velocity is much smaller than the average molecule velocity. If the velocity of the piston is on the order of the average molecule velocity, thermodynamic description fails: the relation between the force and the gas volume becomes dependent on the details of the molecule motion.

The gas-piston system may be viewed as a mechanical system consisting of the balls and the piston. Mass of the piston is much greater than the molecule masses:

V.L. Berdichevsky, *Variational Principles of Continuum Mechanics*,
Interaction of Mechanics and Mathematics, DOI 10.1007/978-3-540-88467-5_2,

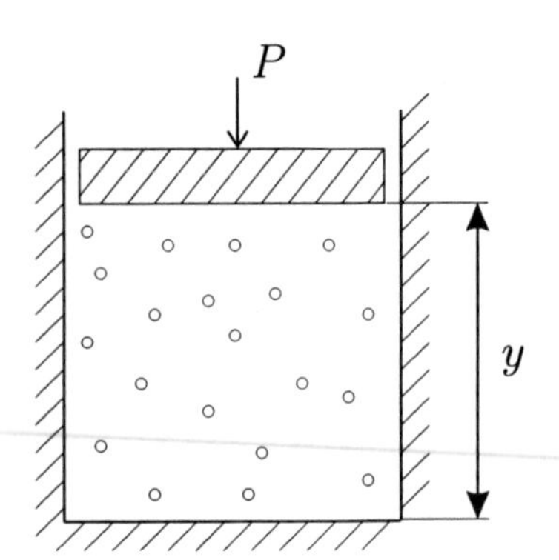

Fig. 2.1 Gas under piston

this is the reason why the coordinate of the piston, y, changes slowly.[1] In case of elastic collisions, the system is Hamiltonian. One may say that the thermodynamic description of the system "gas under piston" corresponds to elimination of the fast degrees of freedom from the governing Hamiltonian equations. In fact, this situation is generic: classical thermodynamics is a theory of slow variables for a Hamiltonian system which governs micromotion.

Thermodynamics is concerned with systems possessing at least two well separated time scales and, thus, characterized by fast and slow variables. Thermodynamic theory is a theory of slow variables for such systems – this was a major Boltzmann's insight. One can say that thermodynamic equations are the equations that are obtained by elimination of fast variables from the governing equations.

Why should macromotion obey the first and second laws of thermodynamics? Clearly, this must be caused by some special features of microdynamics. It turns out that these features are ergodicity, mixing and the Hamiltonian structure of the underlying microdynamics. The meaning of the terms ergodicity and mixing is discussed in the next section. The absence of ergodicity or Hamiltonian structure would prevent the existence at macrolevel temperature and entropy. The absence of mixing would yield the violation of the second law.[2]

We call the laws of thermodynamics obtained by elimination of fast degrees of freedom from Hamiltonian equations primary thermodynamics. The characteristic features of primary thermodynamics are the appearance of two new slow variables, temperature and entropy, and the dissipation of energy of slow variables (the total energy of fast and slow variables is conserved in isolated systems).

The first and the second laws of thermodynamics are the constraints which must be obeyed by any macroscopic theory. There are additional independent constraints,

---

[1] In fact, the slow change of y is accompanied by fast oscillations of small magnitude due to the collisions of the piston with the molecules. In thermodynamic description, these oscillations can be neglected. They are studied in the theory of thermodynamic fluctuations (Sect. 2.4).

[2] These statements will be rectified further in one respect: in fact, to have the laws of thermodynamics on macrolevel, the microequations might possess slightly more general structure than the Hamiltonian one.

which are sometimes called the third law of thermodynamics, Onsager's reciprocal relations (they are considered further in Sect. 2.6). Are there other constraints of a similar level of universality? Yes, there are. It turns out that, if dissipation is negligible, the governing equations of some slow variables must possess a Hamiltonian structure. That indicates the existence of quite peculiar "Hamiltonian reciprocities" in macrophysical interactions.

The dissipative equations of primary thermodynamics can also possess two well separated time scales. Elimination of fast degrees of freedom in primary thermodynamics yields the equations of secondary thermodynamics. If the fast variables in primary thermodynamics perform some chaotic motion then, after elimination of fast degrees of freedom and transition to the secondary thermodynamics, two new slow variables appear, "secondary entropy" and "secondary temperature." It is quite plausible that the secondary entropy possesses the features which are similar to the features of the usual thermodynamic entropy.

We touch upon all these issues in this chapter and further in Chap. 17.

2.2 Temperature

If a mechanical system is governed by Hamiltonian equations, and its motion is sufficiently chaotic, one can introduce the notion of temperature. First, the term "sufficiently chaotic" must be explained.

Consider a Hamiltonian system with Hamiltonian $H(p,q)$. Function $H(p,q)$ does not change in the course of motion; its value is called the energy of the system. Let energy have the value E. Any trajectory of the system lies on a surface in phase space defined by the equation

$$H(p,q) = E.$$

This surface is called an energy surface. It is assumed that energy surfaces bound finite regions in phase space.

The system is called ergodic if (almost) any trajectory covers the entire energy surface. That means the following. Let a trajectory start at some point A. Consider a point B with some vicinity of this point $\triangle B$. For ergodic systems, sooner or later, the trajectory will pass through the vicinity, $\triangle B$, of the point B for any choice of B and $\triangle B$ (Fig. 2.2). Since $\triangle B$ can be chosen as small as we wish, the trajectory will be passing closer and closer to B. The time of the next passage can, however, be very large. Such a behavior is observed for almost any trajectory in the sense that the set of points A for which trajectories behave differently has zero area on the energy surface. For example, there might be periodic trajectories on the energy surface of an ergodic Hamiltonian system, but the area covered by such trajectories is zero.

Intuitively, ergodicity is a feature of chaotic motion. There is another feature of chaotic motion, mixing. To define mixing one has to view the trajectories of the Hamiltonian system on an energy surface as the trajectories of particles of some

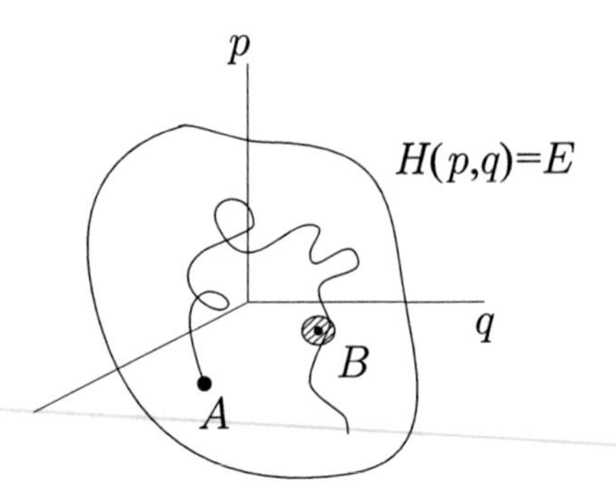

Fig. 2.2 A sketch for the definition of ergodicity; shadowed region is $\triangle B$

media. One can put an ink spot on the energy surface and observe its evolution in the course of motion (Fig. 2.3). If the spot tends to cover densely the entire energy surface, the system is called mixing. It turns out that every mixing system is ergodic. Ergodic systems are not necessarily mixing. Equilibrium thermodynamics discussed in Sects. 2.2–2.5 holds true for ergodic systems. In order the laws of nonequilibrium thermodynamics to be true the underlying Hamiltonian system must be also mixing. The notion of temperature and entropy is based on ergodicity only.

Consider the time average of some function, $\varphi(p,q)$, of generalized coordinates and momenta along a trajectory $p(t), q(t)$:

$$\langle\varphi\rangle = \lim_{\theta\to\infty}\frac{1}{\theta}\int_0^\theta \varphi\left(p(t),q(t)\right)dt.$$

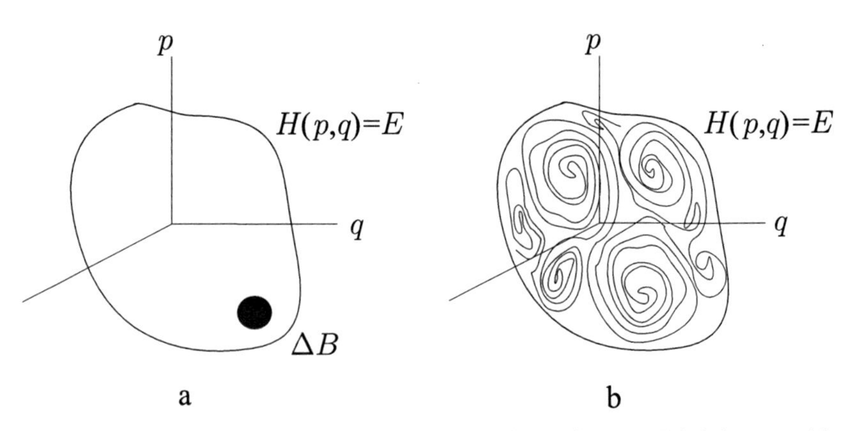

Fig. 2.3 A sketch for the definition of mixing: an initial small spot $\triangle B$ is being spread by the phase flow to cover densely the entire energy surface

As was discovered by Boltzmann, for ergodic systems this time average is the same for (almost) all trajectories on the same energy surface. Moreover, this time average can be computed without knowing the particular trajectory by the formula

$$\langle\varphi\rangle = \frac{\frac{\partial}{\partial E}\int\limits_{H(p,q)\le E}\varphi(p,q)dpdq}{\frac{\partial}{\partial E}\int\limits_{H(p,q)\le E}dpdq}. \tag{2.1}$$

Here $dp = dp_1 \dots dp_n, \ \ dq = dq^1 \dots dq^n$.

This formula was proven with mathematical rigor by Birkhoff and Khinchine, and is called usually the Birkhoff-Khinchine theorem.

Plugging in (2.1) various functions $\varphi(p,q)$ and computing the integrals on the right hand side, one can find their time averages. Functions $p_1\frac{\partial H}{\partial p_1}, \dots, p_N\frac{\partial H}{\partial p_N}$ are of special interest: for systems with the Hamiltonian

$$H(p,q) = \frac{p_1^2}{2m_1} + \dots + \frac{p_N^2}{2m_N} + U(q) \tag{2.2}$$

they are (doubled) kinetic energies of each degree of freedom. Evaluation of the integrals from (2.1) for these functions is simple and yields the so-called equipartition law :

$$\left\langle p_1\frac{\partial H}{\partial p_1}\right\rangle = \dots = \left\langle p_N\frac{\partial H}{\partial p_N}\right\rangle. \tag{2.3}$$

Indeed, let us find $\left\langle p_1\frac{\partial H}{\partial p_1}\right\rangle$. Consider the integral,

$$\int\limits_{H(p,q)\le E} p_1\frac{\partial H}{\partial p_1}dpdq.$$

Using the step function

$$\theta(x) = \begin{cases} 1 & x \ge 0 \\ 0 & x < 0 \end{cases} \tag{2.4}$$

we can write this integral as an integral over the entire phase space,

$$\int p_1\frac{\partial H}{\partial p_1}\theta(E - H(p,q))dpdq.$$

Therefore,

$$\frac{\partial}{\partial E}\int\limits_{H(p,q)\le E} p_1\frac{\partial H}{\partial p_1}dpdq = \int p_1\frac{\partial H}{\partial p_1}\theta'(E - H(p,q))dpdq,$$

where $\theta'(x)$ is the derivative of the step-function (the derivative, $\theta'(x)$, is equal to the δ-function, but this is not essential at the moment). The integrand can be written as

$$p_1 \frac{\partial H}{\partial p_1} \theta'(E - H(p,q)) = -p_1 \frac{\partial}{\partial p_1} \theta(E - H(p,q))$$
$$= -\frac{\partial}{\partial p_1} (p_1 \theta(E - H(p,q))) + \theta(E - H(p,q)) . \tag{2.5}$$

The integral of the first term in the right hand side over the entire phase space is equal to zero due to the divergence theorem and vanishing of $\theta(E - H(p,q))$ at infinity ($H(p,q) > E$ at infinity). Therefore,

$$\frac{\partial}{\partial E} \int_{H(p,q) \leq E} p_1 \frac{\partial H}{\partial p_1} dpdq = \int \theta(E - H(p,q)) dpdq. \tag{2.6}$$

The result (2.6) does not depend on the choice of a particular component of momenta, $p_1, \ldots, p_n$, and the same answer we get for integrals of $p_2 \frac{\partial H}{\partial p_2}, \ldots, p_n \frac{\partial H}{\partial p_n}$. Then the equipartition law follows from (2.1) and (2.6)

For the systems with Hamiltonians (2.2) equipartition law means that the averaged values of kinetic energies of all degrees of freedom are the same. The common value (2.3) is denoted by T and called absolute temperature. We drop the adjective and call it temperature because no other temperatures will appear in our consideration.

The integral in the right hand side of (2.6) is called the phase volume,

$$\Gamma(E) = \int \theta(E - H(p,q)) dpdq = \int_{H(p,q) \leq E} dpdq.$$

The denominator in (2.1) is the derivative of the phase volume, $d\Gamma(E)/dE$, which will also be denoted for brevity $\Gamma_E(E)$.

Finally, temperature can be expressed explicitly in terms of the phase volume $\Gamma(E)$:

$$T = \frac{\Gamma(E)}{d\Gamma(E)/dE}. \tag{2.7}$$

As follows from (2.7), temperature has the dimension of energy. Traditionally, temperature is measured in degrees. The two numbers are linked by Boltzmann's constant k: if T° is the value of temperature in degrees Kelvin, then

$$T = kT^\circ.$$

The constant k is very small:

$$k = 1.38 \times 10^{-16} \text{ erg/deg} .$$

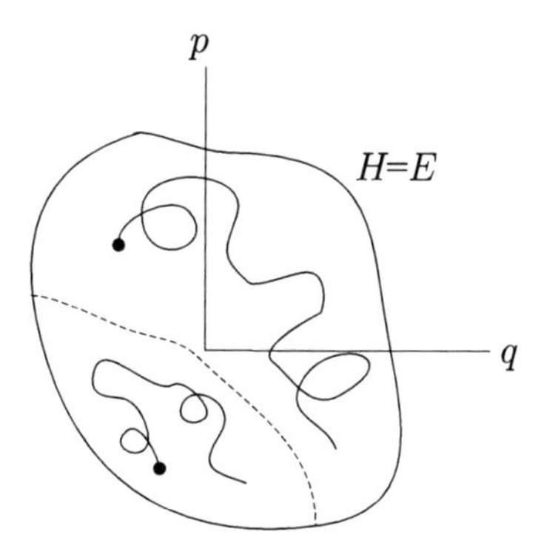

Fig. 2.4 A sketch of energy surface for a two-temperature system

i.e. one degree Kelvin corresponds to energy of about 10^{-16} $\mathrm{g\,cm^2/s^2}$. Temperature becomes on the order of unity if measured in electron-volts. (1 eV is the energy which an electron gains accelerating between the points with the difference of electric potential equal to 1 V; this energy is very small due to the small mass of an electron). Room temperature corresponds to about $1/40$ eV. Energy units for temperature are convenient in all theoretical treatments and will be used here; Boltzmann's constant appears only at the stage of comparing theory and experiments.

Ergodicity is necessary to introduce temperature. If the system is not ergodic, temperature may not exist. For example, consider the system for which the energy surface contains two parts such that trajectory started in one part always remains in that part (see Fig. 2.4). For such systems, formula (2.1) does not hold, and the equipartition law is not true. Under some additional conditions, one may say that the system has two temperatures corresponding to each part of the energy surface, but such a situation is beyond the scope of classical thermodynamics.

Ergodicity yields immediately irreversibility of macromotion. This is seen from the equipartition law. Consider, for example, the gas-piston system. Let the piston be given some initial velocity. After some time, the equipartition of energy sets up in the system. This means that the average kinetic energy of the piston is equal to the average kinetic energy of a molecule. Since the mass of the piston is much larger than the mass of molecules, the velocity of the piston becomes very small, i.e. the piston comes to rest. This clearly demonstrates the irreversible character of piston's motion in spite of reversibility of the underlying microdynamics.

2.3 Entropy

Entropy is a characteristic of slow processes in ergodic Hamiltonian systems.[3]

---

[3] The term "entropy" is widely explored now in many different senses. What we mean here by entropy is, precisely, the thermodynamic entropy introduced by Clausius. The meaning of Clausius' entropy for ergodic Hamiltonian systems was determined by J.W. Gibbs and P. Hertz.

To explain what entropy is, consider first the gas-piston system. Let the position of the piston, y, be changed slowly by some "hard device." This means that we prescribe the function, $y(t)$. Changing y we do some work. Therefore energy of the system changes. To find the dependence of energy on time we note that Hamiltonian of the system,

$$H(p,q,y) = \frac{p_1^2}{2m} + \ldots + \frac{p_n^2}{2m} + U(q_1,\ldots,q_n,y), \tag{2.8}$$

depends explicitly on time through the dependence of potential energy on y. The energy rate is

$$\frac{dE}{dt} = \frac{dH(p,q,y)}{dt} = \frac{\partial H}{\partial p_i}\frac{dp_i}{dt} + \frac{\partial H}{\partial q^i}\frac{dq^i}{dt} + \frac{\partial H}{\partial y}\frac{dy}{dt} = \frac{\partial H(p,q,y)}{\partial y}\frac{dy}{dt}. \tag{2.9}$$

The two terms in (2.9) are canceled due to Hamiltonian equations (1.63). The derivative $\partial H/\partial y$ has the meaning of force which one has to apply in order to make the piston move along the path $y(t)$.[4]

If we know the trajectory of the systems, $p(t), q(t)$, we could find the energy at time t by integrating (2.9):

$$E(t) - E(t_0) = \int_{t_0}^{t} \frac{\partial H(p(\tau),q(\tau),y(\tau))}{\partial y(\tau)}\frac{dy(\tau)}{d\tau}d\tau. \tag{2.10}$$

In principle, one obtains different values of energy at time t for different trajectories $p(\tau), q(\tau)$ and different piston paths $y(\tau)$. Remarkably, for ergodic Hamiltonian systems and a slow change of the parameter y, the final value of energy depends only on the final value of the parameter y, the initial values of energy, $E_0 = E(t_0)$ and the initial value of the parameter y, $y_0 = y(t_0)$, and depends neither on the trajectories, $p(\tau), q(\tau)$, nor on the path, $y(\tau)$:

$$E(t) = \text{function}(y(t), y_0, E_0). \tag{2.11}$$

Moreover, the dependence of the final value of energy, $E(t)$, on y_0 and E_0 is "degenerated": the parameters y_0 and E_0 enter (2.11) only through a combination $S(E_0, y_0)$, where $S(E, y)$ is some function which is determined uniquely by Hamilton function $H(p,q,y)$:

$$E(t) = E(y(t), S_0(E_0, y_0)). \tag{2.12}$$

[4] For the gas-piston system the factor $\partial H(p,q,y)/\partial y$ does not depend on p due to (2.8), but this is not essential in our reasoning and we proceed in a more general setting.

Derivation of (2.12) is quite simple. First, we note that on the right hand side of (2.10) $dy/d\tau$ changes slowly. Therefore, $\partial H/\partial y$ may be substituted by its average over the energy surface $H(p, q, y(\tau)) = E(\tau)$,

$$E(t) - E(t_0) = \int_{t_0}^{t} \left\langle \frac{\partial H}{\partial y} \right\rangle \frac{dy}{d\tau} d\tau,$$

or, after differentiation with respect to t,

$$\frac{dE}{dt} = \left\langle \frac{\partial H}{\partial y} \right\rangle \frac{dy}{dt}. \tag{2.13}$$

The average value $\langle \partial H/\partial y \rangle$ can be computed using the Birkhoff-Khinchine theorem (2.1) as

$$\left\langle \frac{\partial H}{\partial y} \right\rangle = \frac{\dfrac{\partial}{\partial E} \displaystyle\int_{H(p,q,y) \le E} \frac{\partial H}{\partial y} dpdq}{\dfrac{\partial}{\partial E} \displaystyle\int_{H(p,q,y) \le E} dpdq}. \tag{2.14}$$

Introducing the phase volume bounded by the energy surface $H(p, q, y) = E$,

$$\Gamma(E, y) = \int_{H(p,q,y) \le E} dpdq = \int \theta(E - H(p, q, y)) dpdq, \tag{2.15}$$

and differentiating (2.15) with respect to y we find

$$\begin{aligned} \frac{\partial \Gamma(E, y)}{\partial y} &= -\int \theta'(E - H(p, q, y)) \frac{\partial H(p, q, y)}{\partial y} dpdq \\ &= -\frac{\partial}{\partial E} \int \theta(E - H(p, q, y)) \frac{\partial H(p, q, y)}{\partial y} dpdq \\ &= -\frac{\partial}{\partial E} \int_{H(p,q,y) \le E} \frac{\partial H(p, q, y)}{\partial y} dpdq. \end{aligned} \tag{2.16}$$

From (2.14) and (2.16)

$$\left\langle \frac{\partial H}{\partial y} \right\rangle = -\frac{\partial \Gamma(E, y)/\partial y}{\partial \Gamma(E, y)/\partial E}. \tag{2.17}$$

Plugging (2.17) into (2.13) we obtain

$$\frac{\partial \Gamma(E, y)}{\partial E} \frac{dE}{dt} + \frac{\partial \Gamma(E, y)}{\partial y} \frac{dy}{dt} = 0. \tag{2.18}$$

Hence, the phase volume does not change in time, which is also true of any function of the phase volume, $S(\Gamma)$.

Let us specify function $S(\Gamma)$ by the condition that

$$\frac{\partial S(\Gamma(E, y))}{\partial E} = \frac{1}{T}. \tag{2.19}$$

Since

$$\frac{\partial S(\Gamma(E, y))}{\partial E} = \frac{dS}{d\Gamma}\frac{\partial \Gamma}{\partial E},$$

and according to (2.7), $T = \Gamma\,(\partial\Gamma/\partial E)^{-1}$, we obtain for $S(\Gamma)$ the differential equation

$$\frac{dS}{d\Gamma} = \frac{1}{\Gamma},$$

the only solution of which is

$$S(E, y) = \ln \Gamma(E, y) + const. \tag{2.20}$$

This function is called thermodynamic entropy.

Entropy does not change in the process under consideration. Hence,

$$S(E, y) = S_0, \quad S_0 = S(E_0, y_0). \tag{2.21}$$

If $T > 0$, then $\partial S/\partial E > 0$ and we can solve (2.21) with respect to E. We see that energy is determined by the current value of y, while the initial values of energy and y enter in this dependence only through the combination $S_0 = S_0(E_0, y_0)$, i.e. we arrive at (2.12).

In terms of entropy, equation (2.17) for the force, $\langle \partial H/\partial y\rangle$, takes the form

$$\left\langle \frac{\partial H}{\partial y} \right\rangle = -T\frac{\partial S(E, y)}{\partial y}. \tag{2.22}$$

Obviously, our derivation remains valid if the system has a number of slow parameters, $y^1, \dots, y^m$. In this case, y in all previous equations denotes the set $y = (y^1, \dots, y^m)$ while (2.22) is replaced by the equation

$$\left\langle \frac{\partial H}{\partial y^\mu} \right\rangle = -T\frac{\partial S(E, y)}{\partial y^\mu}, \tag{2.23}$$

where μ runs through values $1, \dots, m$.

Equations (2.19) and (2.23) link temperature, T, and generalized forces, $\langle \partial H/\partial y^\mu\rangle$, with the slow characteristics of the system, E and y^μ. They are called constitutive equations. The constitutive equations are specified as soon as entropy is known as a function of E and y^μ. For a given Hamiltonian of the system, $H(p, q, y)$, one can determine the phase volume computing the integral (2.15), and then find entropy $S(E, y)$ from (2.20).

Constitutive equations (2.19) and (2.23) take a simpler form in the terms of function $E(y, S)$:

$$T = \frac{\partial E(y, S)}{\partial S}, \quad \left\langle \frac{\partial H}{\partial y^\mu} \right\rangle = \frac{\partial E(y, S)}{\partial y^\mu}. \tag{2.24}$$

Indeed, the equation

$$S(E(S, y), y) = S \tag{2.25}$$

must be an identity for all values of parameters S and y. Differentiating (2.25) with respect to S and y^μ, we obtain

$$\frac{\partial S}{\partial E} \frac{\partial E(S, y)}{\partial S} = 1, \quad \frac{\partial S}{\partial E} \frac{\partial E(S, y)}{\partial y^\mu} + \frac{\partial S}{\partial y^\mu} = 0. \tag{2.26}$$

Equations (2.24) follow from (2.19), (2.23) and (2.26).

It is worthing emphasize that, if the system were not ergodic or the parameters y were not slow, then the energy at time t computed from differential equation (2.9) depends on the initial values of the generalized coordinates and momenta p_0, q_0, and the entire trajectory $y(\tau)$, $\tau < t$. Therefore, the constitutive equations obtained do not make sense.

Formulas (2.12), (2.19), (2.21), (2.23) and (2.24) are valid for any ergodic Hamiltonian system no matter how many degrees of freedom it has. For example, one can speak of entropy and temperature of a pendulum, which has just one degree of freedom. In this case the energy surfaces are closed curves in the phase space (for sufficiently small E). Each trajectory covers the entire energy surface, thus the system is ergodic. Assuming for definiteness that the length of the pendulum is a slow parameter, one can find entropy of the pendulum from (1.68): in the limit of small energies, $S = \ln(E\sqrt{l/g}) + const$.

Example. Let us find entropy of a gas occupying volume V. We model the gas by a Hamiltonian system of N rigid spheres of radii a and of equal masses m. Each ball has three translational degrees of freedom, so the total number of degrees of freedom, n, is $3N$. Hamilton function is a sum of kinetic energy K and particle-particle and particle-wall interaction energy U:

$$\begin{aligned} H &= K + U, \\ K &= \frac{p_1^2}{2m} + \ldots + \frac{p_n^2}{2m}, \quad U = U(q_1, \ldots, q_n). \end{aligned} \tag{2.27}$$

Interaction energy is zero if particles do not overlap with each other and the wall, otherwise it is equal to infinity. The question of ergodicity of such a system is not elementary. We proceed assuming that the system is ergodic. We have to find

$$\Gamma = \int_{H(p,q,y)\leq E} dpdq.$$

Since $H = +\infty$ if any two particles overlap or a particle overlaps with the wall and, otherwise, $H = K$ and does not depend on q, the phase volume is the product of two integrals:

$$\Gamma = \Gamma_p \Gamma_q, \quad \Gamma_p = \int_{H(p)\leq E} dp, \quad \Gamma_q = \int_{\text{admissible } q} dq. \tag{2.28}$$

The integral, Γ_p, is the volume of the sphere of the radius $\sqrt{2mE}$ in n-dimensional space. If R is the radius of a sphere in n-dimensional space, its volume is

$$V_n(R) = c_n R^n, \quad c_n = \frac{\pi^{n/2}}{\frac{n}{2}\Gamma(\frac{n}{2})},$$

$\Gamma(s)$ being the Γ-function. For an integer s, $\Gamma(s) = (s-1)!$ So,

$$\Gamma_p = c_n (2mE)^{n/2}. \tag{2.29}$$

The integral, Γ_q, can easily be found in the limit when the ball radius tends to zero. In this limit one may neglect overlapping of the balls and take into account only the positions of the balls inside the volume V. Then

$$\Gamma_q = V^N. \tag{2.30}$$

Dropping additive constants we obtain for entropy from (2.20), (2.28), (2.29) and (2.30):

$$S = \ln(E^{3N/2} V^N) = N\left(\frac{3}{2}\ln E + \ln V\right). \tag{2.31}$$

One can find temperature from (2.19) and (2.31) as

$$\frac{1}{T} = \frac{\partial S}{\partial E} = \frac{3}{2}\frac{N}{E}. \tag{2.32}$$

Determining energy in terms of temperature from (2.32), we arrive at the familiar constitutive equation of the perfect gas:

$$E = \frac{3}{2} N T. \tag{2.33}$$

An important consequence of formula (2.31) is an unbounded growth of entropy when the gas volume increases. Such a behavior of entropy changes if a force, P, acts on the piston. Then the Hamiltonian of the system acquires an additional term:[5]

$$H = K + U + P y. \tag{2.34}$$

Calculation of entropy of the system (2.34) is reduced to the previous one for the system (2.27) by changing E to $E - P y$:

$$S = N \left(\frac{3}{2} \ln(E - P y) + \ln V \right).$$

Taking into account that V is a linear function of y: $V = y\Omega$, Ω being the area of the piston, and dropping unessential constants, we obtain

$$\begin{aligned} S(E, y) &= N \ln \left[\frac{3}{2} \ln(E - P y) + \ln(y\Omega) \right] \\ &= N \ln \left[\frac{P y}{E} \left(1 - \frac{P y}{E} \right)^{3/2} \right] + const. \end{aligned}$$

A graph of the entropy per particle, S/N, as a function of the dimensionless distance $y^* = P y / E$ is shown in Fig. 2.5.

The remarkable feature of this graph is that entropy reaches its maximum value. The point where entropy is maximum corresponds to the equilibrium state of the system the reader is invited to check this fact. It turns out that this property of entropy, reaching its maximum value at equilibrium, is generic. The physical origin of this property is explained in Sect. 2.5.

So far we have considered the case when all external forces act only on slow variables as is seen from the energy equation (2.9). Such processes are called adiabatic. If there are some external forces, F_i, acting on the fast coordinates, q_i, then the energy equation gets the additional term

---

[5] Formula (2.34) becomes obvious if one writes first the energy of the entire system "gas+piston" endowing the piston with some mass M :

$$H = K + U + P y + \frac{Y^2}{2M},$$

where Y is the momentum of the piston. In (2.34) we dropped the kinetic energy of the piston which is negligible because, due to the equipartition of energy over all degrees of freedom, near equilibrium it is on the order of kinetic energy of one molecule, K/N. The sign + at $P y$ corresponds to the negative direction of the force acting on the piston for $P \geq 0$ (indeed, the Lagrange function of the piston corresponding to the Hamilton function chosen is $L = \frac{1}{2} M \dot{y}^2 - P y$, thus $M \ddot{y} = -P$).

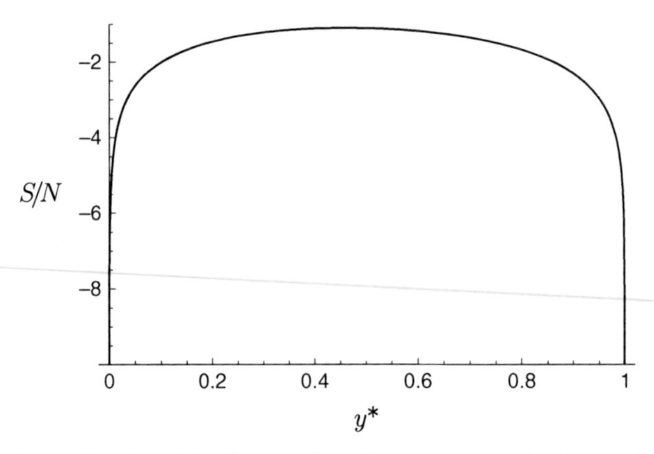

Fig. 2.5 S/N, as a function of the dimensionless distance between the piston and the bottom, $y^* = Py/E$

$$\frac{dE}{dt} = \frac{\partial H}{\partial y^\mu}\frac{dy^\mu}{dt} + \frac{\partial H}{\partial q^i}F^i.$$

Averaging over time and taking into account that E and y change slowly, we have

$$\frac{dE}{dt} = \left\langle \frac{\partial H}{\partial y^\mu} \right\rangle \frac{dy^\mu}{dt} + \frac{dQ}{dt}, \quad \frac{dQ}{dt} = \left\langle \frac{\partial H}{\partial q^i}F^i \right\rangle. \tag{2.35}$$

The additional work of external forces denoted by dQ is called heat supply. Heat supply causes entropy to change. The process is called quasi-equilibrium, if the constitutive equations (2.28) remain valid for $dQ \neq 0$. Then, we have from energy equation (2.35)

$$\frac{\partial E(y,S)}{\partial y^\mu}\frac{dy^\mu}{dt} + \frac{\partial E(y,S)}{\partial S}\frac{dS}{dt} = \left\langle \frac{\partial H}{\partial y^\mu} \right\rangle \frac{dy^\mu}{dt} + \frac{dQ}{dt}. \tag{2.36}$$

The first terms on both sides of (2.36) cancel out due to (2.24), and (2.36) becomes a link between entropy, temperature and heat supply:

$$\frac{dQ}{dt} = T\frac{dS}{dt}. \tag{2.37}$$

If $dQ \neq 0$, entropy may either increase or decrease depending on whether energy is pumped to the system ($dQ > 0$) or taken from the system ($dQ < 0$).

Comparing the two characteristics of the system, energy and entropy, one may say that energy is a more fundamental one: energy makes sense for any system while entropy can be introduced only for slow processes in ergodic systems.

We have seen that the Hamiltonian structure of the equations of micromechanics yields the laws of equilibrium thermodynamics. One may wonder how important it is that the equations of micromechanics are Hamiltonian. In other words: could non-Hamiltonian equations of microdynamics yield the equations of classical equilibrium thermodynamics? It turns out that the class of such equations is indeed slightly wider than the class of Hamiltonian equations (see Appendix E).

2.4 Entropy and Probability

Slow parameters y fluctuate near the equilibrium values. For example, in the system "gas under piston" the distance between the piston and the bottom, y, varies slightly due to the collisions of the piston with the molecules. One may wonder what is the probability density of y, $f(y)$? The answer was found by Einstein: the probability density of slow variables fluctuating near the equilibrium values is determined only by the equilibrium properties of the system, namely, by the function $S(E, y)$, and is given by the formula

$$f(y) = const\ e^{S(E,y)}. \tag{2.38}$$

Formula (2.38) has an asymptotic nature and holds for systems with a very large number of fast degrees of freedom, n. There is a generalization of this formula for systems with any finite n [43, 46].

According to (2.38), the most probable values of y are the values for which entropy takes its maximum value. Since entropy is usually proportional to the number of fast degrees of freedom (for a gas this is seen from (2.31)) which is large, maximum is very sharp and, in fact, the slow variables just fluctuate slightly around the equilibrium values.

2.5 Gibbs Principles

Gibbs suggested use of the maximum property of entropy as the first principle in any modeling of thermodynamic equilibrium.

The first Gibbs principle. *In a state of thermodynamic equilibrium, the entropy S of an isolated system attains its maximum on all possible states of the system with a given energy E.*

The Gibbs principle differs considerably from the variational principles of analytical mechanics. In mechanics, only the particle positions are subject to change, but in the Gibbs principles virtually all characteristics of equilibrium are varied. In the consideration of phase equilibrium, for example, the interphase surface and the masses of both phases are subject to change. An example of the application of the Gibbs principles to the equilibrium of elastic media will be given in Sect. 7.4. Consider here another example.

Let us show that temperatures of two contacting systems are equal if the systems are in thermodynamical equilibrium and isolated from the environment. Denote entropies and energies of the systems by E_1, S_1, and E_2, S_2. Thermodynamical properties of the systems are determined by the functions $S_1(E_1)$ and $S_2(E_2)$. The total entropy of the two systems, by our assumption, is

$$S(E_1, E_2) = S_1(E_1) + S_2(E_2). \tag{2.39}$$

Systems are in contact and may exchange energies: heat may flow from one system to another. The total energy must not change in such a process because the systems are isolated from the environment:

$$E_1 + E_2 = E. \tag{2.40}$$

The total energy E is supposed to be given. In equilibrium, the entropy (2.39) must be maximum on the set of all possible values E_1, E_2 obeying the constraint (2.40). The condition of thermodynamical equilibrium may be obtained, for example, by eliminating the variable E_2,

$$S = S_1(E_1) + S_2(E - E_1),$$

and differentiating entropy with respect to E_1. We get

$$\frac{dS_1(E_1)}{dE_1} - \left.\frac{dS_2(E_2)}{dE_2}\right|_{E_2=E-E_1} = 0$$

or

$$\frac{1}{T_1} = \frac{1}{T_2}$$

as claimed. Similar result holds for many systems in contact. The first Gibbs principle can be put in another form which is often used:

The second Gibbs principle. *In the state of thermodynamic equilibrium, the energy $E(y, S)$ of an isolated system attains its minimum on all states of the system with a given value of entropy S.*

The two Gibbs principles are equivalent except in some degenerate cases.

2.6 Nonequilibrium Processes

Consider an isolated system characterized by a number of slow variables, $y^1, \ldots, y^n$. There are some equilibrium values of y; the system remains in the state with such values of y indefinitely. If the initial values of y differ from the equilibrium values, the slow variables evolve approaching the equilibrium values. The theory of nonequilibrium processes aims to establish equations describing this evolution. In this section we discuss the situation when the governing equations of the evolution to equilibrium are ordinary differential equations.

Clausius found from phenomenological reasoning that entropy of an isolated system may not decrease in the path to equilibrium. This is the so-called second law of thermodynamics. For Hamiltonian systems this feature of entropy was established by Kubo [163]. Hamiltonian systems possess such a feature if the phase flow is mixing. In summary, the first and the second laws of thermodynamics are observed only for one reason: the equations governing micromotion are Hamiltonian and mixing.[6] If micromotion is not Hamiltonian or mixing, one can construct examples showing that the first and/or the second laws are violated.

The governing equations of nonequilibrium thermodynamics are purely phenomenological. They must obey the constraint of the second law: if the system is isolated, its entropy may not decrease. To construct the evolution equations one usually chooses as the main idea that entropy has a maximum value at equilibrium. Then the simplest system of equations warranting the approach to equilibrium is

$$\frac{dy^\mu}{dt} = D^{\mu\nu}(y)\frac{\partial S(E,y)}{\partial y^\nu}, \tag{2.41}$$

where $D^{\mu\nu}(y)$ is some positive semi-definite matrix. Indeed, entropy of the system grows along each trajectory $y(t)$ of the system (2.41):

$$\frac{dS(E,y(t))}{dt} = \frac{\partial S(E,y)}{\partial y^\mu} D^{\mu\nu}(y)\frac{\partial S(E,y)}{\partial y^\nu} \geqslant 0. \tag{2.42}$$

We assumed here that the system is isolated, so the energy remains constant. According to (2.42), $D^{\mu\nu}(y)$ have the meaning of dissipation coefficients, i.e. the coefficients controlling the entropy growth.

If $y(t)$ are close to the equilibrium values, one can use a linearized version of (2.41). To write down the linearized equations (2.41) we accept that the equilibrium corresponds to the zero values of y. Then, expanding S in a Taylor series with respect to y in vicinity of equilibrium, we have

$$S = const - \frac{1}{2}a_{\mu\nu}y^\mu y^\nu,$$

where $a_{\mu\nu}$ is a non-negative symmetric matrix. In linear approximation the coefficients $D^{\mu\nu}$ are some constants, $D^{\mu\nu} = D^{\mu\nu}(0)$. The governing equations take the form a system of linear differential equations,

$$\frac{dy^\mu}{dt} = -D^{\mu\nu}(0)a_{\lambda\nu}y^\lambda. \tag{2.43}$$

---

[6] Up to some refinements of this statement like that made at the end of Sect. 2.3, which, most probably, are of little physical significance.

Equations (2.41) are more a concept than a "Law of Nature": in modeling the evolution to equilibrium, one may try equations of the form (2.41), but, in fact, the physically adequate dynamical equations may have a different form: all one must not violate is the evolution of entropy to its maximum value.

Are there other constraints to the governing equations, which are additional to the first and the second laws of thermodynamics? Yes, there are. They are caused by the Hamiltonian structure of microdynamics. The first such constraint was discovered by Onsager [236]: the dissipative coefficients, $D^{\mu\nu}(0)$, are not arbitrary. They must obey the reciprocal relations

$$D^{\mu\nu}(0) = D^{\nu\mu}(0). \tag{2.44}$$

Onsager's reciprocal relations follow from reversibility of micromotion. The latter takes place if Hamilton function is an even function of momenta, p. There are systems for which Hamilton function is not an even function, like, for example, the systems under action of external constant magnetic field, B: Hamilton function contains the terms of the form, pB, which change the sign if time is reversed . For systems in a magnetic field, the coefficients $D^{\mu\nu}$ may also depend on the magnetic field, and Onsager's reciprocal relations are replaced by

$$D^{\mu\nu}(0, B) = D^{\nu\mu}(0, -B).$$

Note that the coefficients of the linear differential equations (2.43), the product of two symmetric matrices, are not necessarily symmetric.

Onsager's reciprocal relations are independent of the first and the second laws of thermodynamics. They are sometimes called the third law of thermodynamics.

There are also other constraints [44, 50]: if the slow variables are the coordinates and momenta of the underlying Hamiltonian microdynamics, and the dissipation is negligibly small, the equations of slow evolution must possess a Hamiltonian structure with some effective Hamilton function, $H_{\text{eff}}(y, S)$,

$$\frac{dy^\mu}{dt} = \omega^{\mu\nu} \frac{\partial H_{\text{eff}}(y, S)}{\partial y^\nu}. \tag{2.45}$$

Here $\omega^{\mu\nu}$ is the constant tensor defining the Hamiltonian structure

$$\omega^{\mu\nu} = \begin{cases} 1 & \mu \geq n+1,\, \nu = \mu \\ -1 & \mu \leq n,\, \nu = n + \mu \\ 0 & \text{otherwise} \end{cases}.$$

The effective Hamilton function, $H_{\text{eff}}(y, S)$, can be calculated explicitly in terms of the phase volume of the Hamiltonian system. This calculation shows that it has the meaning of the total energy of the system. Entropy in (2.45) is a given constant. In many models of continuum mechanics, the kinematic variables can be viewed as the slow variables of the underlying Hamiltonian system. Therefore, if the dissipation is neglected, the governing dynamics is Hamiltonian, and the corresponding principle of least action must exist.[7] This implies that only quite special interactions with specific "variational reciprocal relations" are possible in continuum mechanics. Most considerations in this book are based on that point.

In general, if the dissipative terms are taken into account, then the slow evolution is governed by the equations

$$\frac{dy^\mu}{dt} = \omega^{\mu\nu}\frac{\partial H_{\text{eff}}(y,S)}{\partial y^\nu} - \frac{1}{T}D^{\mu\nu}(y,S)\frac{\partial H_{\text{eff}}(y,S)}{\partial y^\nu}$$
$$\frac{dS}{dt} = \frac{1}{T^2}D^{\mu\nu}(y,S)\frac{\partial H_{\text{eff}}(y,S)}{\partial y^\mu}\frac{\partial H_{\text{eff}}(y,S)}{\partial y^\nu}, \quad T = \frac{\partial H_{\text{eff}}(y,S)}{\partial S}. \tag{2.46}$$

In these equations, the dissipative processes are characterized by the dissipative coefficients, $D^{\mu\nu}(y,S)$. The dissipative coefficients must be symmetric:

$$D^{\mu\nu}(y,S) = D^{\nu\mu}(y,S). \tag{2.47}$$

Equation (2.47) may be viewed as an extension of Onsager's relations to the non-linear case.

Equations (2.46) form a system of ordinary differential equations for unknown functions $y^\mu(t)$ and $S(t)$. The model is specified by the effective Hamiltonian, $H_{\text{eff}}(y,S)$, and the dissipative coefficients, $D^{\mu\nu}(y,S)$. The effective Hamiltonian is determined by the equilibrium properties of the system since it can be expressed in terms of its phase volume. In contrast, the dissipative coefficients are the characteristics of the nonequilibrium behavior; they describe the mixing properties of the underlying Hamiltonian system. The dissipative coefficients are independent of equilibrium properties: one may envision the systems with the same equilibrium properties but different laws of evolution to equilibrium.

Energy of the system, $H_{\text{eff}}(y,S)$, is conserved in the course of evolution to equilibrium:

$$H_{\text{eff}}(y,S) = E = const, \tag{2.48}$$

as it must be for an isolated system. Indeed,

$$\frac{dH_{\text{eff}}(y,S)}{dt} = \frac{dH_{\text{eff}}(y,S)}{\partial y^\mu}\frac{dy^\mu}{dt} + \frac{dH_{\text{eff}}(y,S)}{\partial S}\frac{dS}{dt}, \tag{2.49}$$

and the right hand side of (2.49) vanishes due to (2.46).

The energy conservation allows us to reduce the order of the system. The resulting equations take a simple form if expressed in terms of the function, $S(E,y)$,

[7] The fact that (2.45) holds for isolated systems is not a constraint in consideration of continuum media: the isolation requirement just specifies the boundary conditions and does not affect the differential equations.

which is the solution of (2.48) with respect to S for a given E. For this function, similarly to (2.26),

$$\frac{1}{T} = \frac{\partial S(E, y)}{\partial E}, \qquad \frac{\partial S(E, y)}{\partial y^\mu} = -\frac{1}{T}\frac{\partial H_{\text{eff}}(y, S)}{\partial y^\mu}.$$

Therefore, the governing differential equations become a system of m differential equations of the first order:

$$\begin{aligned}\frac{dy^\mu}{dt} &= -\omega^{\mu\nu}\frac{1}{\partial S(E, y)/\partial E}\frac{\partial S(E, y)}{\partial y^\nu} \\ &\quad + D^{\mu\nu}(y, S(E, y))\frac{1}{\partial S(E, y)/\partial E}\frac{\partial S(E, y)}{\partial y^\nu}.\end{aligned} \tag{2.50}$$

This system is determined by the functions, $S(E, y)$ and $D^{\mu\nu}(y, S)$, while E is considered as a given parameter.

An important process which is not covered by (2.46) or (2.50) is heat conduction. In this case, the slow variables, y^μ, are the energies of small parts of the body; they are not the coordinates or momenta of the underlying Hamiltonian system. The equations of heat conduction also possess a special structure; the reader is referred to [50] for consideration of this case and for further details regarding the derivation of (2.46) and (2.50).

2.7 Secondary Thermodynamics and Higher Order Thermodynamics

The special structure of thermodynamic equations, comprised of the existence of energy and entropy, the equations of state and the constitutive equations and the special form of equations of nonequilibrium thermodynamics, pertains only to the equations governing the evolution of slow variables of a Hamiltonian system. Such a theory may be called primary thermodynamics. It might happen that the equations of primary thermodynamics also admit two well-separated time scales. One may wonder what are the governing equations for the slow variables of the primary thermodynamics or, in other words, which equations do we get after the elimination of the fast variables in the equations of primary thermodynamics. We call the corresponding theory of slow variables of primary thermodynamics secondary thermodynamics. There are two important examples of secondary thermodynamics: plasticity theory and turbulence theory. Plasticity of crystalline bodies is caused by motion of defects of crystal lattice, such as dislocations. The crystal lattice may be viewed as a Hamiltonian system. The defects are the slow variables of this Hamiltonian system. Therefore the governing dynamical equations for defects are the subject of primary thermodynamics. Accordingly, macroscopic plastic behavior of crystals and polycrystals is the subject of secondary thermodynamics. Another example: turbulence theory. Equations describing fluid motion are the equations of primary thermodynamics. In the case of a chaotic fluid motion, turbulence, the motion is characterized by fast and slow variables. Elimination of fast variables and construction of the equations governing the slow variables is the major goal of turbulence theory. For both examples of secondary thermodynamics, the development of a theory of slow variables has not been completed yet. In particular, it is not known whether the equations of secondary thermodynamics must possess a special structure, as do the equations of primary thermodynamics. Most probably, there are no statements of the same level of generality as for primary thermodynamics. In particular, in turbulence theory different flow geometries may yield quite different systems of equations for slow variables. One feature of secondary thermodynamics can be quite general though. If the equations of primary thermodynamics exhibit chaotic behavior, then a new entropy can enter the theory. As in primary thermodynamics, its meaning is two fold: the new entropy characterizes fluctuations of the slow variables and the information on the system lost in the elimination of fast degrees of freedom. In the case of materials with random microstructures, this new characteristic called entropy of microstructure is introduced and studied in Sect. 18.5–18.8. In contrast to thermodynamic entropy, entropy of microstructure should decrease in an isolated system.[8] This feature is caused by the appearance of attractors in the phase space: the phase volume shrinks when the system approaches the equilibrium state.

One may envision the situations when a secondary thermodynamics model possesses two well-separated time scales. Then the elimination of the fast variables yields the equations of tertiary thermodynamics, etc. What will be common for all levels of thermodynamics is the existence of energy and entropy equations as long as energy and entropy remain slow variables. Besides, thermodynamic entropy, once appeared, will remain an increasing function for closed systems.

The existence of entropy is intimately related to the Hamiltonian structure of microdynamics. Such a structure is guaranteed by the classical approximation in quantum mechanics. However, if the quantum mechanics problem has two well-separated time scales, the elimination of the fast variables may yield the dissipative equations already at the level of the quantum mechanics description. On the next level, the classical mechanics description, one would have a system that is not Hamiltonian but instead a system with dissipation. Presumably, entropy can still be introduced, but such a consideration seems not have been pursued yet. The continuum mechanics level of description will then correspond at least to secondary thermodynamics.

[8] An exception is the case of unstable systems. For such systems entropy of microstructure can be generated without external actions.

Chapter 3
Continuum Mechanics

3.1 Continuum Kinematics

Continuum kinematics. A continuum is a system consisting of an infinite number of particles. More precisely, a continuum is defined as a set of particles which is in a one-to-one correspondence with a set of points of some region $\mathring{V}$ in three-dimensional space R_3. Each particle can be assigned its "name": the coordinates of its counterpart in region $\mathring{V}$. We denote these coordinates by capital Latin letters X^1, X^2, X^3, in contrast to the coordinates of the observer's frame, which are denoted by x^1, x^2, x^3. The latter are called Eulerian coordinates, while the coordinates X^1, X^2, X^3 are usually referred to as Lagrangian coordinates or material coordinates.

Positions of the particles at any instant t are described by the functions

$$x^i = x^i\left(X^1, X^2, X^3, t\right).$$

We use the small Latin indices i, j, k, l for numbering Eulerian coordinates, while for Lagrangian coordinates another group of indices, a, b, c, d is employed. We will call these indices observer's indices and Lagrangian or material indices, respectively. The reason for such a distinctive notation will be explained later. So, for the particle trajectories we write

$$x^i = x^i\left(X^a, t\right) \tag{3.1}$$

or, if the indices are suppressed,[1]

$$x = x\left(X, t\right).$$

Distinguishing the points of the continuum and assigning each one a "name," X^a, is a key point in the definition of continuum. For, if we do not identify particles, we

[1] We do not strictly maintain this order of the arguments throughout the book and, in cases when the dynamic effects are of primary concern, we write $x = x(t, X)$.

V.L. Berdichevsky, *Variational Principles of Continuum Mechanics*,
Interaction of Mechanics and Mathematics, DOI 10.1007/978-3-540-88467-5_3,

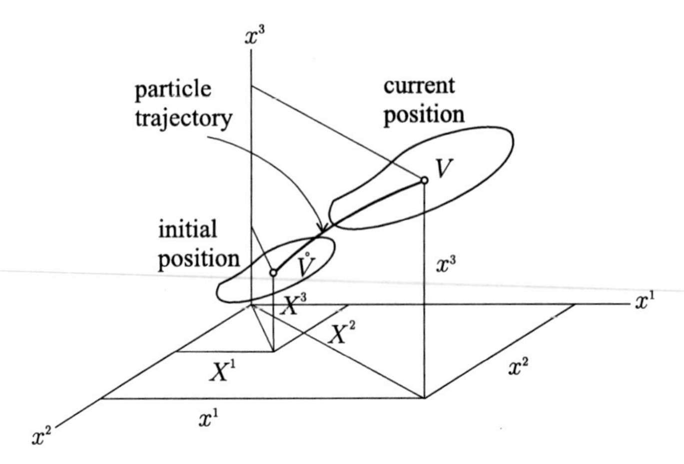

Fig. 3.1 Lagrangian and Eulerian coordinates

are not able to introduce such notions, as, say, velocity (defining velocity, one has to say velocity of which object is considered).

Region $\overset{\circ}{V}$ is usually identified with the initial position of the continuum. The current position of the continuum is denoted by V (Fig. 3.1). If we wish to emphasize that the region V changes in time, we write $V(t)$ instead of V.

For a given observer's space point x and a given instant t, one can consider (3.1) as a system of three (non-linear) equations with respect to Lagrangian coordinates. In principle, this system of equations can be solved, and one can write

$$X^a = X^a(x^i, t) \text{ or } X = X(x, t). \tag{3.2}$$

Equations (3.2) show which particle arrives at the space point x at the time t.

We will call the functions $x(X, t)$ the particle trajectories, and, in case of statics, when these functions do not depend on time, the particle positions. The inverse functions, $X(x, t)$ will be called the Lagrangian coordinate flow.

In addition to functions $x(X, t)$, behavior of continuum may need to be characterized by other functions $u(X, t)$, some "internal degrees of freedom," like temperature, plastic deformations, concentration of chemical species, etc. Selection of the proper set of characteristics is the first step in continuum modeling.

Tensor character of continuum mechanics equations. We have introduced two coordinate frames, the observer frame with the Eulerian coordinates x^i, and the material frame with the Lagrangian coordinates X^a. Even if Eulerian coordinates are Cartesian, Lagrangian coordinates are, in general, curvilinear (Fig. 3.2). Therefore, the technique of curvilinear coordinates appears naturally. Besides, all the relations we consider must not depend on the choice of Eulerian and Lagrangian coordinates. Thus, tensor analysis must be employed. In order to make the text approachable for the reader who is not familiar with tensor analysis and the theory of curvilinear coordinates, the following simplifications are made. First, the text is written in such

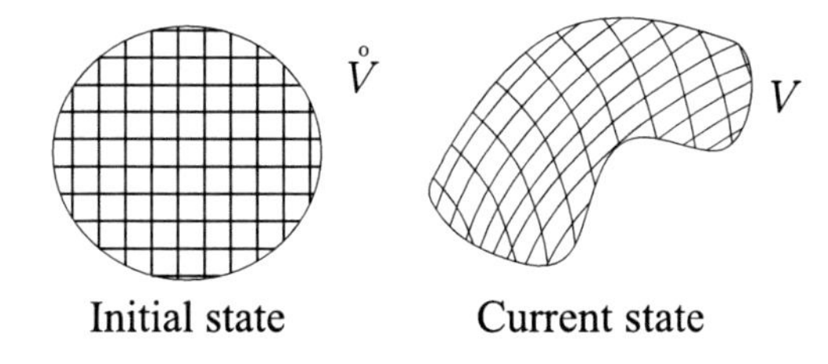

Fig. 3.2 Curvilinear nature of Lagrangian coordinates: material frame in the initial and the current states

a way that all remarks concerning tensor character of the relations derived can be omitted without detriment to understanding the key part of the material. Second, though all equations in the book are written in a proper tensor form distinguishing upper and lower indices, the reader may decline making a difference between upper and lower Eulerian indices i, j, k, l, accepting that the observer frame is Cartesian. Therefore, for example, one can identify $x^1(X^a, t)$ and $x_1(X^a, t)$. Such identification, however, cannot be made for quantities with Lagrangian indices; for example, X^a and X_a must be treated as different quantities. If a quantity with Lagrangian indices is used with upper and lower indices we always provide the link between the two. Third, we use the Cartesian coordinates for the observer's frame; therefore the covariant derivatives do not appear, and the space derivatives with respect to Eulerian coordinates are just $\partial/\partial x^i$. The subsections concerned with the issues on tensor features of continuum mechanics are furnished with the sign $*$ which indicates that the subsection can be omitted by an uninterested reader. The only tensor notation which the reader is supposed to know is the summation convention over repeated indices introduced in Sect. 1.4. One simple statement from tensor analysis, the sum $a^{ij}b_{ij}$ is zero if a^{ij} is symmetric and b_{ij} is anti-symmetric, is also used. Perhaps, the only part of the book where the knowledge of tensor analysis cannot be avoided is the shell theory (Chap. 14) where the coordinate system on the shell middle surface is inevitably curvilinear.

Distortion. Derivatives

$$x_a^i = \frac{\partial x^i(X, t)}{\partial X^a}$$

are called distortions (the term deformation gradient is often used as well). Derivatives

$$X_i^a = \frac{\partial X^a(x, t)}{\partial x^i}$$

are called inverse distortions.

They obey the equations

$$x_a^i X_j^a = \delta_j^i, \quad x_a^i X_i^b = \delta_a^b, \tag{3.3}$$

where δ_j^i is Kronecker's delta: $\delta_j^i = 1$ if $i = j$ and $\delta_j^i = 0$ if $i \neq j$.

The first equation (3.3) can be obtained from the identity

$$x^i\left(X\left(x,t\right),t\right)=x^i, \tag{3.4}$$

which expresses the fact that functions $X\left(x,t\right)$ are solutions of (3.1). Differentiating (3.4) with respect to x^j and using the chain rule, we obtain

$$\frac{\partial x^i\left(X\left(x,t\right),t\right)}{\partial x^j}=\frac{\partial x^i\left(X\left(x,t\right),t\right)}{\partial X^a}\frac{\partial X^a\left(x,t\right)}{\partial x^j}=x^i_a X^a_j=\frac{\partial x^i}{\partial x^j}=\delta^i_j,$$

i.e. the first equation (3.3) holds. The second equation (3.3) is derived similarly from the identity

$$X^a\left(x\left(X,t\right),t\right)=X^a.$$

Matrices will be denoted by double bars: $\left\|x^i_a\right\|$ means the matrix with the components x^i_a. Equations (3.3) mean that the matrix $\left\|X^a_i\right\|$ is inverse to the matrix $\left\|x^i_a\right\|$. Indeed, let the upper indices number the columns, while the lower indices number the rows. Then (3.3) can be written as

$$\left\|x^i_a\right\|\cdot\left\|X^a_j\right\|=\left\|\delta^i_j\right\|,\quad \left\|X^a_i\right\|\cdot\left\|x^i_b\right\|=\left\|\delta^a_b\right\|,$$

i.e. X^a_i are the components of the inverse matrix. This explains the term "inverse distortion" for X^a_i.

Velocity. Time derivative when Lagrangian coordinates are kept constant is denoted by d/dt and called material time derivative. By definition, the particle velocity is

$$v^i=\frac{dx^i(X,t)}{dt}.$$

As has been mentioned, velocity cannot be defined if Lagrangian coordinates are not introduced explicitly: $v^i\left(X,t\right)$ is the velocity of the particle with the Lagrangian coordinate X.

Velocity and distortion obey the compatibility relation

$$\frac{\partial v^i}{\partial X^a}=\frac{dx^i_a}{dt}. \tag{3.5}$$

Inverse matrix. Let the determinant of matrix $\left\|a^i_j\right\|$,

$$a=\det\left\|a^i_j\right\|,$$

be not equal to zero and $a^{(-1)i}_j$ be the components of the matrix which is inverse to the matrix $\left\|a^i_j\right\|$:

$$\left\| a^i_j \right\| \cdot \left\| a^{(-1)j}_k \right\| = \left\| \delta^i_k \right\| .$$

To write this equation in the index form we accept a convention that the upper and the lower indices are the numbers of the row and the column, respectively. Then $a^{(-1)i}_j$ obey to the system of equations

$$a^i_j a^{(-1)j}_{\ \ k} = \delta^i_k, \tag{3.6}$$

Equations (3.6) may be considered as a system of n linear equations with respect to $a^{(-1)j}_{\ \ k}$ assuming that a^i_j are given. It follows from (3.6) that $a^{(-1)i}_j$ obeys also to a system of equations,

$$a^i_j a^{(-1)k}_{\ \ i} = \delta^k_j. \tag{3.7}$$

In equation (3.6) the lower index of a^i_j is involved in summation; in equation (3.7) the upper index does. Equations (3.7) is an index form of the matrix equation,

$$\left\| a^{(-1)k}_{\ \ i} \right\| \cdot \left\| a^i_j \right\| = \left\| \delta^k_j \right\| .$$

To prove (3.7) one can first check that, due to (3.6),

$$(a^i_j a^{(-1)k}_{\ \ i} - \delta^k_j) a^m_k = 0. \tag{3.8}$$

(opening brackets in (3.8) we arrive at an identity, $a^m_k - a^m_k = 0$). Since det $\left\| a^i_j \right\| \neq 0$, equation (3.7) follows from (3.8).
Metric tensor. The distance ds between the points with coordinates x^i and $x^i + dx^i$ is, by definition,

$$ds^2 = g_{ij} dx^i dx^j, \tag{3.9}$$

where $g_{ij} = g_{ji}$ are called the covariant components of metric tensor. Determinant of the matrix $\left\| g_{ij} \right\|$ is denoted by g :

$$g = \det \left\| g_{ij} \right\| . \tag{3.10}$$

In Cartesian coordinates g_{ij} are constant and equal to Kronecker's delta δ_{ij} while $g = 1$.

Contravariant components of metric tensor, g^{ij}, are, by definition, the components of the matrix inverse to the matrix $\left\| g_{ij} \right\|$:

$$g^{ij} g_{kj} = \delta^i_k.$$

In Cartesian coordinates, $g^{ij} = g_{ij} = \delta_{ij}$.

Similarly, the distance between the two material points X^a and $X^a + dX^a$ is

$$ds^2 = g_{ab} dX^a dX^b, \tag{3.11}$$

where g_{ab} are the Lagrangian components of the metric tensor.

Comparing (3.9) and (3.11) and using the equality

$$dx^i = x^i_a dX^a,$$

we obtain the expression of the metric tensor in Lagrangian coordinates in terms of distortion:

$$g_{ab} = g_{ij} x^i_a x^j_b. \tag{3.12}$$

Contravariant components of metric tensor in Lagrangian coordinates, g^{ab}, are the components of the inverse matrix to the matrix $\|g_{ab}\|$. The reader is invited to check that

$$g^{ab} = g^{ij} X^a_i X^b_j, \quad X^a_i = g_{ij} g^{ab} x^j_b. \tag{3.13}$$

The Levi-Civita symbol. A system of numbers $e_{ijk} = e^k_{ij} = e^{ijk}$ where

$$e_{123} = e_{312} = e_{231} = 1, \quad e_{213} = e_{321} = e_{132} = -1,$$

and all other e_{ijk} are equal to zero is called the three-dimensional Levi-Civita symbol. The Levi-Civita symbol is used for explicitly writing determinants and vector products. For any tensor a^i_j, the determinant of the matrix $\left\|a^i_j\right\|$ is

$$\det \left\|a^i_j\right\| = e_{ijk} a^i_1 a^j_2 a^k_3. \tag{3.14}$$

Note also the following relations:

$$e_{ijk} a^i_l a^j_m a^k_n = \det \left\|a^i_j\right\| e_{lmn}, \tag{3.15}$$

$$\det \left\|a^i_j\right\| = \frac{1}{3!} e_{ijk} a^i_l a^j_m a^k_n e^{lmn}. \tag{3.16}$$

In Cartesian coordinates the vector with the components $e_{ijk} a^i b^j$ is called the vector product of the vectors with components a^i and b^j. An important role in operations with vector products is played by the identities

$$e_{ijk} e^{lmk} = \delta^l_i \delta^m_j - \delta^m_i \delta^l_j, \quad e_{ijk} e^{ljk} = 2\delta^l_i. \tag{3.17}$$

In Cartesian coordinates the Levi-Civita symbol coincides with the Levi-Civita tensor, ε_{ijk}, defined in the next subsection.

The Levi-Civita tensor*. It is easy to derive from (3.15) that the quantities ε_{ijk} which are equal to $\sqrt{g}e_{ijk}$ in coordinate systems of one orientation and to $-\sqrt{g}e_{ijk}$ in coordinate systems of the opposite orientation form the components of a covariant tensor of the third order. This tensor is called the Levi-Civita tensor.

One can introduce two Levi-Civita tensors setting $\varepsilon_{123} = 1$ or $\varepsilon_{123} = -1$ in a dextral Cartesian coordinate system (i.e. at right handed orientation). Accordingly, there are two ways to introduce a vector product. The two vector products differ by a sign. In what follows, we assume $\varepsilon_{123} = \sqrt{g}$ in a dextral coordinate system.

Appearance of the Levi-Civita symbol in a theory indicates that left handed and right handed coordinate systems are not equivalent.[2]

Note the relation for contravariant components of the Levi-Civita tensor:

$$\varepsilon^{ijk} = g^{ii'} g^{jj'} g^{kk'} \varepsilon_{i'j'k'} = \frac{1}{\sqrt{g}} e^{ijk}. \tag{3.18}$$

The identities (3.17) yield the corresponding identities for the Levi-Civita tensor:

$$\varepsilon_{ijk}\varepsilon^{lmk} = \delta_i^l \delta_j^m - \delta_i^m \delta_j^l, \quad \varepsilon_{ijk}\varepsilon^{ljk} = 2\delta_i^l. \tag{3.19}$$

In a Cartesian coordinate system, $\varepsilon_{ijk} = \varepsilon^{ijk} = e_{ijk}$.

Formula for inverse matrix. We will employ a useful explicit formula for the components of the inverse matrix in terms of the determinant a.

The determinant a is a certain function of the matrix components: $a = a\left(a_j^i\right)$. The components of the inverse matrix $a^{(-1)}{}^i_j$ are the derivatives of $\ln a$:

$$a^{(-1)}{}^j_i = \frac{1}{a}\frac{\partial a}{\partial a_j^i} = \frac{\partial \ln a}{\partial a_j^i}. \tag{3.20}$$

The reader can easily check the validity of (3.20) for two-dimensional matrices by direct inspection. Since this formula will be used further on many occasions, we give here its proof for 3×3 matrices; the proof in general case is similar. From (3.14),

$$\frac{\partial a}{\partial a_1^i} = e_{ijk} a_2^j a_3^k.$$

Hence,

$$\frac{1}{a}\frac{\partial a}{\partial a_1^i} a_1^i = 1,$$

$$\frac{1}{a}\frac{\partial a}{\partial a_1^i} a_2^i = e_{ijk} a_2^i a_2^j a_3^k = 0.$$

[2] The exceptions are the cases when ε_{ijk} are contained in products such as $\varepsilon_{ijk}\varepsilon^{lmn}$ which are invariant with respect to the sign change of ε_{ijk}.

Here we used that summation of object, e_{ijk}, which is anti-symmetric over ij , and the tensor $a_2^i a_2^j$, which is symmetric over ij, is zero. Similarly,

$$\frac{1}{a}\frac{\partial a}{\partial a_1^i}a_3^i = 0.$$

Therefore, the term $1/a\ \partial a/\partial a_1^i$ obeys (3.7) with

$$a_i^{(-1)1} = \frac{1}{a}\frac{\partial a}{\partial a_1^i}.$$

Other relations (3.7) are verified in the same way.

Jacobian of transformation from Lagrangian to Eulerian variables and related formulas. Denote by Δ the Jacobian of transformation from Lagrangian to Eulerian variables:

$$\Delta = \det \left\| x_a^i \right\|.$$

The following useful consequences of (3.15) hold:

$$e_{ijk}x_a^i x_b^j x_c^k = \Delta e_{abc}, \qquad e_{abc}X_i^a X_j^b X_k^c = \frac{1}{\Delta}e_{ijk}, \tag{3.21}$$

$$e^{abc}x_a^i x_b^j x_c^k = \Delta e^{ijk}, \qquad e^{ijk}X_i^a X_j^b X_k^c = \frac{1}{\Delta}e^{abc}.$$

Contracting the second equation (3.21) with $x_a^i x_b^j$ we obtain:

$$e_{abc}X_k^c = \frac{1}{\Delta}e_{ijk}x_a^i x_b^j. \tag{3.22}$$

Further contraction of (3.22) with e^{abc} and use of the second formula (3.17) yields

$$X_k^c = \frac{1}{2\Delta}e_{ijk}x_a^i x_b^j e^{abc}. \tag{3.23}$$

The same result can be obtained from relation (3.20) by differentiation of Jacobian,

$$X_i^a = \frac{1}{\Delta}\frac{\partial \Delta}{\partial x_a^i}. \tag{3.24}$$

An equivalent form of (3.23) is obtained by contraction of (3.23) with Levi-Civita symbol, $e^{i'j'k}$, and use of (3.17):

$$X_k^c e^{ijk} = \frac{1}{\Delta}x_a^i x_b^j e^{abc}.$$

Similarly to (3.23), distortion can be expressed in terms of inverse distortion:

$$x_c^k = \frac{1}{2\det \left\| X_i^a \right\|} e^{ijk} X_i^a X_j^b e_{abc}. \tag{3.25}$$

Lagrangian components of metric tensor and Levi-Civita tensor. The determinant of the matrix $\det\|g_{ab}\|$ will be denoted by $\hat{g}$. According to (3.20), the contravariant Lagrangian components of metric tensor are related to $\hat{g}$ by the formula

$$g^{ab} = \frac{1}{\hat{g}} \frac{\partial \hat{g}}{\partial g_{ab}}. \tag{3.26}$$

For Lagrangian components of the Levi-Civita tensor we have

$$\varepsilon_{abc} = \sqrt{\hat{g}} e_{abc}, \qquad \varepsilon^{abc} = \frac{1}{\sqrt{\hat{g}}} e^{abc}.$$

Note that

$$\hat{g} = g\Delta^2. \tag{3.27}$$

Therefore formulas for Lagrangian components of the Levi-Civita tensor can be written also as

$$\varepsilon_{abc} = \sqrt{g}\Delta e_{abc}, \qquad \varepsilon^{abc} = \frac{1}{\sqrt{g}\Delta} e^{abc}. \tag{3.28}$$

The initial position/initial state. Consider the position of continuum at some initial instant t_0:

$$x^i = x^i \left(X^a, t_0\right) = \mathring{x}^i \left(X^a\right). \tag{3.29}$$

Lagrangian coordinates can be always chosen to coincide with the Eulerian coordinates at the initial instant. Then the functions (3.29) become

$$\mathring{x}^1 \left(X^a\right) = X^1, \; \mathring{x}^2 \left(X^a\right) = X^2, \; \mathring{x}^3 \left(X^a\right) = X^3. \tag{3.30}$$

Transformations of Lagrangian coordinates change functions (3.30). Therefore, in order to maintain the ability to keep the invariance of all the relations with respect to the transformations of the Lagrangian coordinates, we will describe the initial position of the system by functions (3.29) of general form.

The components of the metric tensor in the initial state are

$$\mathring{g}_{ab} = g_{ab}(X, t_0) = g_{ij} \mathring{x}_a^i \mathring{x}_b^j, \quad \mathring{g}^{ab} = g^{ij} \mathring{X}_i^a \mathring{X}_j^b.$$

Here,

$$\mathring{x}_a^i = \frac{\partial \mathring{x}^i}{\partial X^a}, \quad \mathring{X}_i^a = \frac{1}{\mathring{\Delta}} \frac{\partial \mathring{\Delta}}{\partial \mathring{x}_a^i}, \quad \mathring{\Delta} = \det \left\| \mathring{x}_a^i \right\|.$$

All the initial values are furnished with the symbol $^\circ$.

Transformation from Eulerian to Lagrangian coordinates. Let $f(x, t)$ be a scalar function of Eulerian coordinates and time. From this function, two different functions of Lagrangian coordinates, $f(\mathring{x}(X), t)$ and $f(x(X, t), t)$, can be constructed. In what follows, if not noted otherwise, we use the second way, i.e. the transformation from Eulerian to Lagrangian coordinates and back is done by means of the particle trajectories $x(X, t)$ and the Lagrangian coordinate flow $X(x, t)$.

Two groups of transformations*. Continuum mechanics equations have a tensor character with respect to the two groups of transformation: the group of Eulerian coordinate transformations,

$$x'^i = x'^i(x^j), \tag{3.31}$$

and the group of Lagrangian coordinate transformations,

$$X'^a = X'^a(X^b). \tag{3.32}$$

Now we can give a precise distinction between Eulerian indices which we denoted by the group of letters i, j, k, l and Lagrangian indices for which the letters a, b, c, d were reserved. The group (3.31) causes the transformation of Eulerian indices while the objects with Lagrangian indices are not affected by this group and behave as scalars. Similarly, the group (3.32) yields the transformations of the objects with Lagrangian indices leaving the objects with Eulerian indices unchanged.

The examples are: velocity, v^i, a vector with respect to the group (3.31) and a set of three scalars with respect to the group (3.32); the metric tensor g_{ab}, a set of six scalars with respect to the group (3.31) and a tensor of the second order with respect to the group (3.32); distortion x^i_a, a set of three vectors with respect to the group (3.31) and a set of three vectors with respect to the group (3.32).

Using the first letters of the Latin alphabet a, b, c, d for Lagrangian coordinates is a tribute to the tradition of the nineteenth century to denote the three Lagrangian coordinates by the first letters of the Latin alphabet a, b, c.

Juggling with indices. Indices are moved up and down by the following rule: For an object with upper index, T^i, by definition,

$$T_i = g_{ij} T^j.$$

Accordingly,

$$T^i = g^{ij} T_j.$$

Since we agreed to deal with Cartesian observer's frame, $g_{ij} = g^{ij} = \delta^{ij}$, we have $T^i = T_i$. Nevertheless, in order to have all the relations in tensor form, we keep the metric tensor when necessary.

For the objects with the Lagrangian indices, keeping the metric tensor while juggling with the indices is always necessary because in Lagrangian coordinates we have two different metric tensors, g_{ab} and $\mathring{g}_{ab}$, and, to avoid ambiguities, we have to specify which one is used for juggling.

Strain measures. Distortion $x^i_a(X,t)$ describes an affine deformation of an infinitesimally small element of the continuum in the neighborhood of a point X. There are two ways of extracting strain from the distortion. Strain is a part of affine deformation not affected by the rotation of the element. The first way uses the metric tensor in the Lagrangian coordinate system. It is apparent that the tensor g_{ab} measuring the distances does not depend on the rotation of the material element. The tensor

$$\varepsilon_{ab} = \frac{1}{2}\left(g_{ab} - \mathring{g}_{ab}\right) = \frac{1}{2}\left(g_{ij}x^i_a x^j_b - \mathring{g}_{ab}\right) \tag{3.33}$$

can be used as a measure of strain.

The second way of defining strain is based on splitting the matrix $\left\|x^i_a\right\|$ into the product of a symmetric positive matrix[3] with the components $|x|_{ab}$ and an orthogonal matrix with the components λ^{ib},

$$x^i_a = |x|_{ab}\,\lambda^{ib}. \tag{3.34}$$

The orthogonality of the matrix $\left\|\lambda^{ib}\right\|$ means that

$$g_{ij}\lambda^{ib}\lambda^{jc} = \mathring{g}^{bc}, \quad \mathring{g}_{bc}\lambda^{ib}\lambda^{jc} = g^{ij}. \tag{3.35}$$

Formula (3.34) corresponds to presenting the affine deformation as a superposition of extensions in three directions and orthogonal rotation. Indeed, by rotating coordinates one can put a symmetric matrix, $|x|_{ab}$, into diagonal form. The fibers directed along the axes of the new coordinate system (along the principal directions of the tensor $|x|_{ab}$) elongate or contract depending on whether the corresponding eigenvalue of the tensor $|x|_{ab}$ is greater or smaller than unity. The final distortion is obtained by rotation of the deformed fibers. Presentation of matrices in the form (3.34) is called the polar decomposition.

Tensors $|x|_{ab}$ and g_{ab} are related as

$$g_{ab} = g_{ij}x^i_a x^j_b = \mathring{g}^{cd}\,|x|_{ac}\,|x|_{bd}\,. \tag{3.36}$$

According to (3.36), g_{ab} and $|x|_{ab}$ can be made diagonal by the same orthogonal (with respect to the metric $\mathring{g}_{ab}$) transformation.

Formula (3.36) provides a one-to-one correspondence between g_{ab} and $|x|_{ab}$. Indeed, if $|x|_{ab}$ is known then g_{ab} is determined by (3.36). The principal coordinates of tensors $|x|_{ab}$ and g_{ab} (the coordinates in which $|x|_{ab}$ and g_{ab} are diagonal) coincide

[3] Positiveness of a symmetric matrix means that its eigenvalues are positive.

according to (3.36). If g_{ab} is known, then, to find $|x|_{ab}$, one makes an orthogonal transformation which diagonalizes g_{ab}. Since $|x|_{ab}$ are diagonal in the same coordinate system and the principal values are positive, we have (for $\mathring{g}_{ab} = \delta_{ab}$)

$$|x|_{11} = \sqrt{g_{11}}, \quad |x|_{22} = \sqrt{g_{22}}, \quad |x|_{33} = \sqrt{g_{33}}.$$

Since the tensor $|x|_{ab}$ is a "square root" of the metric tensor, it is called the distortion modulus.

The tensor $|x|_{ab}$ depends on the distortion semilinearly, i.e. if the distortion is multiplied by a positive constant, $|x|_{ab}$ acquires the same factor.

The tensor λ^{ib} can be expressed in terms of x^i_a as

$$\lambda^{ib} = |x|^{(-1)ba} x^i_a,$$

where $|x|^{(-1)ba}$ is the inverse of the tensor $|x|_{ab}$:

$$|x|_{ab} |x|^{(-1)bc} = \delta^c_a.$$

In the system of coordinates in which the tensor $|x|_{ab}$ is diagonal,

$$|x|^{(-1)11} = \frac{1}{|x|_{11}}, \quad |x|^{(-1)22} = \frac{1}{|x|_{22}}, \quad |x|^{(-1)33} = \frac{1}{|x|_{33}}.$$

Therefore, $|x|_{ab}$ and λ^{ib} are uniquely determined in terms of the distortion x^i_a.

Using the tensor $|x|_{ab}$, we can construct a second strain measure:

$$\gamma_{ab} = |x|_{ab} - \mathring{g}_{ab}. \tag{3.37}$$

Note that

$$|\mathring{x}|_{ab} \equiv |x|_{ab}\big|_{t=t_0} = \mathring{g}_{ab}.$$

This equality can be checked by transforming tensor $|\mathring{x}|_{ab}$ into its principal axes.

According to (3.36) and (3.37), the strain tensors ε_{ab} and γ_{ab} are related as

$$\varepsilon_{ab} = \gamma_{ab} + \frac{1}{2}\mathring{g}^{cd}\gamma_{ac}\gamma_{bd}. \tag{3.38}$$

Strain measures ε_{ab} and γ_{ab} are useful if the deviations of g_{ab} from their initial values, $\mathring{g}_{ab}$, are small. Otherwise, it is more convenient to use g_{ab} as the primary characteristics of deformations.

A representation of three-dimensional orthogonal matrices. The matrix with the components λ^{ia}, orthogonal in sense of (3.35), can be orthogonal in the usual sense if it is multiplied by matrix $\left\|\mathring{x}^i_a\right\|$, $\alpha^{ij} = \mathring{x}^i_a\lambda^{ja}$. Indeed from (3.35) we find that the matrix $\left\|\alpha_{ij}\right\|$ is inverse to itself:

$$\alpha^{ij}\alpha_{kj} = \delta^i_k \quad \alpha^{ij}\alpha_{ik} = \delta^j_k. \tag{3.39}$$

A characteristic feature of the orthogonal transformation in three-dimensional space is that an orthogonal transformation keeps one line, the axis of rotation, immobile. Denote the axis of rotation by C and the unit vector directed along it by c_i. Since the transformation does not change c_i,

$$\alpha^{ij} c_j = c^i.$$

Due to the orthogonality of α^{ij} we can also write

$$\alpha^{ij} c_i = c^j.$$

The orthogonal transformation can be given by vector c_i and the angle θ of counterclockwise (if observed from the end of c_i) rotation around the line C. The couples (c_i, θ) and $(-c_i, -\theta)$ define the same transformation. The components of the orthogonal matrix, α^{ij}, are expressed in terms of c^i and θ as

$$\alpha^{ij} = \cos\theta g^{ij} + (1 - \cos\theta) c^i c^j - \sin\theta \varepsilon^{ijk} c_k. \tag{3.40}$$

This can be checked by writing down (3.40) in an orthogonal coordinate system one of the axes of which coincides with C.

Inversely, if the components of matrix α^{ij} are known, the angle θ is found by means of the formula

$$\cos\theta = \frac{1}{2}\left(\alpha_i^i - 1\right), \tag{3.41}$$

It follows from (3.40) by contraction over indices i, j. Then the vector c_i is computed by contracting (3.40) with ε_{ijk} and using $(3.19)_2$:

$$c_k = -\frac{1}{2\sin\theta} \varepsilon_{ijk} \alpha^{ij}.$$

The analogous expression for the orthogonal matrix with the determinant equal to -1 is

$$\alpha^{ij} = \cos\theta g^{ij} - (1 + \cos\theta) c^i c^j - \sin\theta \varepsilon^{ijk} c_k. \tag{3.42}$$

Space and time derivatives. The derivatives with respect to time with Eulerian coordinates held constant will be denoted in several ways depending on convenience:

$$\partial_t \equiv (\cdot)_{,t} \equiv (\cdot)_t \equiv \frac{\partial}{\partial t}.$$

Similarly, for spatial derivatives with respect to Eulerian coordinates we use the notation

$$\frac{\partial}{\partial x^i} \equiv \partial_i \equiv (\cdot)_{,i}$$

while spatial derivatives with respect to Lagrangian coordinates are denoted as

$$\frac{\partial}{\partial X^a} \equiv \partial_a \equiv (\cdot)_{,a}\,.$$

Obviously, for any function of Eulerian coordinates and time, $u(x,t)$,

$$\frac{du}{dt} = \frac{du(x(X,t),t)}{dt} = \frac{\partial u}{\partial t} + \frac{\partial u}{\partial x^i}\frac{dx^i}{dt} = \frac{\partial u}{\partial t} + v^i \frac{\partial u}{\partial x^i}.$$

The order of differentiation with respect to time and space coordinates, ∂_t and ∂_i, can be changed:

$$\frac{\partial}{\partial x^i}\frac{\partial}{\partial t} = \frac{\partial}{\partial t}\frac{\partial}{\partial x^i}.$$

Similarly,

$$\frac{\partial}{\partial X^a}\frac{d}{dt} = \frac{d}{dt}\frac{\partial}{\partial X^a}.$$

However,

$$\frac{\partial}{\partial x^i}\frac{d}{dt} \neq \frac{d}{dt}\frac{\partial}{\partial x^i}, \qquad \frac{\partial}{\partial X^a}\frac{\partial}{\partial t} \neq \frac{\partial}{\partial t}\frac{\partial}{\partial X^a}.$$

The operator ∂_t applied to the components of a tensor with Eulerian indices results in tensor components with a similar index structure. The same is true for the operator d/dt applied to tensor components with Lagrangian indices.

The material time derivative of tensor components with Eulerian indices does not yield a tensor. For example,

$$\frac{dx^i_a}{dt} = \frac{\partial}{\partial X^a}\frac{dx^i}{dt} = \frac{\partial v^i}{\partial X^a} = x^k_a \frac{\partial v^i}{\partial x^k}. \tag{3.43}$$

Quantities dx^i_a/dt do not transform by tensor rules since they are expressed in terms of partial (not covariant) derivatives of velocity.

The strain rate tensor. The components of the strain rate tensor in the Lagrangian coordinate system are defined as

$$e_{ab} = \frac{d\varepsilon_{ab}}{dt}.$$

Differentiating (3.33) with respect to time, we get

$$2e_{ab} = \left(g_{ij}\frac{\partial v^i}{\partial X^a} x^j_b\right) + (a \leftrightarrow b)\,. \tag{3.44}$$

By $(a \leftrightarrow b)$ we denote the previous term in brackets with index a changed to b and index b changed by a.

Equation (3.44) can be rewritten as[4]

$$2e_{ab} = x^i_a x^j_b \left(\frac{\partial v^i}{\partial x^j} + \frac{\partial v^j}{\partial x^i} \right). \tag{3.45}$$

Contracting (3.45) with inverse distortions $X^a_i X^b_j$, we obtain the following expression for the strain rate tensor in the Eulerian coordinate system, $e_{ij} = X^a_i X^b_j e_{ab}$:

$$e_{ij} = \partial_{(i} v_{j)} \equiv \frac{1}{2} \left(\frac{\partial v^i}{\partial x^j} + \frac{\partial v^j}{\partial x^i} \right). \tag{3.46}$$

The parentheses in indices stand for symmetrization:

$$a_{(ij)} \equiv \frac{1}{2} \left(a_{ij} + a_{ji} \right).$$

Let us emphasize that the time derivatives of Eulerian components of the strain tensor, $\varepsilon_{ij} = X^a_i X^b_j \varepsilon_{ab}$, are not equal to e_{ij} :

$$e_{ij} \neq \frac{d\varepsilon_{ij}}{dt}.$$

Rigid motion. Consider the motion of the system when the distortion x^i_a does not depend on the Lagrangian coordinates. In this case, the particle trajectories are

$$x^i = r^i(t) + \alpha^i_a(t) X^a. \tag{3.47}$$

Functions $x^i = r^i(t)$ define the trajectory of the particle with the zero Lagrangian coordinates.

The motion (3.47) is called homogeneous deformation. It is called rigid if the distortion, $x^i_a = \alpha^i_a$, is an orthogonal matrix, i.e.

$$g_{ij} \alpha^i_a \alpha^j_b = \mathring{g}_{ab}, \quad \mathring{g}^{ab} \alpha^i_a \alpha^j_b = g^{ij}. \tag{3.48}$$

---

[4] In curvilinear Euler's coordinates one has to differentiate in (3.44) the metric tensor as well. That yields an additional term $\frac{dg_{ij}}{dt} x^i_a x^j_b$ in (3.45). Since

$$\frac{dg_{ij}\left(x^k(X^a, t)\right)}{dt} = \frac{dg_{ij}}{dx^k} v^k, \quad \frac{\partial g_{ij}}{\partial x^k} = g_{mj} \Gamma^m_{ik} + g_{mi} \Gamma^m_{jk},$$

where Γ^m_{ik} are Kristoffel's symbols, (3.45) gets a tensor form

$$2e_{ab} = x^i_a x^i_b \left(\nabla_i v_j + \nabla_j v_i \right).$$

where ∇_i is the covariant space derivative in Eulerian coordinates.

For $g_{ij} = \delta_{ij}$, $\mathring{g}_{ab} = \delta_{ab}$, the matrix α^i_a is orthogonal in the usual sense.

Note that the second equation (3.48) follows from the first one, and vice versa. Indeed, denote $\mathring{g}^{ab}\alpha^i_a\alpha^j_b$ by $\tilde{g}^{ij}$. Contracting the first equation (3.48) with $\mathring{g}^{ac}\alpha^k_c$, we get the relation $g_{ij}\tilde{g}^{ik}\alpha^j_b = \alpha^k_b$ which, due to the non-degeneracy of the matrix $\|\alpha^i_a\|$, implies that $g_{ij}\tilde{g}^{ik} = \delta^k_j$. Therefore $\tilde{g}^{ik} = g^{(-1)ik} = g^{ik}$, as claimed; similarly, one can prove that the first equation (3.48) follows from the second one.

In rigid motion, according to the first equation (3.48), the strain tensor is equal to zero. The inverse statement is also true, i.e. if the strain tensor is equal to zero at every point of the continuum, the particle trajectories are given by formulas (3.47) where α^i_a is an orthogonal matrix. Hence, the rigid motion can be defined as a motion for which the distance between any two points of the continuum does not change over time.

The continuum which performs only rigid motions is called a rigid body. The velocities of the particles of the rigid body can be found by differentiating (3.47):

$$v^i\left(X^a, t\right) = u^i(t) + \frac{d\alpha^i_a(t)}{dt}X^a, \quad u^i \equiv \frac{dr^i}{dt}. \tag{3.49}$$

Here u^i is the velocity of the point with the zero Lagrangian coordinates.

Let us find the velocity field v^i as a function of Eulerian coordinates. In order to do this, we first need to express the Lagrangian coordinates in terms of Eulerian coordinates from (3.47):

$$X^a\left(x^i, t\right) = \alpha^{ja}(t)\left(x_j - r_j(t)\right). \tag{3.50}$$

It does not matter which metric tensor, $\mathring{g}^{ab}$ or g^{ab}, is used to lift the index in α^i_a because $\mathring{g}^{ab}=g^{ab}$ for a rigid motion. Substituting (3.50) into (3.49), we get

$$v^i\left(x^k, t\right) = u^i(t) + \omega^{ji}(t)\left(x_j - r_j(t)\right). \tag{3.51}$$

where we introduced the notation

$$\omega^{ji} = \alpha^{ja}(t)\frac{d\alpha^i_a}{dt}. \tag{3.52}$$

Tensor ω^{ji} is antisymmetric with respect to i, j. Indeed, differentiating the second equation (3.48) with respect to t and using the definition (3.52), we obtain

$$\frac{d}{dt}\left(\alpha^i_a\alpha^j_b\mathring{g}^{ab}\right) = \omega^{ij} + \omega^{ji} = 0.$$

Antisymmetric tensors of the second order, ω^{ij}, in a three-dimensional space are in one-to-one correspondence with vectors, ω_k, such that

$$\omega^{ij} = \varepsilon^{ijk}\omega_k, \quad \omega_k = \frac{1}{2}\varepsilon_{kij}\omega^{ij}. \tag{3.53}$$

Relations (3.53) follow from (3.19). The vector

$$\omega_k = \frac{1}{2}\varepsilon_{kij}\alpha^{ia}\frac{d\alpha_a^j}{dt}, \tag{3.54}$$

is called the angular velocity vector of a rigid body, ω^{ij} the tensor of angular velocity, $\omega^{ij}dt$ the tensor of infinitesimally small rotation, and $d\varphi_i = \omega_i dt$ the angle of infinitesimally small rotation. The kinematic relation (3.51) in terms of the vector of angular velocity takes the form

$$v^i\left(x^k, t\right) = u^i(t) + \varepsilon^{ikj}\omega_k\left(x_j - r_j(t)\right). \tag{3.55}$$

Time derivative of Lagrangian coordinates. By the definition of the material time derivative,

$$\frac{dX^a}{dt} = 0. \tag{3.56}$$

If one plugs into (3.56) the dependence of X^a on Eulerian coordinates, one gets

$$\frac{dX^a(t,x)}{dt} = \frac{dX^a(t,x(t,X))}{dt} = \frac{\partial X^a(t,x)}{\partial t} + \frac{\partial X^a}{\partial x^i}\frac{dx^i}{dt} = 0$$

or

$$\frac{\partial X^a}{\partial t} + v^i\frac{\partial X^a}{\partial x^i} = 0. \tag{3.57}$$

Equations (3.57) may be considered as a system of three linear equations with respect to velocity. Contracting (3.57) with x_a^j and using the first equation (3.3) we obtain

$$v^i = -x_a^i\frac{\partial X^a}{\partial t} \tag{3.58}$$

Equation (3.58) presents velocity in terms of time and space derivatives of the Lagrangian coordinate flow $X^a(t,x)$ because x_a^i may be viewed as some functions of the derivatives of $X^a(t,x)$, X_i^a.

Time derivative of the inverse distortion. Let us show that

$$\frac{dX_j^a}{dt} = -X_i^a\frac{\partial v^i}{\partial x^j}. \tag{3.59}$$

Indeed, differentiating (3.57) with respect to x^j, we have

$$\frac{\partial}{\partial t}\frac{\partial X^a}{\partial x^j} + v^i\frac{\partial}{\partial x^i}\frac{\partial X^a}{\partial x^j} + \frac{\partial v^i}{\partial x^j}\frac{\partial X^a}{\partial x^i} = 0. \tag{3.60}$$

Since the sum of the first two terms in (3.60) is dX_j^a/dt, we arrive at (3.59).

Invariant integration. In an Eulerian coordinate system, the volume element dV is

$$dV = \sqrt{g}dx^1dx^2dx^3 \tag{3.61}$$

where g is the determinant (3.10). After the coordinate transformation, $x \to X$,

$$dV = \sqrt{g}\,|\Delta|\,dX^1dX^2dX^3,$$

where Δ is the Jacobian of transformation. Since, due to (3.27),

$$\sqrt{\hat{g}} = \sqrt{g}\,|\Delta|\,, \tag{3.62}$$

the volume element in Lagrangian coordinates in the current state is

$$dV = \sqrt{\hat{g}}dX^1dX^2dX^3. \tag{3.63}$$

Similarly, in the initial state,

$$d\mathring{V} = \sqrt{\mathring{g}}dX^1dX^2dX^3, \qquad \mathring{g} = \det\|\mathring{g}_{ab}\|\,. \tag{3.64}$$

It can be proved that dV and $d\mathring{V}$ are scalars with respect to transformations of Lagrangian coordinates which do not change the orientation of the coordinate system.

Integrals

$$\int_V \Phi dV, \quad \int_{\mathring{V}} \Phi d\mathring{V}$$

have the same form in all Lagrangian coordinate systems, X^a, if Φ is a scalar with respect to transformations of Lagrangian coordinates.

The following equality holds:

$$\int_V \Phi\sqrt{\hat{g}}dX^1dX^2dX^3 = \int_V \Phi\sqrt{g}dx^1dx^2dx^3, \tag{3.65}$$

where V is the region occupied by the system in the current state. Equation (3.65) allows one to convert the integrals over Eulerian variables to the integrals over Lagrangian variables and vice versa.

The continuity equation. Denote by ρ the mass density. Then the mass of the material occupying the volume element dV is ρdV. We assume that the mass of each volume element does not change in the course of motion, i.e.

$$\rho dV = \rho_0 d\mathring{V}, \tag{3.66}$$

where ρ_0 is the mass density in the initial state. Then, from (3.63), (3.64), and (3.66), it follows the so-called continuity equation in Lagrangian variables:

$$\rho\sqrt{\hat{g}} = \rho_0\sqrt{\mathring{g}}. \tag{3.67}$$

In the initial state, the mass density ρ_0 and the metric tensor are some given functions of X^a. For given particle trajectories, one can find $\hat{g}$; therefore, the continuity equation can be considered as an equation allowing one to find the mass density, ρ, if the continuum motion is known:

$$\rho = \frac{\rho_0\sqrt{\mathring{g}}}{\sqrt{\hat{g}}}. \tag{3.68}$$

In terms of the determinant of distortion, Δ, according to (3.62),

$$\rho = \frac{\rho_0\sqrt{\mathring{g}}}{\sqrt{g}}\frac{1}{|\Delta|}. \tag{3.69}$$

Hence, the mass density is a function of distortion, x^i_a. Further we assume that $\Delta > 0$, and the sign of the absolute value in (3.69) can be dropped: at the initial state $\Delta > 0$, and, in order to change the sign, Δ must become zero at some space point; vanishing of Δ would mean a collapse of a material volume to a point or a surface – we exclude such cases from consideration.

In many cases it is convenient to consider mass density as a function of Eulerian coordinates. As such, mass density obeys the equation

$$\frac{d\rho}{dt} + \rho\frac{\partial v^i}{\partial x^i} = 0. \tag{3.70}$$

To derive (3.70) we differentiate (3.69) with respect to time while keeping the Lagrangian coordinates constant and use (3.24) and (3.43):

$$\frac{d\rho}{dt} = -\frac{\rho_0\sqrt{\mathring{g}}}{\sqrt{g}\Delta^2}\frac{\partial\Delta}{\partial x^i_a}\frac{dx^i_a}{dt} = -\rho X^a_i\frac{dx^i_a}{dt} = -\rho X^a_i\frac{\partial v^i}{\partial X^a} = -\rho\frac{\partial v^i}{\partial x^i}.$$

The formula for mass density (3.68) may be considered as the solution of the differential equation (3.70) for mass density.

Note the relation for the time derivative of the distortion determinant, Δ, which follows from (3.69) and (3.70):

$$\frac{d\Delta}{dt} = \Delta\frac{\partial v^i}{\partial x^i}. \tag{3.71}$$

Equation (3.66) allows one to convert easily the integrals of densities per unit mass over the Eulerian variables into the integrals over the Lagrangian variables;

for example, in the case of the internal energy density per unit mass, U, one can write

$$\int\limits_{V} \rho U dV = \int\limits_{\mathring{V}} \rho_0 U d\mathring{V}. \tag{3.72}$$

Derivative of the Jacobian with respect to parameters. Formula (3.71) is a special case of a more general relation. Consider a coordinate transformation, $X \to x$, which depends of parameters y^μ, $\mu = 1, ..., m$:

$$x^i = x^i \left(X^a, y^\mu\right).$$

Let Δ be the Jacobian of this transformation:

$$\Delta = \det \left\| \frac{\partial x^i}{\partial X^a} \right\|.$$

Then

$$\left.\frac{\partial \Delta}{\partial y^\mu}\right|_{X=\text{const}} = \Delta \frac{\partial}{\partial x^i} \frac{\partial x^i(X, y)}{\partial y^\mu}. \tag{3.73}$$

Indeed,

$$\left.\frac{\partial \Delta}{\partial y^\mu}\right|_{X=const} = \frac{\partial \Delta}{\partial x^i_a} \frac{\partial x^i_a}{\partial y^\mu} = \Delta X^a_i \frac{\partial}{\partial X^a} \frac{\partial x^i(X, y)}{\partial y^\mu} = \Delta \frac{\partial}{\partial x^i} \frac{\partial x^i(X, y)}{\partial y^\mu}.$$

Some identities. Consider an arbitrary coordinate transformation, $x = x(X)$. For the Jacobian of this transformation an identity holds

$$\frac{\partial}{\partial x^j} \frac{1}{\Delta} x^j_a = 0. \tag{3.74}$$

It can be checked by direct inspection:

$$x^j_a \frac{\partial}{\partial x^j} \frac{1}{\Delta} = -x^j_a \frac{1}{\Delta^2} \frac{\partial \Delta}{\partial x^i_b} \frac{\partial}{\partial x^j} x^i_b$$

$$= -x^j_a \frac{1}{\Delta} X^b_i \frac{\partial x^i_b}{\partial x^j} = -\frac{1}{\Delta} X^b_i \frac{\partial x^i_b}{\partial X^a}$$

$$= -\frac{1}{\Delta} X^b_i \frac{\partial x^i_a}{\partial X^b} = -\frac{1}{\Delta} \frac{\partial x^i_a}{\partial x^i}.$$

The identity (3.74) yields another form of (3.71), which employs the Lagrangian components of velocity, $v^a = X_i^a v^i$,

$$\frac{d\Delta}{dt} = \frac{\partial \Delta v^a}{\partial X^a}. \tag{3.75}$$

Replacing in the identity (3.75) x by X and X by x and taking into account that $\det \|\partial X / \partial x\| = 1/\Delta$, we obtain another identity,

$$\frac{\partial}{\partial X^a}\left(\frac{\partial X^a}{\partial x^i}\Delta\right) = 0. \tag{3.76}$$

Divergence form of (3.73). Note another form of equation (3.73):

$$\left.\frac{\partial \Delta}{\partial y^\mu}\right|_{X=const} = -\frac{\partial}{\partial X^a}\left(\frac{\partial X^a}{\partial y^\mu}\Delta\right). \tag{3.77}$$

It is obtained from (3.73) using the relation,

$$-\frac{\partial x^i(X, y)}{\partial y^\mu} = x_a^i \frac{\partial X^a(x, y)}{\partial y^\mu}, \tag{3.78}$$

which follows from differentiation of the identity

$$x^i\left(X^a\left(x^k, y\right), y\right) = x^i$$

with respect to y^μ. Plugging (3.78) in (3.73) we have

$$\left.\frac{\partial \Delta}{\partial y^\mu}\right|_{X=const} = -\Delta \frac{\partial}{\partial x^i} x_a^i \frac{\partial X^a}{\partial y^\mu}.$$

Employing (3.74) we obtain

$$\left.\frac{\partial \Delta}{\partial y^\mu}\right|_{X=const} = -\Delta \frac{\partial}{\partial x^i}\frac{1}{\Delta} x_a^i \frac{\partial X^a}{\partial y^\mu}\Delta = -x_a^i \frac{\partial}{\partial x^i}\frac{\partial X^a}{\partial y^\mu}\Delta = -\frac{\partial}{\partial X^a}\frac{\partial X^a}{\partial y^\mu}\Delta$$

as claimed.

An identity. In transformations of equations of continuum mechanics the following identity proves to be useful:

$$\frac{\partial(\rho x_a^j)}{\partial x^j} = \frac{\rho}{\rho_0\sqrt{\mathring{g}}}\frac{\partial(\rho_0\sqrt{\mathring{g}})}{\partial X^a}. \tag{3.79}$$

For homogeneous continuum, when $\rho_0 = const$, and for Cartesian Lagrangian coordinates in the initial state (with $\mathring{g} = 1$), the identity (3.79) simplifies to

$$\frac{\partial(\rho x_a^j)}{\partial x^j} = 0. \tag{3.80}$$

Formulas (3.79) and (3.80) are written in Cartesian coordinates x. In curvilinear coordinates partial derivatives with respect to x^i must be replaced by covariant derivatives.

Formula (3.79) follows from (3.68) and (3.74):

$$\frac{\partial \rho x_a^j}{\partial x^j} = \frac{\partial}{\partial x^j}\frac{\rho_0\sqrt{\mathring{g}}}{\Delta}x_a^j = \frac{1}{\Delta}x_a^j\frac{\partial}{\partial x^j}\rho_0\sqrt{\mathring{g}}$$

$$= \frac{1}{\Delta}\frac{\partial \rho_0\sqrt{\mathring{g}}}{\partial X^a} = \frac{\rho}{\rho_0\sqrt{\mathring{g}}}\frac{\partial(\rho_0\sqrt{\mathring{g}})}{\partial X^a}.$$

An identity similar to (3.79) holds for the inverse distortion,

$$\frac{\partial}{\partial X^a}\frac{1}{\rho}X_i^a = -\frac{1}{\rho\rho_0\sqrt{\mathring{g}}}\frac{\partial(\rho_0\sqrt{\mathring{g}})}{\partial x^i}. \tag{3.81}$$

To prove (3.81) we note that

$$\frac{\partial}{\partial X^a}\frac{1}{\rho}X_i^a = x_a^j\frac{\partial}{\partial x^j}\frac{1}{\rho}X_i^a = \frac{1}{\rho}\left(\frac{\partial}{\partial x^j}\rho x_a^j\frac{1}{\rho}X_i^a - \frac{1}{\rho}X_i^a\frac{\partial}{\partial x^j}\rho x_a^j\right) = -\frac{1}{\rho^2}X_i^a\frac{\partial}{\partial x^j}\rho x_a^j.$$

Plugging here (3.79) we arrive at (3.81). In homogeneous media referred to Cartesian Lagrangian coordinates,

$$\frac{\partial}{\partial X^a}\frac{1}{\rho}X_i^a = 0. \tag{3.82}$$

The divergence theorem. In this subsection, several versions of the divergence theorem that will be used later are introduced.

The divergence theorem is a multi-dimensional generalization of Newton's formula: for any function of one variable, $f(x)$,

$$\int_a^b \frac{df(x)}{dx}dx = f(b) - f(a).$$

This formula means that the integral of df/dx does not depend on the value of f inside the integration region, it is determined only by the end values of f. The explanation of such a "paradoxical" feature is simple: the integral is the limit value of a sum,

$$\int_a^b \frac{df(x)}{dx}dx = \sum_{k=0}^{n} \frac{f(a+(k+1)\varepsilon) - f(a+k\varepsilon)}{\varepsilon}\varepsilon, \qquad b = a + (n+1)\varepsilon.$$

Writing down this sum in full,

$$\sum_{k=0}^{n} \frac{f(a+(k+1)\varepsilon) - f(a+k\varepsilon)}{\varepsilon}\varepsilon = f(a+\varepsilon) - f(a) + f(a+2\varepsilon) - f(a+\varepsilon) + \ldots + f(a+(n+1)\varepsilon) - f(a+n\varepsilon)$$

we see that all the values of function f inside the integration region cancel out. It turns out that a similar fact holds for multi-dimensional integrals. We formulate it first for a bounded region V in three-dimensional space R_3. Let $\zeta^\alpha(\alpha = 1, 2)$ be some parameters on the surface ∂V bounding the region V. The parametric equations of ∂V are

$$x^i = r^i(\zeta^\alpha).$$

The functions $r^i(\zeta^\alpha)$ are assumed to be piecewise differentiable. The coordinates ζ^1, ζ^2 are introduced in such a way that adding a third space coordinate, ζ^3, increasing away from ∂V in the outward direction yields a coordinate system, $\zeta^1, \zeta^2, \zeta^3$, which has the same orientation as the coordinate system x^i, i.e. $\det\|\partial x/\partial\zeta\| > 0$.

Consider the object

$$N_i = e_{ijk} r_1^j r_2^k, \quad r_\alpha^j \equiv \frac{\partial r^j}{\partial \zeta^\alpha}. \tag{3.83}$$

which is defined on ∂V. This object is "orthogonal" to ∂V in the sense that the contraction of N_i with the tangent vectors r_α^i is zero:

$$N_i r_\alpha^i = 0.$$

It is directed outside of V.

Let some continuous differentiable functions Ω^i $(i = 1, 2, 3)$ be defined on the closed region V. Then, the following equation holds:

$$\int_V \frac{\partial \Omega^i}{\partial x^i} dx^1 dx^2 dx^3 = \int_{\partial V} \Omega^i N_i d\zeta^1 d\zeta^2. \tag{3.84}$$

Equation (3.84) is called the divergence theorem.

We will also need the divergence theorem in the four-dimensional space-time. Let V_4 be a region in space-time R_4, which is swept by the motion of a three-dimensional

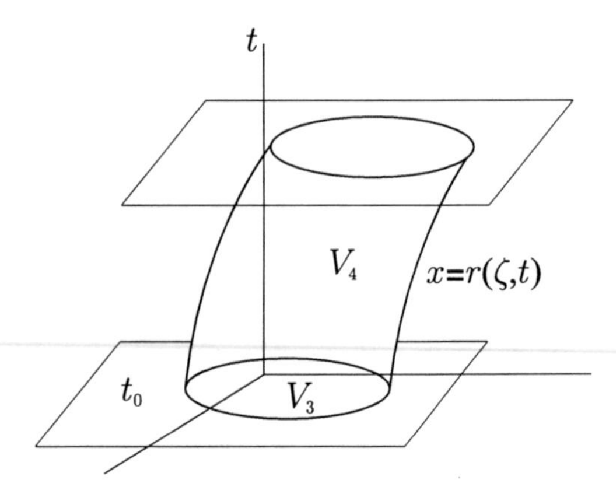

Fig. 3.3 Motion of a three-dimensional region V_3 in four-dimensional space-time

region V_3 in R_3 (Fig. 3.3). Denote as before the parameters on the boundary, ∂V_3, of the region V_3 by ζ^1, ζ^2. Then the "time" part of the boundary ∂V_4 of the region V_4 is described by the parametric equations

$$x^i = r^i\left(\zeta^1, \zeta^2, t\right), \quad x^4 = t, \tag{3.85}$$

where small Latin indices correspond to the projections on the space coordinate axes and the index 4 marks the projection on the time axis. Denote the quantities defined by (3.83) by N_i, where r^i_α are derivatives of the function (3.85) with respect to ζ^α.

For any continuous differentiable functions, Ω, Ω^i, in the closed region V_4, the following equality holds:

$$\int\limits_{V_4}\left(\frac{\partial\Omega}{\partial t}+\frac{\partial\Omega^i}{\partial x^i}\right)dx^1dx^2dx^3dt =$$

$$= \int\limits_{t_0}^{t}\int\limits_{\partial V_3}\left(\Omega^i - \Omega r^i_t\right)N_i d\zeta^1 d\zeta^2 dt + \left[\int\limits_{V_3}\Omega dx^1dx^2dx^3\right]_{t_0}^{t_1}. \tag{3.86}$$

where $r^i_t \equiv \partial r^i/\partial t\big|_{\zeta^\alpha=const}$.

Invariant form of the divergence theorem*. Equation (3.84) can be written in the "invariant" form if we put in (3.84) $\Omega^i = \sqrt{g}\omega^i$, where ω^i are the components of some vector field. In order to do that, we use the relation for the covariant[5] divergence of the vector ω^i,

$$\nabla_i\omega^i = \frac{1}{\sqrt{g}}\frac{\partial\sqrt{g}\omega^i}{\partial x^i}, \tag{3.87}$$

[5] The reader who is not familiar with the notion of covariant derivatives can find the necessary definitions in Sect. 4.6.

and the relation between N_i and the unit normal vector n_i,

$$N_i = n_i \frac{\sqrt{a}}{\sqrt{g}}.$$

where a is the determinant of the surface metric tensor (see Sect. 14.1, and, in particular, (14.7)). Introducing the invariant volume elements as $dV = \sqrt{g}dx^1dx^2dx^3$ and the invariant surface elements as $dA = \sqrt{a}d\zeta^1 d\zeta^2$, we can write (3.84) as

$$\int_V \nabla_i \omega^i dV = \int_{\partial V} \omega^i n_i dA. \tag{3.88}$$

Emphasize that, despite its "tensor" appearance, (3.88) has a purely analytical nature and is not related to notions of the covariant differentiation and/or metrics that were used in (3.88).

Choosing $\omega^1 = \varphi$, and $\omega^2 = \omega^3 = 0$, we obtain from (3.88) in Cartesian coordinates,

$$\int_V \frac{\partial \varphi}{\partial x^1} dV = \int_{\partial V} \varphi n_1 dA.$$

Similar relations hold for $\partial\varphi/\partial x^2$ and $\partial\varphi/\partial x^3$. Thus,

$$\int_V \frac{\partial \varphi}{\partial x^i} dV = \int_{\partial V} \varphi n_i dA. \tag{3.89}$$

If in (3.88) x^i are interpreted as the Lagrangian coordinates X^a, g_{ij} as the components of the metric tensor in the initial state, $\mathring{g}_{ab}$, and ω^i as the components of the vector in the Lagrangian coordinate system, ω^a, then (3.88) becomes

$$\int_{\mathring{V}} \mathring{\nabla}_a \omega^a d\mathring{V} = \int_{\partial \mathring{V}} \omega^a \mathring{n}_a d\mathring{A}. \tag{3.90}$$

Similarly, if g_{ij} are interpreted as the components of the metric tensor in not the initial but the current state, g_{ab}, then instead of (3.90) we get

$$\int_V \nabla_a \omega^a dV = \int_{\partial V} \omega^a n_a dA. \tag{3.91}$$

Let us introduce in planes $t = const$ in V_4 a spatial metric tensor, $g_{ij}\left(x^k, t\right)$, and substitute into (3.86):

$$\Omega = \sqrt{g}\omega, \quad \Omega^i = \sqrt{g}\omega^i,$$

$$dV = \sqrt{g} dx^1 dx^2 dx^3, \quad dA = \sqrt{a} d\zeta^1 d\zeta^2,$$

where ω^i are the components of a spatial vector field. Then (3.86) takes the form

$$\int_{V_4} \left(\frac{1}{\sqrt{g}} \frac{\partial \omega \sqrt{g}}{\partial t} + \nabla_i \omega^i \right) dV dt =$$

$$= \int_{t_0}^{t_1} \int_{\partial V_3} \left(\omega^i - \omega r_t^i \right) n_i dA dt + \left[\int_{V_3} \omega dV \right]_{t_0}^{t_1}. \tag{3.92}$$

The quantity $r_t^i n_i$ means the surface velocity component in the direction of the normal to the surface.

Identifying in (3.92) the coordinates, x, with the Lagrangian coordinates, and the metric tensor, g_{ij}, with the initial metric tensor, $\mathring{g}_{ab}$, we get

$$\int_{\mathring{V}_3 \times [t_0, t_1]} \left(\frac{d\omega}{dt} + \mathring{\nabla}_a \omega^a \right) d\mathring{V} dt =$$

$$= \int_{t_0}^{t_1} \int_{\partial \mathring{V}_3} \left(\omega^a - \omega r_t^a \right) \mathring{n}_a d\mathring{A} dt + \left[\int_{\mathring{V}_3} \omega d\mathring{V} \right]_{t_0}^{t_1}. \tag{3.93}$$

The quantity $r_t^a \mathring{n}_a$ can be interpreted as the velocity of the boundary, $\partial \mathring{V}_3$, of the region V_3 over the particles.

If in (3.92) the coordinates, x, are identified with the Lagrangian coordinates, X, while g_{ij} are set to be the current Lagrangian metric tensor, g_{ab}, then (3.92) becomes

$$\int_{V_4} \left(\frac{1}{\sqrt{\hat{g}}} \frac{d}{dt} \omega \sqrt{\hat{g}} + \nabla_a \omega^a \right) dV dt =$$

$$= \int_{t_0}^{t_1} \int_{\partial V_3} \left(\omega^a - \omega r_t^a \right) n_a dA dt + \left[\int_{V_3} \omega dV \right]_{t_0}^{t_1}. \tag{3.94}$$

Another useful form of the divergence theorem is obtained if one substitutes ω in (3.94) by $\rho\omega$. Then, using the continuity equation (3.67), one can write (3.94) as

$$\int_{V_4} \left(\rho \frac{d\omega}{dt} + \nabla_a \omega^a \right) dV dt =$$

$$= \int_{t_0}^{t_1} \int_{\partial V_3} \left(\omega^a - \rho \omega r_t^a \right) n_a dA dt + \left[\int_{V_3} \rho \omega dV \right]_{t_0}^{t_1}. \tag{3.95}$$

In particular, when $\omega^a = 0$, (3.95) takes the form

$$\int_{V_4} \rho \frac{d\omega}{dt} dV dt = -\int_{t_0}^{t} \int_{\partial V_3} \rho \omega r_t^a n_a dA dt + \left[\int_{V_3} \rho \omega dV \right]_{t_0}^{t_1}. \tag{3.96}$$

3.2 Basic Laws of Continuum Mechanics

Momentum equation. The second Newton law for one particle reads: the rate of momentum, mv^i, is equal to the external force acting on the particle

$$\frac{d}{dt} mv^i = F^i.$$

What is an analogy of this law for a continuum? The answer is not self-evident and for the classical continuum models is as follows.

Consider a piece of continuum, some region $\mathring{V}$ in the space of Lagrangian coordinates. For an observer, this piece occupies some moving region $V(t)$. It consists of the same material particles. Momentum of an infinitesimally small piece of material, dV, is, by definition, the product of the mass of the particle, ρdV, and the particle velocity, v^i. We postulate that the momentum of the entire piece of continuum is the sum of momenta of all its infinitesimally small parts:

$$\text{momentum} = \int_{V(t)} \rho v^i dV.$$

Denote by F^i the total force acting on the material in region V. Then we accept the following continuum version of the second Newton law:

$$\frac{d}{dt} \int_{V(t)} \rho v^i dV = F^i. \tag{3.97}$$

The total force, F^i, can be split into the sum of the surface force and the body force:

$$F^i = F^i_{surface} + F^i_{body}. \tag{3.98}$$

Both forces are assumed to possess a force density:

$$F^i_{surface} = \int_{\partial V} f^i dA, \quad F^i_{body} = \int_{V} \rho g^i dV. \tag{3.99}$$

The body force per unit mass, g^i, is a function of x and t; in most applications, the only body force is gravity; the vector g^i, the gravity acceleration, is constant. The surface force per unit area, f^i, can depend not only on the space point of the boundary and time, but also on the geometrical characteristics of the surface ∂V. The crucial assumption of classical continuum mechanics is that the surface force density depends linearly on the normal vector to ∂V:

$$f^i = \sigma^{ij} n_{j,} \tag{3.100}$$

and "the coefficients" σ^{ij} are some functions of x and t only. These coefficients are called the stress tensor. The unit normal vector n_j is directed, by our convention, outside of the region V. Surface force density may depend on the geometry of the surface in a more complex way as we will see later in Sect. 4.5, but for all classical models formula (3.100) holds true.

In principle, the stress tensor may be non-symmetric, but it is assumed symmetric in all classical models:

$$\sigma^{ij} = \sigma^{ji}. \tag{3.101}$$

Equations (3.97), (3.98), (3.99) and (3.100) constitutes the integral form of momentum equation for classical models of continuum media.

Differential form of momentum equation. Let us show that (3.97), (3.98), (3.99) and (3.100) are equivalent to the following equation:

$$\rho \frac{dv^i}{dt} = \frac{\partial \sigma^{ij}}{\partial x^j} + \rho g^i. \tag{3.102}$$

Consider the momentum equation (3.97). We would like to move the time derivative under the integral, but then some additional terms must appear because the limits of integration depend on time. To get around this complication, we write the integral in terms of Lagrangian coordinates by means of (3.66):

$$\int\limits_{V(t)} \rho v^i dV = \int\limits_{\mathring{V}} \rho_0 v^i d\mathring{V}. \tag{3.103}$$

The limits of integration in the right hand side of (3.103) do not depend on time; thus we can move the time derivative under the integral:

$$\frac{d}{dt} \int\limits_{V(t)} \rho v^i dV = \int\limits_{\mathring{V}} \rho_0 \frac{dv^i}{dt} d\mathring{V}.$$

Returning back to Eulerian coordinates by means of (3.66) we obtain

$$\frac{d}{dt}\int\limits_{V(t)} \rho v^i dV = \int\limits_{V(t)} \rho \frac{dv^i}{dt} dV. \tag{3.104}$$

The total force (3.98) can also be written as a volume integral due to (3.100):

$$F^i = \int\limits_V \left(\frac{\partial \sigma^{ij}}{\partial x^j} + \rho g^i \right) dV. \tag{3.105}$$

Here we used the divergence theorem. Equating (3.104) and (3.105) we have

$$\int\limits_V \left(\rho \frac{dv^i}{dt} - \frac{\partial \sigma^{ij}}{\partial x^j} - \rho g^i \right) dV = 0. \tag{3.106}$$

Since the region V is arbitrary and all functions involved are assumed to be continuous, the integrand must be identically equal to zero, otherwise one can find a region V for which the left hand side of (3.106) is not zero. We thus arrive at the momentum equation (3.102).

Piola-Kirchhoff stress tensor. Stress tensor σ^{ij} is also called Cauchy's stress tensor. The projection of this tensor over the second index to Lagrangian coordinates with a factor, ρ_0/ρ,

$$p^{ia} = \frac{\rho_0}{\rho} \sigma^{ij} X_j^a,$$

is usually called Piola-Kirchhoff stress tensor. The Piola-Kirchhoff stress tensor has one observer's and one Lagrangian index, i.e. it behaves as a triad of vectors under transformation of observer's frames and a triad of vectors under transformation of Lagrangian frames. There is convention in tensor analysis to keep the same core letter for the components of the tensor in different coordinate systems. Therefore, one can write for Piola-Kirchhoff stress tensor

$$p^{ia} = \frac{\rho_0}{\rho} \sigma^{ia}$$

or

$$\sigma^{ia} = \frac{\rho}{\rho_0} p^{ia}.$$

Note also here the relation between the Lagrangian and Eulerian components of the stress tensor:

$$\sigma^{ab} = \sigma^{ij} X_i^a X_j^b.$$

Momentum equation in Lagrangian coordinates. Momentum equation in Lagrangian coordinates takes a simple form if we use the Piola-Kirchhoff stress tensor[6]

$$\rho_0 \frac{dv^i}{dt} = \frac{\partial p^{ia}}{\partial X^a} + \rho_0 g^i. \tag{3.108}$$

Another form of this equation,

$$\rho_0 \frac{\partial^2 x^i(t, X)}{\partial t^2} = \frac{\partial p^{ia}}{\partial X^a} + \rho_0 g^i, \tag{3.109}$$

emphasizes that the independent variables in this equation are Lagrangian coordinates and time.

To derive (3.108) from (3.102) one can plug into (3.102) the expression of Cauchy's stress tensor in terms of Piola-Kirchhoff's stress tensor,

$$\sigma^{ij} = \frac{\rho}{\rho_0} p^{ia} x_a^j,$$

and use the identity (3.73).

Energy equation. Consider the total energy of material contained in some region $V(t)$. As we discussed in Chap. 2, energy is a primary characteristic of a mechanical system which always exists. Continuum is an approximate description for mechanical systems consisting of a large number of material particles. Energy is defined as the total energy of these particles. Total energy can be presented as a sum of the kinetic energy of the macroscopic motion,

$$\int_{V(t)} \rho \frac{v^2}{2} dV$$

and the remainder, which is called internal energy. The latter is assumed to have some energy density. The internal energy density per unit mass is denoted by U. So,

---

[6] Here we assume that the Lagrangian coordinates are Cartesian in the initial state, and $\mathring{g} = 1$. Otherwise, the momentum equation is

$$\rho_0 \frac{dv^i}{dt} = \frac{1}{\sqrt{\mathring{g}}} \frac{\partial}{\partial X^a} \left(\sqrt{\mathring{g}} p^{ia} \right) + \rho_0 g^i, \tag{3.107}$$

or, in terms of the covariant derivative in the initial state,

$$\rho_0 \frac{dv^i}{dt} = \mathring{\nabla}_a p^{ia} + \rho_0 g^i.$$

Note that the covariant derivative $\mathring{\nabla}_a$ "acts" only on the Lagrangian index of Piola-Kirchhoff tensor.

$$\text{total energy} = \int_{V(t)} \rho \frac{v^2}{2} dV + \int_{V(t)} \rho U dV.$$

The energy rate is caused by the work of external forces

$$\frac{d}{dt} \int_{V(t)} \rho \left(\frac{v^2}{2} + U \right) dV = P + \frac{dQ}{dt}. \tag{3.110}$$

Here P is the power of external forces (the work of external forces per unit time)

$$P = \int_{\partial V(t)} \sigma^{ij} n_j v_i dA + \int_{V(t)} \rho g^i v_i dV, \tag{3.111}$$

and heat supply dQ is the work additional to the work of external forces. Heat supply is the work done on "microscopic degrees of freedom" as discussed in Chap. 2.

It is assumed in classical continuum models that the heat supply is due only to some energy flux through the surface, and there is a heat supply surface density, q, such that

$$\frac{dQ}{dt} = \int_{\partial V} q \, dA. \tag{3.112}$$

Moreover, the dependence of the heat supply surface density, q, on the geometry of the surface is similar to (3.100):

$$q = -q^i n_i, \tag{3.113}$$

where q^i are some functions of the space points and time. The vector q^i is called the heat flux vector. The minus sign in (3.113) is caused by the convention that energy should decrease if the vector q^i is directed outside of the region V and increase otherwise.

Combining (3.110), (3.111), (3.112) and (3.113) and repeating the reasoning that yields the momentum equation in differential form (3.102) we obtain the energy equation in differential form:

$$\rho \frac{d}{dt} \left(\frac{v^2}{2} + U \right) = \frac{\partial (\sigma^{ij} v_i)}{\partial x^j} + \rho g^i v_i - \frac{\partial q^i}{\partial x^i}. \tag{3.114}$$

This equation may be simplified. Note that the momentum equation (3.102) yields an equation for the kinetic energy rate: contracting (3.102) with v_i we have

$$\rho \frac{d}{dt} \frac{v^2}{2} = \frac{\partial \sigma^{ij}}{\partial x^j} v_i + \rho g^i v_i. \tag{3.115}$$

Deducting (3.115) from (3.114) we obtain the equation for the internal energy rate:

$$\rho \frac{dU}{dt} = \sigma^{ij} \frac{\partial v_i}{\partial x^j} - \frac{\partial q^i}{\partial x^i}. \tag{3.116}$$

Entropy equation. If the processes in the system are sufficiently slow, one can introduce $\mathcal{S}$, the entropy of the system. We assume that entropy of the system is the sum of entropies of all subsystems, i.e. the entropy density per unit mass, S, exists:

$$\mathcal{S} = \int_V \rho S dV.$$

Entropy density has the dimension $(mass)^{-1}$ if the total entropy is taken as dimensionless.

The second law of thermodynamics claims that for an isolated system the total entropy does not decrease:

$$\frac{d\mathcal{S}}{dt} = \int_V \rho \frac{dS}{dt} dV \geq 0. \tag{3.117}$$

Every continuum model must obey this constraint.

3.3 Classical Continuum Models

All continuum models can be roughly split into two categories. The first is comprised of the models that have as the primary characteristics only the particle positions $x(t, X)$ and the entropy $S(t, X)$. Functions $x(t, X)$ and $S(t, X)$ are analogous to the generalized coordinates of classical mechanics. Note that the "generalized coordinates" of continuum mechanics include, in addition to the functions $x(t, X)$, which are a direct analogy of the particle coordinates in classical mechanics, an additional "coordinate" $S(t, X)$. The second type of model is formed by the models which have some additional primary characteristics like plastic deformations, concentrations of chemical species, etc. We call these additional characteristics the internal degrees of freedom. We consider first the models which do not have internal degrees of freedom.

All classical continuum models are based on the assumption that internal energy density, U, is a function of distortion and entropy density only:

$$U = U(x_a^i, S). \tag{3.118}$$

Function $U(x_a^i, S)$ specifies the physical properties of the media. To close the system of three momentum equations (3.102) and energy equation (3.115), one has to prescribe the constitutive equations which link the remaining unknowns in

(3.102) and (3.115), σ^{ij} and q^i, with four functions $x^i(t, X)$ and $S(t, X)$. Note that mass density in (3.102) and (3.115) can be considered as a function of distortion according to the conservation of mass equation (3.69). The constitutive equations for σ^{ij} and q^i cannot be taken arbitrarily: the second law of thermodynamics (3.117) must be satisfied. If σ^{ij} and q^i were known, then, for models with the internal energy (3.118), the rate of entropy can be found from the energy equation (3.116):

$$\rho \frac{\partial U}{\partial x_a^i} \frac{dx_a^i}{dt} + \rho \frac{\partial U}{\partial S} \frac{dS}{dt} = \sigma^{ij} \frac{\partial v_i}{\partial x^j} - \frac{\partial q^i}{\partial x^i}. \tag{3.119}$$

The derivative $\partial U / \partial S$ is absolute temperature

$$\frac{\partial U(x_a^i, S)}{\partial S} = T. \tag{3.120}$$

Equation (3.120) is just the definition of temperature. One may consider (3.120) as an equation linking entropy and temperature and use temperature as the primary unknown function instead of entropy. Then (3.120) will serve as an equation to find entropy.

Note that the first term in the right hand side of (3.119) can be written in different forms:

$$\sigma^{ij} \frac{\partial v_i}{\partial x^j} = \frac{\rho}{\rho_0} p_i^a \frac{\partial v^i}{\partial X^a}, \tag{3.121}$$

or

$$\sigma^{ij} \frac{\partial v_i}{\partial x^j} = \sigma^{ij} e_{ij} = \sigma^{ab} \frac{d\varepsilon_{ab}}{dt}. \tag{3.122}$$

Let us use (3.121). Then from (3.119) the entropy rate is

$$\rho \frac{dS}{dt} = \frac{1}{T} \frac{\rho}{\rho_0} \left(p_i^a - \rho_0 \frac{\partial U}{\partial x_a^i} \right) \frac{\partial v^i}{\partial X^a} - \frac{1}{T} \frac{\partial q^i}{\partial x^i}.$$

The closing equations for p_i^a and q^i must be such that for an isolated system the total entropy does not decrease at any instant:

$$\int_V \rho \frac{dS}{dt} = \int_V \left[\frac{1}{T} \frac{\rho}{\rho_0} \left(p_i^a - \rho_0 \frac{\partial U}{\partial x_a^i} \right) \frac{\partial v^i}{\partial X^a} - \frac{1}{T} \frac{\partial q^i}{\partial x^i} \right] dV \geq 0. \tag{3.123}$$

Isolation of the system requires some special conditions at the boundary. In particular, the heat supply through the boundary must be zero:

$$q^i n_i = 0 \text{ at } \partial V.$$

Integrating the last term of the integral in (3.123) by parts, we put the second law of thermodynamics in the form

$$\int_V \left(\frac{1}{T} \frac{\rho}{\rho_0} \left(p_i^a - \rho_0 \frac{\partial U}{\partial x_a^i} \right) \frac{\partial v^i}{\partial X^a} + q^i \frac{\partial}{\partial x^i} \frac{1}{T} \right) dV \geq 0. \tag{3.124}$$

The consequences of the inequality (3.124) depend on the further assumptions for the stresses and the heat flux.

Heat conductivity. Consider first the case when the continuum does not move. Then the only process remaining is the heat conduction. It is described by the equation

$$\rho \frac{dU}{dt} = -\frac{\partial q^i}{\partial x^i}. \tag{3.125}$$

To close this equation, we need the constitutive equation for the heat flux. It must satisfy the second law of thermodynamics (3.124). Now (3.125) takes the form

$$\int_V q^i \frac{\partial}{\partial x^i} \frac{1}{T} dV \geq 0. \tag{3.126}$$

The simplest relation to obey this inequality is the Fourier heat conduction law:

$$q_i = \varkappa \frac{\partial}{\partial x^i} \frac{1}{T}, \tag{3.127}$$

with some positive parameter, $\varkappa$, the coefficient of heat conductivity.

Let us choose the function $U(X^a, t)$ as the independent thermodynamic variable, and describe the thermodynamic properties of the material by the function $S = S_0(U)$. Then the temperature is linked to U by the formula

$$\frac{1}{T} = \frac{dS_0(U)}{dU}. \tag{3.128}$$

Equations (3.125), (3.128) and (3.127) form a closed system of equations for heat conductivity.

Fourier law is often written in a slightly different form,

$$q_i = -\gamma \frac{\partial T}{\partial x^i},$$

It has an advantage that the coefficient $\gamma = \varkappa / T^2$ only slightly depends on temperature for many materials.

To demonstrate the importance of the inequality (3.126) consider an improvement of heat conductivity theory suggested by Cattaneo. One can derive from the equations obtained that the heat propagates with an infinite velocity. To get rid of this paradox, Cattaneo proposed to replace (3.127) by the equation

$$q_i + \alpha \frac{dq_i}{dt} = \varkappa \frac{\partial}{\partial x^i} \frac{1}{T}, \tag{3.129}$$

The energy equation (3.125) along with equations (3.128) and (3.129) form a closed system of equations. However, these equations contradict the inequality of the second law of thermodynamics (3.126). Indeed, in Cattaneo's theory, q^i and U can be given independently for the initial instant since the energy equation and the equation for q^i (3.129) contain the first derivatives of q^i and U with respect to time. By a specific choice of the initial values, the integrand in (3.126) and, hence, the integral (3.126) can be made negative. Therefore, all the formulas of Cattaneo's theory cannot be left unchanged. The question arises whether it is possible to choose such equations of state for entropy that entropy increases in an isolated system. The negative answer to this question would mean that (3.129) is not consistent with the laws of thermodynamics. It turns out that a modification of the equation of state does exist. One has to change only the constitutive equation for entropy:

$$S = S_0(U) - \frac{\alpha}{2\varkappa} q_i q^i, \qquad \frac{1}{T} = \frac{dS_0(U)}{dU}.$$

The reader can check that the total entropy of the system does increase. The negative sign in the equation for the entropy has a clear meaning: the heat flux induces some order, and the entropy density decreases.

Elastic body. Theory of elasticity is based on the assumption that p_i^a and q^i may depend only on distortion, entropy (or temperature) and temperature gradient.

Besides, these relationships are universal, i.e. they are valid for any values of x_a^i, T and $\partial T/\partial x^i$. Let us take first $T = const$. Then inequality (3.124) simplifies to

$$\int_V \frac{1}{T} \frac{\rho}{\rho_0} \left(p_i^a - \rho_0 \frac{\partial U}{\partial x_a^i} \right) \frac{\partial v^i}{\partial X^a} dV \geq 0. \tag{3.130}$$

The momentum equation contains the time derivative of the velocity field. Therefore, the initial data for velocity must be provided. The initial velocity field can be chosen arbitrarily. Applying inequality (3.130) to the initial instant and using the fact that p_i^a and $\partial U/\partial x_a^i$ do not depend on velocity field, we see that the expression in parenthesis must vanish and thus

$$p_i^a = \rho_0 \frac{\partial U}{\partial x_a^i}. \tag{3.131}$$

Otherwise, one can choose the velocity field in such a way that entropy rate is negative (the reasoning is similar to that of the main lemma of calculus of variation in Chap. 1). Since at the initial instant any distributions of distortion and entropy can be taken, and the dependence $p_i^a(x_b^j, S)$ is universal, (3.131) gives the required constitutive equations for the stress tensor.

Consider now an arbitrary temperature field. Because of (3.131), inequality (3.124) is now a constraint for the heat flux vector (3.126). Any dependence of q^i on the temperature gradient satisfying to (3.126) is possible, in principle. The simplest is the Fourier law, which we write for the general anisotropic case as

$$q^i = D^{ij}\frac{\partial}{\partial x^j}\frac{1}{T}. \tag{3.132}$$

The heat conductivities, D^{ij}, must be positive definite:

$$D^{ij}\xi_i\xi_j \geq 0 \quad \text{for any } \xi_i.$$

The heat conductivities may depend on temperature and distortion. It can be shown that Onsager's relations for heat conduction imply the symmetry of tensor D^{ij},

$$D^{ij} = D^{ji}.$$

The momentum (3.108), energy (3.116), continuity (3.68) and constitutive equations (3.120), (3.131) and (3.132) form the closed system of equations for elastic heat conducting body. The model is specified by prescribing the internal energy $U(x_a^i, S)$ and the heat conductivities D^{ij}.

Stress-strain relations for elastic body. The constitutive equations (3.131) can be further specified if we note that internal energy must be invariant with respect to rigid rotations of an elastic body. If a body with the particle positions $x^i(t, X)$ is rotated, the new particle positions, $x'^i(t, X)$, are

$$x'^i(t, X) = \alpha_j^i\, x^j(t, X)$$

where α_j^i are the components of an orthogonal matrix. Therefore, the distortion in the new position is

$$x_a'^i = \alpha_j^i x_a^j.$$

Internal energy must be the same in both states for any orthogonal matrix α_j^i and any initial distortion x_a^i:

$$U\left(\alpha_j^i x_a^j, S\right) = U\left(x_a^j, S\right). \tag{3.133}$$

One can show that to satisfy (3.133), function U may depend on distortion only through the metric tensor $g_{ab} = g_{ij}x_a^i x_a^j$, or, equivalently, through the strain tensor $\varepsilon_{ab} = \frac{1}{2}(g_{ab} - \mathring{g}_{ab})$. Since

$$\frac{\partial g_{bc}}{\partial x_a^i} = g_{ij}\delta_b^a x_c^j + g_{kj}x_b^k\delta_c^a\delta_i^j = x_{ic}\delta_b^a + x_{ib}\delta_c^a \tag{3.134}$$

we can write the stress-strain relations for elastic body (3.131) as

$$p^{ia} = 2\rho_0\frac{\partial U}{\partial g_{ab}}x_b^i = \rho_0\frac{\partial U}{\partial \varepsilon_{ab}}x_b^i, \tag{3.135}$$

In a Eulerian frame, these relations take the form

$$\sigma^{ij} = \frac{\rho}{\rho_0}p^{ia}x_a^j = \rho x_a^i x_b^j\frac{\partial U(\varepsilon_{ab}, S)}{\partial \varepsilon_{ab}}. \tag{3.136}$$

The stress components σ^{ij} and p^{ia} change if the body is rotated. This does not occur for Lagrangian components of the stress tensor,

$$\sigma^{ab} = \rho\frac{\partial U(\varepsilon_{ab}, S)}{\partial \varepsilon_{ab}}. \tag{3.137}$$

One comment on the meaning of the derivatives, $\partial U(\varepsilon_{ab}, S)/\partial\varepsilon_{ab}$ is now in order. Internal energy is a function of a symmetric tensor, $\varepsilon_{ab} = \varepsilon_{ba}$. Let, say, in a two-dimensional space and for a Cartesian initial metrics, $\mathring{g}^{ab} = \delta^{ab}$, the internal energy is

$$U = \mu\mathring{g}^{ab}\mathring{g}^{cd}\varepsilon_{ac}\varepsilon_{bd} = \mu(\varepsilon_{11}^2 + 2\varepsilon_{12}^2 + \varepsilon_{22}^2).$$

Derivatives of this function over $\varepsilon_{11}, \varepsilon_{12}$ and ε_{22} are

$$\frac{\partial U}{\partial \varepsilon_{11}} = 2\mu\varepsilon_{11}, \qquad \frac{\partial U}{\partial \varepsilon_{12}} = 4\mu\varepsilon_{12}, \qquad \frac{\partial U}{\partial \varepsilon_{22}} = 2\mu\varepsilon_{22}.$$

We see that these derivatives do not form a tensor: there is an extra factor 2 in the 12-compontent. Therefore, we must define the derivatives in a different way. This definition must be such that the equations for the differential of U.

$$dU = \frac{\partial U}{\partial \varepsilon_{ab}}d\varepsilon_{ab},$$

remains true. To have the tensor relations, we will differentiate $U(\varepsilon_{ab})$ as if the tensor ε_{ab} is not symmetric. We set $\varepsilon_{ab} = \varepsilon_{ba}$ only in the final formulas. The reader is invited to check that such a rule yields the relations indeed.

Entropic elasticity. The stress-strain relations contain the internal energy as a function of strains and entropy. If the internal energy were known as a function of strains and temperature then the form of the stress-strain relations would change. To discuss the corresponding changes, it is convenient first to write down the stress-strain relations in terms of entropy as the primary thermodynamic potential. Entropy is a function of internal energy and strains,

$$S = S(U, \varepsilon_{ab}). \tag{3.138}$$

For a given dependence of internal energy on strain and entropy, $U = U(\varepsilon_{ab}, S)$, function (3.138) may be viewed as a solution of the equation

$$U(\varepsilon_{ab}, S) = U.$$

Therefore, the derivatives of $U(\varepsilon_{ab}, S)$ and $S(U, \varepsilon_{ab})$ are linked. To find the corresponding relationships, we differentiate the identity

$$U(\varepsilon_{ab}, S(U, \varepsilon_{ab})) = U, \tag{3.139}$$

with respect to ε_{ab} and U :

$$\frac{\partial U(\varepsilon_{ab}, S)}{\partial \varepsilon_{ab}} + \frac{\partial U}{\partial S}\frac{\partial S(U, \varepsilon_{ab})}{\partial \varepsilon_{ab}} = 0, \qquad \frac{\partial U(\varepsilon_{ab}, S)}{\partial S}\frac{\partial S(U, \varepsilon_{ab})}{\partial U} = 1.$$

Hence,

$$\frac{\partial U(\varepsilon_{ab}, S)}{\partial \varepsilon_{ab}} = -T\frac{\partial S(U, \varepsilon_{ab})}{\partial \varepsilon_{ab}}, \qquad \frac{\partial S(U, \varepsilon_{ab})}{\partial U} = \left(\frac{\partial U(\varepsilon_{ab}, S)}{\partial S}\right)^{-1} = \frac{1}{T}. \tag{3.140}$$

In terms of the thermodynamics function $S(U, \varepsilon_{ab})$, the stress-strain relations take the form

$$\sigma^{ab} = -\rho T\frac{\partial S(U, \varepsilon_{ab})}{\partial \varepsilon_{ab}}. \tag{3.141}$$

Experiments show that some materials, such as rubbers or polymers, exhibit the following feature: their internal energy is a function of temperature only and, practically, does not depend on strain:

$$U = U_0(T).$$

Equivalently, temperature is some function of internal energy:

$$T = T_0(U).$$

According to the second equation (3.140), in this case

$$\frac{\partial S(U, \varepsilon_{ab})}{\partial U} = \frac{1}{T_0(U)}. \tag{3.142}$$

The general solution of (3.142) is

$$S(U,\varepsilon_{ab}) = S_0(U) + S_1(\varepsilon_{ab}). \tag{3.143}$$

where $S_0(U)$ is the solution of the ordinary differential equation

$$\frac{dS_0(U)}{dU} = \frac{1}{T_0(U)}.$$

Finally, for the stress-strain relations we have from (3.141) and (3.143)

$$\sigma^{ab} = -\rho T \frac{\partial S_1(\varepsilon_{ab})}{\partial \varepsilon_{ab}}. \tag{3.144}$$

To see some qualitative peculiarities of such stress-strain relations, consider the one-dimensional case when the stress and strain tensors have just one component, σ and ε. According to (3.144),

$$\sigma = -\rho T \frac{\partial S_1(\varepsilon)}{\partial \varepsilon}.$$

Let the dependence of stresses on strains be linear, i.e.

$$S_1 = -\frac{1}{2}\lambda\varepsilon^2. \tag{3.145}$$

Here λ is the material rigidity, $\lambda > 0$; the negative sign in (3.145) roots in Gibbs' variational principle: in equilibrium, i.e. at $\varepsilon = 0$, entropy must be maximum. So,

$$\sigma = \rho T \lambda \varepsilon.$$

We see that the rigidity of material increases if temperature is raised. This is opposite to typical behavior of metals for which the rigidity decreases when temperature rises.

Another outcome of our consideration is that the material properties are uniquely defined only if the internal energy is given in terms of their "native arguments", ε_{ab} and S. If one uses as an argument temperature instead of entropy, then an uncertainty appears in the costitutive equations. The uncertainty can be fixed by providing an additional information about entropy, like (3.140).

The class of elasticity models (3.143) is sometimes called entropic elasticity.

Ideal compressible fluid. The special case of an elastic body when the internal energy depends on distortion through the mass density only,

$$U = U(\rho, S), \tag{3.146}$$

is called an ideal fluid. To derive the constitutive equations in this case we need the relation

$$\frac{\partial \rho}{\partial x^i_a} = -\rho X^a_i. \tag{3.147}$$

It follows from (3.24) and (3.69):

$$\frac{\partial \rho}{\partial x^i_a} = \frac{\partial}{\partial x^i_a}\frac{\rho_0\sqrt{\mathring{g}}}{\sqrt{g}\Delta} = -\frac{\rho_0\sqrt{\mathring{g}}}{\sqrt{g}\Delta^2}\frac{\partial \Delta}{\partial x^i_a} = -\rho X^a_i.$$

Then, from (3.131) and (3.147) we find the components of Piola-Kirchhoff's tensor,

$$p^a_i = -\rho\rho_0\frac{\partial U(\rho, S)}{\partial \rho}X^a_i,$$

and the components of the stress tensor in observer's frame:

$$\sigma^{ij} = -pg^{ij}, \quad p \equiv \rho^2\frac{\partial U(\rho, S)}{\partial \rho}. \tag{3.148}$$

The scalar p is called pressure.

The closed system of equations for an ideal, compressible, heat-conducting fluid takes the form

$$\begin{aligned}
\frac{\partial \rho}{\partial t} + \frac{\partial \rho v^i}{\partial x^i} &= 0,\\
\rho\frac{dv^i}{dt} &= -\frac{\partial p}{\partial x^i} + \rho g_i,\\
\rho\frac{dU(\rho, S)}{dt} &= -p\frac{\partial v^i}{\partial x^i} - \frac{\partial q^i}{\partial x^i},\\
q_i &= \varkappa\frac{\partial}{\partial x^i}\frac{1}{T},\\
p = \rho^2\frac{\partial U(\rho, S)}{\partial \rho}&, \quad T = \frac{\partial U(\rho, S)}{\partial S}.
\end{aligned} \tag{3.149}$$

Function $U(\rho, S)$ in (3.149) is assumed to be given. The special case,

$$U(\rho, S) = a\rho^{\gamma-1}e^{S/c_v}, \tag{3.150}$$

where a, γ and c_v are some positive constants, and $\gamma > 1$, is called a perfect gas. The reader is invited to check that, for a perfect gas,

$$U = c_v T \text{ and } p = \rho RT$$

where $R \equiv (\gamma - 1)c_v$, is the so-called gas constant, and c_v the heat capacity for constant volume.

The constant a in (3.150) is nonessential: by changing a one only shifts entropy by a constant. The model for an ideal gas contains two physical characteristics, c_v and γ.

Incompressible ideal fluid. If "the fluid rigidity" with respect to the volume change tends to infinity, then the variations of the density become negligibly small. One can accept that in the first approximation the density is constant. Then the system of (3.149) splits into two subsystems: the first serves to find velocities and pressure (density ρ is a given constant),

$$\begin{aligned}\frac{\partial v^i}{\partial x^i} &= 0, \\ \rho\frac{\partial v^i}{\partial t} &= -\frac{\partial p}{\partial x^i} + \rho g_i,\end{aligned} \tag{3.151}$$

and the second allows one to determine temperature distribution when fluid motion is known from (3.151):

$$\rho\frac{dU(S)}{dt} = -\frac{\partial q^i}{\partial x^i}, \quad q_i = \varkappa\frac{\partial}{\partial x^i}\frac{1}{T}, \quad T = \frac{dU(S)}{dS}. \tag{3.152}$$

Equations (3.151) are called Euler equations.

Viscous compressible fluid. For a compressible fluid internal energy is, as before, a function of density and entropy. What changes is the assumption made regarding the stress tensor: for viscous fluid the stress tensor depends on the velocity field. Consider again the inequality of the second law of thermodynamics (3.124). We put it in the form

$$\int_V \left(\frac{1}{T}\left(\sigma^{ij} + \rho^2\frac{\partial U(\rho, S)}{\partial \rho}g^{ij}\right)\frac{\partial v_i}{\partial x^j} + q^i\frac{\partial}{\partial x^i}\frac{1}{T}\right)dV \geq 0. \tag{3.153}$$

Let us define tensor τ^{ij} by the equation

$$\sigma^{ij} = -\rho^2\frac{\partial U(\rho, S)}{\partial \rho}g^{ij} + \tau^{ij}. \tag{3.154}$$

Tensor τ^{ij} is symmetric since σ^{ij} and g^{ij} are symmetric. Due to the symmetry of τ^{ij}, the sum $\tau^{ij}\partial v_i/\partial x^j$ can also be be written as

$$\tau^{ij}\frac{\partial v_i}{\partial x^j} = \tau^{ij}e_{ij} \tag{3.155}$$

where e_{ij} is the strain rate tensor (3.46). So,

$$\int_V \left(\frac{1}{T}\tau^{ij}e_{ij} + q^i\frac{\partial}{\partial x^j}\frac{1}{T}\right)dV \geq 0. \tag{3.156}$$

In principle, any dependence of τ^{ij} and q^i on e_{ij} and the temperature gradient which does not violate the inequality (3.156) is acceptable. The simplest one is a linear dependence of τ^{ij} on e_{ij}:

$$\tau^{ij} = \mu^{ijkl} e_{kl} \tag{3.157}$$

with a positive tensor μ^{ijkl} ($\mu^{ijkl} e_{ij} e_{kl} \geq 0$ for any e_{ij}), and Fourier's law for q^i (3.132). This case is called a Newtonian fluid.

For an isotropic Newtonian fluid,

$$\tau^{ij} = \lambda e^k_k g^{ij} + 2\mu e^{ij}. \tag{3.158}$$

In general, one usually assumes that the dissipation, D, defined as

$$D = \tau^{ij} e_{ij} + T q^i \partial_i T^{-1}, \tag{3.159}$$

is a function[7] of e_{ij} and $T^{-1}_{,i}$:

$$D = D\left(e_{ij}, \partial_i T^{-1}\right),$$

and the closing relations are

$$\tau^{ij} = \lambda \frac{\partial D}{\partial e_{ij}}, \quad T q^i = \lambda \frac{\partial D}{\partial \left(\partial_i T^{-1}\right)}. \tag{3.160}$$

Here λ is the parameter determined from (3.159) and (3.160)

$$\lambda \left(\frac{\partial D}{\partial e_{ij}} e_{ij} + \frac{\partial D}{\partial \left(\partial_i T^{-1}\right)} \partial_i T^{-1} \right) = D. \tag{3.161}$$

Newtonian fluids correspond to a quadratic function D. In this case, the parameter λ is equal to $1/2$.

Incompressible viscous fluid. If the fluid is incompressible then, as for incompressible ideal fluids, the determination of the temperature field separates from the determination of fluid motion. The closed system of equations for fluid motion consists of four equations for four unknowns, velocity v^i and pressure p:

$$\frac{\partial v^i}{\partial x^i} = 0 \tag{3.162}$$

---

[7] Dissipation may also depend on temperature and mass density, but this is not emphasized in the notation.

$$\rho \frac{dv_i}{dt} = -\frac{\partial p}{\partial x^i} + \mu \Delta v_i.$$

Here Δ is Laplace's operator:

$$\Delta = \frac{\partial^2}{\partial x_1^2} + \frac{\partial^2}{\partial x_2^2} + \frac{\partial^2}{\partial x_3^2}.$$

Equations (3.162) are called Navier-Stokes equations.

Temperature is determined from the energy equation after velocity field has been found from Navier-Stokes equations:

$$\rho \frac{dU(S)}{dt} = 2\mu e_{ij} e^{ij} - \frac{\partial}{\partial x^i} \varkappa \frac{\partial}{\partial x_i} \frac{1}{T}. \tag{3.163}$$

The first term in the right hand side of (3.163) describes heating of fluid caused by viscosity.

Plastic bodies. Among a wide class of continuum models with internal degrees of freedom we consider the models of classical plasticity theory. They endow the material with additional degrees of freedom – the tensor field of plastic deformation $\varepsilon_{ab}^{(p)}$. Internal energy of a plastic body depends on elastic deformation $\varepsilon_{ab}^{(e)} = \varepsilon_{ab} - \varepsilon_{ab}^{(p)}$, plastic deformation $\varepsilon_{ab}^{(p)}$ and entropy:

$$U = U\left(\varepsilon_{ab}^{(e)},\ \varepsilon_{ab}^{(p)},\ S\right). \tag{3.164}$$

Following the same line of reasoning as for elastic bodies, we first determine the entropy rate from the energy equation (3.116):

$$\rho T \frac{dS}{dt} = \left(\sigma^{ab} - \rho \frac{\partial U}{\partial \varepsilon_{ab}^{(e)}}\right) \frac{d\varepsilon_{ab}}{dt} + \left(\rho \frac{\partial U}{\partial \varepsilon_{ab}^{(e)}} - \rho \frac{\partial U}{\partial \varepsilon_{ab}^{(p)}}\right) \frac{d\varepsilon_{ab}^{(p)}}{dt} - \frac{\partial q^i}{\partial x^i},$$

and then write the inequality of the second law of thermodynamics as

$$\int \left[\frac{1}{T}\left(\sigma^{ab} - \rho \frac{\partial U}{\partial \varepsilon_{ab}^{(e)}}\right) \frac{d\varepsilon_{ab}}{dt} + \frac{1}{T}\left(\rho \frac{\partial U}{\partial \varepsilon_{ab}^{(e)}} - \rho \frac{\partial U}{\partial \varepsilon_{ab}^{(p)}}\right) \frac{d\varepsilon_{ab}^{(p)}}{dt} + q^i \frac{\partial}{\partial x^i} \frac{1}{T}\right] dV \geq 0. \tag{3.165}$$

Consider, for simplicity, the cases when heat flux, q^i, can be neglected. If the tensor σ^{ab} does not depend on strain rates then it necessarily follows from (3.165) that

$$\sigma^{ab} = \rho \frac{\partial U\left(\varepsilon_{ab}^{(e)},\ \varepsilon_{ab}^{(p)},\ S\right)}{\partial \varepsilon_{ab}^{(e)}}. \tag{3.166}$$

The closing equations for plastic deformation are chosen in such a way that

$$\left(\rho\frac{\partial U}{\partial\varepsilon_{ab}^{(e)}}-\rho\frac{\partial U}{\partial\varepsilon_{ab}^{(p)}}\right)\frac{d\varepsilon_{ab}^{(p)}}{dt}\geq 0. \tag{3.167}$$

If the internal energy does not depend explicitly on plastic deformations then inequality (3.167) simplifies to

$$\sigma^{ab}\frac{d\varepsilon_{ab}^{(p)}}{dt}\geq 0.$$

Denote the expression from (3.167) in parenthesis by τ^{ab}, so that

$$\tau^{ab}=\rho\frac{\partial U}{\partial\varepsilon_{ab}^{(e)}}-\rho\frac{\partial U}{\partial\varepsilon_{ab}^{(p)}}=\sigma^{ab}-\rho\frac{\partial U}{\partial\varepsilon_{ab}^{(p)}}. \tag{3.168}$$

The dissipation in plastic flow is

$$\rho T\frac{dS}{dt}=D=\tau^{ab}\dot{\varepsilon}_{ab}^{(p)},\quad \dot{\varepsilon}_{ab}^{(p)}\equiv\frac{d\varepsilon_{ab}^{(p)}}{dt}. \tag{3.169}$$

If one assumes that the dissipation is a function of the plastic strain rate,[8]

$$D=D\left(\dot{\varepsilon}_{ab}^{(p)}\right),$$

then, similarly to (3.160), one postulates that

$$\tau^{ab}=\lambda\frac{\partial D}{\partial\dot{\varepsilon}_{ab}^{(p)}}. \tag{3.170}$$

The parameter λ is determined from (3.169) and (3.170):

$$\lambda=D\bigg/\frac{\partial D}{\partial\dot{\varepsilon}_{ab}^{(p)}}\dot{\varepsilon}_{ab}^{(p)}.$$

If τ^{ij} and $e_{ij}^{(p)}$ are the components of the tensors τ and the plastic strain rate in an Eulerian frame,

$$\tau^{ij}=x_a^i x_b^j\tau^{ab},\quad e_{ij}^{(p)}=X_i^a X_j^b\dot{\varepsilon}_{ab}^{(p)},$$

then, obviously,

[8] Dissipation may depend on distortion as well.

$$\tau^{ij} = \lambda \frac{\partial D}{\partial e_{ij}^{(p)}}. \tag{3.171}$$

For metals, plastic deformation is usually incompressible and

$$g^{ij} e_{ij}^{(p)} = 0.$$

Therefore, dissipation depends only on the deviator part of the plastic strain rate

$$e_{ij}^{\prime(p)} = e_{ij}^{(p)} - \frac{1}{3} e^{(p)k}_{\ \ k} g_{ij}.$$

One can then obtain from (3.171) that the trace of the tensor τ^{ij} is zero:

$$\tau_i^i = 0,$$

while the deviator part of the tensor τ, $\tau^{\prime ij} = \tau^{ij} - 1/3\tau_k^k g^{ij}$, is

$$\tau^{\prime ij} = \lambda \frac{\partial D}{\partial e_{ij}^{\prime(p)}}.$$

The behavior of many metals is well described by the model

$$D = k \left(e_{ij}^{\prime(p)} e^{\prime(p)ij} \right)^{\frac{1}{2}\left(1+\frac{1}{m}\right)}. \tag{3.172}$$

It contains two material characteristics, k and m. The dimensionless parameter, m, is usually very large. In the limit $m \to \infty$, this model transforms into the von Mises model, where

$$D = k\sqrt{e_{ij}^{\prime(p)} e_{ij}^{\prime(p)}}. \tag{3.173}$$

For the von Mises model, $\lambda = 1$, and

$$\tau^{\prime ij} = k \frac{e_{ij}^{\prime(p)}}{\sqrt{e_{mn}^{\prime(p)} e^{\prime(p)mn}}}. \tag{3.174}$$

If $e_{ij}^{\prime(p)} \neq 0$, i.e. the material is deforming plastically, then, from (3.174),

$$\tau^{\prime ij} \tau_{ij}^{\prime} = k^2. \tag{3.175}$$

This equation defines a surface in the six-dimensional space of variables τ_{ij}. The plastic flow occurs only if the stresses lie on this surface.

If there is no plastic deformation, $e_{ij}^{\prime(p)} = 0$, then (3.174) contain an indeterminacy of the type $0/0$. The analysis of the non-degenerated model (3.172) shows that for $e_{ij}^{\prime(p)} = 0$ the tensor $\tau^{\prime ij}$ lies inside the surface (3.175).

Usually, elastic deformation is much smaller than plastic deformation, and one can equate the total and plastic strains. Accordingly, one can identify the total strain rate e_{ij} with the plastic strain rate, $e_{ij}^{(p)}$. Then we obtain for von Mises dissipation:

$$D = k\sqrt{e'_{ij}e^{\prime ij}}. \tag{3.176}$$

If, additionally, the internal energy depends only on elastic deformations, then

$$\sigma^{\prime ij} = \tau^{\prime ij} = k\frac{e^{\prime ij}}{\sqrt{e'_{mn}e^{\prime mn}}}. \tag{3.177}$$

Equations (3.176) and (3.177) characterize the material called an ideal plastic body.

A remarkable feature of the von Mises dissipation is the independence of the total dissipation on the rate of the process. Namely, for the deformation path, $\varepsilon_{ij}(t)$, starting at the strain state, ε_{ij}^1, and ending at the strain state, ε_{ij}^2, the integral

$$\int_{\varepsilon_{ij}^1}^{\varepsilon_{ij}^2} D^{ij}\left(e'_{ij}\right) dt$$

depends only on the path, $\varepsilon_{ij}(t)$, and does not depend on the rate with which this path is passed. Such a feature is characteristic for many materials.

3.4 Thermodynamic Formalism

In this section we briefly describe what is called the thermodynamic formalism in derivation of governing equations. For simplicity, we assume that continuum does not move, and there is no heat conductivity. To incorporate continuum motion and heat conduction one needs to complicate the treatment in accordance with what was said in the previous section.

So, let the state of the continuum be characterized by some fields, $u^{\varkappa}(X,t)$, $\varkappa = 1,\ldots,m$. For example, these could be plastic deformation (or plastic distortion), concentrations of chemical species, etc. We assume that internal energy density is some function of $u^{\varkappa}$, their spatial derivatives and entropy:

$$U = U\left(u^{\varkappa}, u^{\varkappa}_{,a}, S\right). \tag{3.178}$$

This is a crucial assumption. In fact, when we have chosen some set of parameters to model the state of a material, we can never be sure a priori that energy is completely determined by these parameters, and there are no additional "hidden" parameters which affect the value of energy. The validity of (3.178) can be checked only experimentally either by direct measurements of energy, entropy and other parameters involved or by inspecting the validity of the consequences of (3.175). To make further relations simpler, we include the factor, ρ_0, in U, so, in this section U means the internal energy per unit volume of the initial state.

Let $\mathring{V}$ be some piece of the material. The energy of the material confined in region $\mathring{V}$ is

$$E = \int_{\mathring{V}} U\left(u^{\varkappa}, u^{\varkappa}_{,a}, S\right) d\mathring{V}. \tag{3.179}$$

Let the material be adiabatically isolated. Then the first law of thermodynamics states that the energy rate is equal to the power of external forces (recall that, as we have just assumed, heat fluxes are neglected)

$$\frac{d}{dt} \int_{\mathring{V}0} U\left(u^{\varkappa}, u^{\varkappa}_{,a}, S\right) d\mathring{V} = P. \tag{3.180}$$

The structure of the power is controlled by the form of energy. We have seen in Sections 3.2 and 3.3 that formula for the power (3.108) is consistent with the assumption that energy density depends on the first derivatives of displacements. Similarly, we set in the general case

$$P = \int_{\partial\mathring{V}} \sigma^a_{\varkappa} \frac{du^{\varkappa}}{dt} \mathring{n}_a d\mathring{A} + \int_{\mathring{V}} g_{\varkappa} \frac{du^{\varkappa}}{dt} d\mathring{V}. \tag{3.181}$$

Here $\sigma^a_{\varkappa}$ and $g_{\varkappa}$ are some "generalized stresses" and "generalized body force". Transforming the surface integral (3.181) into a volume integral by means of divergence theorem, we write the first law of thermodynamics as

$$\int_{\mathring{V}} \left[\frac{\partial U}{\partial S}\dot{S} + \left(\frac{\partial U}{\partial u^{\varkappa}} - \partial_a \sigma^a_{\varkappa} - g_{\varkappa}\right)\dot{u}^{\varkappa} + \left(\frac{\partial U}{\partial u^{\varkappa}_{,a}} - \sigma^a_{\varkappa}\right)\dot{u}^{\varkappa}_{,a}\right] d\mathring{V} = 0.$$

Since the region $\mathring{V}$ is arbitrary, this equation may be satisfied only if the integrand is zero identically. Employing the notation, T, for the derivative, $\partial U/\partial S$, we have

$$\rho_0 T\dot{S} + \left(\frac{\partial U}{\partial u^{\varkappa}} - \partial_a \sigma^a_{\varkappa} - g_{\varkappa}\right)\dot{u}^{\varkappa} + \left(\frac{\partial U}{\partial u^{\varkappa}_{,a}} - \sigma^a_{\varkappa}\right)\dot{u}^{\varkappa}_{,a} = 0. \tag{3.182}$$

Let us rewrite (3.182) in the form,

$$\rho_0 T\dot{S} = \tau_\varkappa \dot{u}^\varkappa + \tau_\varkappa^a \dot{u}_{,a}^\varkappa, \tag{3.183}$$

where we introduced the notation

$$\tau_\varkappa \equiv \partial_a \sigma_\varkappa^a - \rho_0 \frac{\partial U}{\partial u^\varkappa} + g_\varkappa, \qquad \tau_\varkappa^a \equiv \sigma_\varkappa^a - \rho_0 \frac{\partial U}{\partial u_{,a}^\varkappa}. \tag{3.184}$$

The generalized forces,$\tau_\varkappa$ and $\tau_\varkappa^a$, control the entropy rate and, thus, describe the non-equilibrium features of the model. If $\tau_\varkappa = 0$ and $\tau_\varkappa^a = 0$ and $U\left(u^\varkappa, u_{,a}^\varkappa, S\right)$ is given, then (3.184) form a closed system of equations for $u^\varkappa$. Remarkably, this system of equations has a variational form:

$$\partial_a \frac{\partial U}{\partial u_{,a}^\varkappa} - \frac{\partial U}{\partial u^\varkappa} + g_\varkappa = 0,$$

Using the variational derivative,

$$\frac{\delta U}{\delta u^\varkappa} = \frac{\partial U}{\partial u^\varkappa} - \partial_a \frac{\partial U}{\partial u_{,a}^\varkappa},$$

we can write this system of equations as

$$\frac{\delta U}{\delta u^\varkappa} = g_\varkappa.$$

To obtain a closed system equations for $u^\varkappa$ in general case, one can, in addition to specifying function $U\left(u^\varkappa, u_{,a}^\varkappa, S\right)$, prescribe the dependence of $\tau_\varkappa$, $\tau_\varkappa^a$, and $g_\varkappa$ on $u^\varkappa$ and their derivatives. Then, eliminating $\sigma_\varkappa^a$ from (3.184), we arrive at the following system of equations for $u^\varkappa$:

$$\tau_\varkappa = \partial_a \left(\rho_0 \frac{\partial U}{\partial u_{,a}^\varkappa} + \tau_\varkappa^a\right) - \rho_0 \frac{\partial U}{\partial u^\varkappa} + g_\varkappa. \tag{3.185}$$

As to $g_\varkappa$, usually, there are no "body forces" working on the change of the internal parameters, $u^\varkappa$, i.e. $g_\varkappa = 0$. We accept this in what follows.

The only constraint for the admissible dependencies of $\tau_\varkappa^a$ and $\tau_\varkappa$ on $u^\varkappa$ and their derivatives is the condition that entropy does not decay (or dissipation, D, the right hand side of (3.183) is non-negative):

$$D \equiv \tau_\varkappa \dot{u}^\varkappa + \tau_\varkappa^a \dot{u}_{,a}^\varkappa \geq 0. \tag{3.186}$$

It is often supposed that there exists a function of $u^\varkappa$, $\dot{u}^\varkappa$, and $u_{,a}^\varkappa$, $\mathcal{D}$, such that

$$\tau_\varkappa = \frac{\partial \mathcal{D}}{\partial \dot{u}^\varkappa}, \quad \tau_\varkappa^a = \frac{\partial \mathcal{D}}{\partial \dot{u}_{,a}^\varkappa}. \tag{3.187}$$

Function $\mathcal{D}$ is called the dissipation potential. The potential law is not a "law of Nature", but rather a way to obtain a model with simply controlled mathematical features.

In case of the potential law (3.187) the governing equations (3.185) take especially simple "variational" form:

$$\frac{\delta U}{\delta u^{\varkappa}} = -\frac{\delta \mathcal{D}}{\delta \dot{u}^{\varkappa}}. \tag{3.188}$$

Here $\delta \mathcal{D}/\delta \dot{u}^{\varkappa}$ is the variational derivative:

$$\frac{\delta \mathcal{D}}{\delta \dot{u}^{\varkappa}} = \frac{\partial \mathcal{D}}{\partial \dot{u}^{\varkappa}} - \partial_a \frac{\partial \mathcal{D}}{\partial \dot{u}^{\varkappa}_{,a}}.$$

The dissipative potential, $\mathcal{D}$, is simply related to dissipation if $\mathcal{D}$ is a homogeneous function, i.e. for any λ and some number, m,

$$\mathcal{D}\left(\lambda \dot{u}^{\varkappa}, \lambda \dot{u}^{\varkappa}_{,a}\right) = \lambda^m \mathcal{D}\left(\dot{u}^{\varkappa}, \dot{u}^{\varkappa}_{,a}\right). \tag{3.189}$$

According to Euler's identity for homogeneous functions,

$$\dot{u}^{\varkappa} \frac{\partial \mathcal{D}}{\partial \dot{u}^{\varkappa}} + \dot{u}^{\varkappa}_{,a} \frac{\partial \mathcal{D}}{\partial \dot{u}^{\varkappa}_{,a}} = m\mathcal{D}\left(\dot{u}^{\varkappa}, \dot{u}^{\varkappa}_{,a}\right),$$

the dissipative potential differs from the dissipation by the factor, m :

$$D = m\mathcal{D}. \tag{3.190}$$

In linear theory, D and $\mathcal{D}$ are quadratic functions and $m = 2$. For ideal-plasticity-type models, $m = 1$.

Chapter 4
Principle of Least Action in Continuum Mechanics

As we discussed in Sect. 2.6, for reversible processes the governing equations of mechanics must have a Hamiltonian structure, and accordingly a principle of least action must exist. In another extreme case, when the inertial effects and the internal interactions described by the internal energy can be ignored, variational principles also exist, but they are due to the special structure of the models rather than to the laws of Nature. In this chapter we consider the principle of least action in continuum mechanics of reversible processes and some related issues. The variational principles for dissipative processes are presented in the second part of the book along with the other variational features of the classical continuum models.

4.1 Variation of Integral Functionals

Let a continuum be characterized by some functions $u^{\varkappa}(x^i, t)$, $\varkappa = 1, \ldots, m$, $i = 1, 2, 3$. In what follows the number of the variables x^i is not essential; besides, time does not play a special role. Therefore, we include time in the set of independent variables and write $u^{\varkappa} = u^{\varkappa}(x)$ assuming that $x = \{x^1, \ldots, x^n\}$ is a point of n-dimensional space. We write, for brevity, $u(x)$ for the set $\{u^1(x), \ldots, u^m(x)\}$, when this cannot cause confusion.

Let a functional, $I(u)$, be given, i.e. there is a rule which allows one to compute the number, $I(u)$, for each $u(x)$. The major example of such a functional for us is an integral functional

$$I(u) = \int_V L(x^i, u^{\varkappa}, u^{\varkappa}_{,i}) d^n x, \quad u^{\varkappa}_{,i} \equiv \frac{\partial u^{\varkappa}}{\partial x^i}, \tag{4.1}$$

where V is some region in n-dimensional space of x-variables. For such functionals, function L is called Lagrangian. The case when Lagrangian depends on higher derivatives will be considered further in Sect. 4.4.

We assume that region V is compact to avoid the technicalities caused by the unboundedness of V; some peculiarities of the variational problems for an unbounded domain are considered in Sect. 11.7. All functions involved are assumed

V.L. Berdichevsky, *Variational Principles of Continuum Mechanics*,
Interaction of Mechanics and Mathematics, DOI 10.1007/978-3-540-88467-5_4,

to be sufficiently smooth to make the derivation of the final equations meaningful. In particular, the boundary ∂V of region V is piecewise smooth, and Lagrangian is a smooth function of its arguments.

Physical reasoning to be considered later sets some constraints on the admissible function $u(x)$. A typical constraint is prescribing the values of $u(x)$ at a piece of the boundary, ∂V_u, of the boundary ∂V:

$$u(x) = u_b(x) \quad \text{at } \partial V_u. \tag{4.2}$$

Here $u_b(x)$ are the given boundary values of functions $u(x)$.

The constraints specify the set of admissible functions $u(x)$. We denote this set by $\mathcal{M}$. A description of the set $\mathcal{M}$ also includes the characterization of smoothness of the admissible functions. In order for the integral (4.1) to be sensible, it is sufficient to include in the set $\mathcal{M}$ the functions $u(x)$ which are continuous along with their derivatives in the closed region V. The physically important case of piecewise differentiable functions will be treated in Sect. 7.4.

A variational principle usually states that the true process (or, in statics, the equilibrium state) of a continuum is a stationary point of the functional $I(u)$ on the set $\mathcal{M}$, i.e. variation of this functional, δI, vanishes for all admissible variations δu.

Let $u(x)$ be a stationary point of the functional $I(u)$ and $u(x) = u(x) + \delta u(x)$ be a small disturbance of the function $u(x)$. Keeping only the leading small terms in the difference $I(u + \delta u) - I(u)$, we find the variation of the functional, $I(u)$,

$$\delta I = \int_V \left(\frac{\partial L}{\partial u^\varkappa} \delta u^\varkappa + \frac{\partial L}{\partial u^\varkappa_{,i}} (\delta u^\varkappa)_{,i} \right) d^n x. \tag{4.3}$$

Functions $\delta u^\varkappa$ and $(\delta u^\varkappa)_{,i}$ are not independent because $(\delta u^\varkappa)_{,i}$ are completely determined by $\delta u^\varkappa$. To put (4.3) in the form which contains only independent terms, we integrate the second term in the integrand by parts:

$$\begin{aligned}\delta I &= \int_V \left(\left(\frac{\partial L}{\partial u^\varkappa} - \frac{\partial}{\partial x^i} \frac{\partial L}{\partial u^\varkappa_{,i}} \right) \delta u^\varkappa + \frac{\partial}{\partial x^i} \left(\frac{\partial L}{\partial u^\varkappa_{,i}} \delta u^\varkappa \right) \right) dV \\ &= \int_V \frac{\delta L}{\delta u^\varkappa} \delta u^\varkappa d^n x + \int_{\partial V} \frac{\partial L}{\partial u^\varkappa_{,i}} \delta u^\varkappa n_i dA. \end{aligned} \tag{4.4}$$

Here dA is the area element at ∂V, n_i the unit normal vector at ∂V and a notation for the variational derivative of L is introduced,

$$\frac{\delta L}{\delta u^\varkappa} = \frac{\partial L}{\partial u^\varkappa} - \frac{\partial}{\partial x^i} \frac{\partial L}{\partial u^\varkappa_{,i}}. \tag{4.5}$$

Note that

$$\delta u = 0 \quad \text{at } \partial V_u$$

due to the boundary conditions (4.2). Therefore, the surface integral in (4.4) is, in fact, an integral over $\partial V - \partial V_u$.

The terms in (4.4) are independent because the value of $\delta u^\varkappa$ inside the region V and at its boundary can be changed independently of each other.

Formula (4.4) is the sought expression for the variations of functional $I(u)$. We will often encounter this expression in that or in a similar form. Further, the derivations will be given only in cases when they do not repeat the above one; otherwise, only the result will be stated.

The main lemma of calculus of variations. For the case under consideration the main lemma of calculus of variations takes the form: if

$$\int_V F_\varkappa \bar{u}^\varkappa d^n x = 0, \tag{4.6}$$

where $F_\varkappa$ are some continuous functions of coordinates which are independent of $\bar{u}^\varkappa$, and $\bar{u}^\varkappa$ are arbitrary smooth functions which are zero on the boundary ∂V of region V, then

$$F_\varkappa = 0 \quad \text{in } V.$$

This lemma can be proved in the same way as in Sect. 1.3.

Note the following consequence: if

$$\int_V F_\varkappa \bar{u}^\varkappa d^n x + \int_{\partial V} f_\varkappa \bar{u}^\varkappa d^{n-1} x = 0, \tag{4.7}$$

where F and f are continuous functions of coordinates which are independent of $\bar{u}^\varkappa$, and $\bar{u}^\varkappa$ are arbitrary smooth functions, then

$$F_\varkappa = 0 \quad \text{in } V, \qquad f_\varkappa = 0 \quad \text{on } \partial V. \tag{4.8}$$

To obtain the first equation (4.8) one should consider functions $\bar{u}^\varkappa$ which are equal to zero at ∂V; then the first equation (4.8) follows from the main lemma of calculus of variations. Thus, the first term in (4.7) is zero. Then the second term in (4.7) is also zero and, applying again the main lemma of calculus of variations, one gets the second equation (4.8).

At the stationary point,

$$\delta I = 0. \tag{4.9}$$

From (4.9), (4.4) and the main lemma of variational calculus it follows that the stationary points of the functional (4.1) satisfy the equations

$$\frac{\delta L}{\delta u^\varkappa} \equiv \frac{\partial L}{\partial u^\varkappa} - \frac{\partial}{\partial x} \frac{\partial L}{\partial u^\varkappa_{,i}} = 0 \quad \text{in } V, \tag{4.10}$$

and the boundary conditions

$$\frac{\partial L}{\partial u^{\varkappa}_{,i}} n_i = 0 \quad \text{on } \partial V - \partial V_u. \tag{4.11}$$

Equations (4.10), (4.11) and (4.2) form a closed system of equations and boundary conditions to find the stationary point. As has been already mentioned on a similar occasion in Sect. 1.3, the remarkable feature of the variational approach is that one always[1] gets a proper number of boundary conditions to solve the partial differential equations (4.10).

Note that the inhomogeneous boundary conditions,

$$\frac{\partial L}{\partial u^{\varkappa}_{,i}} n_i = f_{\varkappa} \quad \text{at } \partial V - \partial V_u,$$

where $f_{\varkappa}$ are some prescribed functions at $\partial V - \partial V_u$, are obtained if the surface integral

$$\int\limits_{\partial V - \partial V_u} f_{\varkappa} u^{\varkappa} dA$$

is added to the functional (4.1).

In our consideration, variations $\delta u(x)$ are the infinitesimally small disturbances of the stationary point u. In calculus of variations one sometimes uses a slightly different language which proves to be useful for non-smooth variational problems. One considers a curve in the functional space, $u = u(\varepsilon)$, i.e. a set of functions, $u(x, \varepsilon)$, smoothly depending on parameter ε. The curve passes through the stationary point at $\varepsilon = 0$. Along this curve the functional becomes a function of one variable, $I(u(\varepsilon))$. The derivative of this function must vanish at the stationary point

$$\frac{d}{d\varepsilon} I(u(\varepsilon)) = 0 \quad \text{for } \varepsilon = 0.$$

One can easily check that

$$\frac{d}{d\varepsilon} I(u(\varepsilon)) = I'\left(u(\varepsilon), \frac{du}{d\varepsilon}\right).$$

The notation $I'\left(u, \frac{du}{d\varepsilon}\right)$ is similar to that in (1.13); in the case under consideration $I'(u, \delta u) = \delta I$. For infinitesimally small $d\varepsilon$, then one can set

[1] With the exception of some degenerated cases; further discussion of this issue the reader will find in Sect. 5.5 in connection with the notion of "feeling the constraints."

$$\delta u = \left. \frac{du}{d\varepsilon} \right|_{\varepsilon=0} d\varepsilon. \tag{4.12}$$

Various variations, δu, correspond to different curves passing through the stationary point. The idea of considering the functional on some curves (or on some surfaces) in the functional space is quite fruitful; in fact, the entire finite element ideology is based on this concept. For the purpose of the derivation of equations for the stationary point, the "variation language" is enough, and we will abide with it.

Energy-momentum tensor. For any function, $L\left(x^i, u^\varkappa, u^\varkappa_{,i}\right)$, the following identity holds:

$$\frac{\partial}{\partial x^j} \left(\frac{\partial L}{\partial x^\varkappa_{,j}} u^\varkappa_{,i} - L \delta^j_i \right) = - \frac{\delta L}{\delta u^\varkappa} u^\varkappa_{,i} - \partial_i L. \tag{4.13}$$

Here $\partial_i L$ are the partial derivatives of the function, $L\left(x^i, u^\varkappa, u^\varkappa_{,i}\right)$, with respect to x^i, while $\partial / \partial x^j$ in the left hand side of (4.13) mean the "full derivative," i.e. the derivative taking into account the dependence of all functions involved, including $u^\varkappa$ and $u^\varkappa_{,i}$, on x^i.

The identity (4.13) can be checked by direct inspection.

If the Lagrangian does not depend explicitly on the coordinates, i.e. $L = L\left(u^\varkappa, u^\varkappa_{,i}\right)$, then any solution of Euler equations (4.10) also satisfies the equations

$$\frac{\partial}{\partial x^j} \left(\frac{\partial L}{\partial u^\varkappa_{,j}} u^\varkappa_{,i} - L \delta^j_i \right) = 0. \tag{4.14}$$

If index i runs through four values, three marking the spatial coordinates and one the time, the corresponding equations are the momentum equations and the energy equation, as the inspection shows for various models. The tensor

$$P^j_i = \frac{\partial L}{\partial u^\varkappa_{,j}} u^\varkappa_{,i} - L \delta^j_i \tag{4.15}$$

is called the energy-momentum tensor. For some field variables, $-P^j_i$ has the meaning of the energy-momentum tensor.

The facts mentioned have a deep origin in the invariance of the action functional with respect to translations, but we do not dwell on these issues here.

Now we turn to discussion of the principle of least action in continuum mechanics.

4.2 Variations of Kinematic Parameters

In this section the relations for the variations of key kinematic characteristics of the continuum motion are summarized.

Variation of velocity. If $u^{\varkappa}$ are some functions of Lagrangian coordinates and time, $u^{\varkappa} = u^{\varkappa}(X^a, t)$, then the variations, $\delta u^{\varkappa}$, were defined as derivatives with respect to an auxiliary parameter ε,

$$\delta u^{\varkappa} = \left.\frac{\partial u^{\varkappa}(X^a, t, \varepsilon)}{\partial \varepsilon}\right|_{\varepsilon=0} d\varepsilon. \tag{4.16}$$

Therefore, the operator δ commutes with the operators d/dt and $\partial/\partial X^a$. The interchangeability of the operators δ and d/dt implies that

$$\delta v^i = \frac{d\delta x^i}{dt}.$$

If δx^i are considered as functions of the Eulerian coordinates and time, then

$$\delta v^i = \frac{\partial \delta x^i}{\partial t} + v^k \frac{\partial \delta x^i}{\partial x^k}. \tag{4.17}$$

Variation of the Jacobian Δ. Using the equation for the components of the inverse matrix (3.24) and the interchangeability of the operators δ and $\partial/\partial X^a$, we have

$$\delta \Delta = \frac{\partial \Delta}{\partial x^i_a} \delta x^i_a = \frac{\partial \Delta}{\partial x^i_a} \frac{\partial \delta x^i}{\partial X^a} = \Delta X^a_i \frac{\partial \delta x^i}{\partial X^a}. \tag{4.18}$$

Consider now the variations of the particle trajectories, $\delta x^i(X^a, t)$, as functions of the Eulerian coordinates. Then, the right hand side of (4.18) may be viewed as the chain rule for differentiation of δx^i with respect to x^i:

$$\Delta X^a_i \frac{\partial \delta x^i}{\partial X^a} = \Delta \frac{\partial X^a}{\partial x^i} \frac{\partial \delta x^i}{\partial X^a} = \Delta \frac{\partial \delta x^i}{\partial x^i}.$$

Finally, for the variation of the determinant of the distortion we obtain

$$\delta \Delta = \Delta \frac{\partial \delta x^i}{\partial x^i}. \tag{4.19}$$

Variation of mass density. Taking the variation of (3.69), we find

$$\delta \rho = -\frac{\rho}{\Delta} \delta \Delta.$$

Using equation (4.19), we get

$$\delta \rho = -\rho \frac{\partial \delta x^i}{\partial x^i}. \tag{4.20}$$

Variation of the strain tensor. From (3.33), we have

$$\delta\varepsilon_{ab} = \frac{1}{2}\delta g_{ab} = \frac{1}{2}\delta\left(g_{ij}x_a^i x_b^j\right) = \frac{1}{2}g_{ij}\left(x_a^i\frac{\partial\delta x^j}{\partial X^b} + \frac{\partial\delta x^i}{\partial X^a}x_b^j\right). \tag{4.21}$$

Consider δx^i in (4.21) as functions of Eulerian coordinates, and substitute $\partial\delta x^i/\partial X^a$ by $x_a^k\partial\delta x^i/\partial x^k$. Equation (4.21) take the form

$$\delta\varepsilon_{ab} = \frac{1}{2}g_{ij}\left(x_a^k x_b^j + x_b^k x_a^j\right)\frac{\partial\delta x^i}{\partial x^k} = \frac{1}{2}x_a^i x_b^j\left(\partial_i\delta x_j + \partial_j\delta x_i\right) = x_a^i x_b^j\partial_{(i}\,\delta x_{j)}. \tag{4.22}$$

Variation of inverse distortion. The variations X_i^a are most easily found from the definition of X_i^a as the components of the inverse of matrix x_a^i. Taking the variation of the equation $x_b^j X_i^b = \delta_i^j$, we get

$$x_b^j\delta X_i^b + X_i^b\delta x_b^j = 0. \tag{4.23}$$

Let us contract (4.23) with X_j^a with respect to index j. Since $x_b^j X_j^a = \delta_b^a$, the equality (4.23) becomes

$$\delta X_i^a = -X_j^a X_i^b\delta x_b^j. \tag{4.24}$$

If we use the interchangeability of the operators δ and ∂_b and consider δx^j as functions of the Eulerian coordinates, we obtain

$$\delta X_i^a = -X_j^a\frac{\partial\delta x^j}{\partial x^i}. \tag{4.25}$$

Variation of contravariant components of Lagrangian metrics. The contravariant components of the metric tensor are the components of the inverse matrix to the matrix $\|g_{ab}\|$. Therefore, as in computation of inverse distortion, we vary the equation $g^{ab}g_{bc} = \delta_c^a$. Similarly to (4.24),

$$\delta g^{ab} = -g^{ac}g^{bd}\delta g_{cd} = -2g^{ac}g^{bd}\delta\varepsilon_{cd}. \tag{4.26}$$

From the relations (4.26) and (4.22) we get

$$\delta g^{ab} = X_i^a X_j^b\left(\partial^i\delta x^j + \partial^j\delta x^i\right). \tag{4.27}$$

Variations of particle trajectories of a rigid body. The motion of the rigid body is described by functions $r^i(t)$ and $\alpha_a^i(t)$ (see (3.47)). If there are no additional constraints on the motion, the functions $r^i(t)$ can change independently while functions $\alpha_a^i(t)$ change in such a way as to satisfy the orthogonality condition (3.48). Taking the variation of the second relation (3.48), we get

$$\delta\left(\overset{\circ}{g}{}^{ab}\alpha^{i}_{a}\alpha^{j}_{b}\right)=\alpha^{ia}\delta\alpha^{j}_{a}+\alpha^{ja}\delta\alpha^{i}_{a}=0. \tag{4.28}$$

Equation (4.28) means that infinitesimally small tensor $\delta\varphi^{ij}=\alpha^{ia}\delta\alpha^{j}_{a}$ is antisymmetric with respect to indices i, j, but otherwise is arbitrary. The variation of the particle trajectories is

$$\delta x^{i}=\delta r^{i}(t)+\left(x_{j}-r_{j}(t)\right)\delta\varphi^{ji}. \tag{4.29}$$

If the vector $\delta\varphi_k$ corresponding to the antisymmetric tensor $\delta\varphi^{ji}$ is defined as

$$\delta\varphi_{k}=\frac{1}{2}\varepsilon_{ijk}\delta\varphi^{ij},\quad \delta\varphi^{ij}=\varepsilon^{ijk}\delta\varphi_{k},$$

then (4.29) can be written as

$$\delta x^{i}=\delta r^{i}+\varepsilon^{ijk}\delta\varphi_{j}\left(x_{k}-r_{k}(t)\right). \tag{4.30}$$

Note that $\delta\varphi^{ij}$ and $\delta\varphi_j$ are some infinitesimally small quantities which are not variations of any characteristics of motion. The tensor $\delta\varphi^{ij}$ is analogous to the tensor of an infinitesimally small rotation, and $\delta\varphi_j$ is analogous to an infinitesimally small angle of rotation.

If variation δ coincides with the time increment, d, along the real trajectory, (4.29) and (4.30) transform to (3.51) and (3.55).

Variation of a function of Eulerian coordinates. Functions $u^{\varkappa}$ can be considered as functions of the Eulerian coordinates. Then the corresponding set of the trial functions is $u^{\varkappa}(x,t,\varepsilon)$. The quantity

$$\partial u^{\varkappa}=\left.\frac{\partial u^{\varkappa}(x,t,\varepsilon)}{\partial\varepsilon}\right|_{\varepsilon=0,\ x=const}d\varepsilon$$

is the variation with the Eulerian coordinates held constant; it is denoted by the symbol ∂. The variation δ (for constant X^a) and the variation ∂ (for constant x) are related as

$$\delta u^{\varkappa}=\left.\frac{\partial}{\partial\varepsilon}u^{\varkappa}\left(x^{i}\left(X^{a},t,\varepsilon\right),t,\varepsilon\right)\right|_{\varepsilon=0,\ X^{a}=const}d\varepsilon=\partial u^{\varkappa}+\delta x^{i}\frac{\partial u^{\varkappa}}{\partial x^{i}}. \tag{4.31}$$

The operator ∂ commutes with the operators $\partial/\partial x^i$ and $\partial/\partial t$ and does not commute with the operators $\partial/\partial X^a$ and d/dt.

The formulas establishing the relations between $\partial d/dt$, $\partial\partial/\partial X^a$ and $d\partial/dt$, $\partial\partial/\partial X^a$ are easily obtained by the means of (4.31). For an arbitrary function F,

$$\partial \frac{d}{dt} F = \delta \frac{d}{dt} F - \delta x^i \frac{\partial}{\partial x^i} \frac{d}{dt} F = \frac{d\delta F}{dt} - \delta x^i \frac{\partial}{\partial x^i} \frac{dF}{dt} =$$
$$= \frac{d}{dt} \partial F + \frac{d}{dt} \left(\delta x^i \frac{\partial F}{\partial x^i} \right) - \delta x^i \frac{\partial}{\partial x^i} \frac{dF}{dt},$$

$$\partial \frac{\partial}{\partial X^a} F = \delta \frac{\partial}{\partial X^a} F - \delta x^i \frac{\partial}{\partial x^i} \frac{\partial}{\partial X^a} F =$$
$$= \frac{\partial}{\partial X^a} \partial F + \frac{\partial}{\partial X^a} \delta x^i \frac{\partial}{\partial x^i} F - \delta x^i \frac{\partial}{\partial x^i} \frac{\partial}{\partial X^a} F.$$

Here are some formulas for the variations when the Eulerian coordinates are held constant.

Eulerian variation of velocity. Applying (4.31) we get

$$\partial v^i = \delta v^i - \delta x^k \frac{\partial v^i}{\partial x^k} = \frac{d\delta x^i}{dt} - \delta x^k \frac{\partial v^i}{\partial x^k}. \tag{4.32}$$

Eulerian variation of mass density. Using (4.20) and (4.31), we get

$$\partial \rho = \delta \rho - \delta x^i \frac{\partial \rho}{\partial x^i} = -\rho \partial_i \delta x^i - \delta x^i \frac{\partial \rho}{\partial x^i} = -\partial_i \left(\rho \delta x^i \right). \tag{4.33}$$

Variation of Lagrangian coordinates. Writing the condition for the variation δ being equal to zero in the Lagrangian coordinates,

$$\delta X^a = \partial X^a + \delta x^i \frac{\partial X^a}{\partial x^i} = 0,$$

we obtain, for the variation of the Lagrangian coordinates at constant x

$$\partial X^a = -X_i^a \delta x^i. \tag{4.34}$$

4.3 Principle of Least Action

Continuum mechanics aims to model the behavior of a large number of particles. Dynamics of particles is governed by the principle of least action with the action functional of the form

$$\int_{t_0}^{t_1} (\text{Lagrange function})\, dt$$

In the continuum description, Lagrange function possesses a density, the Lagrangian, L,

$$\text{Lagrange function} = \int\limits_{\overset{\circ}{V}} L\, d\overset{\circ}{V} \tag{4.35}$$

where $\overset{\circ}{V}$ is the region occupied by the continuum in the initial state. The summation over particles is made in two steps: first, $L d\overset{\circ}{V}$ is the total Lagrange function of particles which are in the volume $d\overset{\circ}{V}$; second, summation over particles from different parts of the region $\overset{\circ}{V}$ is replaced by integration over $\overset{\circ}{V}$. The latter procedure assumes that the interaction between particles from different parts of the region $\overset{\circ}{V}$ is negligible, and the total Lagrange function is approximately equal to the sum of Lagrange functions of small parts of the region $\overset{\circ}{V}$.

As in classical mechanics, L is the difference of kinetic and potential energies. Denoting the kinetic and potential energies per unit mass by K and U, we have

$$L = \rho_0 (K - U). \tag{4.36}$$

The classical continuum models correspond to the assumption that the key kinematic characteristics of motion are the particle trajectories, $x^i(X^a, t)$, and entropy, S,

$$K = \frac{1}{2} v_i v^i, \quad U = U\left(x^i_a, S\right). \tag{4.37}$$

Internal energy for inhomogeneous media may depend on Lagrangian coordinates, but we do not mention this explicitly.

Finally, the action functional is

$$I\left(x(X,t), S(t,X)\right) = \int\limits_{t_0}^{t_1} \int\limits_{\overset{\circ}{V}} \rho_0 \left(\frac{1}{2} v_i v^i - U\left(x^i_a, S\right)\right) d\overset{\circ}{V} dt. \tag{4.38}$$

The action functional can also be written as an integral over the current state, the moving region, $V(t)$. According to (3.72),

$$I = \int\limits_{t_0}^{t_1} \int\limits_{V(t)} \rho \left(\frac{1}{2} v_i v^i - U\left(x^i_a, S\right)\right) dV dt. \tag{4.39}$$

Let the dissipation be negligible. Then in each particle, entropy, S, does not change and becomes a certain function of Lagrangian coordinates, $S = S(X)$, known from the initial conditions. Thus, entropy drops out of the set of unknown functions.

Suppose that, as in Hamilton variational principle, the initial and the final positions of the particles are prescribed:

$$x(t_0, X) = \overset{0}{x}(X), \quad x(t_1, X) = \overset{1}{x}(X). \tag{4.40}$$

Let us show that then the following variational principle holds.

Principle of least action. *The true motion of continuum is a stationary point of the action functional* (4.38) *on the set of all particle trajectories obeying the constraint* (4.40).

The action functional (4.38) has the form (4.1) with x and u replaced by (X, t) and $x(X, t)$, respectively. Thus, we may apply the formulas of Sect. 4.1. If function L is an arbitrary function of v^i and x^i_a, v^i and x^i_a being derivatives of $x^i(X, t)$ with respect to t and X^a, then from (4.10),

$$\frac{d}{dt}\frac{\partial L}{\partial v^i} + \frac{\partial}{\partial X^a}\frac{\partial L}{\partial x^i_a} = 0. \tag{4.41}$$

For the function

$$L = \rho_0\left(\frac{1}{2}v_i v^i - U\left(x^i_a, S\right)\right)$$

these equations take the form

$$\rho_0 \frac{d}{dt}v_i = \frac{\partial}{\partial X^a}\rho_0\frac{\partial U}{\partial x^i_a} \tag{4.42}$$

or, recalling the definition of the Piola-Kirchhoff tensor (3.131),

$$\rho_0 \frac{d}{dt}v_i = \frac{\partial}{\partial X^a}p^a_i. \tag{4.43}$$

For Cauchy's stress tensor, $\sigma^{ij} = \frac{\rho}{\rho_0}x^j_a p^{ia}$, (4.43), as shown in Sect. 3.2, takes the usual form of the momentum equations

$$\rho\frac{dv^i}{dt} = \frac{\partial \sigma^{ij}}{\partial x^j}. \tag{4.44}$$

Constraints (4.40) vanish the variations of $x(X, t)$ at the initial and finite times. The variations of $x(X, t)$ at the boundary of the body are arbitrary. Therefore, the boundary conditions of Sect. 4.1, (4.11), read

$$p^a_i \mathring{n}_a = 0 \tag{4.45}$$

where $\mathring{n}_a$ are the Lagrangian components of the unit normal vector of the surface $\partial\mathring{V}$.

These formulas can be derived directly from the principle of least action using the relationships for variations from Sect. 4.2.

To include boundary tractions, one has to add into the action functional the surface integral

$$\int_{t_0}^{t_1}\int_{\partial\mathring{V}} f_i(X,t)x^i(X,t)\,d\mathring{A}dt. \tag{4.46}$$

To take into account the body forces, the volume integral must be added:

$$\int_{t_0}^{t_1}\int_{\mathring{V}} \rho_0 g_i(X,t)x^i(X,t)\,d\mathring{V}dt. \tag{4.47}$$

Accordingly, (4.42) and (4.45) change to

$$\rho_0\frac{d}{dt}v_i = \frac{\partial}{\partial X^a}\rho_0\frac{\partial U}{\partial x_a^i} + \rho_0 g_i, \quad p_i^a\mathring{n}_a = f_i \quad \text{at } \partial\mathring{V}. \tag{4.48}$$

The origin of the additional terms (4.46) and (4.47) is apparent: on the microlevel, the external forces act on the particles and cause the terms in Hamiltonian functions like $f_i q^i$; (4.46) and (4.47) are the homogenized form of these contributions.

4.4 Variational Equations

In general, the equations of continuum mechanics cannot be obtained from a variational principle. But they can always be derived from a variational equation, i.e. the statement that the variation of the action functional is equal not to zero but to some functional, δA, which is a linear functional of variations:

$$\delta I = \delta A. \tag{4.49}$$

Variational equations lose the major feature of the variational principle, the special structure of the governing equations, because any equation can be obtained from the variational equation, unless the possible functionals δA are somehow constrained. Indeed, let I be functional of a required function u, while δA has the form

$$\delta A = \int_V F\delta u d^n x.$$

Then it follows from the variational equation that

$$\frac{\delta L}{\delta u} = F. \tag{4.50}$$

Any equation for function u, $G = 0$, can be written in the form (4.50): it is enough to set $F = G + \delta L/\delta u$. Therefore, the information on the admissible forms of the functional δA is what actually defines the variational equation. Here is a brief review of the suggestions made.

The principle of virtual displacements. Historically, the first variational equation was the "golden rule" of mechanics – the principle of virtual displacement. Its formulation for a lever was given in Aristotle's "Physics" (fourth century BC). Further significant development of the principle was made by Stevin and Galileo. The modern form of the principle was essentially obtained by Johann Bernoulli. The formulation of the principle of virtual displacement for a system of material points that is subject to some kinematical constraints is as follows. Let $\mathbf{x}_{(s)}$ be the position vector of particle s, $\mathbf{F}_{(s)}$ the force acting on particle s, and $\delta\mathbf{x}_{(s)}$ the infinitesimally small displacements compatible with the constraints.[2] The constraints are assumed to be "ideal"; this term will explained later. The system is in equilibrium if and only if the total work of all forces on possible displacements is zero:

$$\sum_s \mathbf{F}_{(s)} \cdot \delta\mathbf{x}_{(s)} = 0. \tag{4.51}$$

If there are no kinematic constraints, the assertion (4.51) is "trivial," since it is equivalent to equalities $\mathbf{F}_{(s)} = 0$. "Nontrivial" conditions arise as a result of constraints.

In the papers of the nineteenth century, the principle of virtual displacements was also called the principle of virtual work or principle of virtual velocities or principle of virtual powers (the latter two terms are related to the possibility to replace $\delta\mathbf{x}_{(s)}$ by kinematically admissible velocities; up to an infinitesimally small factor, the velocities take the same values as $\delta\mathbf{x}_{(s)}$).

The d'Alambert principle and the energy equation for virtual displacements. In dynamics, the variational equation (4.51) remains true if the inertial forces are added to $F_{(s)}$:

$$\sum_s \left(\mathbf{F}_{(s)} - m_{(s)}\mathbf{a}_{(s)}\right) \cdot \delta\mathbf{x}_{(s)} = 0, \tag{4.52}$$

where $m_{(s)}$ is the mass and $\mathbf{a}_{(s)}$ is the acceleration of the particle s. Equation (4.52) is called the d'Alambert principle.

If there are no interactions between particles, (4.52) yields the second Newton's Law. In the presence of the interaction forces, (4.52) becomes "nontrivial."

In the d'Alambert principle the inertial forces are in some sense as important as the other forces: for the derivation of the basic equations one has to include in the static equation one more force, the force related to the particle acceleration

---

[2] The indices of non-tensor nature are put in parentheses.

with respect to the inertial frame of reference. Note that in the co-moving frame of reference, i.e. a non-inertial deforming frame where all the particles are at rest, the Newton's equations can be considered as equilibrium equations for the system of forces including the inertial forces.

The d'Alambert variational equation (principle) can be taken as the primary postulate of the mechanics of systems with a finite number of degrees of freedom.

L.I. Sedov suggested the idea that the variational equation of mechanics is, in fact, the energy equation written for virtual displacements. Let us show how to transform the d'Alambert principle into the energy equation for virtual motions.

The quantity $\sum_s \mathbf{F}_{(s)}\delta\mathbf{x}_{(s)}$ apparently represents the work done by the external forces through the virtual displacements, $\delta\mathbf{x}_{(s)}$; we denote this quantity by $\delta\mathcal{A}^{(e)}$:

$$\delta\mathcal{A}^{(e)} = \sum_s \mathbf{F}_{(s)} \cdot \delta\mathbf{x}_{(s)}.$$

Consider the expression

$$m\mathbf{a}\cdot\delta\mathbf{x} = m\frac{d\mathbf{v}}{dt}\cdot\delta\mathbf{x}.$$

Let us add and subtract the term $m\mathbf{v}\delta\mathbf{v}$ in the right-hand side of this expression. Then

$$m\mathbf{a}\cdot\delta\mathbf{x} = \delta\left(\frac{1}{2}m\mathbf{v}^2\right) + m\frac{d\mathbf{v}}{dt}\cdot\delta\mathbf{x} - m\mathbf{v}\cdot\delta\frac{d\mathbf{x}}{dt}, \qquad \mathbf{v}^2 \equiv \mathbf{v}\cdot\mathbf{v}$$

Define the kinetic energy as

$$\mathcal{K} = \sum_s \frac{1}{2}m_{(s)}\mathbf{v}_{(s)}^2,$$

and functional $\delta\Omega$ as

$$\delta\Omega = -\sum_s \left(m_{(s)}\frac{d\mathbf{v}_{(s)}}{dt}\cdot\delta\mathbf{x}_{(s)} - m_{(s)}\mathbf{v}_{(s)} \cdot \delta\frac{d\mathbf{x}_{(s)}}{dt}\right). \tag{4.53}$$

Then the d'Alambert principle takes the form

$$\delta\mathcal{K} = \delta\mathcal{A}^{(e)} + \delta\Omega. \tag{4.54}$$

It is seen from (4.53) that the functional $\delta\Omega$ has the following property: for real motion it is equal to zero, i.e. replacing the admissible variation δ by the increment in the real process, d, causes $\delta\Omega$ to vanish, i.e.

$$\delta\Omega|_{\delta=d} = 0.$$

For real motion the variational equation (4.54) becomes the energy equation,

$$d\mathcal{K} = d\mathcal{A}^{(e)}.$$

Variational equation (4.54) can be considered as the first law of thermodynamics for the virtual displacements in the case of a system with a finite number of degrees of freedom. For the virtual displacements, the energy equation has an additional contribution $\delta\Omega$ which is not present in the equation for the real displacements.

Now we can explain the meaning of the term "ideal constraints" used in the formulation of the principle of virtual displacements. As is apparent in (4.54), the variational equation does not include the energy contribution from dissipation. In this case the constraints are called ideal. For non-ideal constraints the variational equation would contain the additional dissipation term.

A variational version of the first law of thermodynamics. Let us move on to the consideration of the variational equation in the context of continuum mechanics. In a real process, the equation of the first law of thermodynamics is

$$dE = d\mathcal{A}^{(e)} + d\mathcal{Q} \tag{4.55}$$

where E is the energy of the system, $d\mathcal{A}^{(e)}$ is the work done by the external macroscopic forces, and $d\mathcal{Q}$ is the energy supply caused by heat and, possibly, by other forms of energy.

If the increment d of the real process in (4.55) is replaced by an increment δ of the arbitrary admissible process,[3] (4.55) will, generally speaking, not hold. Denoting the arising "discrepancy" by $\delta\Omega$, we can write

$$\delta E = \delta\mathcal{A}^{(e)} + \delta\mathcal{Q} + \delta\Omega. \tag{4.56}$$

Functional $\delta\Omega$ is equal to zero for real variations,

$$\delta\Omega|_{\delta=d} = 0. \tag{4.57}$$

If the functionals E, $\delta\mathcal{A}^{(e)}$, $\delta\mathcal{Q}$ and $\delta\Omega$ are defined, then (4.56) becomes the variational equation equivalent to the first law of thermodynamics for the admissible virtual variations. Functional δE is by definition the variation of functional E. Functional $\delta\mathcal{A}^{(e)}$ is also easy to define: usually, $d\mathcal{A}^{(e)}$ is the linear functional of increments $du^{\varkappa}$, and therefore $\delta\mathcal{A}^{(e)}$ means the value of functional $d\mathcal{A}^{(e)}$ on $\delta u^{\varkappa}$.

[3] Here and in what follows we will assume that the increments of the parameters in real processes belong to the set of all admissible variations.

Let dQ be the heat supply. In order to define functional δQ, we need to consider the definition of δQ based on the second law of thermodynamics:

$$dQ = \int_V \rho T dS dV - dQ', \tag{4.58}$$

where dQ' is the so-called uncompensated heat. In classical models dQ' is the linear functional of increments $du^{\varkappa}$. Therefore, the heat supply as defined by (4.58) is also a linear functional of $du^{\varkappa}$. Consequently, the heat supply for any admissible process can be defined as the value of this functional at $\delta u^{\varkappa}$, and

$$\delta Q = \int_V \rho T \delta S dV - \delta Q'. \tag{4.59}$$

Variational equation of the first law of thermodynamics becomes

$$\delta E = \delta \mathcal{A}^{(e)} + \int_V \rho T \delta S dV - \delta Q' + \delta \Omega. \tag{4.60}$$

Note that in the variational statement of the first law of thermodynamics (4.60) the second law of thermodynamics in the form (4.59) was used.

Sedov's variational equation. In the variational equation (4.60) not all functionals are independent: only some terms can be given, while others are determined from (4.60). To distinguish the two kinds of terms, let us first set for definiteness that

$$E = \int_V \rho \left(\frac{1}{2} v^2 + U \right) dV,$$

$$\delta \Omega = \int_V \rho \left(v_i \frac{d\delta x^i}{dt} - \frac{dv_i}{dt} \delta x^i \right) dV,$$

and express the work done by external forces by the sum of surface and body forces:

$$\delta \mathcal{A}^{(e)} = \delta \mathcal{A}^{(e)}_{body} + \delta \mathcal{A}^{(e)}_{surf}.$$

Then the variational equation can be written as

$$\delta \int_V \rho \left(\frac{1}{2} v_i v^i - U \right) dV - \frac{d}{dt} \int_V \rho v_i \delta x^i dV + \delta \mathcal{A}^{(e)}_{body} + \delta \mathcal{A}^{(e)}_{surf} +$$
$$+ \int_V \rho T \delta S dV - \delta Q' = 0.$$

Let us integrate this equation over an arbitrary interval $[t_0, t_1]$. Using the notation

$$L = \rho\left(\frac{1}{2}v_i v^i - U\right),$$

$$\delta W = -\left[\int_V \rho v_i \delta x^i dV\right]_{t_0}^{t_1} + \int_{t_0}^{t_1} \delta \mathcal{A}^{(e)}_{surf} dt,$$

$$\delta W^* = \int_{t_0}^{t_1}\left(\int_V \rho T \delta S dV - \delta Q' + \delta \mathcal{A}^{(e)}_{body}\right) dt,$$

we get

$$\delta \int_{t_0}^{t_1} \int_V L dV dt + \delta W^* + \delta W = 0. \tag{4.61}$$

Variational equation (4.61) holds for an arbitrary volume V and an arbitrary time interval $[t_0, t_1]$, and, therefore, is equivalent to the variational energy equation in the "local" (for any time instant) form.

The functional δW is an integral over the boundary of a four-dimensional region $V \times [t_0, t_1]$ of the linear combination of variations of the parameters involved.

The function L and the functional δW^* are prescribed, while the functional δW can be found from the variational equation (4.61).

The relation between functional δW^* and the uncompensated heat $\delta Q'$ and the postulates of thermodynamics of irreversible processes can be used to prescribe δW^*.

It is clear that the variational equation (4.61) can be also interpreted as the second law of thermodynamics for the virtual processes. Then, U should be considered as an independent thermodynamic variable and S a known function of U and the thermodynamical parameters. The energy equation completes the closed system of equations.

The variational equation (4.61) has two distinctive features. First, it is written not for the entire region occupied by the continuum, but for any arbitrary part of this region; this makes the variational equation very close to the energy equation. Arbitrariness of the region leads to the appearance of the functional δW in the variational equation. This functional describes the interactions of the piece of the material with its surroundings and is also determined by this equation. The calculation of δW corresponds to establishing the equations of state. Second, the variational equation contains the contributions of irreversible processes.

In essence, the variational equation (4.61) expresses in a compact form the laws of thermodynamics if these laws are taken into account in construction of the functional δW^*.

4.5 Models with High Derivatives

In classical continuum models, energy density depends on distortion, the first space derivatives of $x(X,t)$. If energy density depends on higher derivatives, the continuum acquires some new interesting features which we discuss in this section. These features are common in statics and dynamics, therefore we focus here on the static case. We consider a general setting when the state of continuum is described by some field variables, $u^{\varkappa}(x)$, $\varkappa = 1,\dots,m$. We accept for simplicity that the process is adiabatic, and entropy drops out from the set of required functions. The internal energy is assumed to be a function of $u^{\varkappa}$, $u^{\varkappa}_{,i}$ and $u^{\varkappa}_{,ij}$:

$$U = U\left(u^{\varkappa}, u^{\varkappa}_{,i}, u^{\varkappa}_{,ij}\right).$$

By U in this section we mean the energy per unit volume, therefore the total energy is

$$E = \int_V U\left(u^{\varkappa}, u^{\varkappa}_{,i}, u^{\varkappa}_{,ij}\right) dV. \tag{4.62}$$

Let is find the variation of energy. We have

$$\delta E = \int_V \left[\frac{\partial U}{\partial u^{\varkappa}}\delta u^{\varkappa} + \frac{\partial U}{\partial u^{\varkappa}_{,i}}\frac{\partial \delta u^{\varkappa}}{\partial x^i} + \frac{\partial U}{\partial u^{\varkappa}_{,ij}}\frac{\partial^2 \delta u^{\varkappa}}{\partial x^i \partial x^j}\right] dV. \tag{4.63}$$

To extract the independent variations in the integrand we do, as before, integration by parts. Integrating by parts the last term in (4.63) we have

$$\delta E = \int_V \left[\frac{\partial U}{\partial u^{\varkappa}}\delta u^{\varkappa} + \frac{\delta U}{\delta u^{\varkappa}_{,i}}\frac{\partial \delta u^{\varkappa}}{\partial x^i}\right] dV + \int_{\partial V} \frac{\partial U}{\partial u^{\varkappa}_{,ij}}\frac{\partial \delta u^{\varkappa}}{\partial x^i} n_j dA. \tag{4.64}$$

Here $\delta U/\delta u^{\varkappa}_{,i}$ is the variational derivative:

$$\frac{\delta U}{\delta u^{\varkappa}_{,i}} \equiv \frac{\partial U}{\partial u^{\varkappa}_{,i}} - \frac{\partial}{\partial x^j}\frac{\partial U}{\partial u^{\varkappa}_{,ij}}.$$

Integrating by parts the last term in the volume integral we obtain

$$\delta E = \int_V \frac{\delta U}{\delta u^{\varkappa}}\delta u^{\varkappa} dV + \int_{\partial V} \left(\frac{\delta U}{\delta u^{\varkappa}_{,i}}\delta u^{\varkappa} n_i + \frac{\partial U}{\partial u^{\varkappa}_{,ij}}\frac{\partial \delta u^{\varkappa}}{\partial x^i} n_j\right) dA. \tag{4.65}$$

Here

$$\frac{\delta U}{\delta u^{\varkappa}}=\frac{\partial U}{\partial u^{\varkappa}}-\frac{\partial}{\partial x^{i}}\frac{\delta U}{\delta u_{,i}^{\varkappa}}=\frac{\partial U}{\partial u^{\varkappa}}-\frac{\partial}{\partial x^{i}}\frac{\partial U}{\partial u_{,i}^{\varkappa}}+\frac{\partial^{2}}{\partial x^{i}\partial x^{j}}\frac{\partial U}{\partial u_{,ij}^{\varkappa}}.$$

Now we see the most distinctive feature of the models with high derivatives: in addition to the usual work of surface forces on variations $\delta u^{\varkappa}$, some new surface forces appear, which work on the gradient of variations. The variations in the surface integral are still dependent, and we need to make an additional transformation to define the surface forces uniquely. To put the surface integral to a suitable form we split the gradient, $\partial/\partial x^i$ into the sum of normal derivative and tangent derivatives. Let $x^i = r^i(\xi^\alpha)$ be the parametric equations of the surface, ∂V, ξ^α being the parameters on the surface; Greek indices run through the values 1, 2. Then $r_\alpha^i \equiv \partial r^i/\partial \xi^\alpha$ are the components of the two tangent vectors. We introduce two other vectors, r_i^α, by the relation

$$r_\alpha^i r_j^\alpha = \delta_j^i - n^i n_j.$$

The explicit formulas for r_i^α are given further (see (14.5) and (14.15)). Then the following decomposition of the gradient holds true:

$$\frac{\partial}{\partial x^i} = r_i^\alpha \frac{\partial}{\partial \xi^\alpha} + n_i \frac{\partial}{\partial n}, \qquad \frac{\partial}{\partial n} \equiv n_i \frac{\partial}{\partial x^i}. \tag{4.66}$$

Its derivation is given in Sect. 14.1 (see (14.17)).

Note that for a smooth surface and smooth two-dimensional vector, v^α, the divergence theorem has the form

$$\int_{\Sigma} v_{;\alpha}^\alpha dA = \int_{\partial\Sigma} v^\alpha \nu_\alpha ds \tag{4.67}$$

where a semi-colon in indices denotes covariant surface derivative (its definition is given in Sect. 14.1), ν_α is the normal unit vector to the curve, $\partial\Sigma$, which is tangent to Σ, and s is the arc length along $\partial\Sigma$.

If the surface Σ is a closed smooth surface, and v^α a smooth vector field on Σ, then , as follows from (4.67),

$$\int_{\Sigma} v_{;\alpha}^\alpha dA = 0. \tag{4.68}$$

Using (4.66) and (4.68) we can rewrite (4.65) as

$$\delta E = \int_V \frac{\delta U}{\delta u^{\varkappa}} \delta u^{\varkappa} dV + \int_{\partial V} \left(P_{\varkappa} \delta u^{\varkappa} + Q_{\varkappa} \frac{\partial \delta u^{\varkappa}}{\partial n} \right) dA \tag{4.69}$$

where

$$P_{\varkappa} = \frac{\delta U}{\delta u^{\varkappa}_{,i}} n_i - \left(\frac{\partial U}{\partial u^{\varkappa}_{,ij}} r_i^{\alpha} n_j \right)_{;\alpha}, \qquad Q_{\varkappa} = \frac{\partial U}{\partial u^{\varkappa}_{,ij}} n_i n_j. \tag{4.70}$$

Obviously, $\delta u^{\varkappa}$ and $\partial \delta u^{\varkappa} / \partial n$ are independent of ∂V.

For adiabatic processes variation of energy must be balanced with the work of external forces, δA :

$$\delta E = \delta A. \tag{4.71}$$

Equation (4.69) suggests that the work of external forces should have the form

$$\delta A = \int_V F_{\varkappa} \delta u^{\varkappa} dV + \int_{\partial V} \left(f_{\varkappa} \delta u^{\varkappa} + g_{\varkappa} \frac{\partial \delta u^{\varkappa}}{\partial n} \right) dA. \tag{4.72}$$

The work contains additional "higher order" surface forces, $g_{\varkappa}$. If the external "body forces," $F_{\varkappa}$, and "surface forces," $f_{\varkappa}$ and $g_{\varkappa}$, are given we obtain from (4.71) a closed system of equations and boundary conditions for $u^{\varkappa}$:

$$\frac{\delta U}{\delta u^{\varkappa}} = F_{\varkappa} \quad \text{in } V, \tag{4.73}$$

$$\frac{\delta U}{\delta u^{\varkappa}_{,i}} n_i - \left(\frac{\partial U}{\partial u^{\varkappa}_{,ij}} r_i^{\alpha} n_j \right)_{;\alpha} = f_{\varkappa}, \qquad \frac{\partial U}{\partial u^{\varkappa}_{,ij}} n_i n_j = g_{\varkappa} \qquad \text{on } \partial V$$

There are three major differences from the equations of classical continuum models. First, the equations are of higher order, and, therefore, require more boundary conditions. Second, in classical continuum models the surface forces work only on infinitesimally small displacements. In models with the second derivatives an additional surface force appears which works on normal derivatives of infinitesimally small "displacements," $\partial \delta u^{\varkappa} / \partial n$. Third, as we see from (4.73), the "usual surface force," $f_{\varkappa}$, depends on the surface derivatives of the normal and tangent vectors, and, thus, on the geometry of the surface. In classical continuum models, the surface force is linear with respect to the normal vector of the surface (see (3.100)). In models with high derivatives it is not linear and depends on curvatures of the surface. An example of models with high derivatives, theory of elastic plates and shells, will be considered in Chap. 14.

4.6 Tensor Variations

In derivation of the governing equations from the least action principle we did not use the tensorial nature of the characteristics involved. Remarkably, if the Lagrangian is a scalar and its arguments are tensors, the resulting governing equations automatically have the tensor form. This section explains that point, and, additionally discusses an alternative way of deriving the governing equations based on the notion of tensor variations. We begin with a reminder of the basic facts of tensor analysis.

Basic vectors. Consider in a Euclidian three-dimensional space a coordinate system with coordinates, x^i. This system is, in general, curvilinear. Let $\mathbf{r}(x)$ be the position vector of the point x. Then the basic vectors of the coordinates system, $\mathbf{e}_i$, are defined as

$$\mathbf{e}_i = \frac{\partial \mathbf{r}}{\partial x^i}. \tag{4.74}$$

The scalar products of the basic vectors are the covariant components of the metric tensor,

$$g_{ij} = \mathbf{e}_i \cdot \mathbf{e}_j. \tag{4.75}$$

while the components of the matrix inverse to $\|g_{ij}\|$ are the contravariant components of the metric tensor, g^{ij} . The basic vectors with upper indices are introduced by the formula

$$\mathbf{e}^i = g^{ij}\mathbf{e}_j. \tag{4.76}$$

An alternate definition of $\mathbf{e}^i$ follows from (4.76) and (4.75):

$$\mathbf{e}_i \cdot \mathbf{e}^j = \delta_i^j. \tag{4.77}$$

If the point x is shifted to $x + dx$, the basic vectors get the increments $d\mathbf{e}_i$. The increments are proportional to the shift, dx. Expanding the increments over the basic vectors, one can write

$$d\mathbf{e}_i = \Gamma_{ij}^k dx^j \mathbf{e}_k. \tag{4.78}$$

The coefficients Γ_{ij}^k are called Christoffel's symbols of the coordinate frame. It follows from (4.74) and (4.78) that Christoffel's symbols are symmetric over low indices:

$$\Gamma_{ij}^k = \Gamma_{ji}^k.$$

The expression for Christoffel's symbols in terms of metric tensor can be found by differentiating (4.75) and solving the resulting system of equations with respect to Christoffel's symbols. One gets:

$$\Gamma_{ij}^k = \frac{1}{2} g^{km} \left(\frac{\partial g_{mi}}{\partial x^j} + \frac{\partial g_{mj}}{\partial x^i} - \frac{\partial g_{ij}}{\partial x^m} \right). \tag{4.79}$$

Contracting (4.79) over j, k and using the equality

$$g^{ij} = \frac{1}{g}\frac{\partial g}{\partial g_{ij}}, \quad g = \det \left\| g_{ij} \right\|,$$

which follows from (3.20), we obtain an important identity:

$$\Gamma^j_{ij} = \frac{1}{2g}\frac{\partial g}{\partial x^i} = \frac{1}{\sqrt{g}}\frac{\partial \sqrt{g}}{\partial x^i}. \tag{4.80}$$

Differentiating (4.77), one obtains for the differentials of $\mathbf{e}^i$:

$$d\mathbf{e}^i = -\Gamma^i_{kj} dx^j \mathbf{e}^k \tag{4.81}$$

Tensors. Consider a new coordinate system, $x'^i = x'^i \left(x^j\right)$. It is convenient to mark a new coordinate system by putting the prime sign not at the root letter, but at the index: $x^{i'} = x^{i'} \left(x^j\right)$. The basic vectors of new coordinate system, $\mathbf{e}_{i'}$, differ from $\mathbf{e}_i$. From (4.74),

$$\mathbf{e}_{i'} = \frac{\partial \mathbf{r}}{\partial x^{i'}} = \frac{\partial \mathbf{r}}{\partial x^j}\frac{\partial x^j}{\partial x^{i'}}. \tag{4.82}$$

Accordingly,

$$g_{i'j'} = \mathbf{e}_{i'} \cdot \mathbf{e}_{j'} = g_{ij}\frac{\partial x^i}{\partial x^{i'}}\frac{\partial x^j}{\partial x^{j'}}.$$

One can check by direct inspection that, for the inverse matrix,

$$g^{i'j'} = g^{ij}\frac{\partial x^{i'}}{\partial x^i}\frac{\partial x^{j'}}{\partial x^j}$$

and

$$\mathbf{e}^{i'} = g^{i'j'}\mathbf{e}_{j'} = \frac{\partial x^{i'}}{\partial x^i}\mathbf{e}^i.$$

The set of functions $T^{i_1 \ldots i_k}_{j_1 \ldots j_m}$ form the components of a tensors, if in any coordinate system, $x^{i'}$, the new components $T^{i'_1 \ldots i'_k}_{j'_1 \ldots j'_m}$ obey the equality

$$\mathbf{T} = T^{i'_1 \ldots i'_k}_{j'_1 \ldots j'_m}\mathbf{e}_{i'_1} \ldots \mathbf{e}_{i'_k}\mathbf{e}^{j'_1} \ldots \mathbf{e}^{j'_m} = T^{i_1 \ldots i_k}_{j_1 \ldots j_m}\mathbf{e}_{i_1} \ldots \mathbf{e}_{i_k}\mathbf{e}^{j_1} \ldots \mathbf{e}^{j_m}.$$

Covariant derivatives. Consider the tensor $\mathbf{T} = T^{i_1 \ldots i_k}_{j_1 \ldots j_m}\mathbf{e}_{i_1} \ldots \mathbf{e}_{i_k}\mathbf{e}^{j_1} \ldots \mathbf{e}^{j_m}$. By definition, the result of an infinitesimal parallel transport of $\mathbf{T}$ from point $x + dx$ to point x is the tensor $\tilde{\mathbf{T}} = \mathbf{T}(x + dx)$ in which the basic vectors $\mathbf{e}_i\,(x + dx)$ and

$\mathbf{e}^j(x+dx)$ are replaced by $\mathbf{e}_i(x)+\Gamma^k_{ij}dx^j\mathbf{e}_k(x)$, $\mathbf{e}^j(x)-\Gamma^j_{ik}dx^i\mathbf{e}^k(x)$, respectively and only the infinitesimally small terms of the first order are retained. Covariant derivatives $\nabla_i T^{i_1\dots i_k}_{j_1\dots j_m}$ are defined by the equation

$$\tilde{\mathbf{T}}-\mathbf{T}(x)=\nabla_i T^{i_1\dots i_k}_{j_1\dots j_m}\mathbf{e}_{i_1}\dots\mathbf{e}_{i_k}\mathbf{e}^{j_1}\dots\mathbf{e}^{j_m}dx^i.$$

Hence, for example, for vector $T^i\mathbf{e}_i$, we have

$$\nabla_j T^i=\frac{\partial T^i}{\partial x^j}+\Gamma^i_{jk}T^k.$$

Lagrangian covariant derivatives. Let $\mathbf{e}_a$, $\mathbf{e}^a$ and $\mathring{\mathbf{e}}_a$, $\mathring{\mathbf{e}}^a$ be the basic vectors of the Lagrangian coordinate system in the initial and current states:

$$\mathbf{e}_a=x^i_a\mathbf{e}_i,\quad \mathbf{e}^a=X^a_i\mathbf{e}^i,\quad \mathring{\mathbf{e}}_a=\mathring{x}^i_a\mathbf{e}_i,\quad \mathring{\mathbf{e}}^a=\mathring{X}^a_i\mathbf{e}^i,$$

and Γ^c_{ab} and $\mathring{\Gamma}^c_{ab}$ be the respective Christoffel's symbols:

$$\begin{aligned}d\mathbf{e}_a&=\Gamma^c_{ab}dX^b\mathbf{e}_c,\quad & d\mathbf{e}^a&=-\Gamma^a_{cb}dX^b\mathbf{e}^c,\\ d\mathring{\mathbf{e}}_a&=\mathring{\Gamma}^c_{ab}dX^b\mathring{\mathbf{e}}_c,\quad & d\mathring{\mathbf{e}}^a&=-\mathring{\Gamma}^a_{cb}dX^b\mathring{\mathbf{e}}^c.\end{aligned}$$

Christoffel's symbols Γ^c_{ab} and $\mathring{\Gamma}^c_{ab}$ can be expressed in terms of g_{ab} and $\mathring{g}_{ab}$ by means of formulas analogous to (4.79). Also note that Γ^c_{ab} can be written in terms of the particle trajectories as

$$\Gamma^c_{ab}=\Gamma^k_{ij}x^i_a x^j_b X^c_k+\frac{\partial^2 x^i}{\partial X^a\partial X^b}X^c_i. \tag{4.83}$$

It is seen from (4.83) that Γ^c_{ab} are not zeros even if Christoffel's symbols of observer's frame, Γ^k_{ij}, are zeros.

Consider quantities with the Lagrangian indices, T^b. Using T^b, we can construct two vectors:

$$\mathbf{T}=T^b\mathbf{e}_b\ \text{ and }\ \mathring{\mathbf{T}}=T^b\mathring{\mathbf{e}}_b.$$

Accordingly, we may introduce two operators of covariant differentiation, ∇_a and $\mathring{\nabla}_a$:

$$\frac{\partial\mathbf{T}}{\partial X^a}=\nabla_a T^b\mathbf{e}_b,\quad \frac{\partial\mathring{\mathbf{T}}}{\partial X^a}=\mathring{\nabla}_a T^b\mathring{\mathbf{e}}_b.$$

In the case of tensors of higher orders, the number of possibilities to define covariant differential growth drastically. We will use the covariant derivatives ∇_a and $\mathring{\nabla}_a$ corresponding to the differentiation of the tensor

$$T^{a_1\dots a_k}_{b_1\dots b_m}\mathbf{e}_{a_1}\dots\mathbf{e}_{a_k}\mathbf{e}^{b_1}\dots\mathbf{e}^{b_m}\ \text{and}\ T^{a_1\dots a_k}_{b_1\dots b_m}\mathring{\mathbf{e}}_{a_1}\dots\mathring{\mathbf{e}}_{a_k}\mathring{\mathbf{e}}^{b_1}\dots\mathring{\mathbf{e}}^{b_m}.$$

For the operator ∇_a, the parallel transport occurs by means of Christoffel's symbol Γ^c_{ab}, while for $\mathring{\nabla}_a$ by means of $\mathring{\Gamma}^c_{ab}$. In particular,

$$\nabla_a T^b = \frac{\partial T_b}{\partial X^a} + \Gamma^b_{ac}T^c, \quad \mathring{\nabla}_a T^b = \frac{\partial T_b}{\partial X^a} + \mathring{\Gamma}^b_{ac}T^c.$$

Covariant derivatives of metric tensors with respect to the corresponding parallel transport, as one can check by direct inspection, are zero:

$$\nabla_a g_{bc} = 0, \quad \mathring{\nabla}_a \mathring{g}_{bc} = 0, \quad \nabla_k g_{ij} = 0. \tag{4.84}$$

Differentiating the tensors with both Eulerian and Lagrangian indices, we will include the parallel transport over each index. For example,

$$\nabla_b x^i_a = \frac{\partial x^i_a}{\partial x^b} + \Gamma^i_{jk}x^j_a x^k_b - \Gamma^c_{ab}x^i_c.$$

Using (4.83), one can check by inspection that

$$\nabla_b x^i_a = 0.$$

Time derivatives. The tensor time derivative for the constant Lagrangian coordinates, $\tilde{d}/dt$, is defined as follows. Let $u^{\varkappa i_1\dots i_n}_{j_1\dots j_m}$ be the components of a tensor where $\varkappa$ stands for some set of the Lagrangian indices. Define $V^\varkappa$ as

$$V^\varkappa = u^{\varkappa i_1\dots i_n}_{j_1\dots j_m}\mathbf{e}_{i_1}\dots\mathbf{e}_{i_k}\mathbf{e}^{j_1}\dots\mathbf{e}^{j_m}.$$

Derivatives $dV^\varkappa/dt$ transform as a tensor with Lagrangian indices $\varkappa$. To calculate $dV^\varkappa/dt$ we note that

$$\frac{d\mathbf{e}_i}{dt} = \frac{\partial\mathbf{e}_i}{\partial x^k}v^k = \Gamma^l_{ik}\mathbf{e}_l v^k, \quad \frac{d\mathbf{e}^i}{dt} = -\Gamma^i_{lk}\mathbf{e}^l v^\varkappa.$$

Hence,

$$\frac{dV^\varkappa}{dt} =$$

$$= \left(\frac{du^{\varkappa i_1\dots i_n}_{j_1\dots j_m}}{dt} + u^{\varkappa l i_1\dots i_n}_{j_1\dots j_m}\Gamma^i_{lk}v^k + \dots - u^{\varkappa i_1\dots i_n}_{j_1\dots j_{m-1}l}\Gamma^l_{j_m k}v^k\right)\mathbf{e}_{i_1}\dots\mathbf{e}_{i_k}\mathbf{e}^{j_1}\dots\mathbf{e}^{j_m} =$$

$$= \frac{\tilde{d}u^{\varkappa i_1\dots i_n}_{j_1\dots j_m}}{dt}\mathbf{e}_{i_1}\dots\mathbf{e}_{i_k}\mathbf{e}^{j_1}\dots\mathbf{e}^{j_m}.$$

It follows from the last equality that $\tilde{d}u^{\varkappa i_1 \ldots i_n}_{j_1 \ldots j_m}/dt$ are the components of a tensor. For example, the tensor time derivative of the distortion is

$$\frac{\tilde{d}x^i_a}{dt} = \frac{dx^i_a}{dt} + \Gamma^i_{lk} x^l_a v^k = x^k_a \frac{\partial v^i}{\partial x^k} + \Gamma^i_{kl} x^k_a v^l = x^k_a \nabla_k v^i.$$

The usual rules for differentiation of sums and products hold for $\tilde{d}/dt$. In differentiation of the contractions, like $a_i b^i$, the derivatives d/dt and $\tilde{d}/dt$ coincide:

$$\frac{d}{dt}\left(a_i b^i\right) = \frac{da_i}{dt} b^i + a^i \frac{db^i}{dt} = \frac{\tilde{d}a_i}{dt} b^i + a^i \frac{\tilde{d}b^i}{dt} = \frac{\tilde{d}}{dt}\left(a_i b^i\right). \tag{4.85}$$

Tensor Δ^c_{ab}. Christoffel's symbols Γ^c_{ab} and $\mathring{\Gamma}^c_{ab}$ are not tensors. The difference $\Delta^c_{ab} = \Gamma^c_{ab} - \mathring{\Gamma}^c_{ab}$ is a tensor. It follows from the formula

$$\Delta^c_{ab} = \frac{1}{2} g^{cd} \left(\mathring{\nabla}_b g_{ad} + \mathring{\nabla}_a g_{bd} - \mathring{\nabla}_d g_{ab}\right). \tag{4.86}$$

The covariant derivative ∇_a can be expressed in terms of the covariant derivative $\mathring{\nabla}_a$ and tensor Δ^c_{ab}. For example, for the covariant derivatives of a vector the corresponding relation is

$$\nabla_a T^b = \mathring{\nabla}_a T^b + \Delta^b_{ac} T^c.$$

The tensor Δ^c_{ab} can be expressed in terms of gradients of the strain tensor ε_{ab} : from (4.86) and (4.84)

$$\Delta^c_{ab} = g^{cd} \left(\mathring{\nabla}_b \varepsilon_{ad} + \mathring{\nabla}_a \varepsilon_{bd} - \mathring{\nabla}_d \varepsilon_{ab}\right). \tag{4.87}$$

Here tensor g^{cd} can be considered as a function of $\mathring{g}_{ab}$ and ε_{ab}.

Direct tensor notation. This is a widely used notation when, for example, for a vector, one writes $\mathbf{v}$, implying that $\mathbf{v} = v^i \mathbf{e}_i$. Unfortunately, the attractive simplicity of the direct notation is accompanied by some shortcomings. Dealing with the components of a vector we do not know the vector. This is emphasized by the formula $\mathbf{v} = v^i \mathbf{e}_i$: to prescribe a vector one needs to specify both the components, v^i, and the frame, $\mathbf{e}_i$. For the same components of, say, the strain tensor (4.2), one can define three different tensors,

$$\boldsymbol{\varepsilon}_1 = \varepsilon_{ab} \mathbf{e}^a \mathbf{e}^b, \qquad \boldsymbol{\varepsilon}_2 = \varepsilon_{ab} \mathring{\mathbf{e}}^a \mathring{\mathbf{e}}^b, \qquad \boldsymbol{\varepsilon}_3 = \varepsilon_{ab} \mathbf{e}^a \mathring{\mathbf{e}}^b.$$

We are interested in the dependence of energy on the components of the strain tensor, not on the entire tensor itself: it does not matter whether the strain tensor is the tensor $\boldsymbol{\varepsilon}_1$, $\boldsymbol{\varepsilon}_2$ or $\boldsymbol{\varepsilon}_3$. If we, nevertheless, write $U = U(\boldsymbol{\varepsilon}_1)$, we introduce into energy the extra arguments, the basic vectors, on which energy, in fact, does not depend. Mathematically, nothing is wrong: function may be independent on some of the arguments, but physically, this complication does not seem reasonable. Another shortcoming of the formula $U = U(\boldsymbol{\varepsilon}_1)$ is that one has to list all other arguments of energy, and the form of the arguments not mentioned by writing $U = U(\boldsymbol{\varepsilon}_1)$ may affect the actual value of energy. For example, in the case of an isotropic media the additional argument is just a tensor of the second order formed from the metric tensor. We have, however, a number of possibilities:

$$\mathbf{g}_1 = g^{ab}\mathbf{e}_a\mathbf{e}_b, \qquad \mathbf{g}_2 = \overset{\circ}{g}{}^{ab}\mathbf{e}_a\mathbf{e}_b, \qquad \mathbf{g}_3 = g^{ab}\overset{\circ}{\mathbf{e}}_a\overset{\circ}{\mathbf{e}}_b,$$

not to mention a few more. The models with energies, say, $U = U(\boldsymbol{\varepsilon}_1, \mathbf{g}_1)$ and $U = U(\boldsymbol{\varepsilon}_1, \mathbf{g}_2)$ are different. For example, in the case of the linear dependence of energy on the strains,

$$U(\boldsymbol{\varepsilon}_1, \mathbf{g}_1) = const \quad g^{ab}\varepsilon_{ab}, \qquad U(\boldsymbol{\varepsilon}_1, \mathbf{g}_2) = const \quad \overset{\circ}{g}{}^{ab}\varepsilon_{ab}.$$

These are two different functions. Without specifying the additional arguments in energy, the model remains undetermined. Of course, after all necessary specializations, the direct tensor notation makes sense; however, such specializations are needed only because we introduced the artificial argument into energy, the basic vectors. This is why the index notation is employed in the book: it avoids any ambiguities.

The tensor variations. The rules of transformation of required functions and their variations under coordinate transformations are, generally speaking, different. Consider, for example, the particle trajectories, $x^i\,(X^a, t)$. Functions $x^i\,(X^a, t)$ do not transform by the tensor rules. In a different coordinate frame, $x'^i = f^i\left(x^j\right)$, the particle trajectories are given by the function

$$x'^i\left(X^a, t\right) = f^i\left(x^j\left(X^a, t\right)\right).$$

However, the variations of particle trajectories form the vector components. Indeed,

$$\delta x'^i\left(X^a, t\right) = \delta f^i\left(x^j\left(X^a, t\right)\right) = \frac{\partial x'^i}{\partial x^j}\delta x^j.$$

The Christoffel's symbol of the Lagrangian coordinate system, Γ^c_{ab}, is not a tensor, while its variation, $\delta\Gamma^c_{ab}$, is a tensor of the third order: from (4.86) and (4.26)

$$\begin{aligned}\delta\Gamma^c_{ab} &= \delta\Delta^c_{ab} = -g^{ce}\Delta^d_{ab}\delta g_{ed} + \frac{1}{2}g^{cd}\left(\overset{\circ}{\nabla}_b\delta g_{ad} + \overset{\circ}{\nabla}_a\delta g_{bd} - \overset{\circ}{\nabla}_d\delta g_{ab}\right)\\ &= \frac{1}{2}g^{cd}\left(\nabla_b\delta g_{ad} + \nabla_a\delta g_{bd} - \nabla_d\delta g_{ab}\right)\end{aligned} \tag{4.88}$$

Equation (4.25) shows that the opposite is also true: the variations of the components of a tensor are not necessarily a tensor. In general, this does not cause any complications, because, as will be explained further, the final equations automatically possess a tensor nature. Nevertheless, it is worth knowing how to modify the definition of variations in order to deal with tensors at each step of the derivation of the governing equations.

Let $u^{\varkappa}$ be the components of a tensor, where the multi-index $\varkappa$ corresponds to a set of the Lagrangian indices. The variations $\delta u^{\varkappa}$ are the components of the tensor with the index structure similar to that of the tensor $u^{\varkappa}$ because the transformation matrix $\partial X'^{a}/\partial X^{b}$ can be taken outside the variation operator.

Consider a tensor with the components $u_j^{\varkappa i}$ where the multi-index $\varkappa$ still corresponds to a set of Lagrangian indices, and i, j are the observer's indices. The variation $\delta u_j^{\varkappa i}$ defined in the same way as before is no longer a tensor since the transformation matrix $\partial x'^{j}/\partial x^{i}$ depends on x^i and cannot be taken outside the variation operator.

Define $V^{\varkappa}$ as

$$V^{\varkappa} = u_j^{\varkappa i}\mathbf{e}_i\mathbf{e}^j. \tag{4.89}$$

$V^{\varkappa}$ have only the Lagrangian indices, so $\delta V^{\varkappa}$ transform in the same way as the components of the tensor $V^{\varkappa}$. Let us find the relation between $\delta V^{\varkappa}$ and $\delta u_j^{\varkappa i}$. Since

$$\delta\mathbf{e}_i = \frac{\partial\mathbf{e}_i}{\partial x^k}\delta x^k = \Gamma_{ik}^j\mathbf{e}_j\delta x^k, \quad \delta\mathbf{e}^j = -\Gamma_{ik}^j\mathbf{e}^i\delta x^k,$$

taking the variation of (4.89) we obtain

$$\delta V^{\varkappa} = \left(\delta u_j^{\varkappa i} + \Gamma_{lk}^i u_j^{\varkappa l}\delta x^k - \Gamma_{jk}^l u_l^{\varkappa i}\delta x^k\right)\mathbf{e}_i\mathbf{e}^j. \tag{4.90}$$

Formula (4.90) shows that quantities

$$\tilde{\delta} u_j^{\varkappa i} = \delta u_j^{\varkappa i} + \Gamma_{lk}^i u_j^{\varkappa l}\delta x^k - \Gamma_{jk}^l u_l^{\varkappa i}\delta x^k \tag{4.91}$$

are the components of a tensor with respect to both the transformations of the Lagrangian coordinate system and the Eulerian coordinate system. Equation (4.91) can be generalized for the tensors with an arbitrary structure of the Eulerian indices. The variation $\tilde{\delta}$ (4.91) is called the tensor variation.

The tensor variation $\tilde{\delta}$ can also be defined in terms of derivatives with respect to an auxiliary parameter ε like in (4.12). In order to do this, the trial functions $u_j^{\varkappa i}(X^a, t, \varepsilon)$ should be replaced by $\tilde{u}_j^{\varkappa i}(X^a, t, \varepsilon)$ where the tilde stands for a parallel transport of the tensor $u_j^{\varkappa i}$ with the Eulerian indices from the point $x^i(X^a, t, \varepsilon)$ to the point $x^i(X^a, t)$:

$$\tilde{\delta} u_j^{\varkappa i} = \left.\frac{\partial\tilde{u}_j^{\varkappa i}}{\partial\varepsilon}\right|_{\varepsilon=0} d\varepsilon. \tag{4.92}$$

It is easy to check that definitions (4.91) and (4.92) are equivalent.

The expressions for the tensor variations of velocity, distortion and inverse distortion are

$$\tilde{\delta} v^i = \frac{\partial \delta x^i}{\partial t} + v^k \nabla_k \delta x^i = \frac{\tilde{d} \delta x^i}{dt},$$

$$\tilde{\delta} x_a^i = \delta x_a^i + \Gamma_{lk}^i x_a^l \delta x^k = x_a^k \nabla_k \delta x^i,$$

$$\tilde{\delta} X_i^a = \delta X_i^a - \Gamma_{ik}^l X_l^a \delta x^k = -X_j^a \nabla_i \delta x^j. \tag{4.93}$$

Equation (4.93) differ from the (4.17) and (4.25) by having the covariant instead of the partial derivative with respect to x^i. The difference between (4.19) and the equation for the tensor variation of Δ,

$$\tilde{\delta} \Delta = \frac{\partial \Delta}{\partial x_a^i} \tilde{\delta} x_a^i = \Delta X_i^a \tilde{\delta} x_a^i = \Delta \nabla_i \delta x^i, \tag{4.94}$$

is analogous.

Note the relations

$$\delta g_{ab} = \nabla_a \delta x_b + \nabla_b \delta x_a, \quad \delta \varepsilon_{ab} = \nabla_{(a} \delta x_{b)} \tag{4.95}$$

where δx_a are the Lagrangian coordinates of vectors δx^i, $\delta x_a = x_a^i \delta x_i$.

Substituting this in (4.88) we obtain the variations of Christoffel's symbols:

$$\delta \Gamma_{ab}^c = \nabla_a \nabla_b \delta x^c. \tag{4.96}$$

When taking the variations of the covariant derivative of the tensor $u^\varkappa$, the operator δ is apparently interchangeable with the operator $\mathring{\nabla}_a$:

$$\delta \mathring{\nabla}_a u^\varkappa = \mathring{\nabla}_a \delta u^\varkappa.$$

However,

$$\delta \nabla_a u^\varkappa \neq \nabla_a \delta u^\varkappa,$$

since $\delta \Gamma_{ab}^c \neq 0$. The commutator $\delta \nabla_a - \nabla_a \delta$ of the operators δ and ∇_a can easily be found using (4.96).

Consider a tensor with components $u^\varkappa (x^i, t)$ where the multi-index $\varkappa$ corresponds to the set of the Eulerian indices. The variation $\partial u^\varkappa$ with x^i held constant is the tensor with the same index structure as $u^\varkappa$ since $\partial x'^i / \partial x^i$ can be taken outside the operator ∂. For the tensor variation δ with X^a held constant, from the definitions of the operators $\tilde{\delta}$ and ∂ we have

$$\tilde{\delta} u^{\varkappa} = \partial u^{\varkappa} + \delta x^i \nabla_i u^{\varkappa}. \tag{4.97}$$

In particular, equation. (4.97) implies that the tensor variation of the metric tensor $\tilde{\delta} g_{ij}$ is equal to zero, since $\partial g_{ij} = 0$ and $\nabla_k g_{ij} = 0$. Recall that the variation

$$\delta g_{ij} = \frac{\partial g_{ij}}{\partial x^k} \delta x^k$$

in a curvilinear system of coordinates is not zero.

The variation ∂ commutes with the covariant derivative ∇_i,

$$\partial \nabla_i u^{\varkappa} = \nabla_i \partial u^{\varkappa}, \tag{4.98}$$

since $\partial \Gamma^k_{ij} = 0$.

The variation $\tilde{\delta}$ does not commute with ∇_i:

$$\begin{aligned} \tilde{\delta} \nabla_i u^{\varkappa} &= \partial \nabla_i u^{\varkappa} + \delta x^k \nabla_k \nabla_i u^{\varkappa} = \nabla_i \partial u^{\varkappa} + \delta x^k \nabla_k \nabla_i u^{\varkappa} = \\ &= \nabla_i \left(\tilde{\delta} u^{\varkappa} - \delta x^k \nabla_k u^{\varkappa} \right) + \delta x^k \nabla_k \nabla_i u^{\varkappa} = \\ &= \nabla_i \tilde{\delta} u^{\varkappa} - \nabla_k u^{\varkappa} \nabla_i \delta x^k + \delta x^k \left(\nabla_k \nabla_i u^{\varkappa} - \nabla_i \nabla_k u^{\varkappa} \right). \end{aligned} \tag{4.99}$$

In a space where the curvature tensor is equal to zero, $\nabla_k \nabla_i u^{\varkappa} = \nabla_i \nabla_k u^{\varkappa}$, and (4.99) becomes

$$\tilde{\delta} \nabla_i u^{\varkappa} = \nabla_i \tilde{\delta} u^{\varkappa} - \nabla_k u^{\varkappa} \nabla_i \delta x^k. \tag{4.100}$$

The variation of a scalar function. Let the scalar L be a function of a set of tensors with components $u^{\varkappa}$, $L = L(u^{\varkappa})$, where $\varkappa$ corresponds to a set of the observer's indices and the Lagrangian indices. The following equality holds:

$$\delta L = \tilde{\delta} L \tag{4.101}$$

or, more explicitly,

$$\frac{\partial L}{\partial u^{\varkappa}} \delta u^{\varkappa} = \frac{\partial L}{\partial u^{\varkappa}} \tilde{\delta} u^{\varkappa}. \tag{4.102}$$

Indeed, since L does not change in a parallel transport of $u^{\varkappa}$ from the point $x^i(X^a, t, \varepsilon)$ to the point $x^i(X^a, t)$, we can write

$$L\left(u^{\varkappa}\left(x\left(X, t, \varepsilon\right), \varepsilon\right)\right) = L\left(\tilde{u}^{\varkappa}\left(x\left(X, t, \varepsilon\right), \varepsilon\right)\right), \tag{4.103}$$

where the tilde above $u^{\varkappa}$ stands for the parallel transport with respect to the Eulerian indices. Differentiating (4.103) with respect to ε and setting $\varepsilon = 0$, we get (4.102).

Here is an example illustrating (4.102). Let the internal energy density, U, be a function of the distortion component x_a^i. Since one cannot form a scalar only from x_a^i, U must depend also on other tensors with the Eulerian and Lagrangian indices. They characterize the properties of the material and the observer's frame. Assume that the properties of the material are characterized by some tensor with the Lagrangian indices, $K^\varkappa$, and $K^\varkappa$ are given function of the Lagrangian coordinates. Then, $\delta K^\varkappa = \tilde{\delta} K^\varkappa = 0$. Also assume that the metric tensor g_{ij} is the only tensor with the Eulerian indices in U. Then

$$\delta U = \frac{\partial U}{\partial g_{ij}} \delta g_{ij} + \frac{\partial U}{\partial x_a^i} \delta x_a^i. \tag{4.104}$$

where $\delta g_{ij} = g_{ij,k} \delta x^k$. On the other hand, since $\tilde{\delta} g_{ij} = \delta x^k \nabla_k g_{ij} = 0$,

$$\delta U = \frac{\partial U}{\partial x_a^i} \tilde{\delta} x_a^i. \tag{4.105}$$

According to (4.102), variations (4.104) and (4.105) must coincide. Let us derive the equality (4.105) directly from (4.104). The derivation is based on the identity

$$\frac{\partial U}{\partial x_a^i} x_a^k = 2 \frac{\partial U}{\partial g_{mk}} g_{mi}, \tag{4.106}$$

which follows from the scalar nature of U. To obtain it we consider the transformation of the Eulerian coordinates $x^i \to x'^i$, $x^k = b_i^k x'^i$, $x'^i = b_k^{(-1)i} x^k$. Function U, as a scalar, does not change:

$$U\left(g'_{ij}, x'^i_a, K^\varkappa\right) = U\left(g_{ij}, x_a^i, K^\varkappa\right), \tag{4.107}$$

where $g'_{ij} = g_{mn} b_i^m b_j^n$, $x'^i_a = b_j^{(-1)i} x_a^j$. Differentiating (4.107) with respect to b_j^i and setting $b_j^i = \delta_j^i$, we arrive at (4.106).

As follows from (4.106),

$$\frac{\partial U}{\partial g_{mk}} = \frac{1}{2} \frac{\partial U}{\partial x_a^i} x_a^k g^{im}. \tag{4.108}$$

Note an important consequence of this relation: its left hand side is symmetric while the right hand side can be written in terms of the stress tensor:

$$\frac{\partial U}{\partial g_{mk}} = \frac{1}{2\rho} \sigma^{mk}$$

Therefore, the stress tensor in such models is necessarily symmetric

$$\sigma^{mk} = \sigma^{km}.$$

Consider (4.104). We have

$$\delta U = \frac{\partial U}{\partial g_{mk}} g_{mk,s} \delta x^s + \frac{\partial U}{\partial x_a^i} \delta x_a^i.$$

We may replace $g_{mk,s}$ by $g_{mk,s} + g_{sm,k} - g_{sk,m}$ because the addition, $g_{sm,k} - g_{sk,m}$, is anti-symmetric over mk and, thus, its contraction with a symmetric tensor, $\partial U / \partial g_{mk}$, is zero. Substituting $\partial U / \partial g_{mk}$ by the right hand side of (4.108),

$$\delta U = \frac{1}{2} \frac{\partial U}{\partial x_a^i} x_a^k g^{im} (g_{mk,s} + g_{sm,k} - g_{sk,m}) \delta x^s + \frac{\partial U}{\partial x_a^i} \delta x_a^i,$$

we recognize in the first term Christoffel's symbols (see (4.79)). Thus,

$$\delta U = \frac{\partial U}{\partial x_a^i} x_a^k \Gamma_{ks}^i \delta x^s + \frac{\partial U}{\partial x_a^i} \delta x_a^i = \frac{\partial U}{\partial x_a^i} \tilde{\delta} x_a^i,$$

as claimed.

So, depending on convenience, one can use either the usual variation operator δ or the tensor variation operator $\tilde{\delta}$.

Chapter 5
Direct Methods of Calculus of Variations

The variational principles allow one to investigate the properties of the minimizing/stationary elements without the use of differential equations. Such methods of studying the solutions are called direct. The direct methods are especially effective in cases when the functional of the variational problem has only one stationary point, and this stationary point is either the maximum or the minimum of the functional.

There are many direct methods of constructing the approximate solutions, and there are also some direct methods of qualitative analysis, like the analysis of the existence and uniqueness of the solution or the derivation of a priori estimates. This chapter will cover some ideas that form the basis of the direct qualitative methods. As all other reasonings based on the notion of energy, they are simple and almost obvious. However, their application to particular problems often calls for inventiveness, certain skills, and some subtle mathematical techniques.

We begin with the consideration of the existence and uniqueness theory. From the perspective of an engineer or a physicist, it may seem an infringement in extraneous territory as this subject is a classical topic of mathematics. In fact, however, one can hardly understand in depth such issues of mechanics as loss of stability or the proper choice of boundary conditions without being exposed to the simple ideas underlying the existence and uniqueness theory. Such understanding is certainly necessary for everyone involved in construction of new models of continuum media.

The next topic is the deep and powerful theory of dual variational problems. It is based on the notions of convexity and Young-Fenchel transformation, which are discussed in detail.

Another important tool of the direct methods is the asymptotic analysis of the functionals depending on small parameters. We discuss the general scheme and consider a number of examples while most of the applications are spread over the rest of the book.

In this chapter we establish also a link between minimization problems and integration in functional spaces. This relationship is used in Part III for studying some stochastic variational problems. The chapter is concluded with a discussion of several useful tricks, which help to work with variational problems.

V.L. Berdichevsky, *Variational Principles of Continuum Mechanics*,
Interaction of Mechanics and Mathematics, DOI 10.1007/978-3-540-88467-5_5,

5.1 Introductory Remarks

Setting of variational problems. Consider a functional $I(u)$ defined on a set $\mathcal{M}$ of elements u. In continuum mechanics, u are some functions of space variables and, in dynamical problems, time. Since, the sum of two functions, u_1 and u_2, and the multiplication of a function, u, by a real number, are defined, one says that the u-space is a linear space; denote it by $\mathcal{R}$. The space $\mathcal{R}$ is also called a functional space as it is a set of functions. The set $\mathcal{M}$ in which the functional $I(u)$ is defined is a subset of the space $\mathcal{R}$. In continuum mechanics, the set $\mathcal{M}$ is usually specified by the characterization of smoothness of u and the choice of boundary conditions. There is an important special case when the set $\mathcal{M}$ is also linear, i.e. for any two elements of $\mathcal{M}$, u_1 and u_2, the sum $u_1 + u_2$ belongs to $\mathcal{M}$ and the product of any element u of $\mathcal{M}$ by a number λ is also an element of $\mathcal{M}$. An example of such a set $\mathcal{M}$ is a set of continuous functions $u(x)$, defined in a bounded region V, which have zero values on the boundary ∂V of region V:

$$u(x) = 0 \quad \text{at} \quad \partial V. \tag{5.1}$$

Obviously, if $u_1(x)$ and $u_2(x)$ obey (5.1), then $u_1(x) + u_2(x)$ and $\lambda u_1(x)$ do too. If the boundary conditions are inhomogeneous,

$$u(x) = u^{(b)}(x) \quad \text{at} \quad \partial V, \tag{5.2}$$

then the set has the following structure: any element u of $\mathcal{M}$ can be presented in a form

$$u = u_0 + u'$$

where u_0 is some function satisfying the boundary condition (5.2), and u' a function with zero boundary values, i.e. an element of a linear set $\mathcal{M}'$.

Another example of $\mathcal{M}$ is a cone: the set $\mathcal{M}$ is called a cone if, for any element u of $\mathcal{M}$, it also contains the element λu for all positive λ. Cones appear in variational problems with unilateral constraints such as

$$u(x) \geqslant 0.$$

We consider the variational problems of the following type: find the minimum value of the functional $I(u)$ on a given set $\mathcal{M}$. For such a problem we write

$$I(u) \to \min_{u \in \mathcal{M}} \quad \text{or} \quad \min_{u \in \mathcal{M}} I(u) \quad \text{or} \quad \min_{\mathcal{M}} I(u).$$

The minimum and maximum values of $I(u)$ on $\mathcal{M}$ are denoted by $\check{I}$ and $\hat{I}$, respectively:

$$\check{I} = \min_{u \in \mathcal{M}} I(u), \quad \hat{I} = \max_{u \in \mathcal{M}} I(u).$$

If the set $\mathcal{M}$ is selected by an equation like, for example, vanishing of admissible functions at the boundary ∂V of region V,

$$u = 0 \text{ on } \partial V, \tag{5.3}$$

we also use the notation

$$\check{I} = \min_{u \in (5.3)} I(u),$$

or, if the constraint is a short equation,

$$\check{I} = \min_{u \in u|_{\partial V}=0} I(u) .$$

A search for the maximum value of $I(u)$ is equivalent to a search for the minimum value of the functional $-I(u)$; therefore we focus mostly on the minimization problem. In minimization problems we may admit that the functional $I(u)$ takes at some elements of the set $\mathcal{M}$ the positive infinite value: $I = +\infty$ (if it would take also negative infinite values, the solution of the minimization problem is obvious: $\check{I} = -\infty$). That is equivalent to the elimination of such elements from the set $\mathcal{M}$. This way of operating with the inadmissible elements is quite convenient. For example, we may write the original variational problem as a minimization problem for a functional $\tilde{I}(u)$ defined on the entire linear space

$$\tilde{I}(u) = \begin{cases} I(u) & u \in \mathcal{M} \\ +\infty & u \bar{\in} \mathcal{M} \end{cases}$$

Then

$$\check{I} = \min_{u \in \mathcal{R}} \tilde{I}(u).$$

Usually we are interested to find not only the minimum value of the functional $I(u)$ but also the element, $\check{u}$, on which this value is achieved, i.e. such $\check{u}$ that

$$I(\check{u}) = \check{I}. \tag{5.4}$$

Such an element is called the minimizing element or the minimizer. There might be many minimizing elements (example: $\min\limits_{-\infty<u<+\infty} \sin u = -1,\ \check{u} = -\frac{\pi}{2} + 2\pi n$, $n = 0, \pm 1, \pm 2, \ldots$). On the other hand, the element $\check{u}$ may not exist (example: $\min\limits_{0<u<+\infty} \frac{1}{u} = 0$, while $1/u > 0$ at any point). Mathematicians like to emphasize in notation the possibility of non-existence of the minimizing elements. For example, they write $\inf I(u)$ if it is not known whether the existence of the minimizing element is guaranteed and $\min\limits_{\mathcal{M}} I(u)$ only in the case when the minimum value is achieved on

$\mathcal{M}$ (and, accordingly, $\sup I(u)$ instead of $\max I(u)$). We choose rather to use only the symbols min and max; in the usual mathematical notation they correspond to inf and sup.

In variational problems of continuum mechanics the existence of the minimizing element is usually guaranteed by the physics of the problem. Therefore, the variational problem will be called correctly posed (or, briefly, correct) if a minimizing element does exist. For the variational problems of non-physical nature, like the problems of optimal control theory, such terminology is not appropriate because in a typical situation, the minimizing element does not exist. We will not encounter such problems in this book.

On the geometry and dimension of functional spaces. Functional spaces are infinite-dimensional. This means the following. Consider a smooth function of one variable, $u(x)$. It can be approximated by a piece-wise linear function which has at some nodes, x_k, the same values as $u(x)$ (Fig. 5.1). The approximation is determined uniquely by the values of the function at the nodes, $u(x_k)$. To describe the function $u(x)$ completely one has to tend the number of nodes, N, to infinity. In the limit function $u(x)$ is said to have an infinite number of degrees of freedom.

Another way to introduce the "degrees of freedom" of a function is to expand it into a series. For example, a continuous function of one variable $u(x)$, defined on a segment $-a \le x \le a$, can be presented as the Fourier series

$$u(x) = \sum_{k=0}^{\infty} a_k \cos \frac{\pi k x}{a} + \sum_{k=1}^{\infty} b_k \sin \frac{\pi k x}{a}. \tag{5.5}$$

Hence, each function $u(x)$ is specified by an infinite sequence of numbers $\{a_0, a_1, a_2, \ldots, b_1, b_2, \ldots\}$. To select a function, $u(x)$, one has to prescribe these numbers. There is a one-to-one correspondence between all continuous functions

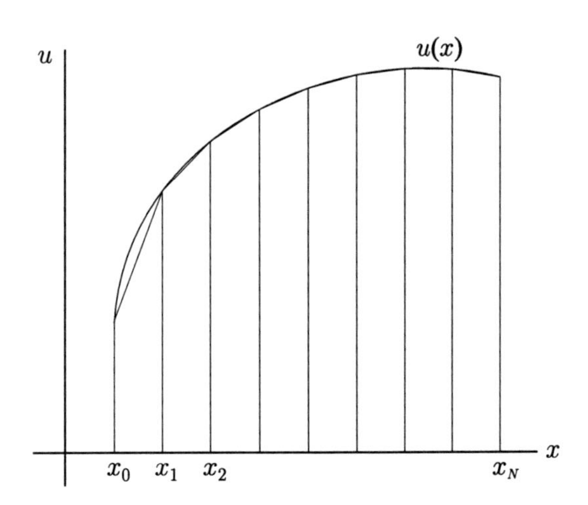

Fig. 5.1 A finite-dimensional truncation in the functional space

$u(x)$ and all sequences[1] $\{a_0, a_1, a_2, \ldots, b_1, b_2, \ldots\}$. In the spirit of vector analysis, one can consider $u(x)$ as a vector in an infinite-dimensional space, with the coordinates $\{a_0, a_1, a_2, \ldots, b_1, b_2, \ldots\}$. The vector $u(x)$ has these coordinates in the basis formed by the vectors (functions)

$$1, \cos\frac{\pi x}{a}, \cos\frac{2\pi x}{a}, \ldots, \sin\frac{\pi x}{a}, \ldots .$$

The same function, $u(x)$, may also be presented in terms of the Taylor series:

$$u(x) = \sum_{k=0}^{\infty} c_k x^k. \tag{5.6}$$

The coefficients $\{c_0, c_1, \ldots\}$ may be viewed as the coordinates of function $u(x)$ in the basis $\{1, x, x^2, \ldots\}$. The coordinates $\{a_0, a_1, a_2, \ldots, b_1, b_2, \ldots\}$ and $\{c_0, c_1, c_2, \ldots\}$ are linked by a linear transformation which can be derived from (5.5) and (5.6).

A functional, $I(u)$, is a function of an infinite number of arguments, $a_0, a_1, a_2, \ldots, b_1, b_2, \ldots$, or $c_0, c_1, c_2, \ldots$. In some cases this functional can be computed explicitly. For example,

$$\begin{aligned} I(u) \equiv \int_{-a}^{a} u^2(x)dx &= \frac{a}{\pi}\int_{-\pi}^{\pi}\left(\sum_0^{\infty} a_k \cos kt + \sum_1^{\infty} b_k \sin kt\right)^2 dt = \\ &= a\left[2a_0^2 + a_1^2 + a_2^2 + \ldots + b_1^2 + b_2^2 + \ldots\right]. \end{aligned} \tag{5.7}$$

Here we used the following property of the harmonic functions:

$$\int_{-\pi}^{\pi} \cos kt \cos mt\, dt = \int_{-\pi}^{\pi} \sin kt \sin mt\, dt = \pi\delta_{km}, \quad \int_{-\pi}^{\pi} \cos kt \sin mt\, dt = 0.$$

One important point is now in order. The infinite dimensionality of the functional space in static problems of continuum mechanics is, in fact, an unnecessary complication. In each application of models of continuum mechanics there is always some characteristic length, l_*, such that the model does not make sense for functions changing on distances less than l_*. For example, if a crystal lattice is modeled by a continuum elastic body, then one can disregard the change of displacements on distances less than the interatomic distance. In terms of the Fourier coefficients, ignoring the changes of the function $u(x)$ on distances less than l_* means dropping in the Fourier series (5.5) all terms with the wavelengths, a/k, smaller than l_*, i.e. keeping only a finite number of coordinates $a_0, a_1, \ldots, a_N, b_1, \ldots, b_N$ with

[1] Which decay fast enough; we do not go into mathematical details.

$N \simeq a/l_*$. Such a dimensional truncation is possible in all problems of continuum mechanics. However, it is also always interesting to see if the results obtained do not depend on the truncation and admit the limit transition $N \to \infty$. The results derived for infinite-dimensional functional spaces may be perceived from this perspective: these results do not depend on the value of the characteristic length l_* and may be viewed as the result corresponding to the limit $l_* \to 0$ (or $N \to \infty$). The idea that the space $\mathcal{R}$ can be considered as finite-dimensional is quite fruitful: then, without loss of any physical feature of the problem, the functional $I(u)$ becomes a function of a finite number of variables.

Linear and quadratic functionals. Two classes of functionals play a very special role in continuum mechanics: linear functionals and quadratic functionals.

Functional $l(u)$ is called linear if, for any u, v and a number λ,

$$l(\lambda u) = \lambda l(u), \quad l(u+v) = l(u) + l(v).$$

This is an example of a linear functional:

$$l(u) = \int_V a(x)u(x)\,dV.$$

Here $a(x)$ is a given smooth function, and $u(x)$ is any integrable function.

Functional $E(u, v)$ is called bilinear if it is linear with respect to each argument, u and v.

Functional $E(u)$ is called quadratic if there is a bilinear symmetric functional, $E(u, v) = E(v, u)$, such that

$$E(u) = E(u, u). \tag{5.8}$$

Boundedness from below. A functional $I(u)$ is called bounded from below on $\mathcal{M}$ if there exists a constant, c, such that for all elements u of $\mathcal{M}$,

$$I(u) \geq c.$$

The fact that functional $I(u)$ has a finite value at a minimizer, $\check{u}$, implies that functional $I(u)$ is bounded from below on $\mathcal{M}$ in a correct variational problem. There are two reasons why a variational problem may not be correct: either the functional is not bounded from below on $\mathcal{M}$, or it is bounded from below but its minimum is not reached at $\mathcal{M}$, i.e. there is no element $\check{u}$ for which $I(\check{u}) = \check{I}$. To determine whether the problem is correct, it is natural to start from the investigation of the boundedness from below of the functional on $\mathcal{M}$. Here are some examples.

Example 1. Linear functional, $l(u)$, is always unbounded on a linear set $\mathcal{M}$ unless it is equal to zero identically. Indeed, if $l(u) \neq 0$ for some u, then $l(\lambda u) = \lambda l(u)$ could be made as large or as small as we wish by the appropriate choice of λ.

Example 2. Let $\mathcal{M}$ be a set of differentiable functions of one variable, defined on the segment $[0, a]$. Consider the functional on $\mathcal{M}$,

$$E(u) = \frac{1}{2}\int_0^a A(x)\left(\frac{du}{dx}\right)^2 dx, \tag{5.9}$$

where $A(x)$ is a positive function. Functional $E(u)$ is obviously quadratic since it can be obtained from a bilinear symmetric functional,

$$E(u, v) = \frac{1}{2}\int_0^a A(x)\frac{du}{dx}\frac{dv}{dx}dx.$$

Functional $E(u)$ is non-negative, and thus, bounded from below by zero.

The functional (5.9) is defined, if $u(x)$ is a differentiable function such that the integral of $A(x)(du/dx)^2$ exists. In what follows we skip such details and use a vague term "smooth function", which assumes that all operations on this function, which are involved in our consideration, are meaningful.

Example 3. Consider the functional

$$I(u) = E(u) - l(u), \tag{5.10}$$

where $E(u)$ is the functional (5.9), and

$$l(u) = \int_0^a g(x)u(x)dx + f_a u(a) - f_0 u(0). \tag{5.11}$$

We assume that functions $A(x)$ and $g(x)$ are smooth, and function $A(x)$ is positive and separated from zero, i.e.

$$A(x) \geq A_0 = const > 0. \tag{5.12}$$

Functional $E(u)$ is bounded from below by zero. Functional $l(u)$ is linear, and, thus, unbounded. The sum $E(u) - l(u)$ can be bounded from below because the linear growth of $l(u)$ is accompanied by the quadratic growth of $E(u)$ and the latter can suppress the negative contribution of $l(u)$. Such "suppression" may not occur if there are functions $\bar{u}$ for which $l(\bar{u}) \neq 0$ while $E(\bar{u}) = 0$. Then $I(\lambda\bar{u}) = -\lambda I(\bar{u})$ and, choosing λ, one can tend $I(\lambda\bar{u})$ to $-\infty$. Functional $E(u)$ is equal to zero for functions $\bar{u} = const$. This suggests that the necessary condition for the boundedness from below of functional $I(u)$ is

$$l(\bar{u}) = 0 \tag{5.13}$$

for all $\bar{u} \equiv const$. The equality (5.13) is equivalent to the following constraint on the entries, $g, \ f_a, \ f_0$, of the functional $l(u)$:

$$\int_0^a g(x)dx + f_a - f_0 = 0. \tag{5.14}$$

To understand the meaning of the constraint (5.14), let us write down the Euler equations for the functional (5.10) assuming that minimizer, $u(x)$, is twice continuously differentiable on $[0, a]$:

$$\frac{d}{dx} A(x) \frac{du}{dx} + g(x) = 0, \quad 0 \leq x \leq a, \tag{5.15}$$

$$A \frac{du}{dx}\bigg|_{x=0} = f_0, \quad A \frac{du}{dx}\bigg|_{x=a} = f_a. \tag{5.16}$$

The boundary value problem (5.15) and (5.16) is not always solvable. To obtain a necessary condition of its solvability, let us integrate (5.15) on $[0, a]$. Using the boundary condition (5.16), we arrive at (5.14). So, the necessary condition of the boundedness from below of the functional $I(u)$ coincides with the necessary condition of the solvability for the boundary value problem (5.15) and (5.16).

The physical meaning of the constraint (5.14) is simple: if the variational problem is interpreted as the equilibrium problem for extension of an elastic beam, and thus, $g(x)$ is the body force while $f_a, \ f_0$ are the end forces, the condition (5.14) means that at equilibrium the sum of all forces acting on the body must be zero.

Let us show that the condition (5.14) is also the sufficient condition for boundedness from below, i.e. if the linear functional $l(u)$ satisfies (5.14), then $I(u)$ is bounded from below.

First, let us note that, if (5.14) holds, functional $I(u)$ is invariant with respect to shifts of u by a constant, $\bar{u}$, since $E(u + \bar{u}) = E(u)$ and $l(u + \bar{u}) = l(u)$. Therefore, the minimizing element of the functional $I(u)$ cannot be determined uniquely: if $\check{u}$ is a minimizing element then $\check{u} + \bar{u}$ is also a minimizing element. We may add a constraint eliminating the shifts for a constant. As such we take, for definiteness,

$$\int_0^a u(x)dx = 0. \tag{5.17}$$

Now we need the following inequality: for smooth functions $u(x)$ satisfying the constraint (5.17),

$$\lambda^2 u^2(x) \leq a \int_0^a \left(\frac{du}{dx}\right)^2 dx, \tag{5.18}$$

where λ is some constant. The factor, a, is included in the right hand side of (5.18) to make the constant λ dimensionless.

The proof is simple. Consider the identity

$$u(x) - u(y) = \int_y^x \frac{du(z)}{dz} dz.$$

Let us integrate it with respect to y from 0 to a. Using (5.17), we get

$$au(x) = \int_0^a dy \int_y^x \frac{du(z)}{dz} dz. \tag{5.19}$$

We square (5.19)

$$a^2 u^2(x) = \left[\int_0^a \left(1 \cdot \int_y^x \frac{du(z)}{dz} dz \right) dy \right]^2$$

and apply the Cauchy inequality: for any two functions, $f(x)$ and $g(x)$,

$$\left(\int f(x)g(x)dx \right)^2 \leq \int f^2(x)dx \int g^2(x)dx \tag{5.20}$$

For further references, note that inequality (5.20) holds true for integrals over multi-dimensional regions as well.[2] We get, using twice the Cauchy inequality,

$$a^2u^2(x) \leq \int_0^a 1^2 dy \int_0^a \left[\int_y^x 1 \cdot \frac{du}{dz} dz \right]^2 dy \leq a \int_0^a \left[\int_y^x 1^2 dz \int_y^x \left(\frac{du}{dz} \right)^2 dz \right] dy. \tag{5.21}$$

Substituting the integration limits, x and y, in (5.21) by 0 and a, respectively, and consequently increasing the value of the right-hand side, we obtain

$$a^2 u^2(x) \leq a^3 \int_0^a \left(\frac{du}{dz} \right)^2 dz. \tag{5.22}$$

[2] The Cauchy inequality follows from the positiveness for any α, β of the quadratic form,

$$\alpha^2 \int f^2 dx + 2\alpha\beta \int fg dx + \beta^2 \int g^2 dx = \int (\alpha f + \beta g)^2 dx \geq 0.$$

Clearly, the equality is achieved only when f and g are proportional.

Hence, the inequality (5.18) holds with the constant λ equal to at least unity (the "best" value for λ, i.e. the largest value of λ for which the inequality (5.18) remains true, is greater than unity).

The following inequalities are the obvious consequences of (5.18):

$$\lambda^2 u^2(0) \leq a \int_0^a \left(\frac{du}{dx}\right)^2 dx, \quad \lambda^2 u^2(a) \leq a \int_0^a \left(\frac{du}{dx}\right)^2 dx, \tag{5.23}$$

$$\mu^2 \int_0^a u^2 dx \leq a^2 \int_0^a \left(\frac{du}{dx}\right)^2 dx. \tag{5.24}$$

with a constant μ which is not less than unity. Inequality (5.24) is also called the Wirtinger inequality.

The elimination of shifts in $u(x)$ by a constant is essential for the inequalities (5.23) and (5.24) to be true: if one ignores the constraint (5.17), then, putting $u(x) = const$, we get zero in the right hand sides and non-zero in the left hand sides. The shift can be excluded by different constraints. The values of the best constants, λ and μ, in (5.23) and (5.24) depend on the choice of the constraint.

The best constant in the Wirtinger inequality is a solution of the following variational problem:

$$\mu^2 = \min_{u(x)\in(5.17)} \frac{a^2 \int_0^a \left(\frac{du}{dx}\right)^2 dx}{\int_0^a u^2 dx}.$$

One can find (for example, by expanding $u(x)$ in Fourier series) that the best value for μ is 2π. Similarly, the best constant λ can be defined by the corresponding variational problem. We will see in Sect. 8.3 that the best constants in the Wirtinger-type inequalities relate to the minimum eigen-frequencies of some mechanical systems.

The best constants depend on the constraints imposed to eliminate the shift. If the constraint (5.17) is replaced by the constraint

$$u(0) = 0 \text{ or } u(0) = u(a) = 0, \tag{5.25}$$

then the best value for μ becomes $\pi/2$ and π, respectively.

Let us return to proving the sufficiency of conditions (5.14) for the boundedness from below of the functional (5.10). Using the inequality

$$|u + v| \leq |u| + |v|$$

and the Cauchy inequality (5.20), we get the following estimate of the linear functional $l(u)$

$$|l(u)| = \left| \int_0^a g(x)\, u\,(x) dx + f_a u\,(a) - f_0 u\,(0) \right| \le$$

$$\le \left| \int_0^a g(x)\, u\,(x) dx \right| + |\, f_a u\,(a)| + |\, f_0 u\,(0)| \le$$

$$\le \left| \int_0^a \frac{a}{\sqrt{A_0}} g\,(x) \frac{\sqrt{A_0}}{a} u\,(x) dx \right| + |\, f_a u\,(a)| + |\, f_0 u\,(0)| \le$$

$$\le \sqrt{\int_0^a \frac{a^2}{A_0} g^2 dx} \sqrt{\int_0^a \frac{A_0}{a^2} u^2 dx} + \frac{\sqrt{a}\,|\, f_a|}{\sqrt{A_0}} \sqrt{\frac{A_0 u^2\,(a)}{a}} +$$

$$+ \frac{\sqrt{a}\,|\, f_0|}{\sqrt{A_0}} \sqrt{\frac{A_0 u^2\,(0)}{a}}. \tag{5.26}$$

Let us increase the right-hand side of (5.26) by means of (5.23), (5.24), and (5.12). We see that the linear functional does not exceed $\sqrt{E(u)}$:

$$|l(u)| \le c\sqrt{E\,(u)}, \tag{5.27}$$

with a constant c equal to

$$c = \sqrt{2} \left(\frac{1}{\mu} \sqrt{\int_0^a \frac{a^2}{A_0} g^2 dx} + \frac{\sqrt{a}\,(|\, f_0| + |\, f_a|)}{\lambda \sqrt{A_0}} \right). \tag{5.28}$$

The inequality (5.27) allows us to complete the estimate of $I(u)$:

$$I\,(u) = E\,(u) - l\,(u) \ge E\,(u) - |l(u)| \ge E\,(u) - c\sqrt{E\,(u)}$$
$$= \left(\sqrt{E\,(u)} - \frac{c}{2} \right)^{2.} - \frac{c^2}{4} \ge -\frac{c^2}{4}$$

This justifies the boundedness of the functional $I\,(u)$ from below.

Equation (5.28) shows that the integrability of $g^2\,(x)$ is sufficient for boundedness from below.

In the example considered, there is a group of transformations, G, acting on $\mathcal{M}$, such that application of the transformation g from G to each element of $\mathcal{M}$ gives again the set $\mathcal{M}$, $g\mathcal{M} = \mathcal{M}$, $g \in G$ (in the example G is the group of shifts of u by a constant). This group leaves the functional E unchanged, $E\,(gu) =$

$E(u)$, $g \in G$, $u \in \mathcal{M}$. If the group G contains more elements than just the identity transformation, the functional E is said to have a kernel. In statics, E is usually the energy, and G is the transformation group corresponding to the rigid motions of the body.

In the same way, as in Example 3, the following statement can be proven: Let the invariance group of $E(u)$, G, be a group of translations with respect to u, $gu = u + \bar{u}_g$, and the set $\mathcal{K}$ of elements $\bar{u}_g$ is a cone, i.e. for every element $\bar{u} \in \mathcal{K}$ and any positive constant λ, $\lambda\bar{u} \in \mathcal{K}$. Let the functional $I(u)$ of the form (5.10) also be bounded from below. Then for all $\bar{u} \in \mathcal{K}$ the linear functional must satisfy the inequality

$$l(\bar{u}) \leq 0, \tag{5.29}$$

If the cone $\mathcal{K}$ contains with every element $\bar{u}$ the element $-\bar{u}$, then, as follows from (5.29), the necessary condition for the boundedness from below of the functional $E(u) - l(u)$ is

$$l(\bar{u}) = 0 \tag{5.30}$$

for all $\bar{u} \in \mathcal{K}$.

The property of $E(u)$ of having a kernel depends on the set on which it is considered. This is illustrated by the following example.

Example 4. Consider the set $\mathcal{M}$ of smooth functions $u(x)$ which consists of all differentiable functions on $[0, a]$, satisfying the condition

$$u(0) = 0. \tag{5.31}$$

In this case, the functional $E(u)$ (5.9) does not have a kernel: shifts for a constant are forbidden by the condition (5.31). The functional $I(u)$ is bounded from below for any $g(x)$ and f_a, since on the set of functions satisfying (5.31), the inequality (5.18) and, consequently, the inequality (5.27) hold true.

The physical interpretation of this fact for an elastic beam is as follows: if one end of the beam is clamped, the equilibrium exists for any forces, not only for forces with the zero resultant.

In the next example, the functional does not have a kernel.

Example 5. Let us change the functional $E(u)$ (5.9) by adding the integral of u^2:

$$E(u) = \frac{1}{2}\int_0^a \left(A(x)\left(\frac{du}{dx}\right)^2 + B(x)u^2 \right) dx, \quad B(x) \geq B_0 > 0. \tag{5.32}$$

The functional $E(u)$ (5.32) is not invariant with respect to shifts. Functional $E(u) - l(u)$ is bounded from below. The proof of the boundedness from below is the same as in Example 3, except that instead of inequalities (5.23), the inequalities

$$\lambda^2 u^2(0) \le a \int_0^a \left(\frac{du}{dx}\right)^2 dx + \frac{1}{a}\int_0^a u^2 dx,$$

$$\lambda^2 u^2(a) \le a \int_0^a \left(\frac{du}{dx}\right)^2 dx + \frac{1}{a}\int_0^a u^2 dx. \tag{5.33}$$

should be used; they are derived similarly to (5.23). It may be checked that, for $B \to 0$, the minimum value of the functional $E(u) - l(u)$ stays bounded if the condition (5.14) is satisfied. Otherwise, $E(u) - l(u) \to -\infty$ when $B \to 0$.

Example 6. Consider the functional $I(u) = E(u) - l(u)$, defined on the functions $u(x)$ of several variables $x = \{x_1, \ldots, x_n\}$, $x \in R_n$, with $E(u)$ and $l(u)$ given by the formulas

$$E(u) = \frac{1}{2}\int_V A^{ij}(x) u_{,i} u_{,j}\, dV,$$

$$l(u) = \int_V g(x) u\, dV + \int_{\partial V} f(x) u\, dA. \tag{5.34}$$

The quadratic form $A^{ij}u_i u_j$ is assumed to be positive: $A^{ij}u_i u_j \ge A_0 u_i u_j$, $A_0 = const > 0$.

The functional $E(u)$ is invariant with respect to shifts by constant functions $u(x) = \bar{u} = const$. The necessary condition of the boundedness from below of the functional $I(u)$ (5.34) is equivalent to the following relation for given functions $g(x)$ and $f(x)$

$$\int_V g(x)\, dV + \int_{\partial V} f(x)\, dA = 0. \tag{5.35}$$

Let us write the Euler equations of the functional $E(u) - l(u)$:

$$\frac{\partial}{\partial x^i} A^{ij}(x) \frac{\partial u}{\partial x^j} + g = 0 \quad \text{in} \quad V, \tag{5.36}$$

$$A^{ij} \frac{\partial u}{\partial x^i} n_j = f \quad \text{on} \quad \partial V. \tag{5.37}$$

Integrating (5.36) over V and using the boundary conditions (5.37) on ∂V, we get (5.35). Hence, this equation is the necessary condition of solvability for the boundary value problem (5.36) and (5.37). In the case of the Laplace equation ($A^{ij} = \delta^{ij}$, $f = 0$) it transforms into the well-known condition of solvability of the von Neuman problem.

The proof of the sufficiency of the condition (5.35) for the boundedness from below of the functional $I(u)$ is based on the Poincaré-Steklov-Fridrichs-Erlich inequalities

$$\lambda^2 \int\limits_{\partial V} u^2 dA \leq \int\limits_V u_{,i} u^{,i} dV, \quad \mu^2 \int\limits_V u^2 dV \leq \int\limits_V u_{,i} u^{,i} dV. \tag{5.38}$$

It is obtained in the same way as in Example 3. The equation

$$\int\limits_V u dV = 0 \tag{5.39}$$

Minimizing sequences. Let the functional $I(u)$ be bounded from below on $\mathcal{M}$. Then there exist sequences $\{u_n\}$ in $\mathcal{M}$ such that $I(u_n) \to \check{I}$ when $n \to \infty$. They are called the minimizing sequences.

Example 7. Let $\mathcal{M}$ be a set of smooth functions $u(x)$ of one variable defined on $[0, a]$ which take the unity value at $x = 0$: $u(0) = 1$. Consider the functional

$$I(u) = \int\limits_0^a u^2 dx.$$

on $\mathcal{M}$. The functional is bounded from below by zero. Let us show that min $I(u) = 0$. Consider a sequence of functions $u_n(x)$ which are linear on the segment $[0,a/n]$, take the values unity and zero at the points $x = 0$ and $x = a/n$, respectively, and equal to zero outside the segment (Fig. 5.2). It is also obvious that $I(u_n) \to 0$ when $n \to \infty$. Hence, the constructed sequence is minimizing. It is obvious that there is no minimizing element on the set of continuous functions because there is no continuous function which is equal to unity at the point $x = 0$ and for which the integral of u^2 is equal to zero.

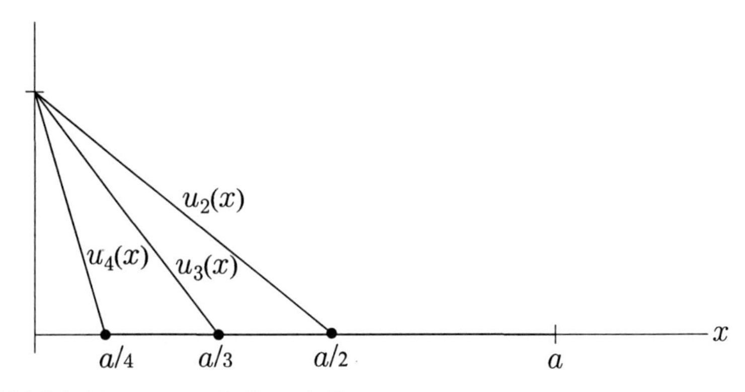

Fig. 5.2 Minimizing sequence in Example 7

As was already mentioned, the incorrectness of the variational problems caused by the absence of the minimizing element is encountered in applications. To solve such a problem means to find $\check{I}$ and to construct a minimizing sequence. In the variational problems of physics and mechanics the minimizing element usually does exist.[3]

5.2 Quadratic Functionals

Quadratic functionals, $E(u)$, were defined as functionals which can be presented in the form

$$E(u) = E(u, u),$$

where $E(u, v) = E(v, u)$ is a bilinear symmetric functional. An immediate consequence of this definition is a "quadratic expansion":

$$E(u + v) = E(u) + 2E(u, v) + E(v). \tag{5.40}$$

Besides, the quadratic functionals grow quadratically along the lines,[4] $u = \lambda u_0$:

$$E(\lambda u_0) = \lambda^2 E(u_0).$$

An identity. The quadratic functionals, as follows from (5.40), obey a useful identity:

$$E\left(\frac{u+v}{2}\right) + E\left(\frac{u-v}{2}\right) = \frac{1}{2}\left(E(u) + E(v)\right). \tag{5.41}$$

Alternative notation. As we have already discussed, in physical problems without loss of generality, the functional space can be replaced by a finite-dimensional space, R_m, of a large dimension, m. Then u becomes a vector in R_m, $E(u)$ a quadratic function in R_m, and $l(u)$ a linear function. Any linear function in R_m can be viewed as a scalar product of some vector, l, and vector u : $l(u) = (l, u)$. By (l, u) we denote the scalar product in R_m : if l_i and u^i are the components of the vectors, l and u,

$$(l, u) = l_i u^i.$$

[3] There are some exceptions such as the problem of thermodynamic equilibrium of two-phase materials when one neglects the interface energy. However, the absense of the minimizer is caused by an oversimplification of the energy functional. To some extent, the situation is similar to the appearance of paradoxes in fluid mechanics when viscosity is neglected.

[4] For a given element u_0, the equation $u = \lambda u_0$ may be interpreted as the parametric equation of the line in $\mathcal{R}$ passing through the origin and the point u_0.

A quadratic function, $E(u)$, may be presented as a scalar product of two vectors, Au and u, where Au is the product of a $m \times m$ matrix, A, and the vector, u :

$$E(u) = (Au, u).$$

The bilinear form, $E(u, v)$, is, obviously, (Au, v). If $m \to \infty$, Au transforms into an operator acting on u. In what follows we will use for quadratic and linear functionals both notations introduced: $E(u) = (Au, u)$ and $l(u) = (l, u)$.

Example. Let $E(u)$ be a functional,

$$E(u) = \int_V \frac{\partial u}{\partial x^i} \frac{\partial u}{\partial x_i} dV,$$

where V is a region in an $n-$dimensional space, and u is a smooth function vanishing on ∂V. Integrating by parts we can write

$$E(u) = \int_V (-\triangle u) u dV,$$

where $\triangle$ is Laplace operator. Setting the scalar product of two functions, u and v, as

$$(u, v) \equiv \int_V uv dV,$$

we see that in this example $A = -\triangle$. Note that the operator, $-\triangle$, is positive in the sense that $(Au, u) \geq 0$ for any u.

Linear problems. Many linear problems of continuum mechanics can be formulated as the minimization problems for a functional which is a sum of quadratic and linear functionals:

$$I(u) = E(u) - l(u) \to \min_{\mathcal{M}} \tag{5.42}$$

Variations of such functionals can easily be found:

$$\begin{aligned} \delta E(u) &= E(u + \delta u) - E(u) = E(u + \delta u, u + \delta u) - E(u) \\ &= 2E(u, \delta u) \\ \delta l(u) &= l(u + \delta u) - l(u) = l(\delta u) \\ \delta I(u) &= 2E(u, \delta u) - l(\delta u) \end{aligned} \tag{5.43}$$

The minimizing element, $\check{u}$, must obey the equation:

$$\delta E(\check{u}) = \delta l(\check{u})$$

for any admissible δu. Thus, for any admissible δu,

$$2E(\check{u}, \delta u) = l(\delta u). \tag{5.44}$$

Another form of this equation is obtained if we assume that $l(u)$ is a scalar product of some element, l, and u : $l(u) = (l, u)$, while $E(u) = \frac{1}{2}(Au, u)$. Then the Euler equation is

$$A\check{u} = l.$$

If the solution of this equation is unique, it can be written as

$$\check{u} = A^{-1}l,$$

where A^{-1} is the inverse operator. The inverse operator can be considered as the infinite-dimensional limit of the inverse symmetric matrix, A_m^{-1}, where A_m is an $m \times m$-matrix corresponding to an $m-$dimensional truncation of the quadratic functional, (Au, u).

Clapeyron theorem. If the set $\mathcal{M}$ is linear, then $\delta u = u' - \check{u}$ belongs to $\mathcal{M}$. We can choose, in particular, $\delta u = \check{u}\varepsilon$, ε being an infinitesimally small number. Plugging it in (5.44) we arrive at the so-called Clapeyron's theorem:

$$2E(\check{u}) = l(\check{u}). \tag{5.45}$$

An important consequence of Clapeyron's theorem is the link between the minimum value of the functional and the value of the quadratic functional at the minimizing element,

$$I(\check{u}) = -E(\check{u}), \tag{5.46}$$

which follows from (5.42) and (5.45). Another useful form of this equation is

$$I(\check{u}) = -\frac{1}{2}l(\check{u}). \tag{5.47}$$

Various forms of the minimization problem for quadratic functionals. The minimization problem for the functional $I = E(u) - l(u)$ may be presented in a number of equivalent forms that are sometimes useful.

Let the kernel of the quadratic functional $E(u)$ be eliminated. One can assign to each element, u, its "length," $\sqrt{E(u)}$. Then the unit sphere in the functional space is defined by the equation $E(u) = 1$.

Let us assign to every element of the functional space, u, an element on the unit sphere, v:

$$v = \frac{u}{\sqrt{E(u)}}, \qquad E(v) = 1.$$

The functional space can be split into rays, $u = \lambda v,\ 0 \le \lambda < +\infty$. The functional $I(u)$ can be considered as a function of two arguments, v and λ. The minimum of the functional $I(u)$ can be searched successively, by first finding the minimum along each ray $u = \lambda v$, and then finding the minimum on the sphere $E(v) = 1$. The search for the minimum along each ray is reduced to minimizing the function $\lambda^2 - \lambda l(v)$ of one variable, λ (v is fixed). The minimum is equal to $-\frac{1}{4}(l(v))^2$. Minimizing this expression with respect to v is equivalent to maximizing the linear functional $l(v)$ on the unit sphere, and we arrive at the variational problem

$$\check{I} = -\frac{1}{4}\hat{l}^2, \quad \hat{l} = \max_{E(v)=1} l(v). \tag{5.48}$$

The variational problem (5.48) can be written in two other equivalent ways: as a maximization problem of the linear functional on a ball:

$$\hat{l} = \max_{v:\ E(v)\le 1} l(v), \tag{5.49}$$

or as a maximization problem over the entire space of a functional of the zeroth order of homogeneity:

$$\hat{l} = \max_{u} \frac{l(u)}{\sqrt{E(u)}}. \tag{5.50}$$

The equivalence of (5.48) and (5.49) can be justified by the following reasoning. Let us expand the set of the admissible elements in the variational problem for $\hat{l}$, (5.48), allowing all elements inside the unit ball, $E(v) \le 1$. The maximum can only increase:

$$\max_{v:\ E(v)\le 1} l(v) \ge \max_{v:\ E(v)=1} l(v). \tag{5.51}$$

Suppose that (5.51) is a strict inequality:

$$\max_{v:\ E(v)\le 1} l(v) > \max_{v:\ E(v)=1} l(v) \equiv \hat{l}.$$

Then there exists a sequence $\{v_n\}$ such that $l(v_n) > \hat{l}$ and $E(v_n) \le 1$. Define the numbers λ_n by the equations: $E(\lambda_n v_n) = 1$. Since $E(\lambda_n v_n) = \lambda_n^2 E(v_n) = 1$ and $E(v_n) \le 1$, the numbers λ_n are not less than unity. Therefore, $l(\lambda_n v_n) = \lambda_n l(v_n) > \lambda_n \hat{l} \ge \hat{l}$, i.e. $l(\lambda_n v_n) > \hat{l}$. This contradicts the definition of $\hat{l}$ (5.48), because the sequence $\{\lambda_n v_n\}$ lies on the sphere $E(v) = 1$. Thus, only the equality sign is possible in (5.51).

Equation (5.50) follows from the fact that the functional $l(u)/\sqrt{E(u)}$ does not change along the rays $u = \lambda v$.

5.3 Existence of the Minimizing Element

In calculus of variations the theorems warranting the existence of the minimizing element are usually the generalizations to the infinite-dimensional case of the theorem which states that a continuous function in a finite-dimensional space reaches both its maximum and minimum values on any closed set $\mathcal{M}$. The analysis of this theorem shows that it is based on three elements: the notion of the convergence of the elements on $\mathcal{M}$, the notion of the continuity of the function (functional) on $\mathcal{M}$ and the structure of the set $\mathcal{M}$: any infinite sequence of elements from $\mathcal{M}$ has to contain a subsequence that converges to an element of $\mathcal{M}$.

The notion of the continuity of the functional (the functional is continuous at point u_0 if for any sequence $\{u_n\}$ converging to u_0, $I(u_n)$ converges to $I(u_0)$), and the above-mentioned property of the set $\mathcal{M}$, which is called compactness, are both based on the way in which the convergence of elements of $\mathcal{M}$ is introduced.

For given functional $I(u)$ and a set $\mathcal{M}$ the convergence of elements of $\mathcal{M}$ can be defined differently. With respect to some convergences functional $I(u)$ can be continuous and the set $\mathcal{M}$ – compact; with respect to the other ones, this may not be true. If it is possible to introduce such convergence of the elements that the set $\mathcal{M}$ is compact and the functional $I(u)$ is continuous with respect to that convergence, then $I(u)$ has a minimizing element on $\mathcal{M}$. This statement is proven similarly to the corresponding theorem for a function of one variable in calculus.

One can check that the assumption of the functional continuity can be relaxed to a weaker assumption of its semi-continuity below. The functional is called semi-continuous below at point u_0 if for any sequence $\{u_n\}$ converging to u_0, for which the sequence $\{I(u_n)\}$ converges,

$$I(u_0) \leq \lim_{n\to\infty} I(u_n).$$

The key types of the convergence of elements in $\mathcal{M}$ are as follows. Usually, the set $\mathcal{M}$ can be considered as a subset of a Banach space. A Banach space is a linear space B, for which a norm, a non-negative functional, $\|u\|$, is defined, where $\|u\|$ is homogeneous ($\|\lambda u\| = |\lambda| \, \|u\|$), non-degenerate ($\|u\| = 0$ if and only if $u = 0$), convex ($\|u + v\| \leq \|u\| + \|v\|$) functional, and the space B is complete with respect to the norm. The latter means that any fundamental sequence $\{u_n\}$ (i.e. the sequence with $\|u_n - u_m\| \to 0$ as $n, m \to \infty$) converges to an element of B.

For a Banach space, the notion of convergence with respect to the norm arises naturally; it is also called strong convergence. One says that the sequence $\{u_n\}$ strongly converges to u_0 if $\|u_n - u_0\| \to 0$ as $n \to \infty$. Strong convergence is not very useful for theorems of the existence of the minimizing element since bounded closed sets in functional spaces (i.e. closed sets contained in a ball $\|u\| \leq R$ of a finite radius R) turn out to be non-compact. However, the bounded closed sets in a Banach space can be compact with respect to the so-called weak convergence: the sequence $\{u_n\}$ weakly converges to u_0 if for any linear continuous functional[5] $l(u)$, $l(u_n) \to l(u_0)$ as $n \to \infty$. Such Banach spaces are often encountered in applications; they include Hilbert spaces (linear spaces for which the scalar product (u, v) is defined, and the norm is introduced as $\|u\| = (u, u)^{1/2}$).

So the existence of the minimizing element is warranted for functionals $I(u)$ which are semi-continuous below with respect to weak convergence on a bounded closed set in the Banach space. The case of the unbounded set $\mathcal{M}$ (for example, when $\mathcal{M}$ is a cone) reduces to the case of the bounded set if the functional $I(u)$ satisfies an additional condition of $I(u) \to +\infty$ when $\|u\| \to \infty$.

One can prove that the functional $\|u\|$ is semi-continuous below with respect to the weak convergence. This property warrants the existence of the minimizing element for the functionals $I(u)$ of the form

$$I(u) = E(u) - l(u)$$

if for some α, $0 < \alpha < 1$, the functional $[E(u)]^\alpha$ has the properties of the norm[6] and $l(u)$ is a linear functional. For quadratic functionals, $\alpha = 1/2$.

The functional space with the norm $\|u\| = \sqrt{E(u)}$ plays a special role in variational problems with quadratic functionals. It is called energy space because $E(u)$ usually has the meaning of energy. To make the energy space a Banach space, one has to include in this space the limits of all fundamental sequences $\{u_n\}$, i.e. such sequences that $\|u_n - u_m\| = \sqrt{E(u_n - u_m)} \to 0$ for $n, m \to \infty$. The minimizing element is an element of the energy space.

In the examples considered above $E(u)$ does not have the properties of the norm since it has a kernel (i.e. it is equal to zero for $u \neq 0$). However, by eliminating the kernel by means of the constraints like (5.17) or (5.25), $\sqrt{E(u)}$ acquires the properties of the norm, and the minimizing element does exist.

The elements of the energy space do not have the smoothness expected of the solution of physical problems. Hence, an independent problem of studying the smoothness properties of the minimizing element arises. It is not as simple as the problem of the existence of the minimizing element in the energy space.

5.4 Uniqueness of the Minimizing Element

The fundamental criterion for the uniqueness of the minimizing element is based on the notion of convexity. We begin its consideration with the definition of convex sets.

---

[5] Continuity of linear functionals with respect to strong convergence is implied.

[6] This condition can be weakened by replacing it by the conditions that the functional $E(u)$ be convex (see below), continuous with respect to some norm $\|u\|$, and coercive, i.e. $E(u)/\|u\| \to \infty$ when $\|u\| \to \infty$.

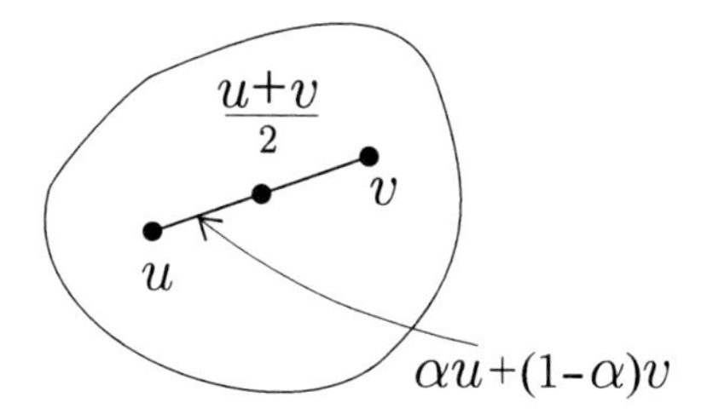

Fig. 5.3 The segment connecting the points u and v

For any two points of a finite-dimensional space, u and v, the sum $\frac{1}{2}(u+v)$ is the center of the segment connecting the points u and v, while any point of this segment can be presented in the form $\alpha u + (1-\alpha)v$, $0 \le \alpha \le 1$. The ends of the segment, u and v, correspond to $\alpha = 1$ and $\alpha = 0$, respectively (Fig. 5.3). We apply this terminology to the functional spaces calling the one-parametric set of elements, $\alpha u + (1-\alpha)v$, $0 \le \alpha \le 1$, the segment connecting the elements u and v.

The set $\mathcal{M}$ is called convex, if, for any two elements u and v, it contains the segment $\alpha u + (1-\alpha)v$, $0 \le \alpha \le 1$ connecting those two elements.

The functional $I(u)$ defined on the convex set $\mathcal{M}$ is called convex if, for any two elements u and v in $\mathcal{M}$ and any $0 \le \alpha \le 1$,

$$I(\alpha u + (1-\alpha)v) \le \alpha I(u) + (1-\alpha)I(v). \tag{5.52}$$

The functional is strictly convex, if for $u \ne v$,

$$I(\alpha u + (1-\alpha)v) < \alpha I(u) + (1-\alpha)I(v), \quad 0 \le \alpha \le 1.$$

Sometimes, in the definition of convex functionals one uses the inequality

$$I\left(\frac{u+v}{2}\right) \le \frac{1}{2}(I(u) + I(v)). \tag{5.53}$$

It follows from (5.52) for $\alpha = \frac{1}{2}$. One can show that the conditions (5.52) and (5.53) are equivalent for functionals which are semi-continuous below. The geometrical meaning of the conditions (5.52) and (5.53) can be seen from Fig. 5.4.

Example 1. Consider the integral functional

$$I(u) = \int_V L\left(x^i, u^\varkappa, \frac{\partial u^\varkappa}{\partial x^i}\right) dV.$$

We are going to show that this functional is convex (strictly convex) if L is a convex (strictly convex) function of $u^\varkappa$ and $\partial u^\varkappa / \partial x^i$. Indeed,

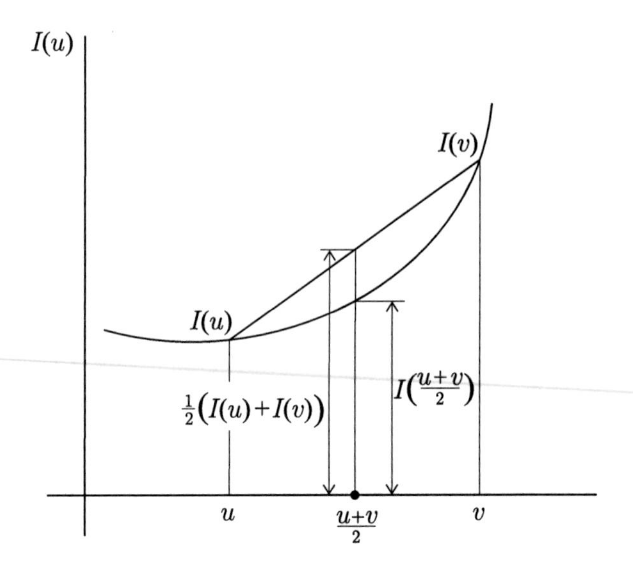

Fig. 5.4 Definition of the convex functional

$$I\left(\alpha u+(1-\alpha)v\right)=$$
$$=\int_V L\left(x^i,\alpha u^\varkappa+(1-\alpha)v^\varkappa,\alpha\frac{\partial u^\varkappa}{\partial x^i}+(1-\alpha)\frac{\partial v^\varkappa}{\partial x^i}\right)dV\leq$$
$$\leq\int_V\left[\alpha L\left(x^i,u^\varkappa,\frac{\partial u^\varkappa}{\partial x^i}\right)+(1-\alpha)L\left(x^i,v^\varkappa,\frac{\partial v^\varkappa}{\partial x^i}\right)\right]dV=$$
$$=\alpha I(u)+(1-\alpha)I(v),$$

as claimed.

A strictly convex functional cannot have more than one minimizing element on a convex set. Indeed, suppose that a strictly convex functional has two minimizing elements, u and v, on $\mathcal{M}$. Then $I(u)=I(v)=\check{I}$. The element $\frac{1}{2}(u+v)$ belongs to $\mathcal{M}$ due to the convexity of the set. According to the strict convexity of the functional $I(u)$, its value at the point $\frac{1}{2}(u+v)$ is less than the minimum value on $\mathcal{M}$,

$$I\left(\frac{u+v}{2}\right)<\frac{1}{2}\left(I(u)+I(v)\right)=\check{I},$$

which contradicts the definition of the minimum value.

Any linear functional obeys (5.52):

$$l(\alpha u+(1-\alpha)v)=\alpha l(u)+(1-\alpha)l(v),$$

and, thus, is convex. The sum of a convex functional and a linear functional is obviously convex. Moreover, the sum of any two convex functionals is convex.

Positive quadratic functionals with zero kernel are strictly convex. Indeed, consider the identity (5.41):

$$E\left(\frac{u+v}{2}\right)+E\left(\frac{u-v}{2}\right)=\frac{1}{2}\left(E\left(u\right)+E\left(v\right)\right). \tag{5.54}$$

By our assumption, functional $E(u)$ has zero kernel, i.e. $E(u)=0$ only for $u=0$. Therefore, for $u\neq v$, $E\left(\frac{u-v}{2}\right)>0$. Dropping $E\left(\frac{u-v}{2}\right)$ in (5.54), we decrease the left hand side and arrive at the inequality

$$E\left(\frac{u+v}{2}\right)<\frac{1}{2}\left(E\left(u\right)+E\left(v\right)\right)$$

which indicates the strict convexity of $E(u)$.

Examples 2–5 of Sect. 5.1 (continued). The quadratic functionals $E\left(u\right)$ in Examples 2–5 are positive. Thus, if the kernels are excluded, these functionals are strictly convex, and the functional $I\left(u\right)=E\left(u\right)-l\left(u\right)$ is strictly convex as well. The set $\mathcal{M}$ in Examples 1–4 is convex. Hence, there is only one minimizing element in the corresponding variational problems.

A convex, but not necessarily strictly convex, functional can have several minimizing elements (the corresponding example is illustrated in Fig. 5.5: the set of the minimizing points of the function is the straight segment).

The set of minimizing elements of a convex functional is convex. Indeed, if u and v are two minimizing elements, $I\left(u\right)=\check{I}$ and $I\left(v\right)=\check{I}$, then

$$I(\alpha u+(1-\alpha)\,v)=\check{I}$$

because

$$\check{I}\leq I\left(\alpha u+(1-\alpha)\,v\right)\leq\alpha I\left(u\right)+(1-\alpha)\,I\left(v\right)=\check{I}.$$

Consequently, all the points on the segment $\alpha u+(1-\alpha)\,v$, $0\leq\alpha\leq1$, are the minimizing elements of $I(u)$.

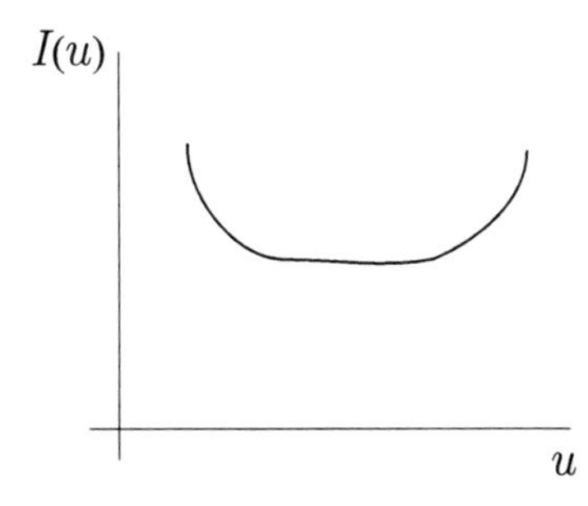

Fig. 5.5 Convex functional with many minimizing elements

5.5 Upper and Lower Estimates

In many problems, the information about the minimum value of the functional $I(u)$ is very important. Sometimes it is of the primary interest (many examples are given further). In particular, the minimum value may be related to the energy of the system.

The estimates of the minimum value are usually based on some auxiliary variational problems which are easier to investigate than the initial variational problem. In constructing the auxiliary problem, either the set $\mathcal{M}$ or the functional $I(u)$ are changed.

Changing the set of the admissible functions. Consider two sets, $\mathcal{M}_1$ and $\mathcal{M}_2$, such that $\mathcal{M}_2 \subset \mathcal{M} \subset \mathcal{M}_1$. Suppose that the functional $I(u)$ can be defined on the set $\mathcal{M}_1$. Denote by $\check{I}_1$ and $\check{I}_2$ the minimum values of the functional $I(u)$ on $\mathcal{M}_1$ and $\mathcal{M}_2$, respectively. It is obvious that

$$\check{I}_1 \leq \check{I} \leq \check{I}_2. \tag{5.55}$$

If one takes as the set $\mathcal{M}_2$ an element, u, of $\mathcal{M}$, one gets an estimate

$$\check{I} \leq I(u) \tag{5.56}$$

In the case of the quadratic functionals of the form $E(u) - l(u)$, minimized on a linear set, an upper estimate of the minimum value, according to the Clapeyron theorem (5.46), corresponds to an estimate of energy from below, while a lower estimate of the minimum value gives an upper energy estimate.

The Rayleigh-Ritz method. The most common method of obtaining an approximate solution of a variational problem is the Rayleigh-Ritz method. The method is as follows. One selects a k-dimensional subset, $\mathcal{M}_k$, of the set $\mathcal{M}$. It is comprised of the elements of the form $a^1 u_1 + \ldots + a^k u_k$, where $u_1, \ldots, u_k$ are some fixed elements of $\mathcal{M}$, while $a^1, \ldots, a^k$ are numerical parameters. The functional I becomes a function of k variables, $a^1, \ldots, a^k$. Denote by $\check{a}^1, \ldots, \check{a}^k$ the minimizer of this function, and by $\check{u}_k$ the element $\check{a}^1 u_1 + \ldots + \check{a}^k u_k$. This element can be considered as an approximation of the minimizing element of the functional I on $\mathcal{M}$; moreover, $\check{I} \leq I(\check{u}_k)$, and if $\check{u}_k$ is the minimizing element of I on $\mathcal{M}_k$, this estimate is the best among all elements of the set $\mathcal{M}_k$.

Increasing the dimensionality of the subset $\mathcal{M}_k$, $\mathcal{M}_1 \subset \mathcal{M}_2 \subset \mathcal{M}_3 \subset \ldots$, we obtain better and better approximations of the minimizing elements and the minimum value. The convergence of this process is proved for a large class of variational problems.

The method of constraint unlocking. Let the set $\mathcal{M}$ be defined by some system of equations or inequalities. By dismissing some of those equations, we obtain a larger set $\mathcal{M}_1$, and $\check{I}_1 \leq \check{I}$. This way of making the lower estimates is called the method of constraint unlocking.

Changing the functional. Another way of making estimates employs a change of the functional. One constructs the functionals I_1 and I_2 in such a way that $I_1 < I < I_2$ on $\mathcal{M}$. Then

$$\check{I}_1 \leq \check{I} \leq \check{I}_2. \tag{5.57}$$

This method is effective if the minimum values $\check{I}_1$ and $\check{I}_2$ of the functionals I_1 and I_2 can be found on $\mathcal{M}$.

Let, for example, the functional I be a sum of two functionals:

$$I = I' + I''$$

where I' and I'' are both bounded from below on $\mathcal{M}$ and their minimum values can be found. Then one can take as a functional I_1 the functional

$$I_1 = I' + \min_{\mathcal{M}} I''$$

The low estimate (5.57) takes the form

$$\min_{\mathcal{M}} I' + \min_{\mathcal{M}} I'' \leq \min_{\mathcal{M}} I. \tag{5.58}$$

Example. Consider the minimization problem for the functional

$$I(u) = E(u) - l(u), \quad E(u) = \int_V L\left(x^i, u^\varkappa, \frac{\partial u^\varkappa}{\partial x^i}\right) dV,$$

$$l(u) = \int_V g_\varkappa(x) u^\varkappa dV + \int_{\partial V_f} f_\varkappa(x) u^\varkappa dA,$$

where ∂V_f is a part of the boundary ∂V where the "external surface forces," $f_\varkappa$, are given. The minimum is sought over all functions $u^\varkappa$, taking the assigned values, on the surface $\partial V_u = \partial V - \partial V_f$:

$$u^\varkappa = u^\varkappa_{(b)}. \tag{5.59}$$

Let us divide the region V into two subregions V' and V'' by the surface Σ. The surface Σ divides each of the surfaces ∂V_u and ∂V_f into two parts, $\partial V'_u$ and $\partial V''_u$, and $\partial V'_f$ and $\partial V''_f$; see Fig. 5.6. In this figure the part of the boundary with "hard" boundary conditions, i.e. the conditions that yield zero variations of the required

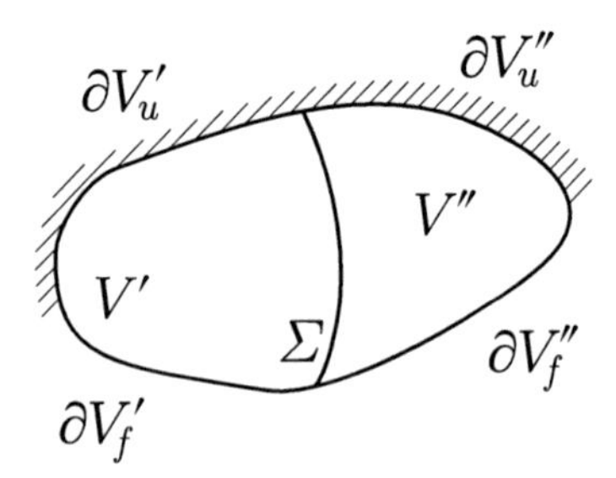

Fig. 5.6 Notation to inequality (5.62)

functions, is shadowed. The boundary of the surface V' comprises surfaces $\partial V'_f$, $\partial V'_u$ and Σ, and the boundary of the surface V'' comprises $\partial V''_u$, $\partial V''_f$, and Σ. Let us define on Σ some functions $p_{\varkappa}(x)$ and define the functionals

$$I'(u) = E'(u) - l'(u), \quad I''(u) = E''(u) - l''(u); \tag{5.60}$$

$$E'(u) = \int_{V'} L dV, \quad E''(u) = \int_{V''} L dV,$$

$$l'(u) = \int_{V'} g_{\varkappa} u^{\varkappa} dV + \int_{\partial V'_f} f_{\varkappa} u^{\varkappa} dA + \int_{\Sigma} p_{\varkappa} u^{\varkappa} dA,$$

$$l''(u) = \int_{V''} g_{\varkappa} u^{\varkappa} dV + \int_{\partial V''_f} f_{\varkappa} u^{\varkappa} dA - \int_{\Sigma} p_{\varkappa} u^{\varkappa} dA, \tag{5.61}$$

Obviously,

$$I' + I'' = I.$$

According to (5.58), for I we get the estimate

$$\check{I}' + \check{I}'' \leq \check{I}. \tag{5.62}$$

If the minimum value of the functional $I(u)$ is related to the energy of the system, the inequality (5.62) provides an estimate of energy in terms of energies of its parts.

Essential and inessential constraints. The estimates of the minimum value are closely related to the important notion of essential and inessential constraints. The constraints removing of which does not change the minimum value of the functional are called inessential. Otherwise, the constraint is essential. We discuss this notion for the following example.

Consider the minimization problem for the functional

$$I(u) = \int_0^a \left(\frac{du}{dx}\right)^2 dx \to \min. \tag{5.63}$$

Minimum is sought on the set of all smooth functions defined on the segment $[0, a]$, which satisfy the conditions

$$u(0) = 1, \quad u(a) = 0. \tag{5.64}$$

The minimum value is positive. Indeed, any function $u(x)$ satisfying the constraints (5.64) can be written as

$$u(x) = \frac{a - x}{a} + v(x), \tag{5.65}$$

where the function $v(x)$ obeys the homogeneous constraints,

$$v(0) = v(a) = 0 \tag{5.66}$$

Changing the argument of the functional I, $u \to v$, we obtain

$$I(v) = \frac{1}{a} + \int_0^a \left(\frac{dv}{dx}\right)^2 dx. \tag{5.67}$$

Hence, the minimum value of the functional $I(u)$ is bounded from below by $1/a$, and the constraints (5.64) are essential to $I(u)$.

Note that $\check{I} = 1/a$. It follows from the inequalities

$$\frac{1}{a} \leq \check{I} \leq I(v)|_{v=0} = \frac{1}{a}.$$

The minimum of the functional $I(u)$ is equal to zero if the first constraint (5.64) is removed. It is achieved on the function $u(x) \equiv 0$. Therefore, the first constraint is essential. The second constraint (5.64) is also essential: if one removes this constraint , then the minimum value of the functional is zero again; it is reached on the functions $u(x) \equiv 1$.

Consider now a variational problem where, in addition to (5.64), one sets a constraint for the boundary value of the derivative,

$$\frac{du}{dx} = c_0 \quad \text{at } x = 0. \tag{5.68}$$

Let us show that this constraint is inessential. After the change of functions, $u \to v$, (5.65), the constraints takes the form

$$v(0)=0,\quad v(a)=0,\quad \left.\frac{dv}{dx}\right|_{x=0}=c\equiv c_0+a^{-1}. \tag{5.69}$$

We assume, for definiteness, that $c>0$. Consider a sequence of functions $\{v_n(x)\}$,

$$v_n(x)=\begin{cases} cx, & 0\le x\le \frac{a_n}{c}\\ -cx+2a_n, & \frac{a_n}{c}\le x\le \frac{2a_n}{c}\\ 0, & \frac{2a_n}{c}\le x\le a\end{cases}$$

where $\{a_n\}$ is a sequence of positive numbers converging to zero (see Fig. 5.7).

Functions $v_n(x)$ satisfy the constraints (5.69), and $I(v_n)=1/a+2ca_n\to 1/a$ as $n\to\infty$. Therefore, the minimum value of the functional $I(u)$ does not change if one prescribes the value of the derivative of $u(x)$ at the boundary: it is still equal to $1/a$.

One may say that the functional itself chooses the constraints which it cannot violate (essential constraints) and which it ignores (inessential constraints). The functional "feels" the essential constraints and does not feel the inessential constraints.

It is also instructive to consider in the above example the case $a=+\infty$. Then, the second constraint (5.64) is replaced by the condition $u(x)\to 0$ for $x\to\infty$. If $a=\infty$, the constraints (5.64) become inessential. To make sure that it is the case, consider the sequence $u_n(x)$ (Fig. 5.8):

$$u_n(x)=\begin{cases} 1-a_nx, & 0\le x\le \frac{1}{a_n}\\ 0, & \frac{1}{a_n}\le x\end{cases}\qquad \{a_n\}\to 0 \text{ for } n\to\infty.$$

For this sequence, $I(u_n)=a_n\to 0$ as $n\to\infty$. Hence, the minimum value is the same as if there are no constraints at all.

Estimates of the closeness of the minimizing element and its approximations. Let a sufficiently narrow fork,

$$a\le \check{I}\le b,$$

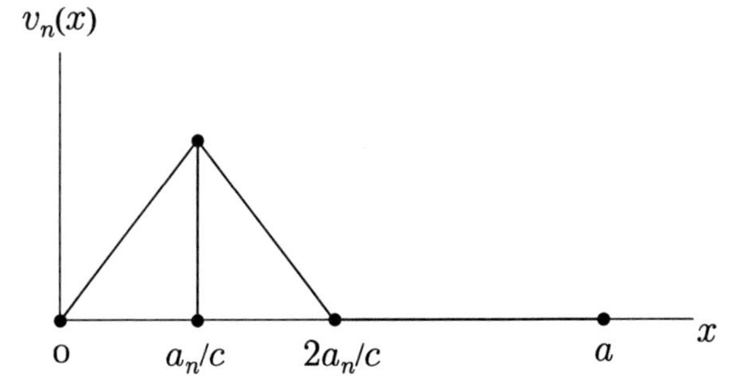

Fig. 5.7 Minimizing sequence in the variational problem with constraints (5.69)

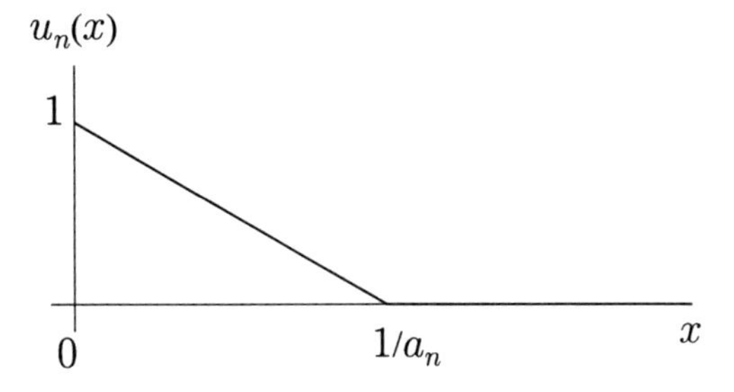

Fig. 5.8 Minimizing sequence for $a = \infty$

be constructed for the minimum value $\check{I}$ of a strictly convex functional $I(u)$, and let an element u be found such that the value of the functional for this element is also within the same fork:

$$a \leq I(u) \leq b.$$

What can be said of the closeness of the element u to the minimizing element $\check{u}$? The difference between u and $\check{u}$ can be estimated in the case of strongly convex functionals. Functional $I(u)$ is called strongly convex if there exists such a positive functional $B(u)$ that for any two elements, u and v,

$$I\left(\frac{u+v}{2}\right) + B\left(\frac{u-v}{2}\right) \leq \frac{1}{2}\left(I(u) + I(v)\right). \tag{5.70}$$

Quadratic positive functionals are strongly convex due to the identity (5.41). It follows from this identity that $B(u) = E(u)$.

In the case of strongly convex functionals, the closeness of the elements u and $\check{u}$ can be characterized in terms of the functional $B(u)$. Indeed, putting in (5.70), $v = \check{u}$, we have

$$I\left(\frac{u+\check{u}}{2}\right) + B\left(\frac{u-\check{u}}{2}\right) \leq \frac{1}{2}\left(I(u) + I(\check{u})\right). \tag{5.71}$$

We can reduce the left-hand side of (5.71) substituting $I\left(\frac{u+\check{u}}{2}\right)$ by the smaller number, $I(\check{u})$. We obtain

$$B\left(\frac{u-\check{u}}{2}\right) \leq \frac{1}{2}\left(I(u) - I(\check{u})\right). \tag{5.72}$$

Hence, if the values of the functional for an approximate solution, u, and for the minimizer are within the same fork, and the fork is narrow, then measure of the difference between the approximate and the exact solution, $B\left(\frac{u-\check{u}}{2}\right)$, is small:

$$B\left(\frac{u-\check{u}}{2}\right) \leq \frac{b-a}{2}. \tag{5.73}$$

For a quadratic functional, $B\left(\frac{u-\check{u}}{2}\right)=E\left(\frac{u-\check{u}}{2}\right)$, and (5.73) provides the energy estimate for the error of the approximate solution:

$$E\left(\frac{u-\check{u}}{2}\right)\leq\frac{b-a}{2}.$$

5.6 Dual Variational Principles

The general scheme. It has long been known that the same system of differential equations can be the system of Euler equations for different functionals. For example, the equations of analytical mechanics for the systems with a finite number of degrees of freedom can be obtained by means of two different variational principles, the Hamilton principle in the phase space and the Lagrange principle. In other areas of mechanics, different principles were also proposed for the same system of equations: the Dirichlet and the Thompson principles in the mechanics of ideal incompressible fluid and in electrostatics, the Lagrange, Castigliano and Reissner principles in the elasticity theory, the Pontrjagin maximum principle in the variational problems with constraints, etc. It turns out that one simple common idea, the idea of duality, lies in the basis of all such principles. This section is concerned with the discussion of this idea.

Consider the minimization problem for the functional $I(u)$ on a set $\mathcal{M}$,

$$\check{I}=\min_{u\in\mathcal{M}}I(u). \tag{5.74}$$

Suppose that it is possible to construct a functional, $\Phi(u,v)$, of two variables, u and v, u being an element of $\mathcal{M}$, and v an element of some other set $\mathcal{N}$, such that

$$I(u)=\max_{v\in\mathcal{N}}\Phi(u,v). \tag{5.75}$$

Then the initial variational problem can be stated as a minimax problem:

$$\check{I}=\min_{u\in\mathcal{M}}\max_{v\in\mathcal{N}}\Phi(u,v). \tag{5.76}$$

Assume that the order of computing of the maximum and minimum values in (5.76) can be changed,[7]

---

[7] If we search the minimum of a function of two variables, $\Phi(u,v)$, we can first minimize the function over u for each fixed v, and then search the minimum of the result over v; or we can search minimum over v for each fixed u and then minimize the result over u. The answer is, obviously, the same:

$$\min_u\min_v\Phi(u,v)=\min_v\min_u\Phi(u,v).$$

$$\check{I} = \max_{v \in \mathcal{N}} \min_{u \in \mathcal{M}} \Phi(u, v). \tag{5.77}$$

Suppose also that the functional $\Phi(u, v)$ is chosen in such a way that $\min_{u \in \mathcal{M}} \Phi(u, v)$ can be easily found. Denote it by $J(v)$:

$$J(v) = \min_{u \in \mathcal{M}} \Phi(u, v). \tag{5.78}$$

Then the initial minimization problem is equivalent to the maximization problem for the functional $J(v)$ on the set $\mathcal{N}$:

$$\check{I} = \max_{v \in \mathcal{N}} J(v). \tag{5.79}$$

The variational problem (5.79) is called dual to the initial variational problem (5.74).

It is possible to construct various dual variational problems, choosing various functionals $\Phi(u, v)$ and sets $\mathcal{N}$. This choice is limited by the two conditions. First, the possibility to change the order of minimization and maximization:

$$\min_{u \in \mathcal{M}} \max_{v \in \mathcal{N}} \Phi(u, v) = \max_{v \in \mathcal{N}} \min_{u \in \mathcal{M}} \Phi(u, v). \tag{5.80}$$

Second, it must be possible to find the functional $J(v)$ (5.78) explicitly. In the variational principles mentioned in the beginning of this section, the latter condition is satisfied by choosing the functional $\Phi(u, v)$ linear with respect to v.

For the validity of (5.80), various conditions of different degrees of generality were suggested. However, it is sometimes easier to check directly whether the order of maximization and the minimization can be changed. Usually, (5.80) can be checked in the following way.

First, let us show that for any functional $\Phi(u, v)$ and for any non-empty sets, $\mathcal{M}$ and $\mathcal{N}$,

$$\max_{v \in \mathcal{N}} \min_{u \in \mathcal{M}} \Phi(u, v) \leq \min_{u \in \mathcal{M}} \max_{v \in \mathcal{N}} \Phi(u, v). \tag{5.81}$$

Indeed,

$$\min_{u \in \mathcal{M}} \Phi(u, v) \leq \Phi(u, v). \tag{5.82}$$

In minimax problems the change of the order of minimization and maximization cannot be always done. This is seen, for example, for a function, $\Phi(u, v) = -u^2 + uv, \quad |u| \leq 1, |v| \leq 1$:

$$\min_{|u| \leq 1} \max_{|v| \leq 1} [-u^2 + uv] = \min_{|u| \leq 1} [-u^2 + |u|] = 0,$$

$$\max_{|v| \leq 1} \min_{|u| \leq 1} [-u^2 + uv] = \max_{|v| \leq 1} \min\{-1 + v, -1 - v\} = -1.$$

If an element u on the right-hand side of (5.82) is fixed, then the inequality (5.82) means that the functional of v, $\min_{u\in\mathcal{M}} \Phi(u, v)$, is not greater than the functional of v, $\Phi(u, v)$. Therefore,

$$\max_{v\in\mathcal{N}} \min_{u\in\mathcal{M}} \Phi(u, v) \leq \max_{v\in\mathcal{N}} \Phi(u, v). \tag{5.83}$$

According to (5.83), the functional of u, $\max_{v\in\mathcal{N}} \Phi(u, v)$ is bounded from below by a constant, $\max_{v\in\mathcal{N}} \min_{u\in\mathcal{M}} \Phi(u, v)$. Consequently, its minimum value, $\min_{u\in\mathcal{M}} \max_{v\in\mathcal{N}} \Phi(u, v)$, is also bounded from below by this constant, and the inequality (5.81) holds true.

The inequality (5.81) can also be written as

$$\max_{v\in\mathcal{N}} J(v) \leq \min_{u\in\mathcal{M}} I(u). \tag{5.84}$$

Denote the minimizing element of the initial variational problem (5.74) by $\check{u}$, and the element v for which

$$\max_{v\in\mathcal{N}} \Phi(\check{u}, v)$$

is achieved by $\hat{v}$. Using (5.84), we get an estimate:

$$J(\hat{v}) \leq \max_{v\in\mathcal{N}} J(v) \leq \min_{u\in\mathcal{M}} I(u) = I(\check{u}). \tag{5.85}$$

In many cases it is possible to compare $J(\check{v})$ and $I(\check{u})$, and, if it turns out that $J(\hat{v}) = I(\check{u})$, then $\max_{v\in\mathcal{N}} J(v) = \check{I}$.

It is essential that the dual variational principle (5.79) allows one easily to obtain the lower estimates of $\check{I}$. Remember that getting the upper estimates of $\check{I}$ is trivial; it is sufficient to calculate the value of functional $I(u)$ at any element of the set $\mathcal{M}$:

$$\check{I} \leq I(u).$$

If a dual variational problem is constructed, then obtaining the lower estimates of $\check{I}$ is just as trivial. It is sufficient to calculate the value of the functional $J(v)$ at any element of the set $\mathcal{N}$:

$$J(v) \leq \check{I}.$$

So, we get a fork

$$J(v) \leq \check{I} \leq I(u).$$

If one can find the elements u and v such that the numbers $I(u)$ and $J(v)$ are close, one gets an approximation for the minimum value without solving any equations.

In some cases, it is hard to find a functional $\Phi(u, v)$ that would satisfy (5.75), but one can construct a functional $\Phi(u, v)$ that obeys the inequality

$$\max_{v\in\mathcal{N}} \Phi(u, v) \leq I(u).$$

Then,

$$\max_{v\in\mathcal{N}} J(v) \leq \check{I}.$$

So, because, as follows from (5.81),

$$\max_{v\in\mathcal{N}} J(v) = \max_{v\in\mathcal{N}} \min_{u\in\mathcal{M}} \Phi(u, v) \leq \min_{u\in\mathcal{M}} \max_{v\in\mathcal{N}} \Phi(u, v) \leq \min_{u\in\mathcal{M}} I(u) = \check{I},$$

the ability to obtain the lower estimates of $\check{I}$ is retained.

An essential role in construction of the functional $\Phi(u, v)$ is played by the Young-Fenchel transformation and other notions of convex analysis to discussion of which we proceed.

5.7 Legendre and Young-Fenchel Transformations

Convex functions. Consider a function $f\left(x^1, \ldots, x^n\right)$ which takes on either a finite value or the value $+\infty$ at the points of n-dimensional space R_n. The set of points in the $(n+1)$-dimensional space R_{n+1} with coordinates $x^1, \ldots, x^n, y$ defined by the condition $y \geq f\left(x^1, \ldots, x^n\right)$, is called the epigraph of the function $f(x)$ and is denoted by epi$f$ (Fig. 5.9). The set of the point in R_n for which $f(x) < +\infty$ is called the effective domain of the function $f(x)$ and is denoted by dom$f$. We will assume that the set dom$f$ is a non-empty subset of R_n, that for all points in dom$f$ the function $f(x)$ is continuous and that $f(x)$ is bounded from below on dom$f$.

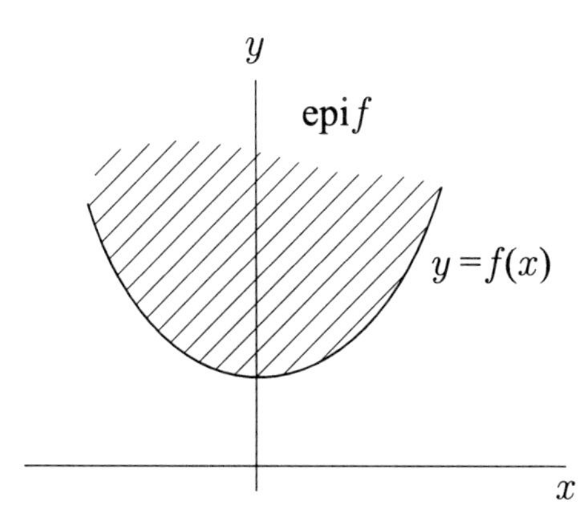

Fig. 5.9 Definition of the epigraph of function $f(x)$

The conditions that epi f is convex and that $f(x)$ is a convex function are equivalent. The continuity of $f(x)$ implies that epi f is a closed set. If $f(x)$ is a convex function, then dom f is a convex set.

If a convex function is defined on some convex set $V \subset R_n$, then it can be extended on the whole set R_n preserving its convexity, setting $f(x) = +\infty$ outside V.

Example 1. The functions $|x|, x^2, e^{\alpha x}$, linear function $a_i x^i$, positive definite quadratic form $a_{ij} x^i x^j$, homogeneous function of the first order $\sqrt{a_{ij} x^i x^j}$, functions

$$f(x) = \begin{cases} \log \frac{1}{x}, & 0 < x, \\ +\infty, & x \leq 0 \end{cases}$$

$$f(x) = \begin{cases} x^p, & 0 \leq x, \; 1 \leq p \leq +\infty \\ +\infty, & x < 0 \end{cases}$$

are all convex.

There are operations on convex functions, which preserve the convexity of those functions:

- The linear combination $a_1 f_1(x) + \ldots + a_m f_m(x)$ of convex functions $f_1(x), \ldots, f_m(x)$ with positive coefficients is a convex function.
- The superposition, $\varphi(f(x))$, of a non-decreasing convex function of one variable, $\varphi(y)$, and a convex function, $f(x)$, is a convex function.
- The function

$$f(x) = \min_{x = x_1 + \ldots + x_m} \left(f_1(x_1) + \ldots + f_m(x_m) \right)$$

  where $f_1(x), \ldots, f_m(x)$ are convex functions and $x \in R_n$, is also convex. It is called the convolution of $f_1, \ldots, f_m$.
- The point-wise maximum of several convex functions,

$$f(x) = \max \{ f_1(x), \ldots, f_m(x) \},$$

is a convex function. Note that point-wise minimum of several convex functions

$$g(x) = \min \{ f_1(x), f_2(x), \ldots, f_m(x) \}$$

is not necessarily convex (see Fig. 5.10).

A convenient criterion for convexity is the following statement: a differentiable function $f(x)$, $x \in R_n$, is convex if and only if for any two points, x_1 and x_2, the inequality holds:

$$0 \leq \left(\left. \frac{\partial f}{\partial x^i} \right|_{x_1} - \left. \frac{\partial f}{\partial x^i} \right|_{x_2} \right) \left(x_1^i - x_2^i \right). \tag{5.86}$$

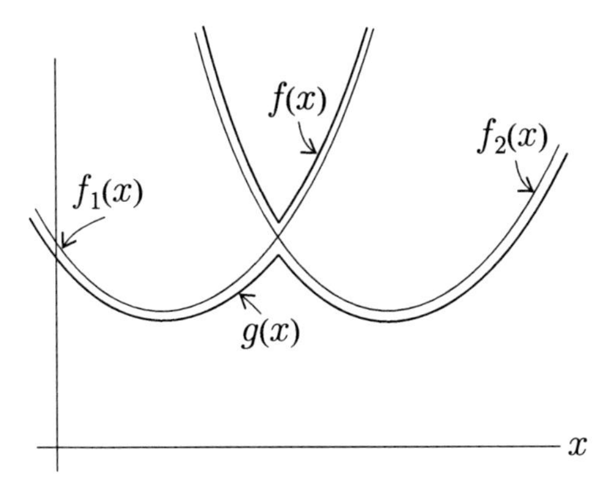

Fig. 5.10 Point-wise maximum of two convex functions, $f(x)$, is always convex; point-wise minimum, $g(x)$, may be not convex

In the proof of this statement, we will assume that the function f is twice continuously differentiable. Consider first a convex, twice differentiable function of one variable, $\varphi(\tau)$,

$$\varphi\left(\frac{\tau_1+\tau_2}{2}\right) \leq \frac{1}{2}\left(\varphi(\tau_1)+\varphi(\tau_2)\right). \tag{5.87}$$

Denote $\tau \equiv \frac{1}{2}(\tau_1+\tau_2)$, $\Delta\tau \equiv \frac{1}{2}(\tau_2-\tau_1)$. Assuming that $\Delta\tau$ is infinitesimally small, we expand the right hand side with respect to $\Delta\tau$, retaining the terms of the order $\Delta\tau$ and $\Delta\tau^2$:

$$\varphi(\tau) \leq \frac{1}{2}\left(\varphi(\tau) - \frac{d\varphi}{d\tau}\Delta\tau + \frac{1}{2}\frac{d^2\varphi}{d\tau^2}(\Delta\tau)^2 + \right.$$

$$\left. +\varphi(\tau) + \frac{d\varphi}{d\tau}\Delta\tau + \frac{1}{2}\frac{d^2\varphi}{d\tau^2}(\Delta\tau)^2\right) = \varphi(\tau) + \frac{1}{2}\frac{d^2\varphi}{d\tau^2}(\Delta\tau)^2.$$

Hence, $0 \leq d^2\varphi/d\tau^2$. Therefore, the derivative of a convex function, $\varphi(\tau)$, is monotonously increasing.

Consider now the function $\varphi(\tau) = f\left(x^i + \tau y^i\right)$ of one variable τ. The function $\varphi(\tau)$ is convex. Since

$$\left.\frac{d\varphi}{d\tau}\right|_{\tau=0} \leq \left.\frac{d\varphi}{d\tau}\right|_{\tau=1},$$

we have

$$\left.\frac{\partial f}{\partial x^i}\right|_x y^i \leq \left.\frac{\partial f}{\partial x^i}\right|_{x+y} y^i. \tag{5.88}$$

Setting in (5.88), $x = x_1,\ y = x_2 - x_1$ we arrive at (5.86). Now we assume that (5.86) is true and prove that function $f(x)$ is convex. For some $0 < \alpha < 1$, consider the expression

$$\Delta = \alpha f(x) + (1-\alpha) f(y) - f(\alpha x + (1-\alpha) y).$$

The argument, $\alpha x + (1-\alpha) y$, can be put in various forms:

$$\alpha x + (1-\alpha) y = y + \alpha (x-y) = x + (1-\alpha)(y-x).$$

Subtracting from Δ the function $\alpha f(x + (1-\alpha)(y-x))$ and adding the same function written as $\alpha f(y + \alpha(x-y))$, we rewrite Δ as

$$\Delta = \alpha (f(x) - f(x + (1-\alpha)(y-x))) + (1-\alpha)(f(y) - f(y + \alpha(x-y))).$$

Applying the Lagrange formula for finite increments, we get

$$\begin{aligned}\Delta = & -\alpha \left.\frac{\partial f}{\partial x^i}\right|_{x+\tau_1(1-\alpha)(y-x)} (1-\alpha)\left(y^i - x^i\right) - \\ & -(1-\alpha) \left.\frac{\partial f}{\partial x^i}\right|_{x+\tau_2\alpha(x-y)} \alpha\left(x^i - y^i\right)\end{aligned}$$

where $0 \le \tau_1 \le 1,\ 0 \le \tau_2 \le 1$. We set $x_1 = x + \tau_1 (1-\alpha)(y-x)$ and $x_2 = y + \tau_2\alpha(x-y)$. Since $x_1 - x_2 = (x-y)[(1-\tau_1)(1-\alpha) + (1-\tau_2)\alpha]$ and the expression in the square brackets is positive,

$$\Delta[(1-\tau_1)(1-\alpha) + (1-\tau_2)\alpha] = \alpha(1-\alpha)\left(\left.\frac{\partial f}{\partial x^i}\right|_{x_1} - \left.\frac{\partial f}{\partial x^i}\right|_{x_2}\right)\left(x_1^i - x_2^i\right),$$

and $\Delta \ge 0$ due to (5.86). This proves the convexity of $f(x)$.

The condition of the strict convexity of the differentiable function $f(x)$ is equivalent to the validity of the strict inequality

$$0 < \left(\left.\frac{\partial f}{\partial x^i}\right|_{x_1} - \left.\frac{\partial f}{\partial x^i}\right|_{x_2}\right)\left(x_1^i - x_2^i\right) \quad \text{for } |x_1 - x_2| \ne 0. \tag{5.89}$$

For applications, another criterion of convexity is also useful: a function, $f(x)$, possessing continuous second derivatives is convex in some convex region, if and only if the quadratic form of the variables $\bar{x}^i$,

$$\frac{\partial^2 f(x)}{\partial x^i \partial x^j}\bar{x}^i\bar{x}^j \ge 0, \tag{5.90}$$

is nonnegative at every point x in this region.

This statement follows from (5.86). Setting $x_2^i = x_1^i + \varepsilon \bar{x}^i$ in (5.86) and tending ε to zero, we get that (5.90) is a necessary condition for convexity. Conversely, let (5.90) be valid. Then

$$\int_0^1 \frac{\partial^2 f\left(x^k + t\bar{x}^k\right)}{\partial x^i \partial x^j} \bar{x}^i \bar{x}^j dt \geq 0.$$

This integral can be written as

$$0 \leq \int_0^1 \frac{d}{dt} \left(\frac{\partial f\left(x^k + t\bar{x}^k\right)}{\partial x^i} \bar{x}^i \right) dt = \left(\frac{\partial f\left(x^k + \bar{x}^k\right)}{\partial x^i} - \frac{\partial f\left(x^k\right)}{\partial x^i} \right) \bar{x}^i. \tag{5.91}$$

Putting in (5.91), $\bar{x}^i = x_2^i - x_1^i$, $x^i = x_1^i$, we obtain (5.86), and, thus, convexity of $f(x)$.

The Legendre transformation. We have already considered the Legendre transformation in Sect. 1.7 when we derived the Hamiltonian equations. Now we discuss it in more detail in the general case.

Let $f(x)$ be a twice continuously differentiable function. Consider a system of non-linear equations with respect to x^i:

$$\frac{\partial f(x)}{\partial x^i} = x_i^* \tag{5.92}$$

where x_i^* are given. If for some values of x_i^* the solution of the system of (5.92) is x^i and at the point x^i the Hessian $\left| \frac{\partial^2 f}{\partial x^i \partial x^j} \right|$ is not zero, then, according to the implicit function theorem, there exists a neighborhood O of this point, for which there is a one-to-one continuously differentiable correspondence between x^i and x_i^* :

$$x^i = x^i\left(x_k^*\right). \tag{5.93}$$

Define the quantity

$$f^{\times} = x_k^* x^k - f(x). \tag{5.94}$$

Let x^i in (5.94) be expressed in terms of x_k^* by means of (5.93). Then $f^{\times}$ becomes a function of x_k^*. It is called the Legendre transformation of the function $f(x)$. The arguments of the Legendre transformation, x_k^*, are called dual variables to x^k.

Example 2. Let us find the Legendre transformation of a quadratic function, $f(x) = \frac{1}{2} a_{ij} x^i x^j$. The system of (5.92) takes the form

$$\frac{\partial f}{\partial x^i} = a_{ij}x^j = x_i^*. \tag{5.95}$$

For a given x_i^*, this is a system of linear equations with respect to x^i. Assuming that $\det\|a_{ij}\| \neq 0$, we can write the solution of (5.95) as

$$x^i = a^{(-1)ij}x_j^*,$$

$a^{(-1)ij}$ being the components of the inverse matrix to the matrix $\|a_{ij}\|$. For $f^\times(x^*)$, we obtain

$$f^\times(x^*) = x_i^*x^i - \frac{1}{2}a_{ij}x^ix^j = \frac{1}{2}x_i^*x^i = \frac{1}{2}a^{(-1)ij}x_i^*x_j^*.$$

We see that the Legendre transformation of the quadratic function is also a quadratic function. One can show that the quadratic function is the only function which possesses such a property.

Example 3. Let x be a number, and

$$f(x) = \frac{1}{r}|x|^r, \quad r > 1.$$

It is easy to check that

$$f^\times(x^*) = \frac{1}{s}|x^*|^s$$

where

$$\frac{1}{r} + \frac{1}{s} = 1.$$

The case $r = 2$ corresponds to quadratic functions. In this case the Legendre transformation is also a quadratic function: $s = 2$. If $1 < r < 2$, then $s > 2$. If $r > 2$, then $1 < s < 2$. For a similar function with a coefficient, a,

$$f = \frac{a}{r}|x|^r, \quad a > 0,$$

its Legendre transformation is

$$f^\times(x^*) = \frac{1}{sa^{s-1}}|x^*|^s = \frac{a}{s}\left|\frac{x^*}{a}\right|^s.$$

A function in n-dimensional space,

$$f(x) = \frac{a}{r}\left(x_ix^i\right)^{\frac{r}{2}},$$

has the Legendre transformation,

$$f^{\times}(x^*) = \frac{1}{sa^{s-1}} \left(x_i^* x^{*i}\right)^{\frac{s}{2}} .$$

For a function,

$$f(x) = \frac{1}{r} \left(a_{ij} x^i x^j\right)^{\frac{r}{2}},$$

the Legendre transformation is

$$f^*(x^*) = \frac{1}{s} \left(a^{-1ij} x_i^* x_j^*\right)^{\frac{s}{2}} .$$

The Legendre transformation is defined only in a small neighborhood of the point x. However, if the function is strictly convex on R_n, then the Legendre transformation can be found for all points in R_n. Indeed, in this case there is a one-to-one correspondence between $\partial f/\partial x^i$ and x^i (otherwise, if there are two different points x_1 and x_2, for which $\partial f/\partial x^i$ are equal, then

$$\left(\left(\frac{\partial f}{\partial x^i}\right)\Big|_{x_1} - \left(\frac{\partial f}{\partial x^i}\right)\Big|_{x_2}\right) \left(x_1^i - x_2^i\right) = 0,$$

which contradicts to (5.89)).

Example 2 shows that the function $f(x)$ can be non-convex, but the Legendre transformation′ is defined on the entire space R_n : such is a quadratic function with the coefficients a_{ij} which have both positive and negative eigenvalues. There are functions for which the Legendre transformation is meaningless.

Example 4. Consider the function $f(x) = |x|$. Equation (5.92) becomes

$$\pm 1 = x^*$$

and does not have a solution for all x^*, except $x^* = \pm 1$. The Legendre transformation′ is meaningless for the function $|x|$.

A generalization of this example is as follows.

Example 5. The Legendre transformation′ is not defined for an arbitrary function of the first order of homogeneity (i.e. function possessing the property $f(\lambda x) = |\lambda| f(x)$) since for this function identically,

$$\det \left\| \frac{\partial^2 f}{\partial x^i \partial x^j} \right\| = 0.$$

Indeed, for the homogeneous functions of the first order Euler identity holds:

$$\frac{\partial f}{\partial x^i} x^i = f.$$

Differentiating the Euler identity with respect to x^i, we get

$$\frac{\partial^2 f}{\partial x^i \partial x^j} x^j = 0. \tag{5.96}$$

Equation (5.96) means that the rows of the matrix $\left\| \frac{\partial^2 f}{\partial x^i \partial x^j} \right\|$ are linearly dependent and thus $\det \left\| \frac{\partial^2 f}{\partial x^i \partial x^j} \right\| = 0$.

For the homogeneous function of first order,

$$f = \sqrt{x_i x^i},$$

the absence of solvability of (5.92) can be seen directly from that equations. Indeed, for this function,

$$\frac{\partial f}{\partial x^i} = \frac{x^i}{\sqrt{x_k x^k}} = x_i^*. \tag{5.97}$$

Equation (5.97) implies that x_i^* lies on the surface of the sphere of unit radius $x_i^* x^{*i} = 1$, and therefore x^i and x_i^* cannot be uniquely related.

The difficulty demonstrated by Examples 4 and 5 is resolved by the Young-Fenchel transformation.

The Young-Fenchel transformation. Consider a function $f(x)$ on R_n and define the function $f^*(x^*)$ as

$$f^*(x^*) = \max_x \left(x_i^* x^i - f(x) \right). \tag{5.98}$$

The function $f^*(x^*)$ is called the Young-Fenchel transformation of the function $f(x)$. Since for any function φ,

$$\max \varphi = -\min(-\varphi),$$

the definition (5.98) can also be written as

$$-f^*(x^*) = \min_x \left(f(x) - x_i^* x^i \right). \tag{5.99}$$

Equation (5.99) has a simple geometric interpretation. Consider in R_{n+1} the graph of the function $f(x)$, $y = f(x)$, and the graph of the linear function $y = x_i^* x^i$ (Fig. 5.11). The quantity, $-f^*(x^*)$, is the minimum vertical distance between these two functions.

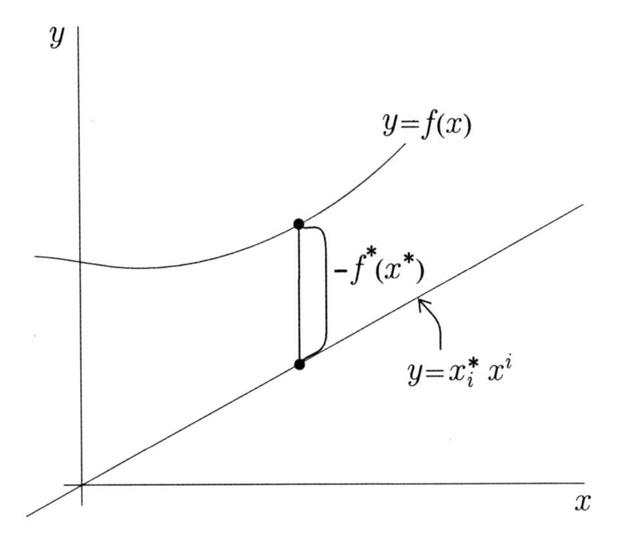

Fig. 5.11 Geometrical interpretation of the Young-Fenchel transform, f^*, of function f

Let the function $f(x)$ be strictly convex and differentiable. Then, due to convexity, the minimum in (5.99) is reached at only one point, and at this point

$$f^* = x_i^* x^i - f(x), \quad \frac{\partial f}{\partial x^i} = x_i^*.$$

Hence, for strictly convex differentiable functions the Young-Fenchel transformation coincides with the Legendre transformation. However, the Young-Fenchel transformation is also valid for functions for which the Legendre transformation is meaningless.

Example 6. Consider the Young-Fenchel transformation for the function $f(x) = |x|$. Let us find

$$\min_x \left(|x| - x^* x\right).$$

It is seen from Fig. 5.12 that $\min_x (|x| - x^* x) = 0$ for $-1 < x^* < 1$; it is reached at $x = 0$. For $x^* = \pm 1$, $\min_x (|x| - x^* x) = 0$; it is reached on the positive semi-axis for $x^* = 1$ and on the negative semi-axes for $x^* = -1$. For $x^* < -1$ and $x^* > 1$, the function $|x| - x^* x$ is not bounded from below; it tends to $-\infty$ for $x \to -\infty$ and $x \to +\infty$. So,

$$f^*\left(x^*\right) = \begin{cases} 0, & |x^*| \leq 1 \\ +\infty, & 1 > |x^*| \end{cases}.$$

A symbolic graph of this function is shown in Fig. 5.13.

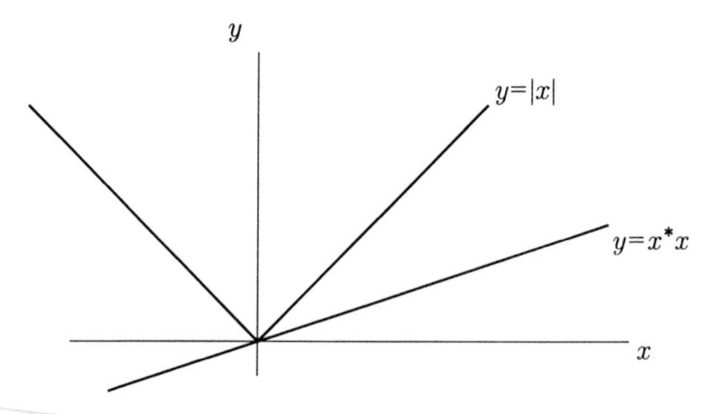

Fig. 5.12 Computation of the Young-Fenchel transform of function $|x|$

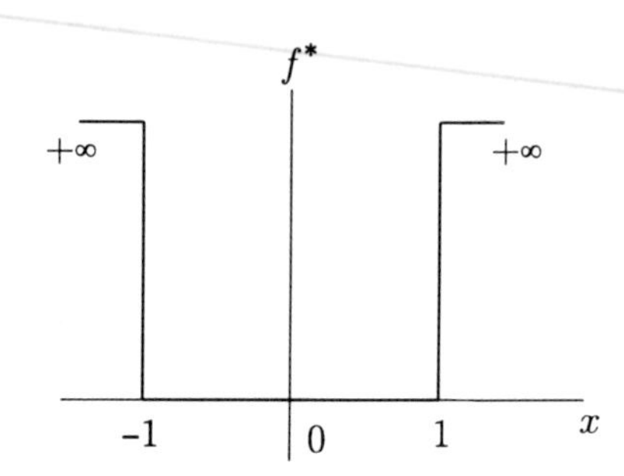

Fig. 5.13 A symbolic graph of the Young-Fenchel transform of the function $|x|$

Example 7. The Young-Fenchel transformation for the homogeneous function of the first order of the form $f = \sqrt{x_i x^i}$ is calculated similarly:

$$f^*\left(x^*\right) = \begin{cases} 0, & x_i^* x^{*i} \leq 1 \\ +\infty, & 1 > x_i^* x^{*i} \end{cases}$$

The Young-Fenchel transformation has a number of remarkable properties.

For any function $f(x)$, its Young-Fenchel transformation is convex. To justify this statement, consider the function $f^*\left(\alpha x^* + (1-\alpha) y^*\right)$, where $0 \leq \alpha \leq 1$. According to the definition (5.98),

$$f^*\left(\alpha x^* + (1-\alpha) y^*\right) = \max_x \left[\left(\alpha x_i^* + (1-\alpha) y_i^*\right) x^i - f(x)\right] =$$
$$= \max_x \left[\alpha \left(x_i^* x^i - f(x)\right) + (1-\alpha)\left(y_i^* x^i - f(x)\right)\right].$$

Using the inequality

$$\max(f+g) \leq \max f + \max g,$$

we get

$$f^{*}\left(\alpha x^{*}+(1-\alpha) y^{*}\right) \leq \alpha \max_{x}\left(x_{i}^{*} x^{i}-f(x)\right)+(1-\alpha) \max_{x}\left(y_{i}^{*} x^{i}-f(x)\right)=$$
$$=\alpha f^{*}\left(x^{*}\right)+(1-\alpha) f^{*}\left(y^{*}\right),$$

i.e. $f^{*}(x^{*})$, is convex.

Example 8. In this example we illustrate the notion of the Young-Fenchel transformation by applying it to thermodynamic functions. Consider the thermodynamic potentials of the ideal compressible gas. The model of the ideal compressible non-heat-conducting gas is specified by its internal energy density, $U(\rho, S)$. Assume that the Cauchy problem for the system of equations of the ideal compressible gas is correct, i.e. small disturbances of initial conditions yield small variations of the solution. It is known that the system of equations possesses such a property if it is hyperbolic. It is not difficult to check that hyperbolicity holds when pressure, $p=\rho^{2}\, \partial U(\rho, S) / \partial \rho$, increases if the density increases. Hence, the internal energy density has to satisfy the condition,

$$\frac{\partial}{\partial \rho} \rho^{2} \frac{\partial U(\rho, S)}{\partial \rho}>0. \tag{5.100}$$

There are two possible interpretations of the inequality (5.100). Define the specific volume as $\vartheta=1 / \rho$. Multiplying (5.100) by ρ^{2}, we can write (5.100) in terms of the derivative with respect to ϑ as

$$\frac{\partial^{2} U(\vartheta, S)}{\partial \vartheta^{2}}>0. \tag{5.101}$$

The inequality (5.101) means that the function $U(\vartheta, S)$ is a convex function of the specific volume ϑ for every fixed value of S. The inequality (5.100) can also be written as

$$\frac{\partial^{2}}{\partial \rho^{2}} \rho U(\rho, S)>0. \tag{5.102}$$

Here we used an identity which holds for any function $f(\rho)$:

$$\frac{1}{\rho} \frac{d}{d \rho} \rho^{2} \frac{d f(\rho)}{d \rho}=\frac{d^{2}}{d \rho^{2}} \rho f(\rho).$$

According to the inequality (5.102), the function $\rho U(\rho, S)$ is convex with respect to ρ for every fixed value of entropy S.

As an example, let us take the internal energy density of the ideal gas,

$$U=a \rho^{\gamma-1} e^{S / c_{v}}, \tag{5.103}$$

where a, γ, c_{v} are positive constants and $\gamma>1$. Entropy in gas dynamics is determined up to an additive constant. Therefore, the factor a can be eliminated. We keep it, however, for dimension reasoning. Function (5.103) is not convex with respect to ρ for $\gamma < 2$, because

$$\frac{\partial^2 U}{\partial \rho^2} = a(\gamma - 1)(\gamma - 2)\rho^{\gamma-3} e^{S/c_v} < 0 \text{ for } \gamma < 2.$$

However, the function $\rho U(\rho, S)$ is a convex function of ρ for all $\gamma > 1$.

In terms of variables ϑ, S the internal energy density of a ideal gas is

$$U = a\frac{1}{\vartheta^{\gamma-1}} e^{S/c_v}$$

It is a convex function of ϑ, as any function of the form b/x^r, for $r > 0, b > 0, x > 0$.

It is natural to require that energy increases monotonously with increase of S (that is equivalent to positiveness of absolute temperature $T = \partial U/\partial S$), and temperature increases monotonously with increase of S ($\partial T/\partial S = \partial^2 U/\partial S^2 > 0$). Then, for every fixed ϑ, U will be a convex function of S. Without loss of generality one can assume that entropy is non-negative. The region on which ϑ and S change is a convex set, $\vartheta \geq 0$, $S \geq 0$. One can extend U to all ϑ, S preserving the convexity property and prohibiting the negative values of ϑ and S by setting

$$\bar{U}(\vartheta, S) = \begin{cases} U(\vartheta, S), & \vartheta \geq 0, \ S \geq 0, \\ +\infty & \text{otherwise} \end{cases}$$

Let us define the free energy $F(\vartheta, T)$ as

$$F(\vartheta, T) = \min_{S \geq 0} (U(\vartheta, S) - TS).$$

This definition can also be written in terms of minimum of the function $\bar{U}$ over all values of S:

$$F(\vartheta, T) = \min_{S} \left(\bar{U}(\vartheta, S) - TS\right).$$

Denote the Young-Fenchel transformation of the function $\bar{U}(\vartheta, S)$ with respect to S by $\bar{U}^*(\vartheta, S^*)$, the Young-Fenchel transformation with respect to ϑ by $\bar{U}^*(\vartheta^*, S)$, and the Young-Fenchel transformation with respect to ϑ and S by $\bar{U}^*(\vartheta^*, S^*)$. The thermodynamic potentials are expressed in terms of these functions as follows. The free energy differs from the Young-Fenchel transformation of internal energy with respect to entropy by the sign:

$$F(\vartheta, T) = -U^*\left(\vartheta, S^*\right)\Big|_{S^*=T}.$$

Therefore $F(\vartheta, T)$ is a concave[8] function of temperature. An important consequence of this fact is that in the increment of free energy, ΔF, resulting from an incremental growth of temperature, ΔT, the coefficient at $(\Delta T)^2$ must be negative due to the concavity of F. In the case of the ideal gas, the free energy is

$$F(\vartheta, T) = c_v T\left(1 - \ln\left(\frac{c_v}{a}\vartheta^{\gamma-1}T\right)\right).$$

This function is convex with respect to specific volume ϑ and concave with respect to temperature, hence its graph in the neighborhood of every point is a saddle.

The enthalpy is defined by the formula

$$i(p, S) = \min_{\vartheta}\left(\bar{U}(\vartheta, S) + p\vartheta\right).$$

This is "almost" the Young-Fenchel transformation: pressure differs from the dual variable to specific volume by the sign, while enthalpy is negative Young-Fenchel transformation of internal energy:

$$i(\rho, S) = -U^*(\vartheta^*, S)\big|_{\vartheta^*=-p}.$$

Since U is usually a decreasing function of ϑ and $\vartheta^* = \partial U/\partial\vartheta < 0$, pressure $p = -\partial U/\partial\vartheta$ is positive. According to its definition, enthalpy is a concave function of pressure. For the ideal gas,

$$i(\rho, S) = \gamma\left(\frac{\gamma'}{\gamma}\right)^{\frac{1}{\gamma'}}\left(ae^{\frac{S}{c_v}}\right)^{\frac{1}{\gamma}} p^{\frac{1}{\gamma'}}, \quad \frac{1}{\gamma} + \frac{1}{\gamma'} = 1.$$

and enthalpy is convex with respect to S. So, in the vicinity of each point the graph of enthalpy is a saddle.

The Gibbs thermodynamic potential (chemical potential) is defined as

$$\mu(p, T) = \min_{\vartheta, S}\left(\bar{U}(\vartheta, S) + p\vartheta - TS\right). \quad (5.104)$$

Chemical potential also differs by the sign from the Young-Fenchel transformation of internal energy:

$$\mu(p, T) = -U^*(\vartheta^*, S^*)\big|_{\substack{S^*=T \\ \vartheta^*=-p}}.$$

Chemical potential is a concave function with respect to both p and T. In the case of the ideal gas,

---

[8] The function $f(x)$ is concave if the negative of this function, $-f(x)$, is convex.

$$\mu(p,T) = c_v T \left\{ \ln \left[\frac{a}{c_v T} \left(\frac{p}{c_v T} \right)^{\gamma-1} \right] + \gamma - (\gamma-1)\ln(\gamma-1) \right\}.$$

If function $F(\vartheta, T)$ is strictly convex with respect to ϑ, pressure $p = -\partial F/\partial\vartheta$ is in one-to-one correspondence with the specific volume, ϑ, for every value of the temperature T. There is an important class of models in which non-convex functions $F(\vartheta, T)$ appear. These are the models which describe the phase transitions. In such models, one value of p can correspond to several values of ϑ. A typical example: the van der Waals gas, for which

$$F(\vartheta, T) = f(T) - \frac{c}{\vartheta} - RT \ln\left(\frac{\vartheta}{b} - 1\right),$$

where c, R, b are positive constants, and $f(T)$ is some function of temperature. A sketch of the free energy of the van der Waals gas is shown in Fig. 5.14. For large values of T, the free energy is convex with respect to ϑ and $\partial^2 F/\partial\vartheta^2 > 0$ (the top curve). If temperature is decreased, then, for some value of temperature T_{cr}, these appears an inflection point A, at which the derivative $F_{\vartheta\vartheta}$ is equal to zero (the middle curve). For $T < T_{cr}$ the free energy is non-convex, and, depending upon the value of pressure p, there are one or three corresponding values of ϑ. The left part of the curve (to the left of point C) on Fig. 5.15 corresponds to the liquid phase, and the right part of the curve (to the right of point D) to the fluid phase. The state corresponding to the segment CD of the curve is in some sense not realized.

We continue the consideration of the properties of the Young-Fenchel transformation.

For any function $f(x)$ and any x and x^*, the following inequality holds:

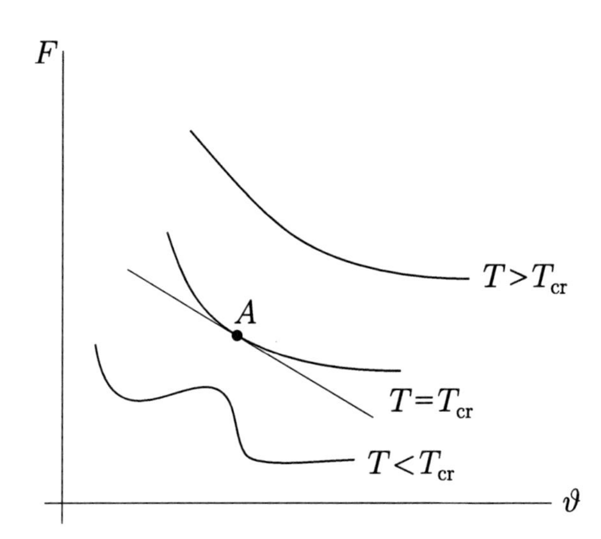

Fig. 5.14 A sketch of free energy of the van der Waals gas

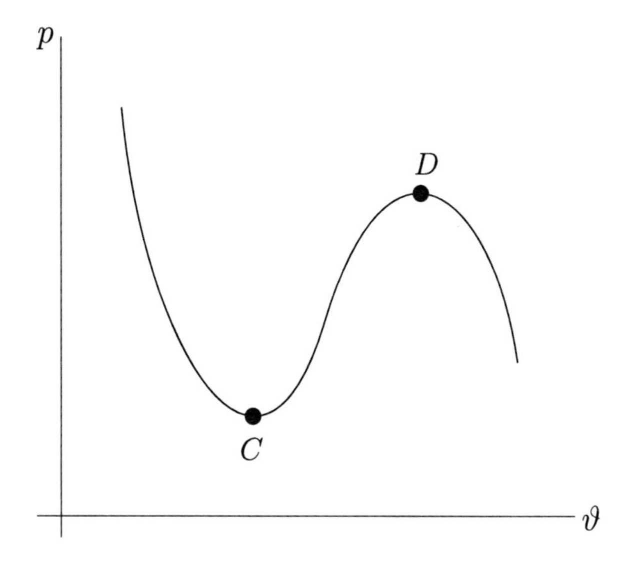

Fig. 5.15 Dependence of pressure on specific volume for van der Walls gas

$$x_i^* x^i \leq f(x) + f^*(x^*) \tag{5.105}$$

This inequality follows directly from the definition (5.98). It is called the Young-Fenchel inequality.

The Young-Fenchel transformation of $f^*(x^*)$ is a function of the dual variable to x^*, i.e. x. Denote this function by $f^{**}(x)$.

For any function $f(x)$ the following inequality holds:

$$f^{**}(x) \leq f(x). \tag{5.106}$$

The proof is simple: the definition of $f^{**}(x)$ in terms of $f(x)$ can be written as

$$f^{**}(x) = \max_{x^*} \left(x^i x_i^* - f^*(x^*)\right) = \max_{x} \left[x^i x_i^* + \min_{z} \left(f(z) - x_i^* z^i\right)\right]$$

If we increase $\min_z$ by substituting it by the value of the function $f(z) - x_i^* z^i$ at the point $z^i = x^i$, we get

$$f^{**}(x) \leq \max_{x^*} f(x) = f(x)$$

as claimed.

The equality

$$f^{**}(x) = f(x) \tag{5.107}$$

holds if and only if the function $f(x)$ is convex and semi-continuous below.

We will prove this statement for the case of continuous functions $f(x)$. If $f^{**}(x) = f(x)$, then $f(x)$ is a convex function as the Young-Fenchel transfor-

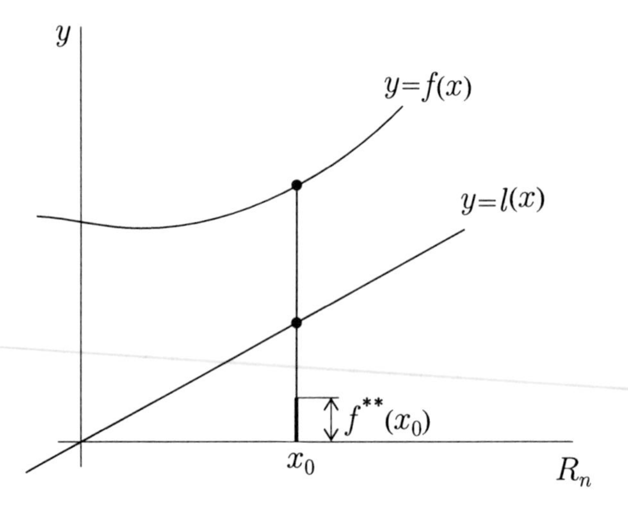

Fig. 5.16 To the proof of the inequality (5.107)

mation of some function. Conversely, let $f(x)$ be convex. According to (5.106), $f^{**}(x) \leq f(x)$. Suppose that there exists a point x_0 for which

$$f^{**}(x_0) < f(x_0). \tag{5.108}$$

Define a linear function $l(x)$ in such a way that $l(x_0) > f^{**}(x)$ and $f(x) > l(x)$ (see Fig. 5.16)[9] Let us write $l(x)$ as

$$l(x) = l(x_0) + z_i^* \left(x^i - x_0^i\right).$$

Since

$$f(x) > l(x) = l(x_0) + z_i^* \left(x^i - x_0^i\right) > f^{**}(x_0) + z_i^* \left(x^i - x_0^i\right),$$

we have

$$z_i^* x^i - f(x) < z_i^* x_0^i - f^{**}(x_0).$$

Consequently,

$$f^*\left(z^*\right) = \max_x \left(z_i^* x^i - f(x)\right) < z_i^* x_0^i - f^{**}(x_0). \tag{5.109}$$

[9] The existence of such a linear function is guaranteed by the separability theorem: for any closed convex set $\mathcal{M}$ in a finite-dimentional space and any point A not belonging to $\mathcal{M}$ there exists a plane separating them or, equivalently, there exists a linear function $l(x)$ taking on positive values on $\mathcal{M}$ and taking on a negative value at the point A. This theorem should be applied in R_{n+1} to the convex set epi f and the point $(x_0, f^{**}(x))$.

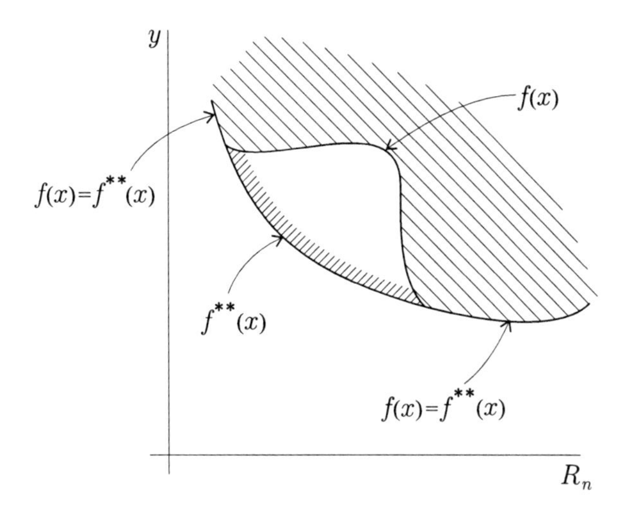

Fig. 5.17 Double Young-Fenchel transformation, $f^{**}(x)$, of a non-convex function $f(x)$

The inequality (5.109) contradicts the Young-Fenchel inequality for the functions f^* and f^{**}:

$$x_i^* x^i \le f^*\left(x^*\right) + f^{**}(x).$$

Therefore, the initial assumption (5.108) is false and $f^{**}(x) = f(x)$ for all x. *For an arbitrary (not necessarily convex) function $f(x)$, the function $f^{**}(x)$ is the maximum convex function which is not greater than $f(x)$.*

The maximum convex function can be found as follows: consider the epigraph, epi f, of the function f (Fig. 5.17, the epigraph is shaded), and construct the smallest possible convex set in the (y, R_n)-space containing epi f; it will be the epigraph of the maximum convex function not greater than f. This set is obtained by "filling out" of all the concave regions of the epigraph of the function $f(x)$. This set is the epigraph of the function $f^{**}(x)$.

The proof will be given again only for a continuous function, $f(x)$. Denote the maximum convex function not greater than f by $\tilde{f}(x)$. Let us show that $\tilde{f}^* = f^*$. Let us fix an arbitrary point x^*. Since $\tilde{f}(x) \le f(x)$, we have $\tilde{f}^*(x^*) \ge f^*(x^*)$. Consequently, if $f^*(x^*) = +\infty$, then $\tilde{f}^*(x^*) = +\infty$ and $\tilde{f}^*(x^*) = f^*(x^*)$. Now let the value of the function f^* at a point x^* be finite. By definition,

$$-f^*\left(x^*\right) = \min_x \left(f(x) - x_i^* x^i\right).$$

Denote the point at which the minimum is reached[10] by x_0. Then

$$-f^*\left(x^*\right) = f(x_0) - x_i^* x_0^i. \tag{5.110}$$

[10] There can be many such points.

It follows from the Young-Fenchel inequality, $x_i^* x^i - f^*(x^*) \leq f(x)$, and (5.110) that

$$x_i^* \left(x^i - x_0^i\right) + f(x_0) \leq f(x). \tag{5.111}$$

We wish to show that $\tilde{f}(x_0) = f(x_0)$. Suppose the contrary,

$$\tilde{f}(x_0) < f(x_0) \tag{5.112}$$

According to (5.112), the value of the function $\tilde{f}(x)$ at x_0 is less than the value of the function $x_i^* \left(x^i - x_0^i\right) + f(x_0)$ at this point. Therefore, there exists a region D in which $\tilde{f}(x) < x_i^* \left(x^i - x_0^i\right) + f(x_0)$, and outside of this region $\tilde{f}(x) \geq x_i^* \left(x^i - x_0^i\right) + f(x_0)$. Define a convex function

$$\tilde{\tilde{f}}(x) = \begin{cases} x_i^*(x^i - x_0^i) + f(x_0), & x \in D \\ \tilde{f}(x), & x \bar{\in} D \end{cases}$$

Inequality (5.111) and the inequality $\tilde{f}(x) \leq f(x)$ imply that $\tilde{\tilde{f}}(x) \leq f(x)$. Moreover, $\tilde{f}(x) < \tilde{\tilde{f}}(x)$ on D. This contradicts the fact that $\tilde{f}$ is the maximum convex function not greater than f. The proof that the graph of the function $\tilde{f}(x)$ cannot be below the line $y = x_i^* \left(x^i - x_0^i\right) + f(x_0)$ is analogous.

Since

$$\begin{aligned} -\tilde{f}^* &= \min_x \left(\tilde{f}(x) - x_i^* x^i\right) \geq \min_x \left(x_i^* \left(x^i - x_0^i\right) + f(x_0) - x_i^* x^i\right) \\ &= f(x_0) - x_i^* x^i = -f^*\left(x^*\right), \end{aligned}$$

and

$$-\tilde{f}^* = \min_x \left(\tilde{f}(x) - x_i^* x^i\right) \leq \min_x \left(f(x) - x_i^* x^i\right) = -f^*\left(x^*\right)$$

we have

$$\tilde{f}^*\left(x^*\right) = f^*\left(x^*\right)$$

Applying the Young-Fenchel transformation to this equality, we get $\tilde{f}^{**} = \tilde{f} = f^{**}$, which completes the proof.

Example 9. Consider a mixture of liquid and its vapor bubbles. The liquid and the vapor are two states of the same matter. This matter is characterized by the free energy $F(\vartheta, T)$, which is non-convex with respect to ϑ, or by the corresponding function $p(\vartheta, T)$, which is shown in Fig. 5.18a for a fixed value of T. The mixture can also be modeled by a continuum. Thus the macro-continuum could be in three states, pure liquid, pure vapor or a mixture of liquid with vapor bubbles. For such macro-continuum, Maxwell proposed the following rule for determining the dependence of the pressure on the specific volume: on the graph of the "true" dependence

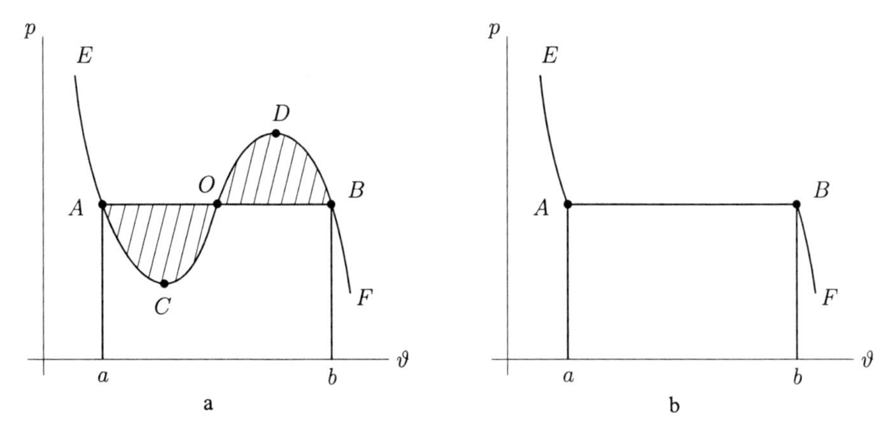

Fig. 5.18 The Maxwell rule

of p on ϑ, Fig. 5.18a, a straight line AB should be drawn in such a way, that the shaded areas ACO and ODB are equal; then, the graph $p(\vartheta, T)$ for the mixture consists of three branches EA, AB and BF (Fig. 5.18b). They correspond to pure liquid, mixture of liquid with vapor bubbles and vapor bubbles, respectively.

Let us show that the Maxwell rule is equivalent to the following statement: the free energy of the mixture is equal to $F^{**}(\vartheta, T)$.

The function $F(\vartheta, T)$ is shown qualitatively in Fig. 5.19. The function $F^{**}(\vartheta, T)$ coincides with $F(\vartheta, T)$ for $\vartheta < a$ and $\vartheta > b$; for $a \leq \vartheta \leq b$, it is linear. Due to the smoothness of $F(\vartheta, T)$, the line AB is tangent to the graph of the function $F(\vartheta, T)$. Therefore,

$$\left.\frac{\partial F}{\partial \vartheta}\right|_{\vartheta=a} = \left.\frac{\partial F}{\partial \vartheta}\right|_{\vartheta=b}.$$

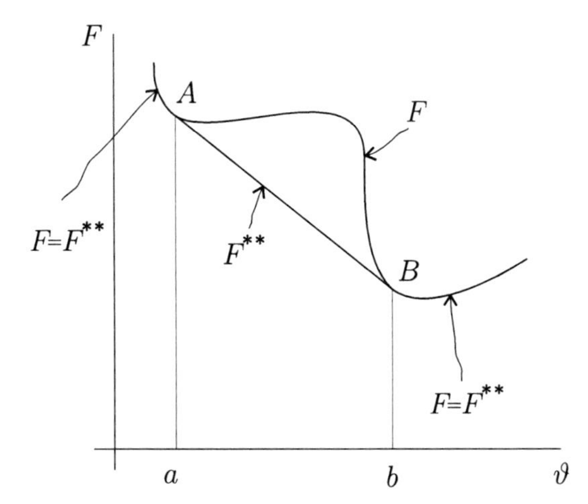

Fig. 5.19 The Maxwell rule in terms of the Young-Fenchel transformation

The figure shows that

$$F(b,T) - F(a,T) = \left.\frac{\partial F}{\partial \vartheta}\right|_{\vartheta=a}(b-a).$$

Besides,

$$F(b,T) - F(a,T) = \int_a^b \frac{\partial F}{\partial \vartheta} d\vartheta.$$

Since $p = -\partial F/\partial \vartheta$, we have

$$\int_a^b p d\vartheta = p|_{\vartheta=a}(b-a). \tag{5.113}$$

The integral in (5.113) represents the area under the curve $p(\vartheta, T)$ on the segment $[a, b]$ and $p|_{\vartheta=a}(b-a)$ represents the area of the rectangle with its top side being AB. Therefore, (5.113) is equivalent to the Maxwell rule.

The Young-Fenchel transformation allows one to present the convex function $f(x)$ as

$$f(x) = \max_{x^*}\left(x_i^* x^i - f^*(x^*)\right). \tag{5.114}$$

If $f(x)$ is non-convex, then the Young-Fenchel transformation provides the best low estimate of $f(x)$ in terms of a convex function

$$f(x) \geq f^{**}(x) = \max_{x^*}\left(x^i x_i^* - f^*(x^*)\right). \tag{5.115}$$

Let the variables $y = (y_1, \ldots, y_m)$ be the parameters of function f, $f = f(x, y)$. Then its Young-Fenchel transformation with respect to x will also depend on parameters, y: $f^* = f^*(x^*, y)$. What can be said about the dependence of f^* on y? *If the function $f = f(x, y)$ is linear with respect to y, then the Young-Fenchel transformation of this function with respect to x is a convex function of y.*

This statement follows from a chain of inequalities:

$$\begin{aligned} f^*\left(x^*, \alpha y_1 + (1-\alpha) y_2\right) &= \max_{x^i}\left[x_i^* x^i - f(x, \alpha y_1 + (1-\alpha) y_2)\right] = \\ &= \max_{x^i}\left[x_i^* x^i - \alpha f(x, y_1) - (1-\alpha) f(x, y_2)\right] = \\ &= \max_{x^i}\left[\alpha\left(x_i^* x^i - f(x, y_1)\right) + (1-\alpha)\left(x_i^* x^i - f(x, y_2)\right)\right] \leq \\ &\leq \alpha f^*\left(x^*, y_1\right) + (1-\alpha) f^*\left(x^*, y_2\right), \end{aligned}$$

as claimed.

This almost obvious property of the Young-Fenchel transformation implies some non-trivial consequences.

Example 10. Let us show that for positive matrices, the diagonal components of the inverse matrix are convex functions of the original matrix components. Consider a positive quadratic form $f = \frac{1}{2}a_{ij}x^i x^j$. Then $f^*(x^*) = \frac{1}{2}a^{(-1)ij}x_i^* x_j^*$ where $a^{(-1)ij}$ are the components of the inverse matrix. Function f depends linearly on the parameters a_{ij}. As it has been established, f^* is a convex function of a_{ij}. Let us set $x_1^* = 1, x_2^* = 0, \ldots, x_n^* = 0$. Then, $2f^* = a^{(-1)11}$ is a convex function of a_{ij}. Similarly, $a^{(-1)22}, \ldots, a^{(-1)nn}$, as well as all linear combinations of $a^{(-1)ij}$ which are obtained by choosing various x_i^*, are convex functions of a_{ij}.

Our next topic is the construction of the dual variational problems by means of Young-Fenchel transformations.

5.8 Examples of Dual Variational Principles

The Dirichlet and von Neuman problems. Consider the minimization problem for the so-called Dirichlet functional,

$$E(u) = \int_\Omega \frac{1}{2}\left[\left(\frac{\partial u}{\partial x}\right)^2 + \left(\frac{\partial u}{\partial y}\right)^2\right] dxdy, \tag{5.116}$$

on the set of all functions of two variables $u(x, y)$ defined on some two-dimensional region Ω with piece-wise smooth boundary $\partial\Omega$ and taking on the assigned values at the boundary:

$$u(x, y)|_{\partial\Omega} = g(s). \tag{5.117}$$

Here $g(s)$ is a continuous function of the arc length, s, along $\partial\Omega$. The minimizing element of the variational problem (5.116), (5.117) is the solution of the Dirichlet problem

$$\frac{\partial^2 \breve{u}}{\partial x^2} + \frac{\partial^2 \breve{u}}{\partial y^2} = 0 \quad \text{in } \Omega, \qquad \breve{u}|_{\partial\Omega} = g(s). \tag{5.118}$$

We are going to construct a dual variational problem. Let us write the integrand in the Dirichlet functional in the form

$$\frac{1}{2}\left[\left(\frac{\partial u}{\partial x}\right)^2 + \left(\frac{\partial u}{\partial y}\right)^2\right] = \max_{p_x, p_y}\left[p_x\frac{\partial u}{\partial x} + p_y\frac{\partial u}{\partial y} - \frac{1}{2}\left(p_x^2 + p_y^2\right)\right].$$

Using this relation we can rewrite the initial variational problem as a minimax problem:

$$\min E(u) = \min_{u \in (5.117)} \int_{\Omega} \max_{p_x, p_y} \left[p_x \frac{\partial u}{\partial x} + p_y \frac{\partial u}{\partial y} - \frac{1}{2} \left(p_x^2 + p_y^2 \right) \right] dxdy.$$

Recall that the notation $u \in$(5.117) means that u satisfies the constraint (5.117).

In the integral, maximum over p_x and p_y is taken at each point (x, y). The maximum can be moved outside the integral, if we perform maximization over arbitrary functions of two variables, $p_x(x, y)$ and $p_y(x, y)$:

$$\min E(u) = \min_{u \in (5.117)} \max_{p_x(x,y), p_y(x,y)} \int_{\Omega} \left[p_x \frac{\partial u}{\partial x} + p_y \frac{\partial u}{\partial y} - \frac{1}{2} \left(p_x^2 + p_y^2 \right) \right] dxdy. \tag{5.119}$$

Further, for brevity, we use for this maximum the notation $\max\limits_{p_x, p_y}$, implying that this is maximization over the functions $p_x(x, y)$ and $p_y(x, y)$.

Let us change the order of maximization and minimization in (5.119). According to the inequality (5.81),

$$\max_{p_x, p_y} \min_{u \in (5.117)} \int_{\Omega} \left[p_x \frac{\partial u}{\partial x} + p_y \frac{\partial u}{\partial y} - \frac{1}{2} \left(p_x^2 + p_y^2 \right) \right] dxdy \le \min_{u \in (5.117)} E(u). \tag{5.120}$$

The minimum value in the left hand side of (5.120) can be found explicitly. Suppose that the functions p_x, p_y are continuously differentiable and continuous on the closed region Ω. Integrating by parts (5.120), we obtain

$$\int_{\Omega} \left[p_x \frac{\partial u}{\partial x} + p_y \frac{\partial u}{\partial y} - \frac{1}{2} \left(p_x^2 + p_y^2 \right) \right] dxdy \tag{5.121}$$
$$= \int_{\partial\Omega} (p_x n_x + p_y n_y) u ds - \int_{\Omega} \left(\frac{\partial p_x}{\partial x} + \frac{\partial p_y}{\partial y} \right) u dxdy - \int_{\Omega} \frac{1}{2} \left(p_x^2 + p_y^2 \right) dxdy.$$

Here, n_x, n_y are the components of the outward unit normal vector on $\partial\Omega$ (Fig. 5.20). Note that function u in the first integral in the right hand side of (5.121) can be replaced by $g(s)$ according to the boundary condition (5.117).

For the integrals in the right hand side of (5.121), we will use the notations

$$E^*(p) = \int_{\Omega} \frac{1}{2} (p_x^2 + p_y^2) dxdy$$

and

$$\Phi(u, p) = - \int_{\Omega} \left(\frac{\partial p_x}{\partial x} + \frac{\partial p_y}{\partial y} \right) u dxdy + l(p),$$

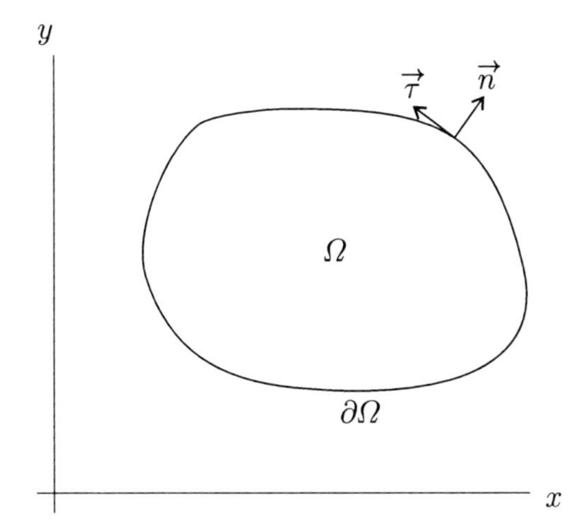

Fig. 5.20 Notation in the Dirichlet problem

where $p \equiv \{p_x, p_y\}$, and $l(p)$ is a linear functional of p:

$$l(p) = \int_{\partial\Omega} \left(p_x n_x + p_y n_y\right) g(s) ds.$$

So the minimization in (5.120) with respect to u is reduced to the minimization of the functional $\Phi(u, p)$, which is linear with respect to u.

Let us show that

$$\min_{u \in (5.117)} \Phi(u, p) = l(p) + \begin{cases} -\infty, & \text{if } \frac{\partial p_x}{\partial x} + \frac{\partial p_y}{\partial y} \neq 0 \\ 0, & \text{if } \frac{\partial p_x}{\partial x} + \frac{\partial p_y}{\partial y} = 0 \end{cases} \tag{5.122}$$

The second case in (5.122) is obvious. To prove (5.122) for the first case, suppose that for some internal point x_0 of the region Ω, the function $\frac{\partial p_x}{\partial x} + \frac{\partial p_y}{\partial y}$ is not zero, and, for definiteness, is greater than zero. Due the continuity of this function, there exists a neighborhood Δ of the point x_0 not intersecting $\partial\Omega$ for which this function is greater than zero. Consider a continuous function $\tilde{u}(x)$ which is positive in Δ, zero on the boundary of Δ, and satisfies the boundary conditions on $\partial\Omega$. Also consider a sequence of functions $u_n(x)$ coinciding with $\tilde{u}(x)$ outside Δ and equal to $n\tilde{u}(x)$ inside Δ. For this sequence, $\Phi(u_n, p) \to -\infty$ as $n \to \infty$, which completes the proof. For the case $\frac{\partial p_x}{\partial x} + \frac{\partial p_y}{\partial y} < 0$ at some point x_0, the proof is analogous.

From (5.121) and (5.122), we get

$$\min_{u\in(5.117)} \int_{\Omega} \left[p_x \frac{\partial u}{\partial x} + p_y \frac{\partial u}{\partial y} - \frac{1}{2}\left(p_x^2 + p_y^2\right)\right] dxdy =$$

$$= \begin{cases} l(p) - E^*(p), & \text{if } \frac{\partial p_x}{\partial x} + \frac{\partial p_y}{\partial y} = 0 \\ -\infty, & \text{if } \frac{\partial p_x}{\partial x} + \frac{\partial p_y}{\partial y} \neq 0 \end{cases}$$

After finding the minimum in (5.120), the maximum is to be found. Therefore, all values of p_x, p_y for which the minimum is equal to $-\infty$ should be excluded. So,

$$\max_p \left(l(p) - E^*(p)\right) \leq \check{E}, \tag{5.123}$$

where the maximum is calculated over all functions p_x, p_y satisfying the conditions

$$\frac{\partial p_x}{\partial x} + \frac{\partial p_y}{\partial y} = 0 \quad \text{in } \Omega. \tag{5.124}$$

Let us show that, actually, there is the equality sign in (5.123):

$$\max_p \left(l(p) - E^*(p)\right) = \check{E}.$$

To this end, consider the functions

$$\check{p}_x = \frac{\partial \check{u}}{\partial x}, \quad \check{p}_y = \frac{\partial \check{u}}{\partial y}, \tag{5.125}$$

where $\check{u}$ is the solution of the boundary value problem (5.118).[11] Due to the first equation (5.118), the functions (5.125) satisfy the constraint (5.124) and, hence, are the admissible functions. Therefore,

$$l(\check{p}) - E^*(\check{p}) \leq \max_p \left(l(p) - E^*(p)\right)$$

and, according to (5.123),

$$l(\check{p}) - E^*(\check{p}) \leq \max_p \left(l(p) - E^*(p)\right) \leq \check{E} = E(\check{u}). \tag{5.126}$$

The numbers $l(\check{p}) - E^*(\check{p})$ and $E(\check{u})$ are equal:

[11] We are assuming that $\check{u}$ is a smooth function and, in particular, the derivatives (5.125) are defined. This assumption makes the proof of the coincidence of the maximum value of $l(p) - E^*(p)$ and the minimum value of $E(u)$ almost trivial; a complete proof requires a more sophisticated technique.

$$l(\check{p}) - E^*(\check{p}) = \int_{\partial\Omega} (\check{p}_x n_x + \check{p}_y n_y)\, \check{u} ds - \frac{1}{2} \int_{\Omega} (\check{p}_x^2 + \check{p}_y^2) dx dy$$
$$= \int_{\Omega} \left[\frac{\partial(\check{p}_x \check{u})}{\partial x} + \frac{\partial(\check{p}_y \check{u})}{\partial y} \right] dx dy - \frac{1}{2} \int_{\Omega} (\check{p}_x^2 + \check{p}_y^2) dx dy$$
$$= \frac{1}{2} \int_{\Omega} \left[\left(\frac{\partial \check{u}}{\partial x} \right)^2 + \left(\frac{\partial \check{u}}{\partial y} \right)^2 \right] dx dy = E(\check{u}).$$

Here we transformed the integral over $\partial\Omega$ into an integral over Ω by means of the divergence theorem and used (5.124).

Since $l(\check{p}) - E^*(\check{p}) = E(\check{u})$, (5.126) implies that

$$l(\check{p}) - E^*(\check{p}) = \max_p \left(l(p) - E^*(p) \right) = E(\check{u}).$$

Hence, the order of maximization and minimization in (5.119) can indeed be changed, and we obtain the dual variational problem:

$$\check{E} = \max_{p \in (5.124)} (l(p) - E^*(p)).$$

The dual variational problem can be presented in a different form if we note that the general solution of (5.124) can be easily written:

$$p_x = \frac{\partial \psi}{\partial y}, \quad p_y = -\frac{\partial \psi}{\partial x}. \tag{5.127}$$

Here $\psi(x, y)$ is an arbitrary function. In a simply connected region, $\psi(x, y)$ must be single-valued; in a multiply connected region it might be multi-valued. For now, we will only consider simply connected regions Ω.

To write down the linear functional $l(p)$ in terms of function ψ, we note that the tangent vector to $\partial\Omega$, $\vec{\tau}$, has the components

$$\tau_x = -n_y, \quad \tau_y = n_x. \tag{5.128}$$

We choose the arc length increasing counter-clockwise; therefore

$$\tau_x = \frac{dx(s)}{ds}, \quad \tau_y = \frac{dy(s)}{ds} \tag{5.129}$$

where $x = x(s), \; y = y(s)$ are the parametric equations of $\partial\Omega$. Hence, we have

$$p_x n_x + p_y n_y = \frac{\partial \psi}{\partial y} \tau_y + \frac{\partial \psi}{\partial x} \tau_x = \frac{\partial \psi}{\partial x} \frac{dx}{ds} + \frac{\partial \psi}{\partial y} \frac{dy}{ds} = \frac{d\psi}{ds}. \tag{5.130}$$

The linear functional, $l(p)$, becomes a functional of ψ:

$$l(\psi) = \int \frac{d\psi}{ds} g(s) ds.$$

Finally, the dual variational problem is the maximization problem for the functional

$$J(\psi) = l(\psi) - E^*(\psi) = \int_{\partial\Omega} \frac{d\psi}{ds} g ds - \int_{\Omega} \frac{1}{2} \left(\left(\frac{\partial\psi}{\partial x} \right)^2 + \left(\frac{\partial\psi}{\partial y} \right)^2 \right) dx dy \quad (5.131)$$

on the set of all functions ψ.

The duality means that

$$\max_{\psi} J(\psi) = \min_{u \in (5.117)} E(u).$$

The functional $J(\psi)$ can be put in a slightly different form by integrating by parts in the first integral in (5.131):

$$J(\psi) = -\int_{\partial\Omega} \psi \frac{dg}{ds} ds - \int_{\Omega} \frac{1}{2} \left(\left(\frac{\partial\psi}{\partial x} \right)^2 + \left(\frac{\partial\psi}{\partial y} \right)^2 \right) dx dy.$$

The maximizing element of the dual variational problem is the solution of the von Neuman problem

$$\Delta\psi = 0, \qquad \frac{d\psi}{dn} = -\frac{dg}{ds}.$$

The functional $J(\psi)$ is invariant with respect to shifts of the function ψ for a constant because function $g(s)$ is continuous. For the same reason the necessary condition for the solvability of this problem,

$$\int_{\partial\Omega} \frac{\partial\psi}{\partial n} ds = 0,$$

is satisfied.

One can say that the von Neuman problem is dual to the Dirichlet problem.

Discontinuity conditions. Solutions of the Dirichlet problem (5.118) are smooth, if $\partial\Omega$ and $g(s)$ are smooth. Therefore, in the dual variational problem the trial fields, p_x and p_y, can be assumed smooth. However, for more complex energy functionals, like, for example, the functional of a heterogeneous medium,

$$E(u) = \int_{\Omega} \frac{1}{2} a(x, y) \left[\left(\frac{\partial u}{\partial x} \right)^2 + \left(\frac{\partial u}{\partial y} \right)^2 \right] dxdy,$$

with the material characteristics, $a(x, y)$, that is discontinuous on some line, Γ, the minimizer may be non-smooth. If the functional is minimized on the set of functions, $u(x, y)$, which are continuous on Γ but may have discontinuous derivatives on Γ, then the variation of the functional gets a contribution,

$$\int_{\Gamma} \left(\left[a \frac{\partial u}{\partial x} \right] n_x + \left[a \frac{\partial u}{\partial y} \right] n_y \right) \delta u \, ds,$$

where $[\varphi]$ denotes the jump of a function, φ, on Γ,

$$[\varphi] = \varphi_+ - \varphi_-,$$

and indices $\pm$ mark the limit values of φ on two sides of Γ; n_x, n_y are the components of the unit normal vector on Γ directed from the side $-$ to the side $+$. Hence, the minimizer should obey the additional condition,

$$\left[a \frac{\partial \breve{u}}{\partial x} \right] n_x + \left[a \frac{\partial \breve{u}}{\partial y} \right] n_y = 0 \text{ on } \Gamma.$$

In the dual variational principle, one should admit the discontinuous trial functions, p_x and p_y. Therefore, integrating by parts in (5.121), one obtains the additional term

$$\int_{\Gamma} \left([p_x] n_x + [p_y] n_y \right) u d s.$$

Accordingly, the trial discontinuous functions in the dual variational principle, p_x and p_y, must obey the condition

$$[p_x] n_x + [p_y] n_y = 0 \text{ on } \Gamma. \tag{5.132}$$

The Dirichlet and Thompson principles

Let V be a region in three-dimensional space which is an exterior of some bounded region with the boundary Ω (Fig. 5.21). For a smooth function, $u(x)$, the Dirichlet functional, $E(u)$, is defined as

$$E(u) = \frac{1}{2} \int_V u_{,i} u^{,i} dV.$$

For the Dirichlet functional, consider a minimization problem on the set of functions, $u(x^i)$, selected by the conditions

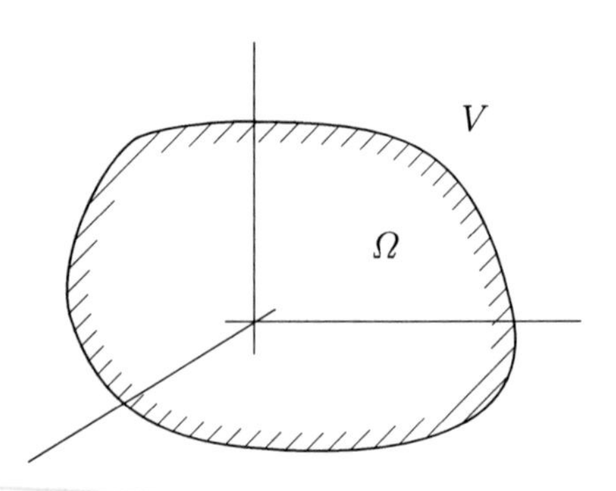

Fig. 5.21 Notation for the Dirichlet and Thompson principles

$$u = 1 \quad \text{on } \Omega,$$

$$u \sim \frac{c_1}{r} + \frac{c_2}{r^2} + \ldots \quad \text{as } r \to \infty, \quad r^2 = x_i x^i. \tag{5.133}$$

The quantity

$$\frac{1}{2\pi} \min_{u \in (5.133)} E(u),$$

has the physical meaning of the electric capacity of the surface Ω. The formulated variational principle for the electric capacity is called the Dirichlet principle. The minimizing function $\check{u}(x)$ satisfies the Laplace equation in the region V,

$$\Delta \check{u} = 0 \quad \text{in } V, \tag{5.134}$$

and the boundary conditions (5.133).

Note that the condition at infinity, $u \to 0$ as $r \to \infty$, is essential: without this condition the minimizing function would be just $\check{u} \equiv 1$, and $E(\check{u}) = 0$. The functional $E(u)$ feels the condition at infinity, as can be seen from the following reasoning. Denote the spherical coordinates in region V by r, φ, θ; $r \geq R(\varphi, \theta)$, $r = R(\varphi, \theta)$ being the parametric equation of the surface Ω. Suppose $u(r, \varphi, \theta)$ tends to some limit value, u_∞, as $r \to \infty$.

Then, for each ray $\varphi = const$, $\theta = const$,

$$u_\infty - u|_{r=R(\varphi,\theta)} = u_\infty - 1 = \int_R^\infty u_{,r} dr = \int_R^\infty r u_{,r} \frac{1}{r} dr.$$

Applying the Cauchy inequality (5.20) to the latter integral, we get

$$(u_\infty - 1)^2 \leq \int_R^\infty r^2 u_{,r}^2 dr \int_R^\infty \frac{dr}{r^2} = \frac{1}{R} \int_R^\infty r^2 u_{,r}^2 dr.$$

Since in spherical coordinates

$$(\nabla u)^2 = u_{,r}^2 + \frac{1}{r^2} u_{,\theta}^2 + \frac{1}{r^2 \sin^2 \theta} u_{,\varphi}^2,$$

we have

$$u_{,r}^2 \leq (\nabla u)^2.$$

Therefore,

$$(u_\infty - 1)^2 \leq \frac{1}{R_{\min}} \int_R^\infty r^2 \nabla u^2 dr \tag{5.135}$$

where $R_{\min}$ is the minimum distance from Ω to the origin. Integrating (5.135) over φ and θ with the weight of the spherical coordinates, $\sin\theta$, we obtain

$$4\pi\, (u_\infty - 1)^2 \leq \frac{1}{R_{\min}} 2E(u).$$

Hence, the change of u_∞ affects the minimum value of the functional.

In the case when $u(r, \varphi, \theta)$ tends to different values, $u_\infty(\varphi, \theta)$, along different rays, the inequality (5.135) is replaced by the inequality

$$\int_0^{2\pi} \int_0^{\pi} (u_\infty(\varphi, \theta) - 1)^2 \sin\theta \, d\varphi d\theta \leq \frac{1}{R_{\min}} 2E(u).$$

We see that energy feels the limit values of u at infinity for each ray.

The electric capacity of surface Ω corresponds to $u_\infty = 0$.

Let us find the variational principle dual to the Dirichlet principle. Writing the integrand in the Dirichlet functional in the form

$$\frac{1}{2} u_{,i} u^{,i} = \max_p \left(p^i u_{,i} - \frac{1}{2} p^i p_i \right),$$

we present the initial variational problem as a minimax problem:

$$\check{E} = \min_{u \in ((5.133))} \frac{1}{2} \int_V u_{,i} u^{,i} dV = \min_{u \in ((5.133))} \max_p \int_V \left(p^i u_{,i} - \frac{1}{2} p^i p_i \right) dV. \tag{5.136}$$

Following the general scheme, we change the order of calculating the maximum and minimum values and calculate the minimum with respect to u.

Suppose the functions p^i are continuously differentiable and continuous on the closed region V and tend to zero as $r \to \infty$. Suppose also that p^i can be expanded in Taylor series with respect to $1/r$ as

$$p \sim \frac{a_2}{r^2} + \frac{a_3}{r^3} + \dots \quad \text{for } r \to \infty. \tag{5.137}$$

For a slower decay, the integral of $p^i p_i$ in (5.136) would diverge.

Integrating by parts, and using (5.133) and (5.137), for the integral in (5.136) we obtain

$$\int_V \left(p^i u_{,i} - \frac{1}{2} p^i p_i \right) dV = \Phi(u, p) - E^*(p), \tag{5.138}$$

$$E^*(p) = \frac{1}{2} \int_V p^i p_i dV, \quad \Phi(u, p) = l(p) - \frac{1}{2} \int_V p^i_{,i} u dV, \quad l(p) = \int_\Omega p^i n_i dA.$$

Minimization of the functional (5.138) with respect to u is reduced to minimizing the linear functional $\Phi(u, p)$. As in the previous example,

$$\min_{u \in (5.133)} \Phi(u, p) = l(p) + \begin{cases} -\infty, & \text{if } \ p^i_{,i} \neq 0 \\ 0, & \text{if } \ p^i_{,i} = 0 \end{cases}. \tag{5.139}$$

Combining (5.139) and (5.138), we have

$$\min_{u \in (5.133)} \int_V \left(p^i u_{,i} - \frac{1}{2} p^i p_i \right) dV = \begin{cases} l(p) - E^*(p), & \text{if } \ p^i_{,i} = 0 \\ -\infty, & \text{if } \ p^i_{,i} \neq 0 \end{cases}.$$

Since we need to calculate the maximum value after calculating the minimum value with respect to u, all functions p for which $\min_u$ is equal to $-\infty$ should be dropped and only functions p^i satisfying the equation

$$p^i_{,i} = 0 \quad \text{in } V, \tag{5.140}$$

should be retained.

As in the previous example, one can check that the order of maximization and minimization in the minimax problem (5.136) can be changed and we arrive at the following dual variational problem:

$$\check{E} = \max_{\substack{p \in (5.140) \\ (5.137)}} \left(l(p) - E^*(p) \right). \tag{5.141}$$

So the dual variational problem is the maximization problem for the functional $l(p) - E^*(p)$ over all vector field satisfying the "incompressibility" condition (5.140).

At first glance, the formulation of the dual variational problem is surprising: we have seen that the quadratic functional of the form

$$\int_a^b u^2 dx$$

does not feel the values of the function $u(x)$ at the boundary (and, in fact, at any fixed point of the segment $[ab]$). Similarly, the functional

$$\int_V p_i p^i dV$$

should not feel the values of p_i at ∂V and, thus, the values of the linear functional, $l(p) = \int_\Omega p^i n_i dA$. The resolution of this "paradox" is in the constraints (5.140) and (5.137): the admissible functions $p^i(x)$ are not arbitrary. Due to these constraints, the linear functional $l(p)$ can be presented as

$$l(p) = \int_V p^i g_{,i} dV$$

with some smooth function g, which is equal to unity at $\partial\Omega$ and decays at infinity as $1/r$.

Therefore,

$$|l(p)|^2 \leq \int_V p_i p^i dV \int_V g_{,i} g_{,i} dV \leq const\ E(p),$$

i.e. energy does feel the values of the functional $l(p)$.

According to (5.141) and (5.50), the dual variational problem can also be stated as the maximization problem of the functional

$$\frac{[l(p)]^2}{4E^*(p)}.$$

The variational principle

$$\max_{\substack{p\in(5.140)\\(5.137)}} \frac{[l(p)]^2}{4E^*(p)} = \check{E}$$

is called the Thompson principle.

The case of general integral functional. Consider the functional $I(u)$ of the form $I(u) = E(u) - l(u)$

$$E(u) = \int_V L(x^i, u^\varkappa, u^\varkappa_{,i}) dV, \qquad u^\varkappa_{,i} \equiv \frac{\partial u^\varkappa}{\partial x^i}$$

$$l(u) = \int_V g_\varkappa u^\varkappa dV + \int_{\partial V_f} f_\varkappa u^\varkappa dA, \tag{5.142}$$

where $u^\varkappa$ are smooth functions of the variables $x_1, \ldots, x_n$, in a closed bounded region V in R_n, ∂V_f is a part of the boundary ∂V of the region V, and $g_\varkappa$ and $f_\varkappa$ are given functions in V and on ∂V_f. The energy density L is a strictly convex function of the variables $u^\varkappa$ and $u^\varkappa_{,i}$.

The original minimization problem is to find the minimum of the functional $I(u)$ on the set of all functions $u^\varkappa$ which take on the assigned values on the surface $\partial V_u = \partial V - \partial V_f$:

$$u^\varkappa = u^\varkappa_{(b)} \quad \text{on } \partial V_u. \tag{5.143}$$

To construct the dual variational problem, we present $L\left(x^i, u^\varkappa, u^\varkappa_i\right)$ using the Young-Fenchel transformation as

$$L\left(x^i, u^\varkappa, u^\varkappa_{,i}\right) = \max_p \left(p_\varkappa u^\varkappa + p^i_\varkappa u^\varkappa_{,i} - L^*\left(x^i, p_\varkappa, p^i_\varkappa\right)\right). \tag{5.144}$$

Here maximization is done over all $p = \left\{p_\varkappa, p^i_\varkappa\right\}$, and $L^*\left(x^i, p_\varkappa, p^i_\varkappa\right)$ is the Young-Fenchel transformation of the function $L\left(x^i, u^\varkappa, u^\varkappa_{,i}\right)$ with respect to the variables $u^\varkappa$ and $u^\varkappa_{,i}$:

$$L^*\left(x^i, p_\varkappa, p^i_\varkappa\right) = \max_{u^\varkappa, u^\varkappa_{,i}} \left(p_\varkappa u^\varkappa + p^i_\varkappa u^\varkappa_{,i} - L\left(x^i, u^\varkappa, u^\varkappa_{,i}\right)\right).$$

Rewriting the initial variational problem as a minimax problem by means of (5.144), we have

$$\check{I} = \min_{u \in (5.143)} \max_p \left[\int_V \left(p_\varkappa u^\varkappa + p^i_\varkappa u^\varkappa_{,i} - L^*\left(x^i, p_\varkappa, p^i_\varkappa\right)\right) dV - l(u)\right]. \tag{5.145}$$

Reversing the order of maximization and minimization in (5.145) yields the relation,

$$\max_p \min_{u \in (5.143)} \left(\Phi(u, p) - E^*(p)\right) \leq \check{I}. \tag{5.146}$$

Here,

$$\Phi(u, p) = \int_V \left(p_\varkappa u^\varkappa + p^i_\varkappa u^\varkappa_{,i} - g_\varkappa u^\varkappa\right) dV - \int_{\partial V_f} f_\varkappa u^\varkappa dA,$$

$$E^*(p) = \int_V L^*\left(x^i, p_\varkappa, p^i_\varkappa\right) dV. \tag{5.147}$$

Let us present $u^\varkappa$ as a sum of some fixed smooth function $h^\varkappa$, taking on the values $u^\varkappa_{(b)}$ on ∂V_u, and the function $u'^\varkappa$, which is equal to zero on ∂V_u:

$$u^\varkappa = h^\varkappa + u'^\varkappa, \quad u'^\varkappa = 0 \quad \text{on } \partial V_u. \tag{5.148}$$

Then, $\Phi(u, p)$ can be written as

$$\Phi(u, p) = l(p) + \Phi\left(u', p\right),$$

where $l(p)$ is a linear functional on the set of functions p defined by the equation

$$l(p) = \Phi(h, p).$$

Calculation of $\min\limits_u$ in (5.146) is reduced to calculation of $\min\limits_{u'} \Phi\left(u', p\right)$:

$$\begin{aligned} \min_{u \in (5.143)} \Phi(u, p) &= \min_{u' \in (5.148)} \left(l(p) + \Phi\left(u', p\right)\right) = \\ &= l(p) + \min_{u' \in (5.148)} \Phi\left(u', p\right). \end{aligned}$$

Since $\Phi(u, p)$ is linear with respect to u,

$$\min_{u' \in (5.148)} \Phi\left(u', p\right) = 0,$$

if for any u'

$$\Phi\left(u', p\right) = 0, \tag{5.149}$$

and

$$\min_{u' \in (5.148)} \Phi\left(u', p\right) = -\infty \tag{5.150}$$

if $\Phi\left(u', p\right) \neq 0$ for at least one element u'. This can be proved in the same way as (5.122). Since we need to find the maximum over p, we have to exclude from consideration all functions p for which $\min\limits_u$ is equal to $-\infty$. As a result, we arrive at the maximization problem for the functional $l(p) - E^*(p)$ on the set of all functions p which satisfy the constraint (5.149). Due to (5.149), the values of the linear functional $l(p)$ do not depend on the specific choice of functions $h^\varkappa$.

If $p_\varkappa$ and $p^i_\varkappa$ are differentiable and continuous functions in the region V, then the constraint (5.149) is reduced to

$$p_\varkappa = \frac{\partial p^i_\varkappa}{\partial x^i} + g_\varkappa \quad \text{in } V, \quad p^i_\varkappa n_i = f_\varkappa \quad \text{on } \partial V_f, \tag{5.151}$$

and the functional $l(p)$ is

$$l(p) = \int_{\partial V_u} p^i_\varkappa n_i u^\varkappa_{(b)} dA. \tag{5.152}$$

Equation (5.146) takes the form,

$$\max_{p\in(5.151)} \left(l(p) - E^*(p)\right) \leq \check{I}. \tag{5.153}$$

Suppose that the functions $\check{u}^\varkappa(x^i)$ which minimize the functional $I(u)$ are twice continuously differentiable. Then, we are going to show that

$$\max_{p\in(5.151)} \left(l(p) - E^*(p)\right) = \check{I}. \tag{5.154}$$

Define

$$\check{p}_\varkappa = \left.\frac{\partial L}{\partial u^\varkappa}\right|_{u^\varkappa=\check{u}^\varkappa}, \quad \check{p}^i_\varkappa = \left.\frac{\partial L}{\partial u^\varkappa_i}\right|_{u^\varkappa=\check{u}^\varkappa}.$$

The Euler equations of the initial variational problem,

$$\frac{\partial L}{\partial u^\varkappa} = \frac{\partial}{\partial x^i}\frac{\partial L}{\partial u^\varkappa_i} + F_\varkappa \quad \text{in } V, \quad \frac{\partial L}{\partial u^\varkappa} n_i = f_\varkappa \quad \text{in } \partial V,$$

imply that the functions $\check{p}_\varkappa$ and $\check{p}^i_\varkappa$ satisfy the constraints (5.151) and, consequently, are the admissible functions. Therefore,

$$l(\check{p}) - E^*(\check{p}) \leq \max_{p\in(5.151)} \left(L(p) - E(p)\right). \tag{5.155}$$

On the other hand,

$$l(\check{p}) - E^*(\check{p}) = \check{I}.$$

Indeed,

$$l(\check{p}) - E^*(\check{p}) = \int\limits_{\partial V_u} \check{p}^i_\varkappa n_i u^\varkappa_{(b)} dA - \int\limits_V L^*\left(x^i, \check{p}_\varkappa, \check{p}^i_\varkappa\right) dV =$$

$$= \int\limits_V \left[\frac{\partial\left(\check{p}^i_\varkappa \check{u}^\varkappa\right)}{\partial x^i} - L^*\left(x^i, \check{p}_\varkappa, \check{p}^i_\varkappa\right)\right] dV - \int\limits_{\partial V_f} \check{p}^i_\varkappa n_i \check{u}^\varkappa dA =$$

$$= \int\limits_V \left[\frac{\partial \check{p}^i_\varkappa}{\partial x^i} \check{u}^\varkappa + \check{p}^i_\varkappa \check{u}^\varkappa_{,i} - \left(\check{p}_\varkappa \check{u}^\varkappa + \check{p}^i_\varkappa \check{u}^\varkappa_{,i} - L\left(x^i, \check{u}^\varkappa, \check{u}^\varkappa_{,i}\right)\right)\right] dV -$$

$$- \int\limits_{\partial V_f} f_\varkappa \check{u}^\varkappa dA = \int\limits_V L\left(x^i, \check{u}^\varkappa, \check{u}^\varkappa_{,i}\right) dV - \int\limits_V g_\varkappa \check{u}^\varkappa dV - \int\limits_{\partial V_f} f_\varkappa \check{u}^\varkappa dA = \check{I}. \quad (5.156)$$

Therefore (5.154) holds true.

So, the maximization problem for the functional $l(p) - E^*(p)$ on the set of functions (5.151) is dual to the initial variational problem.

Consider the specific case, when L does not depend on $u^\varkappa$. Denote by $L^*\left(x^i, p^i_\varkappa\right)$ the Young-Fenchel transformation of the function $L\left(x^i, u^\varkappa_{,i}\right)$ with respect to $u^\varkappa_{,i}$:

$$L^*\left(x^i, p^i_\varkappa\right) = \max_u \left(p^i_\varkappa u^\varkappa_{,i} - L\left(x^i, u^\varkappa_{,i}\right)\right).$$

Function $L^*\left(x^i, p_\varkappa, p^i_\varkappa\right)$ introduced in the beginning of this section is linked to $L^*\left(x^i, p^i_\varkappa\right)$ as

$$L^*\left(x^i, p_\varkappa, p^i_\varkappa\right) = \max_u \left(p_\varkappa u^\varkappa + p^i_\varkappa u^\varkappa_{,i} - L\left(x^i, u^\varkappa_i\right)\right) =$$
$$= \begin{cases} L^*\left(x^i, p^i_\varkappa\right) & \text{for } p_\varkappa = 0 \\ +\infty & \text{for } p_\varkappa \neq 0 \end{cases}$$

If $L^*\left(x^i, p_\varkappa, p^i_\varkappa\right) = +\infty$, then $l(p) - E^*(p)$ is equal to $-\infty$. So in the dual variational problem all non-zero functions $p_\varkappa$ should be excluded from the set of admissible functions. For $p_\varkappa = 0$, (5.151) becomes

$$\frac{\partial p^i_\varkappa}{\partial x^i} + g_\varkappa = 0 \quad \text{in } V, \quad p^i_\varkappa n_i = f_\varkappa \quad \text{on } \partial V_f. \quad (5.157)$$

The dual variational principle (5.154) will be considered further for a number of continuum models.

The Legendre and Young-Fenchel transformation can be used for constructing the "dual" variational problems even in the cases when the functional is not convex. An example is the Hamilton variational principle in the phase space (Sect. 1.6).

Lower estimates of non-convex functionals by means of the dual problem. Consider the minimization problem for the integral functional of the form (5.142) on the set of functions (5.143). Let $L\left(x^i, u^\varkappa, u^\varkappa_{,i}\right)$ be a non-convex function of $u^\varkappa$ and $u^\varkappa_{,i}$. Let us calculate $L^{**}\left(x^i, u^\varkappa, u^\varkappa_{,i}\right)$ – the second Young-Fenchel transformation of the function $L\left(x^i, u^\varkappa, u^\varkappa_{,i}\right)$ with respect to $u^\varkappa, u^\varkappa_{,i}$. We know that

$$L^{**}\left(x^i, u^\varkappa, u^\varkappa_{,i}\right) \leq L\left(x^i, u^\varkappa, u^\varkappa_{,i}\right). \tag{5.158}$$

Therefore, for all fields $u^\varkappa$,

$$I^{**}(u) = \int_V L^{**} dV - l(u) \leq \int_V L dV - l(u) = I(u).$$

Consequently,

$$\check{I}^{**} \leq \check{I}. \tag{5.159}$$

Since the functional $I^{**}(u)$ is convex,

$$\max_p \left(l(p) - \int_V L^*\left(x^i, p_\varkappa, p^i_\varkappa\right) dV \right) = \check{I}^{**}. \tag{5.160}$$

Computing the functional in (5.160) on the admissible elements p, one obtains the lower estimates of $\check{I}^{**}$ and, therefore, $\check{I}$. For some problems with non-convex functionals $\check{I}^{**}$ coincides with $\check{I}$ (see [95]).

5.9 Hashin-Strikman Variational Principle

Linear case. As we have seen in Sect. 5.8, the trial fields of the dual variational problems must obey some differential constraints and, possibly, the discontinuity conditions. While the differential constraints can be easily resolved, the discontinuity conditions are not so easy to deal with if the geometry of discontinuities is complex. Hashin and Strikman suggested a trick, which allows one to overcome this difficulty. They obtained a dual variational principle with the admissible fields that must not obey the differential constraints. We explain the trick first for a simple case of the variational problem

$$E(u) = \int_V \frac{1}{2} a^{ij}(x) \frac{\partial u}{\partial x^i} \frac{\partial u}{\partial x^j} dV \to \min_u, \tag{5.161}$$

where $a^{ij}(x)$ may have discontinuities on surfaces with a complex geometry, and the minimum is sought over all functions, $u(x)$, with the prescribed boundary values,

$$u = u_{(b)} \text{ on } \partial V. \tag{5.162}$$

Further we consider more general cases.

Let us rewrite $E(u)$ by adding and deducting the quadratic functional

$$\int_V \frac{1}{2} a_0 \frac{\partial u}{\partial x^i} \frac{\partial u}{\partial x_i} dV,$$

where a_o is a constant. We have

$$E(u) = \int_V \frac{1}{2} a_0(x) \frac{\partial u}{\partial x^i} \frac{\partial u}{\partial x_i} dV + \int_V \frac{1}{2} \left(a^{ij} - a_0 \delta^{ij}\right) \frac{\partial u}{\partial x^i} \frac{\partial u}{\partial x^j} dV.$$

Let us choose a_o in such a way that the quadratic form

$$\frac{1}{2} \left(a^{ij} - a_0 \delta^{ij}\right) u_{,i} u_{,j}, \qquad \left(u_{,i} \equiv \frac{\partial u}{\partial x^i}\right)$$

be positive definite. Then it can be presented as the Young-Fenchel transformation of the dual quadratic form

$$\frac{1}{2} \left(a^{ij} - a_0 \delta^{ij}\right) u_{,i} u_{,j} = \max_{p^i} \left[p^i u_{,i} - \frac{1}{2} b_{ij} p^i p^j \right], \tag{5.163}$$

where b_{ij} is the inverse tensor to $a^{ij} - a_0 \delta^{ij}$,

$$\left(a^{ij} - a_0 \delta^{ij}\right) b_{jk} = \delta^i_k, \tag{5.164}$$

and the quadratic form, $b_{ij} p^i p^j$, is positive definite. Hence, $E(u)$ is the maximum value in the variational problem

$$E(u) = \max_{p^i(x)} \int_V \left[\frac{1}{2} a_0 u_{,i} u^{,i} + p^i(x) u_{,i} - \frac{1}{2} b_{ij}(x) p^i(x) p^j(x) \right] dV. \tag{5.165}$$

The original variational problem takes the form

$$\min_{u=u_{(b)} \text{ at } \partial V} \max_{p^i(x)} \int_V \left[\frac{1}{2} a_0 u_{,i} u^{,i} + p^i(x) u_{,i} - \frac{1}{2} b_{ij}(x) p^i(x) p^j(x) \right] dV.$$

The order of minimization and maximization can be changed (we do not pause to prove that; it can be done in the same way as for Dirichlet and Neuman problems in Sect. 5.8). We arrive at the relation

$$\min_{u=u_{(b)} \text{ at } \partial V} E(u) = \max_{p^i(x)} \left[\tilde{J}(p) - \int_V \frac{1}{2} b_{ij}(x) p^i(x) p^j(x) dV \right], \tag{5.166}$$

where

$$\tilde{J}(p) = \min_{u=u_{(b)} \text{ at } \partial V} \int_V \left(\frac{1}{2} a_0 u_{,i} u^{,i} + p^i(x) u_{,i} \right) dV. \tag{5.167}$$

So, we get a dual variational principle with the dual variables, $p^i(x)$, which are not constrained. The cost is that the functional $\tilde{J}(p)$ appears which is to be computed. Formula for $\tilde{J}(p)$ can be simplified a bit. Denote by $\mathring{u}(x)$ be the minimizer in the variational problem,

$$\min_{u=u_{(b)} \text{ at } \partial V} \int_V \frac{1}{2} a_0 u_{,i} u^{,i} dV. \tag{5.168}$$

Let E_0 be the minimum value in (5.168):

$$E_0 = \frac{1}{2} \int_V a_0 \mathring{u}_{,i} \mathring{u}^{,i} dV.$$

We set in (5.167)

$$\begin{aligned} u &= \mathring{u} + u', \\ u' &= 0 \text{ on } \partial V. \end{aligned} \tag{5.169}$$

Then

$$\begin{aligned} \tilde{J}(p) &= E_0 + \int_V p^i(x) \mathring{u}_{,i}(x) dV + J(p), \\ J(p) &\equiv \min_{u'=0 \text{ at } \partial V} \int_V \left(\frac{1}{2} a_0 u'_{,i} u'^{,i} + p^i(x) u'_{,i} \right) dV \end{aligned} \tag{5.170}$$

The variational problem (5.170) is well-posed: the functional is obviously bounded from below:

$$\begin{aligned} \int_V \left(\frac{1}{2} a_0 u_{,i} u^{,i} + p^i u_{,i} \right) dV &= \int_V \frac{1}{2} a_0 \left(u_{,i} + p_i \right) \left(u^{,i} + p^i \right) dV - \\ &- \int_V \frac{1}{2} a_0 p_i p^i dV \geqslant -\frac{1}{2} a_0 \int_V p_i p^i dV. \end{aligned} \tag{5.171}$$

As we see from (5.171), it is enough for the boundedness from below that $p^i(x)$ be square integrable functions. In particular, they can be piece-wise continuous functions. We will further obtain the explicit dependence of J on $p^i(x)$. Irrespectively on the explicit form of $J(p)$, we get.

Hashin-Strikman variational principle. *The minimum value, $\check{E}$, of $E(u)$ can be found by solving the variational problem*

$$\check{E} = E_0 + \max_{p^i(x)} \left[\int_V \left(p^i(x)\mathring{u}_{,i}(x) - \frac{1}{2} b_{ij}(x) p^i(x) p^j(x) \right) dV + J(p) \right]. \quad (5.172)$$

Here $p^i(x)$ do not obey any constraints.

Explicit form of $J(p)$ for smooth p. Let us now find the dependence of J on $p^i(x)$. First, we assume that $p^i(x)$ are some smooth functions. Then the minimizer in u' (5.170) obeys the boundary value problem,

$$a_0 \Delta u' = -\frac{\partial p^i}{\partial x^i}, \qquad u'|_{\partial V} = 0 \quad (5.173)$$

Its solution can be written in terms of Green's function, $G(x, y)$, defined as the solution of the boundary value problem

$$\Delta_x G(x, y) = -\delta(x - y), \qquad G(x, y)\big|_{y \in \partial V} = 0. \quad (5.174)$$

Here $\delta(x)$ is the three-dimensional delta-function and Δ_x Laplace's operator acting on x-variables. We take for granted that the solution of this problem is symmetric with respect to x, y,

$$G(x, y) = G(y, x), \quad (5.175)$$

non-negative,

$$G(x, y) \geqslant 0,$$

and smooth everywhere except at the point $x = y$ where it has the singularity of the form

$$G(x, y) = \frac{1}{4\pi |x - y|} + \text{bounded function} \quad \text{as } y \to x.$$

Multiplying (5.174) by $\partial p^i / \partial y^i$ and integrating over y, we obtain

$$\Delta_x \int_V G(x, y) \frac{\partial p^i(y)}{\partial y^i} dV_y = -\int_V \delta(x - y) \frac{\partial p^i(y)}{\partial y^i} dV = -\frac{\partial p^i(x)}{\partial x^i}. \quad (5.176)$$

Comparing (5.173) and (5.176) we see that

$$u'(x) = \frac{1}{a_0} \int_V G(x, y) \frac{\partial p^i(y)}{\partial y^i} dV_y. \tag{5.177}$$

The singularity in (5.177) is integrable. To deal further with discontinuous functions, p^i, we move the derivative from $p^i(y)$ to $G(x, y)$ by integrating by parts. The singularity becomes stronger because $\partial G(x, y)/\partial y^i \backsim 1/|x - y|^2$ but is still integrable. To weaken the singularity of the integrand, before integrating by parts in (5.177), we replace $p^i(y)$ by $p^i(y) - p^i(x)$. That, obviously, does not change (5.177). After integration by parts, we obtain

$$u'(x) = -\frac{1}{a_0} \int_V \frac{\partial G(x, y)}{\partial y^i} \left(p^i(y) - p^i(x)\right) dV_y. \tag{5.178}$$

The differentiation of (5.178) with respect to x^i yields an integral with absolutely integrable singularity; therefore the differentiation is possible and we obtain the derivatives of the minimizer [12]:

$$\frac{\partial u'(x)}{\partial x^i} = -\frac{1}{a_0} \int_V \frac{\partial^2 G(x, y)}{\partial x^i \partial y^j} \left(p^j(y) - p^j(x)\right) dV_y. \tag{5.179}$$

Finally, from Clapeyron's theorem (5.47),

$$J(p) = \frac{1}{2} \int_V p^i(x) u'_{,i} dV =$$

$$= -\frac{1}{2a_0} \int_V \int_V p^i(x) \frac{\partial^2 G(x, y)}{\partial x^i \partial y^j} \left(p^j(y) - p^j(x)\right) dV_x dV_y. \tag{5.180}$$

Formula (5.180) can also be written in a more symmetric form:

$$J(p) = \frac{1}{4a_0} \int_V \int_V \frac{\partial^2 G(x, y)}{\partial x^i \partial y^j} \left(p^i(x)p^j(x) + p^i(y)p^j(y) - 2p^i(x)p^j(y)\right) dV_x dV_y. \tag{5.181}$$

Indeed, from (5.175):

$$\frac{\partial^2 G(x, y)}{\partial x^i \partial y^j} = \frac{\partial^2 G(y, x)}{\partial y^j \partial x^i}. \tag{5.182}$$

---

[12] The contribution to the derivative that is due to differentiation of $p^i(x)$,

$$\frac{1}{a_0} \int_V \frac{\partial G(x, y)}{\partial y^j} \frac{\partial p^j(x)}{\partial x^i} dV_y,$$

is equal to

$$\frac{1}{a_0} \int_{\partial V} G(x, y) n_j dA_y \frac{\partial p^j(x)}{\partial x^i},$$

and is zero because $G(x, y) = 0$ when $y \in \partial V$.

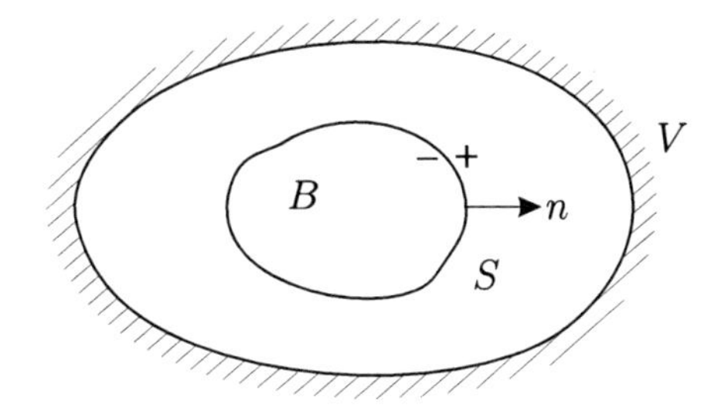

Fig. 5.22 Notation for the variational problem

Therefore, changing in (5.180) x by y, y by x, i by j and j by i, and taking into account (5.182), we get

$$J(p) = -\frac{1}{2a_0} \iint p^j(y) \frac{\partial^2 G(y,x)}{\partial y^i \partial x^j} \left(p^i(x) - p^i(y)\right) dV_x dV_y. \tag{5.183}$$

Due to (5.182) the kernels in (5.180) and (5.183) coincide. Summing up (5.180) and (5.183) we obtain (5.181).

Explicit form of $J(p)$ for discontinuous p. Let us show that (5.181) also holds for piece-wise smooth functions that may have discontinuities on some surfaces. To this end we need an auxiliary statement. Let S be a surface bounding some subregion, B, of V (Fig. 5.22). Consider a variational problem,

$$\int_V \frac{1}{2} u_{,i} u^{,i} dV - \int_S \sigma u dA \rightarrow \min_{u:u=0 \text{ at } \partial V} \tag{5.184}$$

where σ is some function on S. The admissible functions, u, are supposed to be continuous on $S : [u] = 0$. It is shown further in Example 11 of Sect.5.11 that the minimizer of this problem is

$$\check{u}(x) = \int_S G(x,y) \sigma dA. \tag{5.185}$$

Consider now the variational problem (5.170) for functions $p^i(x)$ that have a discontinuity on S. The functional can be rewritten as

$$\int_V \frac{1}{2} a_0 u_{,i} u^{,i} dV - \int_V \frac{\partial p^i}{\partial x^i} u dV - \int_S \left[p^i\right] n^i u dA.$$

The minimizer, $\check{u}$, is the solution of the boundary value problem

$$a_0 \Delta \check{u} = -\frac{\partial p^i}{\partial x^i} \text{ in } V, \quad \check{u} = 0 \text{ on } \partial V, \quad a_0 \left[\check{u}_{,i}\right] n^i = -\left[p^i\right] n_i \text{ on } S.$$

Here $\partial p^i / \partial x^i$ is a piece-wise smooth function. The solution is a sum of two functions, u_1 and u_2, which are the solutions of the following boundary value problems:

$$a_0\Delta u_1 = -\frac{\partial p^i}{\partial x^i} \text{ in } V, \quad u_1 = 0 \text{ on } \partial V, \quad \left[\frac{\partial u_1}{\partial x^i}\right] n^i = 0 \text{ on } S,$$

$$a_0\Delta u_2 = 0 \text{ in } V, \quad u_2 = 0 \text{ on } \partial V, \quad \left[\frac{\partial u_2}{\partial x^i} n^i\right] = -\left[p^i\right] n_i \text{ on } S.$$

The function of x,

$$\frac{1}{2} a_0 \int_V G(x, y) \frac{\partial p^i(y)}{\partial y^i} dV_y,$$

has continuous first derivatives at S if $\partial p^i / \partial y^i$ is piece-wise continuous.[13] Thus,

$$u_1(x) = \frac{1}{a_0} \int_V G(x, y) \frac{\partial p^i(y)}{\partial y^i} dV_y.$$

Function u_2 is equal to the function (5.185) with the factor, $1/a_0$ and $\sigma = \left[p^i\right] n_i$. So

$$\breve{u}(x) = \frac{1}{a_0} \int_V G(x, y) \frac{\partial p^i(y)}{\partial y^i} dV_y + \frac{1}{a_0} \int_S G(x, y) \left[p^i\right] n_i dA. \tag{5.186}$$

Integrating by parts in the volume integral, we see that the surface terms cancel out:

$$\breve{u}(x) = -\frac{1}{a_0} \int_V \frac{\partial G(x, y)}{\partial y^i} p^i(y) dV_y.$$

This formula can also be written as

$$\breve{u}(x) = -\frac{1}{a_0} \int_V \frac{\partial G(x, y)}{\partial y^i} \left(p^i(y) - p^i(x)\right) dV_y,$$

because

$$\int_V \frac{\partial G(x, y)}{\partial y^i} dV_y = 0,$$

due to the boundary condition, $G(x, y) = 0$ at ∂V.

The derivatives of the minimizer can be found by differentiating (5.186): for a point, x, which is strictly inside B or $V - B$,

$$\frac{\partial \breve{u}}{\partial x^i} = \frac{1}{a_0} \int_V \frac{\partial G(x, y)}{\partial x^i} \frac{\partial p^j(y)}{\partial y^j} dV_y + \frac{1}{a_0} \int_S \frac{\partial G(x, y)}{\partial x^i} \left[p^j\right] n_j dA.$$

---

[13] See, e.g., [81], vol. 2, p. 245.

Putting $\partial\left(p^i(y) - p^j(x)\right)/\partial y^j$ instead of $\partial p^j(y)/\partial y^j$ and integrating by parts we arrive at (5.179) for discontinuous p^i. Hence, (5.181) also holds for discontinuous p^i.

Minimum principle. One remark is now in order. If the parameter, a_0, is so big that the quadratic form, $\left(a^{ij} - a_0\delta^{ij}\right)u_{,i}u_{,j}$ is negative definite, then we introduce tensor b^{ij} as the inverse tensor to $a_0\delta^{ij} - a^{ij}$, and formula (5.163) is replaced by the relation

$$\frac{1}{2}\left(a^{ij} - a_0\delta^{ij}\right)u_{,i}u_{,j} = \min_{p^i}\left[\frac{1}{2}b_{ij}p^ip^j - p^iu_{,i}\right].$$

Therefore,

$$E(u) = \min_{p^i(x)}\int_V\left[\frac{1}{2}a_0u_{,i}u^{,j} - p^iu_{,i} + \frac{1}{2}b_{ij}p^ip^j\right]dV,$$

and the Hashin-Strikman principle takes the following form.
Hashin-Strikman variational principle. *The minimum value, $\check{E}$, of E(u) can be found by solving the variational problem*

$$\check{E} = E_0 + \min_{p^i(x)}\left[\int_V\left(-p^i(x)\mathring{u}_{,i} + \frac{1}{2}b_{ij}(x)p^i(x)p^j(x)\right)dV + J(p)\right].$$

Emphasize that the Hashin-Strikman variational principle holds for any value of the parameter, a_0. If one makes estimates by plugging trial fields, then the estimates involve a_0, and an additional optimization of the estimates over a_0 can be done.

Nonlinear case. This construction is easily extended to arbitrary convex functionals,

$$E(u) = \int_V L(x, u_{,i})dV,$$

in the cases when the function

$$\Lambda(x, u_{,i}) = L(x, u_{,i}) - \frac{1}{2}a_0u_{,i}u^{,i}$$

is convex for some a_0. Then $\Lambda(x, u_i)$ can be presented in terms of its Young-Fenchel transformation, $\Lambda^*(x, p^i)$,

$$\Lambda(x, u_{,i}) = \max_{p^i}\left[p^iu_{,i} - \Lambda^*(x, p^i)\right],$$

and

$$E(u) = \max_{p^i}\int_V\left(\frac{1}{2}a_0u_{,i}u^{,i} + p^iu_{,i} - \Lambda^*(x, p^i)\right)dV.$$

From the same line of reasoning we obtain the following

Hashin-Strikman variational principle. *If function, $\Lambda(x, u_{,i})$, is strictly convex, then the variational problem*

$$\check{E} = \min_{u \in (5.162)} \int_V L(x, u_{,i}) dV \tag{5.187}$$

is equivalent to the variational problem

$$\check{E} = E_0 + \max_{p^i(x)} \left[\int_V \left(p^i \mathring{u}_{,i} - \Lambda^*(x, p^i) \right) dV + J(p) \right]$$

where $\mathring{E}$ and $\mathring{u}$ are the minimum value and the minimizer in the variational problem (5.168), and J(p) is the functional (5.181).

If $\Lambda(x, u_{,i})$ is not convex, but the function

$$\Lambda_1(x, u^i) = \frac{1}{2} a_0 u_{,i} u^{,i} - L(x, u_i)$$

is strictly convex, then

$$\frac{1}{2} a_0 u_{,i} u^{,i} - L(x, u_i) = \max_{p^i} \left[p^i u_{,i} - \Lambda^*(x, p^i) \right] = -\min_{p^i} \left[\Lambda^*(x, p^i) - p^i u_{,i} \right],$$

and the original variational problem is equivalent to the variational problem

$$\check{E} = E_0 + \min_{p^i} \int_V \left(\Lambda^*(x, p^i(x)) - p^i(x) \mathring{u}_{,i} \right) dV + J(p).$$

Similar reasoning holds for functionals depending on many required functions. We will discuss this in the case of elastic bodies in Sect. 6.7.

For some functions, like, e.g., $L = \left(a u_i u^i \right)^{\frac{1}{2}}$, neither $L - \frac{1}{2} a_0 u_i u^i$ nor $\frac{1}{2} a_0 u_i u^i - L$ are convex. Therefore, the transformation of the variational problems to the Hashin-Strikman form is not possible. For functions which vanish at some subregion of V like $L = \frac{1}{2} a(x) u_i u^i$ with $a(x) = 0$ somewhere inside V, $L - \frac{1}{2} a_0 u_i u^i$ is not convex for all x, while $\frac{1}{2} a_0 u_i u^i - L$ can be convex. In such cases, the Hashin-Strikman transformation yields only the minimization problem.

5.10 Variational Problems with Constraints

Lagrange multipliers. Usually, the set of admissible functions is defined by the constraints written in the form of equations or inequalities. Some of those constraints, like the boundary conditions, are easy to satisfy, and they do not cause any difficulties in solving the variational problem. For non-local constraints or constraints containing derivatives it is sometimes difficult to find a sufficiently large set of admissible functions. "Getting rid" of such constraints is possible by means of so-called Lagrange multipliers. We explain the idea of Lagrange multiplier method by the following example. Consider the minimization problem for the functional

$$I(u) = \int\limits_V L\left(x^i, u^\varkappa, \frac{\partial u^\varkappa}{\partial x^i}\right) dV$$

with the constraint

$$\mathcal{F}(u) = \int\limits_V F\left(x^i, u^\varkappa, \frac{\partial u^\varkappa}{\partial x^i}\right) dV = 0. \tag{5.188}$$

It is assumed that there exists at least one element u which satisfies the constraint (5.188), besides the value of the functional $I(u)$ is finite on this element.

Consider the following minimax problem:

$$\min_u \max_\lambda \left[I(u) + \lambda \mathcal{F}(u) \right]. \tag{5.189}$$

Here, the minimum is sought over all functions u, while the maximum is sought over all real numbers λ.

Let us show that the problem (5.189) is equivalent to the initial one. Indeed, consider the element u for which $\mathcal{F}(u) \neq 0$. For definiteness, let $\mathcal{F}(u) > 0$. Then, tending λ to $+\infty$, we get

$$\max_\lambda \left[I(u) + \lambda \mathcal{F}(u) \right] = +\infty. \tag{5.190}$$

If $\mathcal{F}(u) < 0$, then tending $\lambda \to -\infty$, we also arrive at (5.190). Therefore, for all u for which $\mathcal{F}(u)$ is not equal to zero, (5.190) holds. After calculating the maximum value over λ, the minimum value must be found. The elements u for which the function $I(u)$ is finite and $\mathcal{F}(u) = 0$ exist by our assumption. Hence, when the maximum value is sought, all u for which (5.190) holds should be excluded and

$$\min_{u \in (5.188)} I(u) = \min_u \max_\lambda \left[I(u) + \lambda \mathcal{F}(u) \right]. \tag{5.191}$$

The auxiliary variable in the minimax problem, λ, is called the Lagrange multiplier for the constraint (5.188).

Suppose that the order of calculation of the maximum and minimum values in (5.191) can be changed:

$$\min_{u \in (5.188)} I(u) = \max_\lambda \min_u \left[I(u) + \lambda \mathcal{F}(u) \right]. \tag{5.192}$$

The problem

$$\min_u \left[I(u) + \lambda \mathcal{F}(u) \right] \tag{5.193}$$

does not have any constraints for u. However, instead of one variational problem with a constraint, a one-parametric set of variational problems depending on the parameter λ needs to be solved. The minimum value in (5.193) is a function of λ. Denote it by $J(\lambda)$. The problem of finding the maximum value of $J(\lambda)$ with respect to λ is the dual one to the initial variational problem:

$$\min_{u\in(5.188)} I(u) = \max_{\lambda} J(\lambda).$$

Suppose that the variational problem (5.193) has a unique minimizing element $\check{u}(\lambda)$, and the function $J(\lambda)$ has a unique maximizing element $\hat{\lambda}$, and that the derivatives of $I(u)$ and $\mathcal{F}(u)$ with respect to u and the derivatives of $J(\lambda)$ and $\check{u}(\lambda)$ with respect to λ exist. Let us show that the solution of the initial variational problem is $\check{u}\left(\hat{\lambda}\right)$. By definition,

$$J(\lambda) = I(\check{u}(\lambda)) + \lambda\mathcal{F}(\check{u}(\lambda)). \tag{5.194}$$

The function $J(\lambda)$ is defined in a line, $-\infty < \lambda < +\infty$. According to our assumption, it is differentiable and reaches its maximum at the point $\hat{\lambda}$. Consequently,

$$\left.\frac{dJ(\lambda)}{d\lambda}\right|_{\lambda=\hat{\lambda}} = I'\left.\left(\check{u}(\lambda), \frac{d\check{u}}{d\lambda}\right)\right|_{\lambda=\hat{\lambda}} + \hat{\lambda}\mathcal{F}'\left.\left(\check{u}(\lambda), \frac{d\check{u}}{d\lambda}\right)\right|_{\lambda=\hat{\lambda}} + \mathcal{F}\left(\check{u}\left(\hat{\lambda}\right)\right) = 0. \tag{5.195}$$

Due to the Euler equations of the variational problem (5.193),

$$I'(\check{u}(\lambda), \bar{u}) + \lambda\mathcal{F}'(\check{u}(\lambda), \bar{u}) = 0 \tag{5.196}$$

for any λ and any function $\bar{u}$. In particular, putting $\lambda = \hat{\lambda}$ and $\bar{u} = d\check{u}/d\lambda$ in (5.196), we get

$$I'\left(\check{u}(\lambda), \frac{d\check{u}}{d\lambda}\right) + \hat{\lambda}\mathcal{F}'\left.\left(\check{u}(\lambda), \frac{d\check{u}}{d\lambda}\right)\right|_{\lambda=\hat{\lambda}} = 0. \tag{5.197}$$

It follows from (5.195) and (5.197) that the function $\check{u}\left(\hat{\lambda}\right)$ satisfies the constraint (5.188):

$$\mathcal{F}\left(\check{u}\left(\hat{\lambda}\right)\right) = 0.$$

Choosing $\hat{\lambda}$ and $\check{u}\left(\hat{\lambda}\right)$ as the trial elements of the dual and the initial variational problems, we get the estimate

$$J\left(\hat{\lambda}\right) \le \max_{\lambda} J(\lambda) = \min_{u\in(5.188)} I(u) \le I\left(\check{u}\left(\hat{\lambda}\right)\right).$$

According to (5.194), $J\left(\hat{\lambda}\right) = I\left(\check{u}\left(\hat{\lambda}\right)\right)$. Hence the upper and the lower estimates coincide, and we obtain,

$$I\left(\check{u}\left(\hat{\lambda}\right)\right) = \min_{u \in (5.188)} I(u),$$

i.e. $\check{u}\left(\hat{\lambda}\right)$ is indeed the minimizing element of the variational problem.

The procedure described reduces the solution of the problem with constraints to solution of a set of the variational problems without constraints.

Consider the changes which need to be made if instead of a constraint of the equality type (5.188) the variational problem involves a constraint of the inequality type:

$$\mathcal{F}(u) = \int_V F\left(x^i, u^\varkappa, \frac{\partial u^\varkappa}{\partial x^i}\right) dV \leq 0. \tag{5.198}$$

In this case, we construct a minimax problem:

$$\min_u \max_{\lambda \geq 0} \left[I(u) + \lambda \mathcal{F}(u)\right], \tag{5.199}$$

where the minimum is sought over all functions u, while the maximum is sought over all nonnegative number λ. The minimax problem (5.199) is equivalent to the original minimization problem with the constraint (5.198): Indeed, if, for some u, the functional $\mathcal{F}(u)$ is positive, then

$$\max_{\lambda \geq 0} \left[I(u) + \lambda \mathcal{F}(u)\right] = +\infty,$$

and in the subsequent calculation of the minimum with respect to u all such u must be disregarded.

If the order of calculation of maximum and minimum in (5.199) can be reversed,

$$\min_{u \in (5.198)} I(u) = \max_{\lambda \geq 0} \min_u \left[I(u) + \lambda \mathcal{F}(u)\right],$$

then the initial variational problem with the constraint is replaced by a family of variational problems without constraints which depend on the parameter λ,

$$J(\lambda) = \min_u \left[I(u) + \lambda \mathcal{F}(u)\right],$$

with the subsequent maximization of the function $J(\lambda)$ on the positive semi-axis, $\lambda \geq 0$.

Let $\mathcal{F}(\check{u}(\lambda)) < 0$ at the minimizer, $\check{u}(\lambda)$, of $I(u) + \lambda \mathcal{F}(u)$. For $\mathcal{F}(\check{u}(\lambda)) < 0$, functional $I(\check{u}) + \lambda \mathcal{F}(\check{u})$ reaches maximum over λ for $\lambda = 0$. If $\mathcal{F}(\check{u}(\lambda)) = 0$, then, in general, the Lagrange multiplier is not zero. Hence, the equality holds,

$$\lambda \mathcal{F}(\check{u}) = 0.$$

It is called the condition of complementary softness.

In the case of the point-wise constraints, the minimax problem is constructed by an analogous scheme. Suppose, functions $u^{\varkappa}$ obey to s equations,

$$f_{\alpha}\left(x^{i}, u^{\varkappa}, \frac{\partial u^{\varkappa}}{\partial x^{i}}\right)=0, \quad \alpha=1, \ldots, s, \tag{5.200}$$

and m inequalities

$$g_{\beta}\left(x^{i}, u^{\varkappa}, \frac{\partial u^{\varkappa}}{\partial x^{i}}\right) \leq 0, \quad \beta=1, \ldots, m. \tag{5.201}$$

Then the Lagrange multipliers, λ^{α} and μ^{β}, are functions of x, and

$$\min_{u \in(5.200),(5.201)} I(u)=$$

$$=\min_{u} \max_{\substack{\lambda^{\alpha}(x), \\ \mu^{\beta}(x) \geq 0}}\left\{I(u)+\int_{V}\left[\lambda^{\alpha}(x) f_{\alpha}\left(x^{i}, u^{\varkappa}, \frac{\partial u^{\varkappa}}{\partial x^{i}}\right)+\mu^{\beta} g_{\beta}\left(x^{i}, u^{\varkappa}, \frac{\partial u^{\varkappa}}{\partial x^{i}}\right)\right] d V\right\}.$$

One proceeds further in the same way as in the previous case by changing the order of minimum and maximum and obtaining an unconstrained variational problem.

In cases when one searches not the minimum but the stationary points of the functional $I(u)$ with some constraints, the general rule of the Lagrange multipliers is as follows: the functional $\Phi(u, \lambda)$ is constructed, the varying of which with respect to λ provides the given constraints, while $\Phi(u, \lambda)=I(u)$ if those constraints are satisfied. Hence, the search of the stationary points of the functional $I(u)$ is reduced to the search of the stationary points of the functional $\Phi(u, \lambda)$. For example, for the constraints of the type (5.200), the functional $\Phi(u, \lambda)$ is

$$\Phi(u, \lambda)=I(u)+\int_{V} \lambda^{\alpha}(x) f_{\alpha}\left(x^{i}, u^{\varkappa}, \frac{\partial u^{\varkappa}}{\partial x^{i}}\right) d V.$$

Consider how the Lagrange multiplier method works in the following example.

Example: The minimum drag body in a hypersonic flow. Consider a body in a hypersonic gas flow in a Cartesian coordinate system, x, y, z (Fig. 5.23). The gas is moving along the z axis in the negative z-direction. The boundary of the body in the half-space $z \geq 0$ (the head part of the body) is given by the equation

$$z=u(x, y).$$

The force exerted by the flow on the body is given by the Newton formula[14]

[14] Here is a brief derivation of this formula. In the case of hypersonic flow, the shock wave is very close to the body. If the body has a smooth shape, then the shock wave practically repeats the form

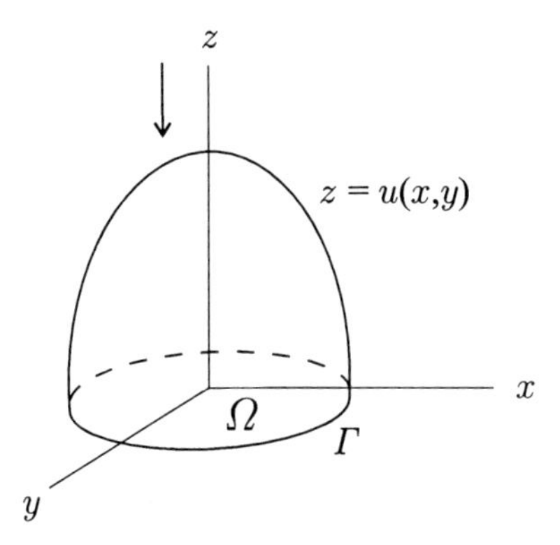

Fig. 5.23 Notation to the hypersonic flow problem

$$\mathcal{F}(u) = \int_{\Omega} \frac{2}{1 + u_x^2 + u_y^2} dxdy, \qquad (5.202)$$

Ω being the cross-section of the body by the plane $z = 0$, and u_x, u_y denote partial the derivatives $u_x = \partial u/\partial x$, $u_y = \partial u/\partial y$. The region Ω is assumed to be simply connected. We will also assume that the function $u(x, y)$ is continuous and almost everywhere differentiable, and that

$$u(x, y) = 0 \quad \text{on } \Gamma. \qquad (5.203)$$

The problem is to find a function $u(x, y)$ for which the drag force $\mathcal{F}(u)$ takes its minimum value.

---

of the body (at least near the head part of the body). According to the conservation of momentum on the shock wave in the ideal gas, $\left[p + \rho v_n^2\right] = 0$, where $[\varphi]$ denotes the difference of the values of the function φ on the two sides of the shock wave, v_n is the normal velocity on the shock wave. Due to the impermeability condition at the body, $v_n \approx 0$ after the shock wave. Therefore, from the momentum conservation the pressure at the body surface is $p = p_1 + \rho_1 v_{n1}^2$ (index 1 marks the values before the shock wave). We assume that $\rho_1 = const$, $v_{n1} = vn_z$, where v, the gas speed far away from the body, is a constant, and n_z is the projection of the normal on the z-axis. The projection of the force on the z-axis is

$$\int pn_z dA = \int p_1 n_z dA + \rho_1 \int v_{n1}^2 n_z dA.$$

The first term (under some conditions) is balanced by the pressure in the tail part of the body. The second term, divided by the constant $\rho_1 v^2/2$ coincides with (5.202) if we take into account the relations linking n_z and the area element dA with the shape of the body:

$$n_z = \left(1 + u_x^2 + u_y^2\right)^{-\frac{1}{2}}, \quad dA = \left(1 + u_x^2 + u_y^2\right)^{-\frac{1}{2}} dxdy.$$

This derivation contains an assumption on the closeness of the shock wave and the body surface. This assumption is acceptable not for all body shapes, and, thus, imposes some implicit constraints on the shape of the body. These constraints are hard to formalize. In what follows we ignore these constraints and consider the minimization problem for the functional (5.202) as a pure mathematical problem.

If the functional $\mathcal{F}(u)$ reaches its minimum value $\check{\mathcal{F}}$ for a smooth function $u(x, y)$, then, according to (5.202), $\mathcal{F}$ is greater than zero. Therefore, it makes sense to call the minimization problem for the functional $\mathcal{F}(u)$ correctly posed, if there exits such a positive constant c, that

$$\mathcal{F}(u) \geq c. \tag{5.204}$$

The minimization problem for the functional $\mathcal{F}(u)$ as stated is not correctly posed. Indeed, consider a conic surface of the height h,

$$u = h\left(1 - \frac{r}{\rho(\theta)}\right), \tag{5.205}$$

where r, θ are polar coordinates in the x-y plane, and $r = \rho(\theta)$ is the parametric equation of the contour Γ. The functional $\mathcal{F}(u)$ for the functions (5.205) becomes

$$\mathcal{F}(u) = \frac{1}{h^2}\int_0^{2\pi} \frac{\rho^4}{1+\frac{\rho^2}{h^2}+\frac{\rho_\theta^2}{\rho^2}} d\theta \leq \frac{1}{h^2}\int_0^{2\pi} \frac{\rho^4}{1+\frac{\rho_\theta^2}{\rho^2}} d\theta, \tag{5.206}$$

where $\rho_\theta \equiv d\rho/d\theta$. It follows from (5.206) that, along with the increase of the height h of the cone, the force on the cone decreases to zero.

To exclude the loss of correctness caused by the possible increase of the size of the body, it is natural to bound above one of the parameters: the surface area of the body

$$\int_\Omega \sqrt{1 + u_x^2 + u_y^2}\, dxdy \leq S, \tag{5.207}$$

the height of the body

$$\max_\Omega u(x, y) \leq h, \tag{5.208}$$

or the volume of the body

$$\int_\Omega u(x, y) dxdy \leq V. \tag{5.209}$$

Consider first the minimization problem on the set of all bodies with the bounded surface area. This problem can be presented as a minimax problem:

$$\check{\mathcal{F}} = \min_{u \in (5.207),(5.203)} \mathcal{F}(u) =$$

$$= \min_{u \in (5.203)} \max_{\lambda \geq 0} \left[\int_\Omega \frac{2}{1+u_x^2+u_y^2} dxdy + \lambda \left(\int_\Omega \sqrt{1+u_x^2+u_y^2} dxdy - S \right) \right].$$

Let us reverse the order of minimization and maximization. According to (5.81), this cannot increase the result:

$$\max_{\lambda \geq 0} \min_{u \in (5.203)} \left[\int_\Omega \frac{2}{1+u_x^2+u_y^2} dxdy + \lambda \left(\int_\Omega \sqrt{1+u_x^2+u_y^2}\, dxdy - S \right) \right] \leq \check{\mathcal{F}}. \tag{5.210}$$

Let us define the function $\sigma(x, y)$ by the equation,

$$\sigma(x, y) = \sqrt{1+u_x^2+u_y^2}, \quad u|_\Gamma = 0, \tag{5.211}$$

and rewrite (5.210) as

$$\max_{\lambda \geq 0} \min_{\sigma \in (5.211)} \left[\int_\Omega \frac{2}{\sigma^2} dxdy + \lambda \left(\int_\Omega \sigma\, dxdy - S \right) \right] \leq \check{\mathcal{F}}. \tag{5.212}$$

The notation $\sigma \in$(5.211) means that for any admissible function $\sigma(x, y)$ there exists a function $u(x, y)$ satisfying (5.211). Let us expand the set of admissible functions $\sigma(x, y)$, replacing the constraint (5.211) by $\sigma \geq 0$. The minimum in (5.212) can only decrease:

$$\max_{\lambda \geq 0} \min_{\sigma \geq 0} \left[\int_\Omega \frac{2}{\sigma^2} dxdy + \lambda \left(\int_\Omega \sigma\, dxdy - S \right) \right] \leq \check{\mathcal{F}}. \tag{5.213}$$

The minimum in (5.213) can easily be found. Its determination is reduced to minization of the function

$$\varphi(\sigma) = \frac{2}{\sigma^2} + \lambda\sigma \quad \text{for } \sigma \geq 0.$$

The function $\varphi(\sigma)$ for $\sigma \geq 0$ is strictly convex and has only one minimum. Solving the equation $d\varphi/d\sigma = 0$, we get $\sigma = (4/\lambda)^{\frac{1}{3}}$; $\varphi(\check{\sigma}) = \min \varphi(\sigma) = \frac{3}{2}(2\lambda)^{\frac{2}{3}}$. Hence,

$$J(\lambda) = \min_\sigma \left[\int_\Omega \left(\frac{2}{\sigma^2} + \lambda\sigma \right) dxdy - \lambda S \right] = \frac{3}{2}(2\lambda)^{\frac{2}{3}} |\Omega| - \lambda S.$$

The function $J(\lambda)$ is strictly concave, and its maximum is reached at the unique point $\hat{\lambda}$; from the equation $\left(\frac{dJ}{d\lambda}\right)\Big|_{\lambda=\hat{\lambda}} = 0$, we find $\hat{\lambda}$: $\hat{\lambda} = 4|\Omega|^3/S^3$, $J(\hat{\lambda}) = 2|\Omega|^3/S^2$. So we obtain a lower estimate for the minimum drag force,

$$2\frac{|\Omega|^3}{S^2} \leq \check{\mathcal{F}}. \tag{5.214}$$

This estimate demonstrates that the minimization problem for $\mathcal{F}(u)$ on the set of all bodies with bounded surface area is correctly posed.

Let us find an upper estimate of the drag force. For the extremum value of the parameter λ, $\hat{\lambda} = 4|\Omega|^3/S^3$, the function $\check{\sigma}(x, y)$ is constant and is equal to $(4/\hat{\lambda})^{\frac{1}{3}} = S/|\Omega|$. Consider the function $u(x, y)$, which is the solution of the boundary value problem

$$u_x^2 + u_y^2 = \frac{S^2}{|\Omega|^2} - 1, \quad u|_\Gamma = 0. \tag{5.215}$$

It is obtained by substituting $\check{\sigma} = S/|\Omega|$ into (5.211). The boundary value problem is meaningful because $S > |\Omega|$.

For the solution $u(x, y)$ of the boundary value problem (5.215), we have

$$\int_\Omega \sqrt{1 + u_x^2 + u_y^2}\, dxdy = S,$$

and therefore, $u(x, y)$ belongs to the set of admissible functions and

$$\check{\mathcal{F}} \leq \mathcal{F}(u).$$

The value of the functional $\mathcal{F}$ for the function u is easily calculated and we arrive at the estimate

$$\check{\mathcal{F}} \leq 2\frac{|\Omega|^3}{S^2}. \tag{5.216}$$

It follows from (5.214) and (5.216) that

$$\check{\mathcal{F}} = 2\frac{|\Omega|^3}{S^2}. \tag{5.217}$$

The minimum drag shape of the body is defined by the solution of the boundary value problem (5.215).

The solutions of (5.215) have the following structure: the lines, $u = const$, are parallel to the contour Γ, and the derivative $\partial u/\partial n$ along the normal to the contour line of the function $u(x, y)$ takes on the values $\pm\sqrt{\frac{S^2}{|\Omega|^2} - 1}$. It is possible to construct many solutions of (5.215), choosing between various contour lines $\frac{\partial u}{\partial n} = +\sqrt{\frac{S^2}{|\Omega|^2} - 1}$ or $\frac{\partial u}{\partial n} = -\sqrt{\frac{S^2}{|\Omega|^2} - 1}$. The solutions for which $\partial u / \partial n$ changes its sign are not meaningful, since the Newton equation for the force (5.202) is not applicable to such bodies. Therefore, we can single out the unique solution of the boundary problem (5.216) by means of an additional condition $\partial u / \partial n \geq 0$.

If Ω is a circle then, among all shapes having Ω as a base and having a surface area not greater than S, the least drag shape is the cone with the surface area S. The minimum drag shape in the case of Ω being an ellipse is shown in Fig. 5.24; it looks like a screwdriver head.

The minimum drag force $\check{\mathcal{F}}$ is determined only by the area of the base $|\Omega|$ and the surface area S, and does not depend on the shape of the base (the minimum drag shape does, of course, depend on the shape of the base). Hence, we also solved another problem: we found the minimum drag shape with the condition that the cross-section area of the body is $|\Omega|$, and the head surface area is not greater than S. The above-mentioned implies that this problem has many solutions (at least one for every region Ω), and the least possible drag force is given by (5.217).

Consider two cones of the same height and the same base area (and therefore, of the same volume), where the shape of the base of one cone is a circle, and of the other a star (Fig. 5.25). Then the star-shaped cone provides less drag than the circular one.

Indeed, the surfaces of both cones are the solutions of (5.215), and (5.217) holds. It only remains to note that the surface area of the star-shaped cone is greater than that of the circular one. It follows, for example from the solution of the following variational problem: among all contours $r = \rho(\theta)$, bounding the regions with a given area $|\Omega|$,

$$\int_0^{2\pi} \frac{1}{2}\rho^2(\theta)d\theta = |\Omega|,$$

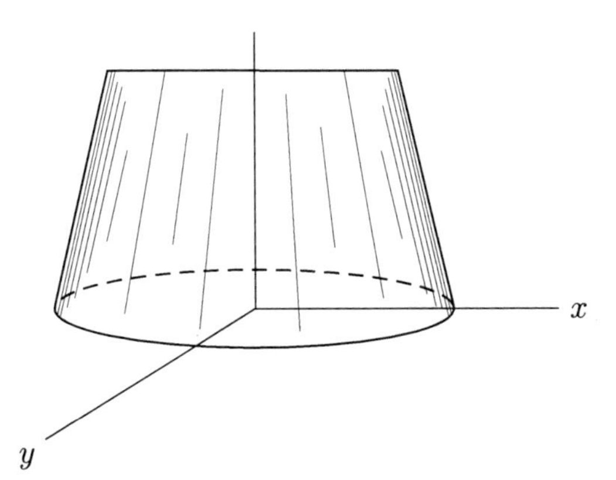

Fig. 5.24 The minimum drag body when the base is an ellipse

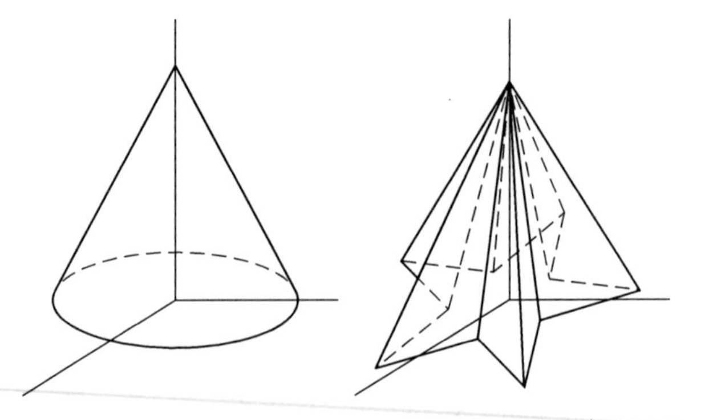

Fig. 5.25 Star-shaped cone provides less drag than a circular cone

find the contour for which the cone of the height h having this region as a base, has the minimum surface area

$$\frac{1}{2}\int_{0}^{2\pi}\sqrt{h^2+\rho^2(\theta)}\rho(\theta)d\theta \rightarrow \min.$$

Here is the solution to this problem. Introducing the Lagrange multiplier λ, we obtain the functional

$$\frac{1}{2}\int_{0}^{2\pi}\left[\rho(\theta)\sqrt{h^2+\rho^2(\theta)}-\lambda\left(\rho^2(\theta)-\frac{|\Omega|}{\pi}\right)\right]d\theta.$$

Its Euler equation for the function $\rho(\theta)$ is $\alpha+\alpha^{-1}=2\lambda$, $\alpha\equiv\check{\rho}/\sqrt{h^2+\check{\rho}^2}$. Consequently, $\check{\rho}\equiv$ *const*, and the stationary value is reached for the cone with the circular base of area $|\Omega|$ and height h. The stationary point is unique since the circular cone is uniquely defined by h and $|\Omega|$. It only remains to show that this is the minimum point. Adding an increment $\varepsilon\bar{\rho}(\theta)$ to $\check{\rho}$, let us calculate the area change up to the terms of order ε^2,

$$\frac{1}{2}\left(h^2+\check{\rho}^2\right)\left(\alpha^2+1\right)\int_{0}^{2\pi}\varepsilon\bar{\rho}d\theta+\frac{1}{2}\left(h^2+\check{\rho}^2\right)\left[\frac{1}{2}\alpha(1-\alpha)+\alpha\right]\int_{0}^{2\pi}\varepsilon^2\bar{\rho}^2d\theta.$$

The first integral is equal to zero due the constraint for the area of the base, while the second integral is positive since $\alpha<1$. Therefore, the circular cone has the minimum surface area among the cones with the same height and the same base area.

The above results can easily be generalized to the case of the region Ω bounded by two closed parallel contours Γ and Γ'. Let the function $u(x,y)$ be equal to zero on Γ and

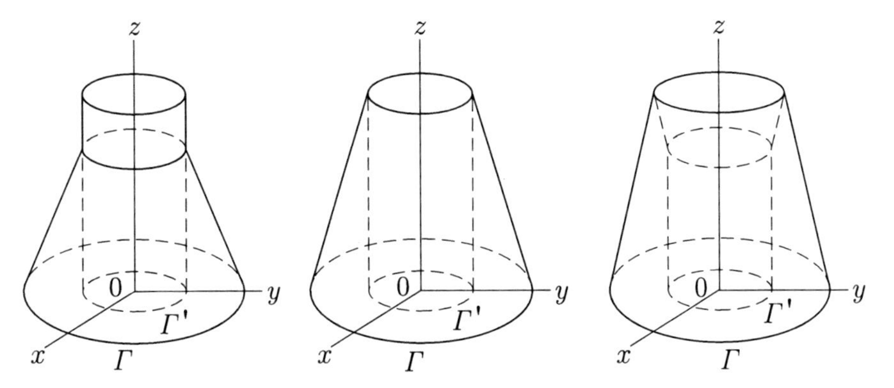

Fig. 5.26 Minimum drag body with the base bounded by contours Γ and Γ′

$$u(x, y) = u_0 = const \quad \text{on } \Gamma'.$$

As in the case of the simply connected region, it can be proven that the sought surface obeys (5.215). The drag force is given by (5.217).

In the case of the circular base, the solution is constructed from conic surfaces and is shown in Fig. 5.26 for different ratios of u_0 and S.

Let us move on the minimization problem for the functional $\mathcal{F}(u)$ with constraints on the height (5.208) and volume (5.209) of the body. First, let us show that this problem is incorrectly posed in the absence of additional conditions. Consider an axis-symmetric body, the cross-section of which in the plane $\theta = const.$ is shown in Fig. 5.27. We will assume that the width of each tooth is ρ/n, where n is an integer, and the height of the prong is equal to ε. Then, $|\partial u/\partial r| = 2\varepsilon n/\rho$, and the functional has the value

$$\mathcal{F}(u) = \int\limits_0^\rho \int\limits_0^{2\pi} \frac{2r}{1+u_r^2} d\theta dr = \frac{2\pi\rho^2}{1+4\frac{\varepsilon^2 n^2}{\rho^2}}. \tag{5.218}$$

Therefore, $\mathcal{F}(u) \to 0$ as $n \to \infty$.

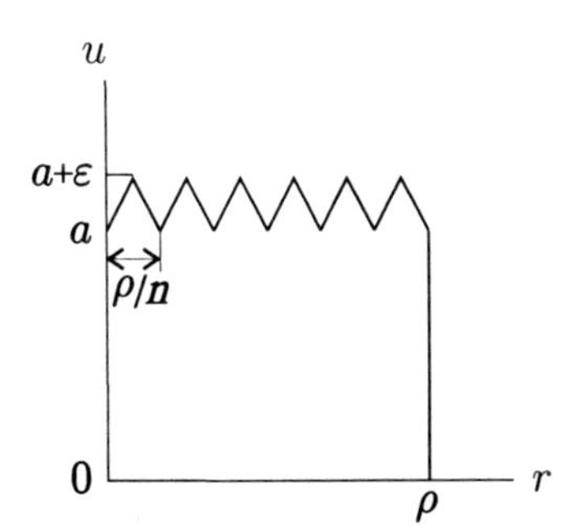

Fig. 5.27 A minimizing sequence for $\mathcal{F}(u)$

By appropriately choosing the constants, a and ε, we can make the height and the volume of the body satisfy the constraints. If the parameter ε is considered to be dependent on n and tends to zero, then the sequence of surfaces will converge to a cylinder closed by a flat cover $z = a$. If we apply an additional constraint of $\varepsilon n \to \infty$ as $n \to \infty$, then, according to (5.218), the corresponding drag force tends to zero.

The existence of a sequence of surfaces for which $\mathcal{F} \to 0$ shows that the minimization problem for the functional $\mathcal{F}(u)$ on the set of bodies with limited height and volume is incorrectly posed.

On the other hand, bounding the surface area of the body makes the problem correctly posed, as we have obtained. Therefore, it makes sense to narrow down the set of admissible functions $u(x, y)$ in such a way that for this set the boundedness of the height or the volume of the body would imply the boundedness of the head surface area. That means that on the set of admissible functions, the inequality

$$\int_{\Omega} \sqrt{1 + u_x^2 + u_y^2} dxdy - \Omega \leq l \max_{\Omega} u(x, y) \tag{5.219}$$

holds in the case of the bounded height or the inequality

$$\int_{\Omega} \sqrt{1 + u_x^2 + u_y^2} dxdy - \Omega \leq k \int_{\Omega} udxdy \tag{5.220}$$

holds in the case of the bounded volume.

The inequalities (5.219) and (5.220) are true, for example, for concave functions $u(x, y)$ (the head part of the body is convex). Indeed, for a concave function $u(x, y)$, the surface area of its graph, $z = u(x, y)$, can be estimated above by the surface area of a cylinder with the cross-section Ω and the height $\max_{\Omega} u(x, y)$. The constant l in (5.220) can be interpreted as the length of contour Γ.

The inequality (5.220) for concave functions can be proved as follows. Consider the cone K, obtained by connecting the farthest point of the body from the plane $z = 0$ with the points on the contour Γ. Due to the convexity of the body, the cone K is completely inside the body. The volume of the cone K, $\frac{1}{3}\Omega \max_{\Omega} u(x, y)$, is the lower estimate of the body's volume,

$$\frac{1}{3}\Omega \max_{\Omega} u(x, y) \leq \int_{\Omega} udxdy \tag{5.221}$$

The inequality (5.220) where k is equal to $3l/\Omega$ follows from the inequalities (5.221) and (5.219).

Due to (5.219) and (5.220), the problem of finding the shape of a convex body having a minimum drag in hypersonic flow is correctly posed on the set of all bodies with bounded height or bounded volume. The solutions of these problems can be obtained only numerically.

Pontrjagin's maximum principle. Consider the variational problem

$$I(x,u)=\int_{t_0}^{t_1} L\big(t,x^i(t),u^\varkappa(t)\big)dt \to \min \tag{5.222}$$

with the constraints

$$\frac{dx^i}{dt}=f^i\left(t,x^i(t),u^\varkappa(t)\right),\quad i=1,\dots,n,\quad \varkappa=1,\dots,m, \tag{5.223}$$

$$x^i(t_0)=x_0^i, \tag{5.224}$$

$$u^\varkappa(t)\in U, \tag{5.225}$$

where U is a set in R_m. The variational problem (5.222), (5.223), (5.224) and (5.225) has differential constraints (5.223). If the functions $u^\varkappa(t)$ satisfying the condition (5.225) are given, then the system of differential equations (5.223) with the initial values (5.224) will determine the functions $x^i(t)$. Having $x^i(t)$ and $u^\varkappa(t)$, we can calculate the value of the functional $I(x,u)$. The minimization of the functional $I(x,u)$ means its minimization with respect to all admissible functions $u^\varkappa(t)$.

The variational problem with the constraints (5.223), (5.224) and (5.225) is an example of the problems in optimal control theory. Let us find the dual variational problem in that case.

First, we rewrite the variational problem by introducing the Lagrange multipliers, $p_i(t)$, as

$$\check{I}=\min_{\substack{x\in(5.224)\\ u\in(5.225)}}\max_{p}\int_{t_0}^{t_1}\left[L\left(t,x^i,u^\varkappa\right)+p_i(t)\left(\frac{dx^i}{dt}-f^i\left(t,x^j,u^\varkappa\right)\right)\right]dt.$$

Suppose that the order of the minimization and maximization can be reversed:

$$\check{I}=\max_{p}\min_{\substack{x\in(5.228)\\ u\in(5.229)}}\int_{t_0}^{t_1}\left[L\left(t,x^i,u^\varkappa\right)+p_i(t)\left(\frac{dx^i}{dt}-f^i\left(t,x^j,u^\varkappa\right)\right)\right]dt. \tag{5.226}$$

Now the minimization problem in (5.226) does not contain differential constraints.

Consider the function

$$\mathcal{H}(t,p,x,u)=p_if^i\left(t,x^j,u^\varkappa\right)-L\left(t,x^i,u^\varkappa\right)$$

which is called the Pontrjagin function. In terms of this function (5.226) takes the form

$$\check{I} = \max_{p} \min_{\substack{x \in (5.224) \\ u \in (5.225)}} \int_{t_0}^{t_1} \left[p_i(t) \frac{dx^i}{dt} - \mathcal{H}(t, p, x, u) \right] dt. \tag{5.227}$$

The minimization with respect to u in (5.227) is reduced to an algebraic problem of maximizing the Pontrjagin function $\mathcal{H}$ with respect to u. Denote its maximum value with respect to u by $H(t, p, x)$:

$$H(t, p, x) = \max_{u \in U} \mathcal{H}(t, p, x, u).$$

Hence, after the calculation of the maximum value with respect to u, the variational problem (5.227) transforms into the Hamilton principle:

$$\check{I} = \max_{p} \min_{x \in (5.224)} \int_{t_0}^{t_1} \left[p_i(t) \frac{dx^i}{dt} - H(t, p, x) \right] dt.$$

This exposition represents the algorithmic part of Pontrjagin's maximum principle. The proof of the principle and its complete formulation can be found in the papers cited in the Bibliographic Comments. It is interesting that changing of the order of minimum and maximum in (5.226) is possible even if functions L and f^i are non-convex.

Integral constraints for derivatives. Consider the minimization problem for the functional

$$I(u) = \int_V L\left(x, u_{,i}\right) dV$$

on the set of functions $u(x)$, $x \in R_n$, with assigned values, $u_{(b)}$, on ∂V. Functional is invariant with respect to the shifts of u for a constant. Therefore, the minimum value does not change if we replace $u_{(b)}$ by $u_{(b)} + const$. Let us try to transform this problem into the minimization problem for the functional

$$\int_V L(x, u_i) dV$$

on the set of n functions $u_i(x)$, where $u_i(x)$ are subject to the constraints warranting that $u_i(x)$ are the derivatives with respect to x^i, of some function $u(x)$, which are equal to $u_{(b)}$ on ∂V. Surely, to be the derivatives of some function, u_i are to satisfy the differential constraints in the region V:

$$u_{i,j} = u_{j,i}.$$

However, we are interested in constraints of another type - the integral constraints. Namely, we are going to show that $u_i(x)$ are the derivatives of some function $u(x)$ equal to $u_{(b)} + const$ on the boundary ∂V of some simply connected region V, if and only if the following equality holds:

$$\int_V p^i u_i dV = \int_{\partial V} p^i n_i u_{(b)} dA, \tag{5.228}$$

where $p^i(x)$ is any smooth field satisfying the equation

$$p^i_{,i} = 0. \tag{5.229}$$

In other words, the following variational principle holds:
Variational principle. *Minimization of the functional*

$$\int_V L(x, u_{,i}) dV$$

on the set of functions $u(x)$ taking the boundary values $u_{(b)} + const$ is equivalent to the minimization problem for the functional

$$\int_V L(x, u_i) dV$$

on the set of functions $u_i(x)$ satisfying the constraint (5.228).

Clearly, this is a version of the dual variational principle.

First we show that (5.229) holds if and only if there exist twice continuously differentiable functions $\psi^{ij}(x)$ such that

$$p^i = \partial_j \psi^{ij}, \quad \psi^{ij} = -\psi^{ji}. \tag{5.230}$$

We prove this fact for the case of $V = R_n$. It is obvious that (5.229) follows from (5.230). Now, let us show that (5.230) follows from (5.229). For any functions $p^i(x)$ satisfying (5.229), there exists the solution $\psi^i(x)$ of equations

$$\Delta \psi^i(x) - \partial^i \partial^j \psi_j(x) = p^i(x). \tag{5.231}$$

Indeed, applying the Fourier transformation to (5.231), we get

$$-k^2 \psi^i(k) + k^i k^j \psi_j(k) = p^i(k), \quad k^2 \equiv k_i k^i \tag{5.232}$$

where $p^i(k)$, $\psi^i(k)$ are the Fourier transformations[15] of the functions $p^i(x)$ and $\psi^i(x)$, respectively, and k^i are the wave numbers (the Fourier transformation parameters). Equations (5.232) form a system of linear equations for $\psi^i(k)$. The rank of the matrix of this system $\left\|-k^2\delta^{ij}+k^ik^j\right\|$ is equal to $n-1$. The necessary condition of the solvability of (5.232) is $p^i(k)k_i=0$ and it is satisfied due to (5.229). From (5.232), we find that $\psi^i(k)=-k^{-2}p^i(k)+\psi(k)k^i$, where $\psi(k)$ is an arbitrary function. Hence, for any functions $p^i(k)$ there exists a function $\psi^i(k)$ satisfying (5.231). Equations (5.231) transform to (5.230) for $\psi^{ij}=\partial^j\psi^i-\partial^i\psi^j$.

To prove (5.228) we note that, if $u_i=u_{,i}$, and $u=u^{(b)}+const$ on ∂V, then (5.228) is satisfied. Now, let (5.228) holds for any solutions of (5.229). According to (5.230), the equality (5.228) can be written as

$$\int_V u_i\partial_j\psi^{ij}dV=\int_{\partial V}u^{(b)}n_i\partial_j\psi^{ij}dA.$$

Integrating by parts, we get

$$\int_V \psi^{ij}u_{[i,j]}dV=\int_{\partial V}\left(u_{(b)}n_i\partial_j\psi^{ij}-u_i\psi^{ij}n_j\right)dA. \tag{5.233}$$

The square brackets in the indices denote the alteration $a_{[i,j]}\equiv\frac{1}{2}\left(a_{ij}-a_{ji}\right)$.

Let $x^i=r^i(\xi^\alpha)$, $\alpha=1,\ldots,n-1$, be the equations of the surface ∂V. Using the notation introduced for the surfaces in R_3 (see Sect. 14.1) and the decomposition of the gradient (14.17), the contraction $n_i\partial_j\psi^{ij}$ is

$$n_i\partial_j\psi^{ij}=n_ir_j^\alpha\psi^{ij}_{,\alpha}. \tag{5.234}$$

Hence, the contraction (5.234) contains only the derivatives ψ^{ij} along the surface ∂V. Assume that $\psi^{ij}=0$ on ∂V. Then the integral on the right-hand side of (5.233) is zero. Due to the arbitrariness of ψ^{ij}, applying the main lemma of calculus of variations to (5.233), we find that $u_{i,j}=u_{j,i}$ in V. Consequently, there exists function $u(x)$ such that $u_i=u_{,i}$. Due to continuity, this also holds at ∂V.

For any functions $u(x)$ and $\psi^{ij}(x)=-\psi^{ji}(x)$, the identity

$$\int_{\partial V}\left(u_{,i}\psi^{ij}-u\psi^{ji}_{,i}\right)n_jdA=0. \tag{5.235}$$

holds. It can be checked by transforming this integral to the volume integral by means of the divergence theorem. Thus, from (5.233) and (5.235),

[15] The basic features of Fourier transformation are considered further in Sect. 6.7.

$$\int_{\partial V} \left(u_{(b)} - u\right) n_i \partial_j \psi^{ij} dA = 0 \tag{5.236}$$

for any functions ψ^{ij} on ∂V. The contraction $n_i r_j^\alpha \psi^{ij}_{,\alpha}$ is

$$n_i r_j^\alpha \psi^{ij}_{,\alpha} = \left(n_i r_i^\alpha \psi^{ij}\right)_{|\alpha} \tag{5.237}$$

where the vertical bar in indices denote the surface covariant derivative along ∂V. This follows from the equalities

$$\left(n_i r_j^\alpha\right)_{|\alpha} = \left(n_j r_i^\alpha\right)_{|\alpha} \tag{5.238}$$

which can be established by means of (14.33) and (14.31). Therefore, $u_{(b)} = u + \ const.$

Unlocking the integral constraints. Let us retain a finite number of constraints corresponding to some set of particular solutions $p^i_{(1)}, \ldots, p^i_{(m)}$ of (5.229), discarding the rest. Then the set of admissible functions will expand, and the minimum of the functional will decrease. The minimum can be found by introducing the Lagrange multipliers $\lambda^1, \ldots, \lambda^m$; we get the following variational problem:

$$\min_{u_i} \max_{\lambda^1,\ldots\lambda^m} \int_V \left(L(x, u_i) - \left(\lambda^1 p^i_{(1)} + \ldots + \lambda^m p^i_{(m)}\right) u_i\right) dV +$$

$$+ \int_{\partial V} (\lambda^1 p^i_{(1)} + \ldots + \lambda^m p^i_{(m)}) n_i u_{(b)} dA \le \check{I}.$$

after reversing the order of calculating of maximum and minimum, finding the minimum with respect to u_i is reduced to calculating the Young-Fenchel transformation of the function L with respect to u_i. Therefore,

$$\max_{\lambda^1,\ldots\lambda^m} \left(\int_{\partial V} p^i n_i u^{(b)} dA - \int_V L^* (x, p_i) dV \right) \le \check{I},$$

where p^i denotes the sum $\lambda^1 p^i_{(1)} + \ldots + \lambda^m p^i_{(m)}$.

We see that relaxing of integral constraints is equivalent to the Rayleigh-Ritz method for the dual problem, and the integral constraint method is based on the same ideas as the duality theory methods.

The above consideration is easily generalized to the integral functional of the form

$$\int_V L\left(x^i, u^\varkappa, u_i^\varkappa\right) dV. \tag{5.239}$$

using the following statement:

For the functions $u^\varkappa(x)$ and $u_i^\varkappa(x)$ to satisfy the relations $u_i^\varkappa = u_{,i}^\varkappa$ and $u^\varkappa = u_{(b)}^\varkappa$ on ∂V_u, it is necessary and sufficient that the relation

$$\int_V \left(p_\varkappa u^\varkappa + p_\varkappa^i u^\varkappa\right) dV = \int_{\partial V_u} p_\varkappa^i n_i u_{(b)}^\varkappa dA \tag{5.240}$$

holds for all the solution of the equations

$$p_\varkappa - p_{\varkappa,i}^i = 0 \qquad \text{in } V, \quad p_\varkappa^i n_i = 0 \qquad \text{on } \partial V_f = \partial V - \partial V_u. \tag{5.241}$$

This statement becomes obvious if using (5.241) we write the equality (5.240) as

$$\int_V \left(p_{\varkappa,i}^i u^\varkappa + p_\varkappa^i u^\varkappa\right) dV = \int_{\partial V_u} p_\varkappa^i n_i u_{(b)}^\varkappa dA \tag{5.242}$$

and take into account that $p_\varkappa^i(x)$ are arbitrary vector fields with zero normal components on ∂V_f.

So, the minimization problem for the functional (5.239) can be written also as the minimization problem for the functional

$$\int_V L\left(x^i, u^\varkappa, u_i^\varkappa\right) dV$$

on the set of all functions $u^\varkappa(x)$ and $u_i^\varkappa(x)$ satisfying the relations (5.240).

The cross-section principle. In the problems without constraints, it is sometimes useful to introduce constraints by means of the following construction.

Let us present the set $\mathcal{M}$ of elements u as a union of non-intersecting subsets $\mathcal{M}_v$, $\mathcal{M} = \cup \mathcal{M}_v$, where v are elements of some set $\mathcal{N}$. The subsets $\mathcal{M}_v$ will be called the cross-sections of the set $\mathcal{M}$.

The minimization problem for functional $I(u)$ on $\mathcal{M}$ can be solved in two steps. First, the minimum of the functional $I(u)$ is sought on the set $\mathcal{M}_v$. The minimum value J of the function $I(u)$ on $\mathcal{M}_v$ depends on v:

$$J(v) = \min_{u \in \mathcal{M}_v} I(u).$$

Then the minimum of the functional $J(v)$ on $\mathcal{N}$ is sought. One may expect that

$$\check{I} = \min_{v \in \mathcal{N}} J(v). \tag{5.243}$$

Let us prove (5.243). Denote $\min J(v)$ by $\check{J}$. Obviously, for any v, $\check{I} \leq J(v)$. Suppose that $\check{I} < \check{J}$. By the definition of the minimum value, there exists a sequence u_n, for which $I(u_n) \to \check{I}$. Denote by v_n the element v, for which $u_n \in \mathcal{M}_v$. Since $\check{I} \leq J(v_n) \leq I(u_n)$, $J(v_n) \to \check{I}$. This refutes the supposition.

The statement (5.243) is called the cross-section principle.

5.11 Variational-Asymptotic Method

In many problems of mechanics and physics there are small and large parameters. These can be geometric parameters like the thickness of a plate or a shell, the diameter of a beam cross-section, the average crystallite size in a polycrystal, the magnitude of deformation or displacements of a continuum, the wave length, the number of gas molecules in a container, or they can be physical parameters: the viscosity or heat conductivity of a fluid, the oscillation frequency of a rigid body, etc. To investigate properly such problems various asymptotic approaches were developed. It is clear that for the problems which allow a variational formulation and, consequently, possessing a special structure, there should exist a direct variational approach based on direct asymptotic analysis of the corresponding functionals. It should automatically take into account the variational structure of the equations and those properties of the solutions, which are dictated by this structure.

A method of asymptotic analysis of functionals will be considered in this section. This method is of heuristic character. It does not have a strict mathematical foundation. However, for problems admitting exact solutions or studied by different methods, there is a complete accordance of the results.

The method of asymptotic analysis of functionals, which will be called the variational-asymptotic method, allows one to consider the minimization problems for functions of a finite number of variables and the problems for differential equations possessing the variational structure from a common point of view.

When applied to differential equations, the variational-asymptotic method has a number of advantages compared to the widely used asymptotic approaches mainly due to the simplicity of analysis, which is more noticeable as the complexity of the system of differential equations increases. The reason for this is that in the variational-asymptotic method only one function, the Lagrangian – instead of a system of differential equations, is the subject of the investigation.

Another advantage of the variational-asymptotic method is that the approximate equations always possess a variational structure while the direct asymptotic analysis of differential equations may yield the approximate equations that do not have such a structure. This can be seen from the analysis of a system of two linear algebraic equations, involving a small parameter, ε:

$$x + \varepsilon (x - y) = 1, \quad y + \varepsilon (y - x) = 0.$$

This system can be interpreted as the equilibrium equations of two interacting springs: x and y are the displacements of the springs from the equilibrium position, and the first spring is subject to the external force of unit magnitude. The interaction is characterized by parameter ε. Let us pose the following problem: for $\varepsilon \to 0$, construct an approximate system of equations the solution of which coincides with the solution of the original system in terms of the order of 1 and ε. Since $x \approx 1$ and $y \approx \varepsilon$, the only possible "simplification" is disregarding the term εy, which is of the order of ε^2. We have

$$x + \varepsilon x = 1, \quad y - \varepsilon x = 0.$$

The initial system of equations is a system of equations of variational type: the left-hand sides of the equations are derivatives of the function

$$f(x, y, \varepsilon) = \frac{1}{2}\left(x^2 + \varepsilon (x - y)^2 + y^2\right),$$

and the equilibrium equations are the equations for the minimizing point of the function $f(x, y, \varepsilon) - x$. For the approximate system of equations, this is not true, there is no function $g(x, y, \varepsilon)$, such that

$$\frac{\partial g(x, y, \varepsilon)}{\partial x} = x + \varepsilon x, \quad \frac{\partial g(x, y, \varepsilon)}{\partial y} = y - \varepsilon x$$

because

$$\frac{\partial}{\partial y}(x + \varepsilon x) \neq \frac{\partial}{\partial x}(y - \varepsilon x).$$

This situation is typical in asymptotic analysis of differential equations with small parameters. The violation of the variational structure of the approximate equations may result in even qualitatively wrong results. For example, the exact self-conjugated equations possess only real eigenvalues, while the approximate equations acquire complex eigenvalues, i.e., physically, the system in an approximate description may become unstable while it is stable in the exact description.

The variational approach in asymptotic analysis of partial differential equations is important also due to the necessity of short-wave extrapolation. This issue will be discussed further in Sect. 14.5.

The variational-asymptotic method is based on the idea of neglecting the small terms in energy. This idea is often used in physics. However, to apply it in a systematic way one has to learn how to recognize small terms, how to treat the situations when neglecting small terms results in the loss of the uniqueness or the existence of the solution, and, moreover, understand how the small terms affect the next approximations and how the iteration procedure could be settled.

The variational-asymptotic method will be formulated as a set of rules, the application of which will be illustrated for a number of examples. Other applications will be encountered in later chapters.

Let the functional $I(u, \varepsilon)$ be defined on some set of elements $\mathcal{M}$ and depend on small parameter ε. We will assume that the functional $I(u, \varepsilon)$ has a finite or a countable number of stationary points. The stationary points will be denoted by $\check{u}$, and, in order to avoid cumbersome notation, will not be numbered. The stationary points $\check{u}$ are functions of ε. This will be emphasized by placing ε in the index, $\check{u}_\varepsilon$.

Let $\check{u}_\varepsilon$ tend to $\check{u}_0$ for $\varepsilon \to 0$. The following questions arise: how can we construct a functional which has $\check{u}_0$ as stationary points? How can we construct an approximate functional, the stationary points of which are the approximations of $\check{u}_\varepsilon$ of a given accuracy?

It is clear that the answers to these questions are related to the simplification of the functional by disregarding small insignificant terms. It is natural to start with investigation of the functional from which all small terms dropped, i.e. the functional $I_0(u) = I(u, 0)$. The following situations may be encountered:

1. $I_0(u)$ has the isolated stationary points
2. $I_0(u)$ has the non-isolated stationary points
3. $I_0(u)$ does not have the stationary points or it is meaningless, i.e. the functional $I(u, \varepsilon)$ is not defined for $\varepsilon = 0$

Let us begin with the first case.

Case 1: $I_0(u)$ has the isolated stationary points

It could be expected that the stationary points of the functional $I_0(u)$ are the first approximations of the stationary points of the initial functional.

Example 1. Consider the behavior of the stationary points of the function of one variable,

$$f(u, \varepsilon) = u^2 + u^3 + 2\varepsilon u + \varepsilon u^2 + \varepsilon^2 u, \tag{5.244}$$

for $\varepsilon \to 0$. The function $f_0(u) = u^2 + u^3$ has two stationary points, $u = 0$ and $u = -\frac{2}{3}$. It is not difficult to check that they are indeed the limits for $\varepsilon \to 0$ for the stationary points of the function $f(u, \varepsilon)$.

The second asymptotic term will be sought in the following way. Let us present u as $u = \check{u}_0 + u'$, where $\check{u}_0$ is a stationary point of the functional $I_0(u)$, and $u' \to 0$ for $\varepsilon \to 0$. We will also keep the leading terms containing u' in the functional $I\left(u_0 + u', \varepsilon\right)$. Thus we obtain the functional $I_1\left(u', \varepsilon\right)$. We expect that the stationary point with respect to u' of this functional is the next asymptotic term.

Example 1 (continued). Consider the stationary point of the function (5.244) in a neighborhood of zero. Thus u is to be small. Let us keep only the leading terms with respect to u in $f(u, \varepsilon)$. Since εu^2 and u^3 are small compared to u^2, and $\varepsilon^2 u$ is small compared to $2u\varepsilon$, we have

$$f_1(u, \varepsilon) = u^2 + 2\varepsilon u.$$

Terms u^2 and $2u\varepsilon$ need to be kept since there are no terms in comparison with which they could be neglected. The function $f_1(u, \varepsilon)$ has the stationary point $u =$

$-\varepsilon$. Hence, the asymptotic approximation of the stationary point of the function $f(u, \varepsilon)$ in the neighborhood of zero is

$$\check{u}_\varepsilon = -\varepsilon + o(\varepsilon).$$

At this point the function $f(u, \varepsilon)$ is the same as $f_1(u, \varepsilon)$, i.e. it has a minimum.

Consider the stationary point of the function $f(u, \varepsilon)$ in the neighborhood of the point $u = -2/3$. Setting $u = -\frac{2}{3} + u'$ and keeping the leading terms with respect to u' in the function $f\left(-\frac{2}{3} + u', \varepsilon\right)$, we arrive at the function[16]

$$f_1(u, \varepsilon) = -u'^2 + \frac{2}{3}\varepsilon u'.$$

Its stationary point is $u' = \frac{1}{3}\varepsilon$ and the asymptotic approximation of the second stationary point is

$$\check{u}_\varepsilon = -\frac{2}{3} + \frac{1}{3}\varepsilon + o(\varepsilon);$$

the function $f(u, \varepsilon)$ has a local maximum at this point.

The next asymptotic terms are constructed analogously. For example, for the stationary point of $f(u, \varepsilon)$ in the neighborhood of zero, we seek u in the form $u = -\varepsilon + u''$, $u'' = o(\varepsilon)$. Keeping only the leading terms with respect to u'' in the function

$$f(u, \varepsilon) = \left(-\varepsilon + u''\right)^2 + \left(-\varepsilon + u''\right)^3 + 2\varepsilon\left(-\varepsilon + u''\right) + \varepsilon\left(-\varepsilon + u''\right)^2 + \varepsilon^2\left(-\varepsilon + u''\right),$$

we get the function

$$f_2(u, \varepsilon) = u''^2 + 2\varepsilon^2 u''.$$

Therefore, $u'' = -\varepsilon^2$ and

$$\check{u}_\varepsilon = -\varepsilon - \varepsilon^2 + o\left(\varepsilon^2\right).$$

Note that the order of the "smallness" of u' is not supposed a priori, but is determined.

Also note that the terms that were insignificant in constructing the leading term of the asymptotic expansion may become important in determining the next terms of the asymptotic expansion.

The example considered demonstrates that the main issue in asymptotic analysis of functionals is to recognize the leading terms and the negligible ones. Usually,

---

[16] From now on, additive constants will be disregarded without mentioning.

this is the most important and the most difficult point of the analysis. However, there are cases when the recognition of the negligible terms does not cause difficulties. Here are two conditions upon satisfying which the terms may be considered negligible.

Let the functional $I(u, \varepsilon)$ includes a sum of two terms, $A(u, \varepsilon)$ and $B(u, \varepsilon)$, and

$$\lim_{\varepsilon\to 0}\max_{u\in\mathcal{M}}\left|\frac{B(u,\varepsilon)}{A(u,\varepsilon)}\right| = 0. \tag{5.245}$$

Then $B(u, \varepsilon)$ is negligible in comparison with $A(u, \varepsilon)$ for all stationary points. We will call such terms globally secondary terms.

Let $\check{u}_\varepsilon \to 0$ for $\varepsilon \to 0$, and for any sequence $\{u_n\}$ converging to $u = 0$:

$$\lim_{\substack{n\to\infty\\ \varepsilon\to 0}}\left|\frac{B(u_n,\varepsilon)}{A(u_n,\varepsilon)}\right| = 0. \tag{5.246}$$

Then the term $B(u, \varepsilon)$ is negligible compared to $A(u, \varepsilon)$ for the stationary point $\check{u}_\varepsilon$. The convergence $u_n \to 0$ is understood in the same sense as the convergence of $\check{u}_\varepsilon \to 0$. The term $B(u, \varepsilon)$ will be called locally secondary compared to $A(u, \varepsilon)$. The case of $\check{u}_\varepsilon \to \check{u}_0$, $\check{u}_0 \neq 0$ is reduced to the case of $\check{u}_\varepsilon \to 0$ by the substitution $u \to u'$: $u = \check{u}_0 + u'$.

In the definitions (5.245) and (5.246), the additive constants in $A(u, \varepsilon)$ are to be excluded. This can be done, for example, by setting the condition

$$A(0, \varepsilon) = 0.$$

In Example 1, the term εu^2 is globally secondary with respect to u^2, $\varepsilon^2 u$ is globally secondary with respect to $2\varepsilon u$ while u^3 is locally secondary with respect to u^2 in the neighborhood of the point $u = 0$.

Example 2. Consider the function

$$f(u, \varepsilon) = \ln(1 + 2u) - \frac{1+2u}{1+2u^2} + \varepsilon u, \quad 1 + 2u > 0. \tag{5.247}$$

The function

$$f_0(u) = \ln(1 + 2u) - \frac{1+2u}{1+2u^2}$$

has a stationary point at $u = 0$. The first terms of expansion $f_0(u)$ with respect to u are $-1 + \frac{20}{3}u^3$. The other terms are locally secondary with respect to u^3. Therefore, the investigation of the function $f(u, \varepsilon)$ can be replaced by the investigation of the function $\frac{20}{3}u^3 + \varepsilon u$ in the first approximation. It has two stationary points $\pm\sqrt{-\frac{\varepsilon}{20}}$ for $\varepsilon \leq 0$ and does not have any stationary points for $\varepsilon > 0$. Consequently, the function $f(u, \varepsilon)$ has the same behavior in the neighborhood of zero.

For any secondary term, $B(u,\varepsilon)$, of the type (5.245) or (5.246), there exists a term of the functional, $A(u,\varepsilon)$, compared to which $B(u,\varepsilon)$ is small. Along with such terms, the functional may have terms which cannot be easily pinpointed as small because the corresponding "large" terms are not present in the functional. Consider, for example, the function

$$f(u,\varepsilon) = u^3 + \varepsilon u^2 + \varepsilon^3 u. \tag{5.248}$$

The function $f_0(u) = f(u,0) = u^3$, and one may expect that all stationary points of the function $f(u,\varepsilon)$ are in the vicinity of $u = 0$. Could we neglect the term $\varepsilon^3 u$? This is not clear because we do not know the order of $\check{u}_\varepsilon$. If, for example, $\check{u}_\varepsilon \sim \varepsilon$, then the first two terms are of the same order, ε^3, while the third, of the order ε^4, may be neglected. If $\check{u}_\varepsilon$ is, say, of the order ε^2, then the last two terms must be kept (they are both of the order ε^5) while the second one, which is of the order ε^6, may be dropped. If we cannot say about a term whether it is the leading one or negligible, we call such a term doubtful. All the terms of function (5.248) are doubtful.

A recipe for studying a functional with doubtful terms is simple: let the stationary points of the functional be hard to find, but the corresponding problem for the functional $\tilde{I}$ with the dropped doubtful terms admits investigation. Then we study the stationary points of the functional $\tilde{I}$ and evaluate the doubtful terms at these stationary points. If the doubtful terms are smaller (in the asymptotic sense) than the kept ones, then $\tilde{I}$ provides the leading asymptotics of the stationary points. Otherwise, the doubtful terms should be kept.

Example 3. Consider the stationary points of the function (5.248) in the neighborhood of zero. There are no clear indications that any one of the terms is small compared to another.

Let us first drop the last term. We obtain the function $u^3 + \varepsilon u^2$, the stationary points of which are $u = 0$ and $u = -\frac{2}{3}\varepsilon$. At $u = 0$ it is not clear whether or not the last term can be disregarded. At $u = -\frac{2}{3}\varepsilon$ it can be done, since the first two terms are of order ε^3, while the last one is of order ε^4.

Let us now discard the second term. We obtain the function $u^3 + \varepsilon^3 u$, which does not have any stationary points for $\varepsilon > 0$, while for $\varepsilon \le 0$, the stationary points are $\pm\sqrt{-\frac{1}{3}\varepsilon^3}$. The first and third terms are of the order of $\varepsilon^{\frac{9}{2}}$, while the second is much larger, of the order of ε^4. Therefore, the second term cannot be discarded.

Let us drop the first term. The function $\varepsilon u^2 + \varepsilon^3 u$ has one stationary point at $-\frac{1}{2}\varepsilon^2$. The first term is of the order of ε^6 and is much smaller than the two retained, which are of the order of ε^5. Therefore, discarding the first term makes sense.

So in the neighborhood of zero we found two stationary points, $\check{u}_\varepsilon = -\frac{2}{3}\varepsilon + o(\varepsilon)$ and $\check{u}_\varepsilon = -\frac{1}{2}\varepsilon^2 + o(\varepsilon^2)$. Since the function (5.248) has only two real stationary points, all the stationary points have been found.

Example 4. **Newton's polygon rule.** To find the leading asymptotic term for the stationary points of the function of the form

$$f(u,\varepsilon)=\sum_{m,n}^{N} a_{mn}\varepsilon^{m}u^{n},$$

which tend to zero for $\varepsilon \to 0$, one can use the method proposed by Newton[17]. We describe this method for the case of the function

$$g(u,\varepsilon)=u^{3}+\varepsilon u^{2}+\varepsilon u^{3}+\varepsilon^{2}u^{4}+\varepsilon^{3}u+\varepsilon^{5}u^{5}.$$

For each term, define a point on a two-dimensional integer lattice with the x-coordinate equal to the order of ε and the y-coordinate equal to the order of u. We get a set of points. For the function $g(u,\varepsilon)$ this set is shown in Fig. 5.28. Then we draw a maximum convex polygon with the vertices from this set, which includes all points of the set. According to Newton's polygon rule, in constructing the leading asymptotic term of the stationary points, it is necessary to keep only those terms the corresponding points of which lie on the sides of the polygon. For $\varepsilon \to 0$, it is only necessary to keep the terms corresponding to the points in the part of the polygon beginning at point A (see Fig. 5.28) and continuing to the right and down and ending at the point C, where the line turns upward. This part of the polygon is shown by a solid line in Fig. 5.28.

Let us derive Newton's polygon rule for the function $g(u,\varepsilon)$ using the general propositions made above. Consider the terms with the power of $\varepsilon \geq 0$ and the power of $u \geq 3$. Since the asymptotic $u \to 0$ is being considered, all the terms except u^{3} can be discarded: they correspond to the terms which are locally secondary compared to u^{3}. For the same reason, among all the terms with the power of $\varepsilon \geq 1$ and the power of $u \geq 2$ only εu^{2} should be kept – the corresponding terms are locally secondary with respect to εu^{2}. Thus, points A, B, and C remain. They correspond to the function (5.248) considered in Example 3. It has two stationary points, and for the construction of the asymptotics of these points the terms corresponding to the points on the edges AB and BC of Newton's polygon should be used.

The derivation of Newton's polygon rule for the general polygon is analogous to the above one.

Case 2: $I_0(u)$ has the non-isolated stationary points

Let us move on to a more interesting for physical applications case when discarding the secondary terms results in appearance of non-isolated stationary points.

A typical finite-dimensional situation can be shown for the function of two variables ($u=\{x,y\}$) of the form

$$f(x,y,\varepsilon)=f_0(x)+\varepsilon g(x,y),\quad -\infty<x<+\infty, -\infty<y<+\infty, \tag{5.249}$$

---

[17] Newton's polygon rule concerns the asymptotic behavior of the roots of polynomials; here it is considered in terms of the problem of finding the stationary points of function $f(u,\varepsilon)$, which is an integral of a polynomial function.

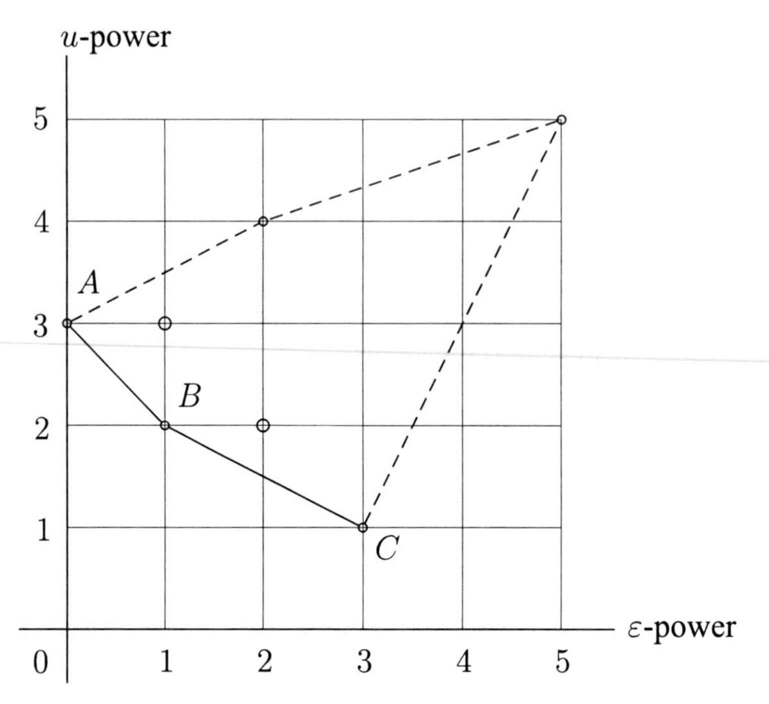

Fig. 5.28 Newton's rule

where the function $f_0(x)$ has some stationary points $\check{x}_0$.

If the presumably small term $\varepsilon g(x, y)$ is dropped, we get the function $f_0(x)$ which has in the (x, y)-plane continuum of stationary points $(\check{x}_0, y)$, $-\infty < y < +\infty$.

Before formulating the general rules for dealing with such cases, let us first consider some examples.

Example 5. Let us find the asymptotic for $\varepsilon \to 0$ of the stationary points of the function

$$f(x, y, \varepsilon) = \cos(x - y) + \varepsilon\left(\frac{1}{x} + y\right), \quad -\infty < x < +\infty, \quad -\infty < y < +\infty.$$

If $\frac{1}{\check{x}_\varepsilon} + \check{y}_\varepsilon$ is bounded for $\varepsilon \to 0$, then the second term is secondary, and $f_0(x, y) = \cos(x - y)$. Denote the set of stationary points of the function f_0, $y = x + \pi k$, $-\infty < x < +\infty$, $k = 0, \pm 1, \pm 2, \ldots$, by $\mathcal{M}_0$. Consider the function $f(x, y, \varepsilon)$ on $\mathcal{M}_0$. The function $f(x, y, \varepsilon)$ on $\mathcal{M}_0$ becomes the function of one variable x (and a number, k): $f = \cos \pi k + \varepsilon\left(\frac{1}{x} + x + \pi k\right)$. The stationary points of f on $\mathcal{M}_0$ with respect to x are $x = \pm 1$. It should be expected that the stationary points of function $f(x, y, \varepsilon)$ converge to $(\pm 1, \pm 1 + \pi k)$ for $\varepsilon \to 0$. The validity of this proposition can be checked directly: it is sufficient to perform the change of variable, $y \to z$: $y = x + z$, and the function f will become a sum of functions which depend only on x and z,

$$f = \cos z + \varepsilon z + \varepsilon \left(\frac{1}{x} + x \right),$$

for which the proposition made is obviously true.

The described procedure of constructing the stationary points made without some precautions can provide wrong results. This is demonstrated in the following example.

Example 6. Let us find the stationary point of the quadratic form

$$f(x, y, \varepsilon) = x^2 - 2x + 4\varepsilon (x - 1) y + \varepsilon^2 y^2 + 2\varepsilon^2 y.$$

The set $\mathcal{M}_0$ of the stationary points of the function $f_0(x, y) = x^2 - 2x$ is the line, $(1, y)$. The function $f(x, y, \varepsilon)$ is a function of y on $\mathcal{M}_0 : f = 1 + \varepsilon^2 \left(y^2 + 2y\right)$. This function has a stationary point at $y = -1$. By analogy with the previous example we could assume that the stationary point of the function $f(x, y, \varepsilon)$ tends to the point $(1, -1)$ for $\varepsilon \to 0$. This, however, is not true. Indeed, writing the stationary condition for the function $f(x, y, \varepsilon)$

$$x - 1 + 2\varepsilon y = 0, \quad 2(x - 1) + \varepsilon y + \varepsilon = 0,$$

we find the stationary points: $\check{x}_\varepsilon = 1 - \frac{2}{3}\varepsilon, \ \check{y}_\varepsilon = \frac{1}{3}$. Why did we get the wrong answer?

Let us fix some point of the set $\mathcal{M}_0$ (defined by assigning a value to y). The search for the stationary point will be carried out by first finding the stationary point with respect to x for a fixed y (the corresponding stationary value of f becomes a function of y only), and then finding the stationary point over all y. Since we suspect that the stationary point is in the neighborhood of set $\mathcal{M}_0$, it is sufficient to consider not all values of x but only those $x = 1 + x'$ for which x' are small. Keeping only the terms important with respect to x' in the function $f\left(1 + x', y, \varepsilon\right)$, we obtain the function $x'^2 + 4\varepsilon x' y$. Consequently, $x' = -2\varepsilon y$. So, we have to consider the function $f(1 - 2\varepsilon y, y, \varepsilon) = -1 - 3\varepsilon^2 y^2 + 2\varepsilon^2 y$ but not $f(1, y, \varepsilon)$ on $\mathcal{M}_0$. This yields $\check{y}_0 = \frac{1}{3}$, in compliance with the correct result. It follows from the above consideration that the stationary point is a saddle point – the function f is maximum with respect to y and minimum with respect to x.

The difference between Examples 5 and 6 is the following. In Example 5, let us fix y and denote the stationary point of the function $f_0(x, y)$ for fixed y by x_0 $(x_0 = y + \pi k)$. Let x' be the first correction to x_0. It is easy to check that the functions $f\left(x_0 + x', y, \varepsilon\right)$ and $f(x_0, y, \varepsilon)$ will differ in terms of the order of ε^2, while to calculate y we need to keep the terms of the order ε. In Example 6, functions $f\left(x_0 + x', y, \varepsilon\right)$ and $f(x_0, y, \varepsilon)$ also differ for terms of the order ε^2. However, to calculate y we need to keep the terms of the order ε^2.

Now we are ready to formulate the general rules for searching the stationary points.

Let $\mathcal{M}_0$ be the set of stationary points of the functional $I_0(u)$ and any element $u \in \mathcal{M}$ can be uniquely presented as $u = u_0 + u'$, $u_0 \in \mathcal{M}_0$, $u' \in \mathcal{M}'$. Fixing u_0 and considering u' to be small, we keep the leading terms with u' in the functional $I\left(u_0 + u', \varepsilon\right)$. We obtain a functional $I_1\left(u_0, u', \varepsilon\right)$. Let us find the stationary points of I_1 with respect to u'. They will depend on u_0.

Suppose that u' is uniquely determined by u_0, $u' = u'(u_0, \varepsilon)$. Representing u as $u = u_0 + u'(u_0, \varepsilon) + u''$, where $u_0 \in \mathcal{M}_0$, $u'' \in \mathcal{M}''$ and u'' is smaller than u' in the asymptotic sense, and keeping the leading terms with respect to u'', we find u'', and so on.

Consider the functionals $I(u_0, \varepsilon)$ and $I\left(u_0 + u'(u_0, \varepsilon), \varepsilon\right)$ on $\mathcal{M}_0$. If their stationary points do not differ significantly, then we can expect that the leading approximation of the stationary points of the initial functional are the stationary points of the functional $I(u_0, \varepsilon)$ on $\mathcal{M}_0$. If the stationary points of the functionals $I(u_0, \varepsilon)$ and $I\left(u_0 + u'(u_0, \varepsilon), \varepsilon\right)$ differ significantly, but the stationary points of the functionals $I\left(u_0 + u'(u_0, \varepsilon), \varepsilon\right)$ and $I\left(u_0 + u'(u_0, \varepsilon) + u''(u_0, \varepsilon), \varepsilon\right)$ do not, then the leading approximation are the stationary points of the functional $I\left(u_0 + u'(u_0, \varepsilon), \varepsilon\right)$. If the stationary points of the functionals $I\left(u_0 + u'(u_0, \varepsilon), \varepsilon\right)$ and $I\left(u_0 + u'(u_0, \varepsilon) + u''\right.$ $\left.(u_0, \varepsilon), \varepsilon\right)$ differ greatly, then the next approximations should be considered.

After the leading term of the asymptotic expansion is found, the next terms are constructed as described above.

The application of this scheme is sometimes hindered by the following obstacle: u' may be not uniquely determined by u_0, and for fixed u_0, runs through some set $\mathcal{M}_1$.

Let $\mathcal{M}_1$ be such that any element $u \in \mathcal{M}$ and be uniquely presented as $u = u_0 + u' + u''$, where $u_0 \in \mathcal{M}_0$, $u' \in \mathcal{M}'$, $u'' \in \mathcal{M}''$. We fix u_0 and u' and, assuming $u'' \ll u'$, we apply the previous scheme to find u''. Usually, after a finite number of steps, no additional variables appear and the expansion $u = u_0 + u' + u'' + \ldots$ can be written as $u = v + w(v, \varepsilon) + w'(v, \varepsilon)$, where v is an element of some set $\mathcal{N}$, and the subsequent terms of w, w', ... are uniquely determined by v. If the stationary points of the functional $I(v + w(v, \varepsilon), \varepsilon)$ and the functional $I(v, \varepsilon)$ on $\mathcal{N}$ are close, then the stationary points of $I(v, \varepsilon)$ are the leading terms of the asymptotic expansion of the stationary points of the initial functional. If the stationary points of the functional $I(v + w(v, \varepsilon), \varepsilon)$ significantly differ from the stationary points of the functional $I(v, \varepsilon)$ but not significantly differ from the stationary points of the functional $I\left(v + w(v, \varepsilon) + w'(v, \varepsilon), \varepsilon\right)$, the leading term is given by the stationary points of the functional $I(v + w(v, \varepsilon), \varepsilon)$. Otherwise, the next approximations should be considered.

Usually, in order to estimate the order of the difference of the two stationary points, it is sufficient to compare the values of the functionals at those points: if the values of the functionals do not differ significantly, their stationary points are in close proximity to each other; corresponding estimates for strictly convex functionals were given in Sect. 5.5.

Before considering more examples we need to introduce the notion of characteristic length.

Notion of characteristic length. To evaluate various terms in the functionals we often need to determine the magnitude of derivatives of required functions. This is closely related to the notion of the characteristic length. The term "characteristic length" may have many different meanings. We will use it in the following sense.

Consider the function $f(x)$ which is defined on the segment $[a,b]$ and can be differentiated a sufficient number of times. Denote the amplitude of the function's oscillation on $[a,b]$ by $\bar{f}$:

$$\bar{f} = \max_{x',x'' \in [a,b]} \left| f\left(x'\right) - f\left(x''\right) \right|.$$

For sufficiently small l, the following inequality holds:

$$\left| \frac{df}{dx} \right| \leq \frac{\bar{f}}{l}. \tag{5.250}$$

The "best" constant in the inequality (5.250) – the greatest l for which the inequality (5.250) holds – will be called the characteristic length of this function. This definition is convenient because the statements about the characteristic length imply the upper estimates of the function's derivatives. Obviously, the characteristic length does not exceed the size of the region:

$$l \leq b - a.$$

Indeed,

$$\left| f\left(x'\right) - f\left(x''\right) \right| = \left| \int_{x'}^{x''} \frac{df}{dx} dx \right| \leq \int_{x'}^{x''} \left| \frac{df}{dx} \right| dx \leq \int_a^b \left| \frac{df}{dx} \right| dx.$$

Therefore,

$$\bar{f} \leq \int_a^b \left| \frac{df}{dx} \right| dx \leq \frac{\bar{f}}{l}(b-a),$$

which yields the statement made.

If we need to estimate higher derivatives, then the corresponding terms are included in the definition of l, and the characteristic length is the best constant in the system of inequalities

$$\left| \frac{df}{dx} \right| \leq \frac{\bar{f}}{l}, \quad \left| \frac{d^2 f}{dx^2} \right| \leq \frac{\bar{f}}{l^2}, \ldots, \left| \frac{d^k f}{dx^k} \right| \leq \frac{\bar{f}}{l^k}. \tag{5.251}$$

If the derivative of the function changes significantly on $[a,$b], the number l might be a two rough characteristic, and a more appropriate one is the local characteristic length, $l(x)$, the best constant in the inequalities (5.251) written for the point x, but not for all points of the segment $[a, b]$.

The definition of the characteristic length of a function of many variables is similar.

Example 7. **Derivation of the Bernoulli-Euler beam theory from Timoshenko's beam theory.** In classical theory of beam bending developed by Bernoulli and Euler, the energy is a functional of the lateral displacement of the beam, $u(x)$,

$$\mathcal{E}_{B-E}(u) = \frac{1}{2}\int_0^l Eh^4 \left(\frac{d^2u}{dx^2}\right)^2 dx \tag{5.252}$$

where E is a coefficient depending on the elastic moduli of the material and on the geometry of the cross-section, h is the diameter of the cross-section, and l is the length of the beam. If, for definiteness, the kinematic boundary conditions are given:

$$u = \frac{du}{dx} = 0 \quad \text{at } x = 0; \quad u = u_l, \quad \frac{du}{dx} = -\psi_l \quad \text{at } x = l, \tag{5.253}$$

the true displacement provide minimum to the functional (5.252) on the set of functions $u(x)$ selected by the conditions (5.253).

Timoshenko proposed a more precise beam theory, in which the kinematics of the beam is described by two functions – the displacement, $u(x)$, and the rotation of the cross-section, $\psi(x)$. Function $\psi(x)$ has the meaning of the angle between the beam cross-sections in the deformed and undeformed states. In Timoshenko's theory, the energy is

$$\mathcal{E}_T(u) = \frac{1}{2}\int_0^l \left[Eh^4\left(\frac{d\psi}{dx}\right)^2 + Gh^2\left(\psi + \frac{du}{dx}\right)^2\right] dx, \tag{5.254}$$

where $G > 0$ is a coefficient determined by the shear modulus of the material and the geometry of the cross-section.

The kinematic conditions corresponding to (5.253) in Timoshenko's theory are

$$u = 0, \quad \psi = 0 \quad \text{at } x = 0, \quad u = u_l, \quad \psi = \psi_l \quad \text{at } x = l. \tag{5.255}$$

The true displacements $u(x)$ and rotations of the cross-section $\psi(x)$ provides minimum to the functional (5.254) on the set of functions $u(x)$, $\psi(x)$ selected by the constraints (5.255).

Let us show that Bernoulli-Euler theory can be considered as the first approximation of Timoshenko's theory for $h \to 0$. We assume that the boundary values u_l and ψ_l and the moduli E and G do not depend on h. We also assume that the minimizing functions have the characteristic length which is much larger than h. This means, in particular, that $h^2\left(\frac{d\psi}{dx}\right)^2 << \psi^2$. Since $Eh^4\left(\frac{d\psi}{dx}\right)^2 << Gh^2\psi^2$, the leading term of Timoshenko's functional is

$$\mathcal{E}_0(u,\psi) = \frac{1}{2}\int_0^l Gh^2\left(\psi + \frac{du}{dx}\right)^2 dx.$$

Minimizing the functional $\mathcal{E}_0(u,\psi)$ and taking into account the conditions (5.255), we obtain that the minimum is equal to zero, and it is reached on the functions u and ψ linked by the relation

$$\psi = -\frac{du}{dx}. \tag{5.256}$$

Consequently, the set $\mathcal{M}_0$ is a set of pairs (u,ψ), where $u(x)$ are arbitrary functions taking on the boundary values (5.253) (due to (5.255) and (5.256)). The functions $\psi(x)$ are calculated from $u(x)$ by means of (5.256).

The relation (5.256) has a simple geometric interpretation. It means that, the cross-section of the beam remains perpendicular to the deformed center line of the beam after the deformation.

If the conditions of the general scheme (u' is uniquely defined by u_0 and $I(u_0,\varepsilon) - I\left(u_0 + u'(u_0,\varepsilon),\varepsilon\right)$ is small) are satisfied, then the functional $\mathcal{E}(u,\psi)$ can be minimized on $\mathcal{M}_0$ in the first approximation. Timoshenko's energy functional and Bernoulli-Euler energy functional coincide on $\mathcal{M}_0$:

$$\mathcal{E}_T = \mathcal{E}_{B-E}.$$

Therefore, in the first approximation, the calculation of u in Timoshenko's theory is reduced to the calculation of u by Bernoulli-Euler theory.

Let us check whether or not these conditions are satisfied. For the following calculations it is convenient to make a change of variables and, instead of ψ, use the function $\varphi = \psi + du/dx$. The function φ is the shear angle, the angle between the deformed cross-section and the cross-section normal to the deformed center line. In terms of u and φ, Timoshenko's energy functional can be written as

$$\mathcal{E}_T(u) = \frac{1}{2}\int_0^l \left[Eh^4\left(\frac{d^2u}{dx^2}\right)^2 - 2Eh^4\frac{d^2u}{dx^2}\frac{d\varphi}{dx} + Eh^4\left(\frac{d\varphi}{dx}\right)^2 + Gh^2\varphi^2\right]dx. \tag{5.257}$$

Fixing the function $u(x)$ (element of the set $\mathcal{M}_0$), we can find the dependence of φ on u. To do this, we keep the leading term in the functional containing φ $\left(Gh^2\varphi^2\right)$, and the term containing both u and φ, $-2Eh^4\frac{d^2u}{dx^2}\frac{d\varphi}{dx}$. The term $Eh^4\left(\frac{d\varphi}{dx}\right)^2$ can be dropped because $h^2\left(\frac{d\varphi}{dx}\right)^2 << \varphi^2$. For determining function φ we get the functional

$$\frac{1}{2}\int_0^l \left(-2Eh^4\frac{d^2u}{dx^2}\frac{d\varphi}{dx} + Gh^2\varphi^2\right)dx. \tag{5.258}$$

After integrating the first term by parts and applying the boundary conditions to φ,

$$\varphi = \frac{du}{dx} \quad \text{for } x = 0, \quad \varphi = \psi_l + \frac{du}{dx} \quad \text{for } x = l, \tag{5.259}$$

the functional (5.258) becomes

$$\frac{1}{2}\int_0^l \left(2Eh^4\frac{d^3u}{dx^3}\varphi + Gh^2\varphi^2\right)dx -$$
$$-2Eh^4\frac{d^2u}{dx^2}\left(\psi_l + \frac{du}{dx}\right)\bigg|_{x=l} + 2Eh^4\frac{d^2u}{dx^2}\frac{du}{dx}\bigg|_{x=0}. \tag{5.260}$$

The non-integral terms are known (since u is given). Therefore, the minimization of the functional (5.260) is reduced to the minimization of the integrand over φ,

$$\varphi = -\frac{Eh^2}{G}\frac{d^3u}{dx^3}. \tag{5.261}$$

Consequently, φ is uniquely defined by u. Moreover, φ is of the order h^2, and the terms related to φ make only a small contribution to the energy compared to Bernoulli-Euler energy ($\mathcal{E}_T - \mathcal{E}_{B-E}$ is small). Hence, Bernoulli-Euler theory is indeed the first approximation of the Timoshenko's theory[18].

The example considered makes the physical meaning of the iteration scheme absolutely transparent. In the first step, the "leading" term of energy is being minimized. If $\psi \neq -du/dx$, the total energy would be of the order of h^2. This is not energetically advantageous since upon satisfying (5.256) the energy is on the order of h^4.

---

[18] Note also how (5.59) is made consistent with the boundary condition (5.259). Denote the solution of the problem in Bernoulli-Euler's theory by u_0. Similarly to the way the function φ was found, the first correction u' to u_0 can be found, $u = u_0 + u' + \ldots$. It turns out that $u' \approx h^2 u_0$. Then, in the first approximation,

$$\varphi = -\frac{Eh^2}{G}\frac{d^3u_0}{dx^3},$$

and the boundary conditions will be satisfied due to the fact that the function $u'(x)$ is subject to constraints

$$\varphi = -\frac{Eh^2}{G}\frac{d^3u_0}{dx^3} = \frac{du'}{dx} \quad \text{for } x = 0, l$$

at the ends of the beam.

It is easy to construct the exact solution of the considered problem according to Timoshenko's theory and to check that the results obtained by the asymptotic analysis of energy correspond to the expansion of the solution of the differential equations in h^2.

Example 8. **Dynamics of systems with non-holonomic constraints.** For some dynamical systems not every trajectory in the phase is admissible. Consider, for example, an ice-skate on a plane. The position of the ice-skate can be described by three numbers, the two coordinates of the point of contact, x and y, and the angle between the ice-skate and x-axis, θ. For given x, y and θ, the velocity components, $\dot{x}$ and $\dot{y}$, cannot be arbitrary because the velocity of the point of contact is directed along the ice-skate. Therefore,

$$\dot{x}\sin\theta - \dot{y}\cos\theta = 0. \tag{5.262}$$

If this condition had the form

$$\frac{d}{dt}\varphi(x, y, \theta) = 0 \qquad \text{or} \qquad \varphi(x, y, \theta) = const, \tag{5.263}$$

we would be able to eliminate one of the coordinates using the equation $\varphi(x, y, \theta) = const$ to obtain a system with two degrees of freedom. Such a constraint is called holonomic. However, (5.262) is not holonomic: it cannot be presented in the form (5.263).

Consider a mechanical system with a non-holonomic constraint:

$$a_i(q)\dot{q}^i = 0. \tag{5.264}$$

Let the system have the kinetic energy $K(q, \dot{q})$ and potential energy $U(q)$. At first glance, to derive the governing dynamical equations of a non-holonomic mechanical system, one has to seek the stationary points of the action functional,

$$I(q) = \int_{t_1}^{t_2} L(q, \dot{q})dt, \qquad L(q, \dot{q}) = K(q, \dot{q}) - U(q),$$

on the set of trajectories selected by the constraint (5.264). In mechanics, however, another recipe was developed: at the real trajectory, $\delta I = 0$ for all variations satisfying the equation

$$a_i(q)\delta q^i = 0. \tag{5.265}$$

The resulting equations are different: in the first case, one obtains the equations

$$\frac{\partial L}{\partial q^i} - \frac{d}{dt}\frac{\partial L}{\partial \dot{q}^i} = \lambda \frac{\partial a_k(q)}{\partial q^i}\dot{q}^k - \frac{d}{dt}(\lambda a_i), \tag{5.266}$$

λ being the Lagrange multiplier for the constraint (5.264), while in the second case the governing equations are

$$\frac{\partial L}{\partial q^i} - \frac{d}{dt}\frac{\partial L}{\partial \dot{q}^i} = \lambda' a_i, \tag{5.267}$$

where λ' is the Lagrange multiplier for the constraint (5.265). Which equations are correct? It turns out that they are both correct but correspond to different physical situations. Equations (5.266) appear if, in a more detailed description, Lagrange function of the system is

$$L(q, \dot{q}) = K(q, \dot{q}) - U(q) + \alpha(a_i(q)\dot{q}^i)^2,$$

where α is a large parameter. According to the general scheme of the variational-asymptotic method, one has to minimize first the leading term. We arrive at the constraint (5.264), and then consider the stationary points of the action functional on the constrained set.

Equations (5.267) appear in another limit: if a system has a dissipative potential, D,

$$D = \frac{1}{2}\mu(a_i(q)\dot{q}^i)^2,$$

and the governing equations are

$$\frac{\partial L}{\partial q^i} - \frac{d}{dt}\frac{\partial L}{\partial \dot{q}^i} = \mu(a_k(q)\dot{q}^k)a_i,$$

then (5.267) correspond to the limit of infinite dissipation, $\mu \to \infty$. Justification of this statement is beyond the scope of this book (see further details in [149–152, 252]).

Example 9. **Pendulum with vibrating suspension point.** Consider a pendulum the suspension point of which is vibrating in a vertical direction. Lagrange function of such a pendulum can be found in the same way as in Example 2 of Sect. 1.6. Instead of (1.49) one has to use the equations

$$x(t) = l \sin q(t), \quad y(t) = l \cos q(t) + a(t),$$

where $a(t)$ is the given y-coordinate of the suspension point. We have

$$L = \frac{1}{2}ml^2\dot{q}^2 - ml \sin q \dot{q}\dot{a} + mgl \cos q. \tag{5.268}$$

Here we keep only the terms depending on $q(t)$.

The "interaction term" between $\dot{q}$ and $\dot{a}$ can also be written as

$$-ml\sin q\dot{q}\dot{a} = ml\frac{d\cos q}{dt}\dot{a} = -ml\cos q\ddot{a} + \frac{d}{dt}ml\cos q\dot{a}.$$

The last term, as a full derivative, does not affect the governing equations, and can be dropped. Thus,

$$L = \frac{1}{2}ml^2\dot{q}^2 + ml\,(g - \ddot{a})\cos q.$$

So, the vibration of the suspension point results in the change of gravity acceleration, g, by the relative acceleration, $g - \ddot{a}$. For our purposes, it will be more convenient to work with the Lagrange function (5.268).

Let the suspension point oscillate very fast. That means that the function $a\,(t)$ can be written as a periodic function of an auxiliary variable, τ,

$$a = a\,(\tau)\,, \quad \tau = \omega t$$

with a large parameter, ω. In physical terms, "large ω" means that ω is much greater than the characteristic frequency of pendulum, $\sqrt{g/l}$. By an appropriate scaling, function $a\,(\tau)$ may be viewed as a 2π-periodic function. Since the Lagrange function depends only on derivatives of a, one can set

$$\langle a \rangle \equiv \frac{1}{2\pi}\int_0^{2\pi} a\,(\tau)\,d\tau = 0. \tag{5.269}$$

The variable τ is called fast time.

We assume that the amplitude of oscillations is small and put

$$a = \frac{1}{\omega}A\,(\tau)$$

where $A\,(\tau)$ is a smooth function of order unity. Then the velocity of the suspension point is of order unity as well:

$$\dot{a} = A_\tau\,(\tau)\,, \quad A_\tau\,(\tau) \equiv \frac{dA\,(\tau)}{d\tau}, \tag{5.270}$$

while acceleration is large:

$$\ddot{a} = \omega A_{\tau\tau}\,(\tau)\,, \quad A_{\tau\tau}\,(\tau) \equiv \frac{d^2A\,(\tau)}{d\tau^2}.$$

If we plug (5.270) in the action functional, we obtain a problem with a large parameter, ω, and we are going to investigate it by the variational-asymptotic method.

Our starting point is the assumption that the large parameter, ω, enters in function $q(t)$ through fast time, τ, i.e.,

$$q = q(t, \tau)$$

and, the dependence of $q(t, \tau)$ on τ is periodic. Function $q(t, \tau)$ may also depend on a small parameter, $1/\omega$, but we do not emphasize this in our notation.

So,

$$\dot{q} = q_t + \omega q_\tau, \quad q_t \equiv \frac{\partial q(t, \tau)}{\partial t}, \quad q_\tau \equiv \frac{\partial q(t, \tau)}{\partial \tau}. \tag{5.271}$$

Plugging (5.271) in (5.268) we obtain

$$\frac{1}{ml} L = \frac{1}{2} l (q_t + \omega q_\tau)^2 - \sin q (q_t + \omega q_\tau) A_\tau + g \cos q. \tag{5.272}$$

To compute the action functional, we have to integrate over time the function of two variables, t and $\tau = \omega t$:

$$I = \int_{t_0}^{t_1} L \, dt.$$

For any function of two variables, $\varphi(t, \omega t)$, for large ω,

$$\int_{t_0}^{t_1} \varphi(t, \omega t) \, dt \approx \int_{t_0}^{t_1} \langle \varphi \rangle \, dt$$

where $\langle \cdot \rangle$ is the average over τ (5.269). Therefore,

$$I = \int_{t_0}^{t_1} \langle L \rangle \, dt. \tag{5.273}$$

For $\omega \to \infty$, the leading term of the Lagrange function (5.272) is

$$\frac{1}{ml} L = \frac{1}{2} l \omega^2 q_\tau^2.$$

Hence, in the first approximation we have to minimize

$$\frac{ml^2 \omega^2}{2} \langle q_\tau^2 \rangle$$

over all periodic functions. The minimum is reached on the functions not depending on τ, $\bar{q}(t)$. We fix a function, $\bar{q}(t)$, and seek the next approximation,

$$q = \bar{q}(t) + q'(t, \tau),$$

where q' is much smaller that $\bar{q}$. Emphasize that we do not yet know the order of q' and are going to find it. In the first approximation,

$$q_t = \bar{q}_t, \quad q_\tau = q'_\tau \equiv \frac{\partial q'(t, \tau)}{\partial \tau}, \tag{5.274}$$

because $\partial q'(t, \tau)/\partial t$ is small compared to $\bar{q}_t$. Plugging (5.274) in (5.272) we have

$$\frac{1}{ml} L = \frac{1}{2} l \left(\bar{q}_t + \omega q'_\tau\right)^2 - \sin\left(\bar{q} + q'\right)\left(\bar{q}_t + \omega q'_\tau\right) A_\tau + g \cos\left(\bar{q} + q'\right).$$

Since $\sin\left(\bar{q} + q'\right) \approx \sin\bar{q} + q'\cos\bar{q}$, $\cos\left(\bar{q} + q'\right) = \cos\bar{q} - q'\sin\bar{q}$, the leading terms of the Lagrange function are

$$\begin{aligned}\frac{1}{ml} L = &\frac{1}{2} l \bar{q}_t^2 + l \bar{q}_t \omega q'_\tau + \frac{1}{2} l \omega^2 q'^2_\tau \\ &- \sin\bar{q}\,\bar{q}_t A_\tau + g\cos\bar{q} - \omega\sin\bar{q} A_\tau q'_\tau.\end{aligned} \tag{5.275}$$

We dropped the interaction terms between $\bar{q}$ and q', $\cos\bar{q}\bar{q}_t q' A_\tau$, $\cos\bar{q}\bar{q}_\tau q' A_\tau$, and $g\sin\bar{q}q'$, which are small in comparison with the leading interaction term $\omega\sin\bar{q} A_\tau q'_\tau$, and the term $\cos\bar{q} q' q'_\tau A_\tau$, which is small compared with the third term. The average over τ of the second and the fourth term in (5.275) is equal to zero due to periodicity of A and q'. So, to find q' we have to minimize the functional

$$\left\langle \frac{1}{2} l \omega^2 q'^2_\tau - \omega \sin\bar{q} A_\tau q'_\tau \right\rangle$$

over all periodic functions q'. This functional can be written as

$$\left\langle \frac{1}{2} l \omega^2 \left(q'_\tau - \frac{1}{l\omega} \sin\bar{q} A_\tau \right)^2 \right\rangle - \frac{1}{2l} \sin^2\bar{q} \left\langle A_\tau^2 \right\rangle.$$

Hence,

$$q'_\tau = \frac{1}{l\omega} \sin\bar{q} A_\tau,$$

and the minimum value of the functional is

$$-g^* \sin^2\bar{q}, \quad g^* \equiv \frac{1}{2l} \left\langle A_\tau^2 \right\rangle.$$

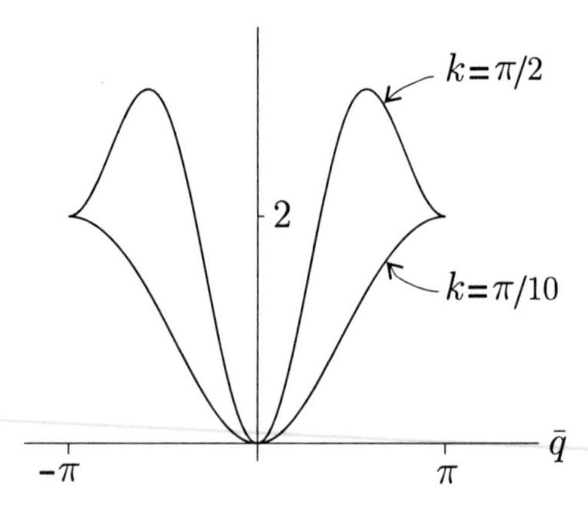

Fig. 5.29 Graphs of function $1 - \cos\bar{q} + k\sin^2\bar{q}$ for $k = \pi/2$ and $k = \pi/10$

Finally,

$$\frac{1}{ml}\langle L\rangle = \frac{1}{2}l\bar{q}_t^2 + g\cos\bar{q} - g^*\sin^2\bar{q}. \tag{5.276}$$

This is the "effective" Lagrange function of the pendulum. The fast vibration of the suspension point results in the change of the gravitation potential energy $g(1-\cos\bar{q})$ by an effective potential energy, $g\left(1-\cos\bar{q}+k\sin^2\bar{q}\right)$, $k = g^*/g$. The graph of this potential energy is shown in Fig. 5.29. For small k, the vibrations are qualitatively similar to vibrations of the usual pendulum, for large k the behavior changes: the potential energy gets an additional local minimum at $\bar{q} = \pi$, i.e. the upper equilibrium position of the pendulum becomes stable[19].

Example 10. **Whitham's method.** Consider a linear homogeneous differential equation with partial derivatives:

$$P\left(\frac{\partial}{\partial t}, \frac{\partial}{\partial x}\right)u = 0, \tag{5.277}$$

where $P(r,s)$ is a polynomial of r, s. This equation always has a solution of the form

$$u = ae^{i(kx-\omega t)}, \tag{5.278}$$

where a, k, ω are some possibly complex constants. For the function $ae^{i(kx-\omega t)}$ to be the solution of (5.277), it is necessary and sufficient for k and ω to satisfy the polynomial equation

$$P(-i\omega, ik) = 0. \tag{5.279}$$

[19] Mathematical justifications and a review can be found in papers by V.I. Yudovich [325] and V. Vladimirov [302].

Solutions of the type (5.278) are called harmonic waves, a the amplitude, k the wave number, ω the frequency, $\theta = kx - \omega t$ the phase (for $\operatorname{Im} a = 0$), and (5.279) the dispersion equation. Harmonic waves are fundamental to the theory of linear equations (5.277), particularly due to the fact that any solution of (5.277) can be presented as a superposition of harmonic waves. The dispersion equation contains all the information about the differential equation and can be used to reconstruct uniquely the differential operator.

The following question arises: what is the analogy of the harmonic waves and the dispersion equation for the non-linear case?

It turns out that the non-linear generalization of the harmonic wave type solutions is

$$u = \psi(a, \theta), \tag{5.280}$$

where a, θ are some functions of x, t,

$$a = a(x, t), \quad \theta = \theta(x, t),$$

and $\psi(a, \theta)$ is a periodic (with a period of 2π) function of θ. Besides, the characteristic lengths of functions $a(x, t), \theta_{,x}(x, t), \theta_{,t}(x, t)$ are much greater[20] than that of $\theta(x, t)$. The functions $a(x, t)$ corresponds to the amplitude, $\theta_{,x}$ to the wave number, and $\theta_{,t}$ to the frequency.

Now we construct the equations to determine the functions ψ, a, and θ, if the governing equation (5.277) is Euler equation of some functional. More specifically, we take (5.277) as the Euler equation of the functional

$$\int_{\Omega} L(u, u_x, u_t)\, dx\, dt \tag{5.281}$$

and look for the solutions of this equation of the form

$$u = \psi(\theta, x, t), \tag{5.282}$$

where θ is some function of x and t, ψ is a periodic (with a period of 2π) function of θ and the characteristic lengths L and T for change of the functions $\theta_{,x}$, $\theta_{,t}$ and $\psi(\theta, x, t)|_{\theta=const}$ over x and t are much greater than the characteristic lengths, l and τ, of the phase θ.

The lengths l and τ can be thought of as the largest constants in the inequalities

$$|\theta_{,x}| \le \frac{2\pi}{l}, \quad |\theta_{,t}| \le \frac{2\pi}{\tau}. \tag{5.283}$$

---

[20] For harmonic waves, the quantities $a, \theta_{,x}, \theta_{,t}$ are constant. Therefore, their corresponding characteristic lengths are infinite.

The definition of l and τ implies the meaning of the function θ as of oscillation phase: the phase increases on 2π on the lengths of the order of l and the time of the order of the oscillation period, τ.

The lengths $\bar{l}$ and $\bar{\tau}$ and the amplitude $\bar{\psi}$ are defined as the largest constants in the inequalities

$$\left|\theta_{,xx}\right| \leq \frac{2\pi}{l\bar{l}}, \quad \left|\theta_{,xt}\right| \leq \frac{2\pi}{l\bar{\tau}}, \quad \left|\theta_{,xt}\right| \leq \frac{2\pi}{\tau\bar{l}}, \quad \left|\theta_{,tt}\right| \leq \frac{2\pi}{\tau\bar{\tau}},$$
$$|\partial_x \psi| \leq \frac{\bar{\psi}}{\bar{l}}, \quad |\partial_t \psi| \leq \frac{\bar{\psi}}{\bar{\tau}}, \quad \left|\psi_{,\theta}\right| \leq \bar{\psi}, \tag{5.284}$$

where $\partial_x \psi$ and $\partial_t \psi$ are the partial derivatives, $\partial\psi/\partial x$ and $\partial\psi/\partial t$, when θ is kept constant.

So, the problem contains two small parameters, $l/\bar{l}$ and $\tau/\bar{\tau}$.

Let us find the derivatives $u_{,x}$, and $u_{,t}$,

$$u_{,x} = \partial_x \psi + \psi_{,\theta}\,\theta_{,x}\,, \qquad u_{,t} = \partial_t \psi + \psi_{,\theta}\,\theta_{,t}\,.$$

Due to the estimates (5.283) and (5.284), in the first approximation

$$u_{,x} = \psi_{,\theta}\,\theta_{,x}\,, \qquad u_{,t} = \psi_{,\theta}\,\theta_{,t}\,.$$

Keeping only the leading terms in the functional, we obtain

$$\iint\limits_{\Omega} L\left(\psi, \psi_{,\theta}\theta_x, \psi_{,\theta}\theta_t\right) dx dt. \tag{5.285}$$

Let us cover the region Σ by the strips, $2\pi k \leq \theta \leq 2\pi\,(k+1)\,,\ k = 0, 1, 2, \ldots$. The integral over Σ can be replaced by the sum of integrals over the strips,

$$\iint\limits_{\Omega} L dx dt = \sum_{\text{strips}} \iint L(\psi, \psi_\theta\theta_x, \psi_\theta\theta_t) \varkappa d\theta d\zeta, \tag{5.286}$$

where ζ is the coordinate along the lines $\theta = const$, $\varkappa$ is the Jacobian of the transformation from the variables x, t to the variables $\theta,\ \zeta$. In the first approximation, since $\theta_{,x}$ and $\theta_{,t}$ change slowly, we may assume that on every strip $\theta_{,x}$, $\theta_{,t}$ do not depend on θ. Therefore, in the first step of the variational-asymptotic method we obtain the same problem for every strip: find the stationary points of the functional

$$\int_0^{2\pi} L\left(\psi, \psi_\theta\theta_{,x}, \psi_\theta\theta_{,t}\right) d\theta \tag{5.287}$$

on the set of periodic functions $\psi(\theta)$. Here $\theta_{,x}$ and $\theta_{,t}$ should be considered as constant parameters. The Euler equation of the functional (5.287) is an ordinary differential equation of the second order. Its solution has two arbitrary constants. One is defined by the periodicity condition[21] $\psi(0) = \psi(2\pi)$. The other can be linked to the amplitude a, for example, by the equation $\max|\psi| = a$. Hence, the set $\mathcal{M}_0$ in the general scheme of the variational-asymptotic method is the set of functions $a(x,t)$ and $\theta(x,t)$.

Denote the value of the functional (5.287) at its stationary point by $2\pi\bar{L}$. The quantity $\bar{L}$ is a function of parameters a, θ_x, θ_t. For small values $l/\bar{l}$ and $\tau/\bar{\tau}$, the sum (5.286) is an integral sum. In the limit $l/\bar{l}, \tau/\bar{\tau} \to 0$, it can be approximated by the integral

$$\int\int_{\Omega} \bar{L}\left(a, \theta_{,x}, \theta_{,t}\right) dx dt. \tag{5.288}$$

Then the equations to determine a and θ are obviously the Euler equations of the functional (5.288):

$$\frac{\partial \bar{L}}{\partial a} = 0, \qquad \frac{\partial}{\partial x}\frac{\partial \bar{L}}{\partial \theta_x} + \frac{\partial}{\partial t}\frac{\partial \bar{L}}{\partial \theta_t} = 0.$$

A more detailed analysis shows that the first of these equations can be interpreted as a non-linear generalization of the dispersion equation. It becomes the dispersion equation of the linear theory in the limit of $a \to 0$. An interesting property of the non-linear dispersion equation is its dependence on the amplitude, which is absent in the linear case.

The generalizations to the case of a large number of required functions and independent arguments are straightforward.

The theory described was suggested by Whitham [316] from some heuristic reasoning.

Example 11. Thin region approximations. Such type of approximations appears in many areas of continuum mechanics. It will be discussed in detail further in Chapters 14 and 15 for elastic plates, shells and beams. Here we consider an example.

Let V be a bounded region in three-dimensional space, and S a closed smooth surface inside V. At each point of S a normal segment is erected of a small length h with the center on S. The segments cover a thin layer ΔB (Fig. 5.30).

Consider the variational problem

$$I(u) = \int_V \frac{1}{2} u_{,i} u^{,i} dV - \int_{\Delta B} \rho u dV \to \min_{u: u=o \text{ on } \partial V}.$$

[21] It is assumed that the periodic solutions exist.

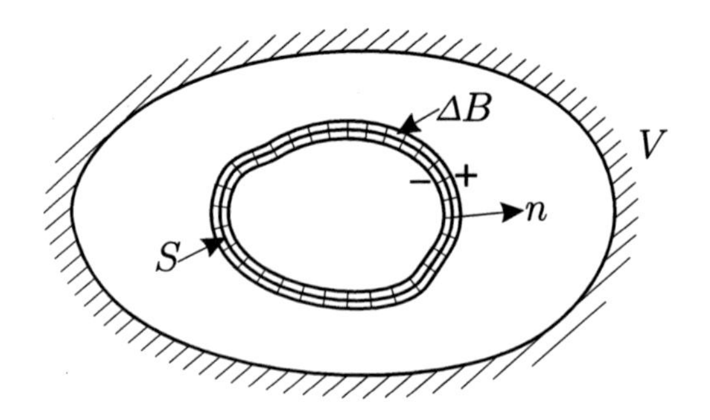

Fig. 5.30 Notation for example 11.

We assume that function ρ is constant along the normal to S and, thus, determines by its values on S. We are going to find the limit solution as $h \to 0$, if ρ tends to infinity as

$$\rho = \frac{\sigma}{h}$$

where σ is a function on S.

Denote by u^+ and u^- the values of u on the two sides of ΔB. Let us seek minimum of the functional $I(u)$ in two steps, first minimizing $I(u)$ over all u with the prescribed values of u^+ and u^-, and then minimizing over u^+,u^-. We split $I(u)$ into the sum of two functionals,

$$I(u) = I_1(u) + I_2(u),$$
$$I_1(u) = \int_{V-\Delta B} \frac{1}{2} u_{,i} u^{,i} dV,$$
$$I_2(u) = \int_{\Delta B} \left(\frac{1}{2} u_{,i} u^{,i} - \rho u \right) dV.$$

For given u^+ and u^- these functionals can be minimized separately. We consider first the minimization of the functional I_2.

We make an assumption that it is enough to consider only smooth functions u^+,u^- with the characteristic length l that is much bigger than h. Moreover, the minimizer of I_2 has the same characteristic length along S as u^+ and u^-. Then in the first term of the integrand, $1/2 u_{,i} u^{,i}$, the derivatives along S can be neglected in comparison with the derivative along the normal, which is of the order $1/h$.

Denote the coordinate along the normal by z, $-h/2 \leqslant z \leqslant h/2$. Then in the leading approximation the variational problem for I_2 is

$$\int_S \int_{-h/2}^{h/2} \left[\frac{1}{2} \left(\frac{\partial u}{\partial z} \right)^2 - \frac{\sigma}{h} u \right] dz \to \min_{\substack{u=u^- \text{ at } z=-h/2 \\ u=u^+ \text{ at } z=h/2}} .$$

The minimizer obeys the boundary value problem

$$\frac{\partial^2 u}{\partial z^2} = -\frac{\sigma}{h}, \quad u = u^- \text{ at } z = -h/2, \quad u = u^+ \text{ at } z = h/2.$$

Thus,

$$u = \frac{u_+ + u_-}{2} + \frac{u_+ - u_-}{h} z - \frac{\sigma}{2h} \left(z^2 - \frac{h^2}{4} \right),$$

and the minimum value of the functional is

$$\check{I}_2 = \int_S \left[\frac{(u_+ - u_-)^2}{2h} - \sigma \frac{u_+ + u_-}{2} - \frac{\sigma^2}{24} h \right] dA. \tag{5.289}$$

The last term in (5.289), being an additive constant, can be dropped.

The minimum value of the functional I_1 with the prescribed boundary values of u, u_+ and u_-, is finite when $h \to 0$. So, in minimization over u_+, u_-, the leading term of the functional as $h \to 0$ is the first term in (5.289),

$$\frac{1}{2h} \int_S (u_+ - u_-)^2 \, dA.$$

The minimum is zero and achieved when $u_+ = u_-$.

Finally, the original variational problem reduces in the leading approximation to the minimization problem for the functional,

$$\int_S \frac{1}{2} u_{,i} u^{,i} dV - \int_S \sigma u dA, \tag{5.290}$$

where minimum is sought over all functions u that are continuous on S and vanish on ∂V.

An important consequence follows from this statement. The minimizer of the original problem can be written in terms of Greens's function (see Sect. 5.9),

$$u(x) = \int_{\Delta B} G(x, y) \rho(y) dV_y.$$

As $h \to 0$, this integral converges to

$$u(x) = \int_S G(x, y) \sigma(y) dA.$$

This must be the minimizer of the variational problem (5.290). On the other hand, the minimizer of the (5.290) obeys the equation on S

$$\left[u_{,i}\right] n^i = -\sigma.$$

We arrive at a well-known property of Green's function:

$$\left[\int_S \frac{\partial G(x,y)}{\partial x^i} \sigma(y) dA\right] n^i = -\sigma(x).$$

Now let us move on to the construction of the next approximations. In physical applications, they are usually referred to as the refined theories. Refined theories are needed in cases when the first approximation is too crude; this can happen if the actual values of the small parameter ε are not sufficiently small. The method of constructing the refined theories consists of two steps: First, a functional is derived which allows one to obtain the corrections to the solution of the next order of magnitude. Second, this refined functional is extrapolated to all, even not small, values of ε.

Usually, to construct a refined theory which takes into account the corrections on the order of $\alpha(\varepsilon)$, one should keep in the functional all the terms on the order of $\alpha(\varepsilon)$ compared to unity. We illustrate this statement by the minimization problem for quadratic functional

$$I(u,\varepsilon) = E(u,\varepsilon) - l(u,\varepsilon),$$

where $E(u,\varepsilon)$ is obtained from a positive symmetric bilinear form $E(u,v,\varepsilon)$

$$E(u,\varepsilon) = E(u,u,\varepsilon)$$

and

$$E(u,v,\varepsilon) = E_0(u,v) + \alpha(\varepsilon) E_1(u,v) + \alpha'(\varepsilon) E_2(u,v) + \dots$$
$$l(u,v) = l_0(u) + \alpha(\varepsilon) l_1(u) + \alpha'(\varepsilon) l_2(u) + \dots, \alpha' = o(\alpha), \quad \alpha(\varepsilon) \to 0 \text{ as } \varepsilon \to 0.$$

The functional is minimized on a linear space $\mathcal{M}$. The limit bilinear form $E_0(u,v)$ is obviously positive. The first approximation u_0 is the minimum point of the functional $I_0 = E_0(u,u) - l_0(u)$. The element u_0 satisfies the Euler equation

$$2E_0(u_0,\bar{u}) - l(\bar{u}) = 0. \tag{5.291}$$

Equation (5.291) is valid for any element $\bar{u} \in \mathcal{M}$.

Let us present u as $u = u_0 + u'$. Keeping the leading terms containing u' in $I(u,\varepsilon)$ and using the equality $2E_0(u_0,u') - l(u') = 0$, which follows from (5.291), we obtain for determining u' the minimization problem

$$I_1\left(u_0, u', \varepsilon\right) = E_0\left(u', u'\right) + \alpha(\varepsilon) E_1\left(u_0, u'\right) - \alpha(\varepsilon) l_1\left(u'\right) \to \min_{u'}.$$

After the substitution $u' \to w$: $u' = \alpha(\varepsilon) w$, it becomes the minimization problem for the functional

$$\alpha'^{-2} I_1 = E_0(w, w) - 2E_1(u_0, w) - l_1(w),$$

which does not depend on small parameter ε. Therefore, $u' \sim \alpha(\varepsilon)$, and, to incorporate the corrections of the order $\alpha(\varepsilon)$, all terms of the functional of the functional of the order $\alpha(\varepsilon)$ should be kept.

The construction of refined theories will be considered in detail in Chap. 14 where the problem of the refinement of the classical shell theory is discussed.

So far, in all cases the functional $I_0(u)$ had the stationary points. It remains to discuss the cases when the functional $I_0(u)$ does not have stationary points or it is meaningless.

Case 3: $I_0(u)$ does not have the stationary points or $I_0(u)$ meaningless
Such a case is not unusual. For example, the function of one variable

$$f(u, \varepsilon) = u + \varepsilon u^2 + \sin \varepsilon u, \tag{5.292}$$

is such a case, because $f_0(u)$ does not have stationary points.

Another example is the minimization problem for the functional

$$\int_{\varepsilon}^{\infty} \frac{1}{r}\left(u^2 + u_{,r}^2\right) dr \tag{5.293}$$

on the set of functions $u(r)$ which are equal to 1 for $r = \varepsilon$. This variational problem arises in the modeling of incompressible flow along an axis-symmetric thin body. The integral (5.293) diverges for $\varepsilon = 0$.

The boundary layer problems also usually belong to this case.

In general, there are no recipes for these cases, except, probably, for the following one: an attempt should be made to reduce these problems to the problems of the type considered earlier by means of a change of the required functions or some other transformations. For example, for the function (5.292), the substitution: $u \to v$: $\varepsilon u = v,\ f \to g : \varepsilon f = g$, yields the function

$$g(v, \varepsilon) = v + v^2 + \varepsilon \sin v$$

which may be studied by the method which has been discussed.

As for the boundary layer problems, the applications of the variational methods are not sufficiently elementary to be presented in an introductory treatment of the subject (the interested reader is referred to [30]).

5.12 Variational Problems and Functional Integrals

In some variational problems, minimization can be replaced by integration. This is a useful trick in studying the stochastic variational problems as will be discussed in Chap. 16. Here we explain the relation between minimization and integration.

First, we need a number of auxiliary facts. Consider in some finite-dimensional space, R_m, a quadratic form,

$$(Au, u) = A_{ij} u^i u^j. \tag{5.294}$$

The form is assumed to be positive:

$$(Au, u) > 0 \quad \text{if } u \neq 0.$$

Then the Gauss formula holds true:

$$\int_{R_m} e^{-\frac{1}{2}(Au,u)} du = \frac{1}{\sqrt{\det A}}. \tag{5.295}$$

Here

$$\det A \equiv \det \left\| A_{ij} \right\|, \quad du = \frac{du_1}{\sqrt{2\pi}} \dots \frac{du_m}{\sqrt{2\pi}}.$$

The Gauss formula can be proved by changing the variables, $u \to \mathring{u}$,

$$u^i = \lambda^i_j \mathring{u}^j, \quad \det \left\| \lambda^i_j \right\| = 1,$$

$\mathring{u}^j$ being the coordinates in which the tensor A_{ij} is diagonal,

$$(Au, u) = A_{ij} \lambda^i_{i'} \lambda^j_{j'} \mathring{u}'^i \mathring{u}'^j = A_1 \left(\mathring{u}^1\right)^2 + \ldots + A_m \left(\mathring{u}^m\right)^2. \tag{5.296}$$

In the new variables,

$$\int_{R_m} e^{-\frac{1}{2}(Au,u)} du = \int_{R_m} e^{-\frac{1}{2}\left(A_1 \left(\mathring{u}^1\right)^2 + \ldots + A_m \left(\mathring{u}^m\right)^2\right)} d\mathring{u}$$
$$= \frac{1}{\sqrt{A_1 \dots A_m}} = \frac{1}{\sqrt{\det A}}.$$

Here we used that[22]

$$\int_{-\infty}^{+\infty} e^{-\frac{1}{2}x^2}dx = \sqrt{2\pi}. \tag{5.297}$$

The Gauss formula admits the following generalization: for any linear function of u, $(l, u) = l_i u^i$,

$$\sqrt{\det A}\int_{R_m} e^{-\frac{1}{2}(Au,u)+(l,u)}du = e^{\frac{1}{2}(A^{-1}l,l)} \tag{5.298}$$

where A^{-1} is the inverse matrix to the matrix A. Formula (5.298) follows from (5.295) and the identity

$$\frac{1}{2}(Au,u) - (l,u) = \frac{1}{2}\left(A\left(u - A^{-1}l\right), \left(u - A^{-1}l\right)\right) - \frac{1}{2}\left(A^{-1}l, l\right). \tag{5.299}$$

Plugging (5.299) in (5.298), changing the variables of integration, $u \to u + A^{-1}l$, and using (5.295), we obtain the right hand side of (5.298).

Consider a quadratic function of a finite number of variables:

$$I(u) = \frac{1}{2}(Au,u) - (l,u), \tag{5.300}$$

$$(Au,u) = A_{ij}u^i u^j, \quad (l,u) = l_i u^i. \tag{5.301}$$

The minimum value of this function is

$$\check{I} = -\frac{1}{2}\left(A^{-1}l, l\right) \tag{5.302}$$

and therefore, formula (5.298) can also be written as

$$e^{-\min\limits_u I(u)} = \sqrt{\det A}\int e^{-I(u)}du. \tag{5.303}$$

---

[22] Note a witty trick suggested by Poisson to find the value of the integral (5.297):

$$\left(\int_{-\infty}^{+\infty} e^{-\frac{1}{2}x^2}dx\right)^2 = \int_{-\infty}^{+\infty} e^{-\frac{1}{2}x^2}dx \int_{-\infty}^{+\infty} e^{-\frac{1}{2}y^2}dy = \int_{-\infty}^{+\infty}\int_{-\infty}^{+\infty} e^{-\frac{1}{2}x^2}e^{-\frac{1}{2}y^2}dxdy$$

$$= \int_{-\infty}^{+\infty}\int_{-\infty}^{+\infty} e^{-\frac{1}{2}(x^2+y^2)}dxdy = \int_0^{+\infty}\int_0^{2\pi} e^{-\frac{1}{2}r^2}rdrd\theta = 2\pi\int_0^{+\infty} e^{-\frac{1}{2}r^2}d\frac{1}{2}r^2 = 2\pi.$$

We see that the computation of the minimum value is reduced to integration. Since any quadratic functional in variational problems of continuum mechanics admits a finite-dimensional truncation, one can write formula (5.303) for a finite-dimensional truncation, and then consider the limit when the dimension of the truncation tends to infinity. In the limit, in the right hand side of (5.303) we obtain what is called the functional integral. We include $\sqrt{\det A}$ in the definition of the "volume element" in the functional space,

$$\mathcal{D}_A u = \sqrt{\det A} du \tag{5.304}$$

and write (5.303) as

$$e^{\min_u I(u)} = \int e^{-I(u)} \mathcal{D}_A u. \tag{5.305}$$

The notation, $\mathcal{D}_A u$, emphasizes that the volume element depends on the operator A.

We will need various generalizations of (5.305) involving non-positive quadratic functionals and complex-valued functionals. We consider them first in the one-dimensional case.

Formula (5.298) in one-dimensional case,

$$\sqrt{A} \int_{-\infty}^{+\infty} e^{-\frac{1}{2}Ax^2 + lx} \frac{dx}{\sqrt{2\pi}} = e^{\frac{1}{2}\frac{l^2}{A}}, \tag{5.306}$$

remains valid if the number l is complex. Indeed, the integral (5.306) may be considered as an integral in the complex z-plane, $z = x + iy$, along the real axis,

$$\sqrt{A} \int_{-\infty}^{+\infty} e^{-\frac{1}{2}Az^2 + lz} \frac{dz}{\sqrt{2\pi}}.$$

The change of variable, $z \to z_1$, $z = z_1 + A^{-1}l$, transforms this integral into the integral over the line in z_1-plane, $-\infty < x < \infty$, $y = y^* \equiv -\operatorname{Im}\left(A^{-1}l\right)$:

$$\sqrt{A} \int_{-\infty}^{+\infty} e^{-\frac{1}{2}Az^2 + lz} \frac{dz}{\sqrt{2\pi}} = \sqrt{A} \int_{-\infty + iy^*}^{+\infty + iy^*} e^{-\frac{1}{2}Az_1^2} \frac{dz_1}{\sqrt{2\pi}} e^{\frac{l^2}{2A}}. \tag{5.307}$$

The integral of $\exp\left[-\frac{1}{2}Az^2\right]$ over the line $[-\infty + iy^*, +\infty + iy^*]$ is equal to the integral over the real axis $[-\infty, +\infty]$. Indeed, compare the integrals of $\exp\left[-\frac{1}{2}Az^2\right]$ over the segments BC and AD (Fig. 5.31). Since the integral of an analytic function, $\exp\left[-\frac{1}{2}Az^2\right]$ over a closed contour $ABCD$ is zero, the difference of these integrals is equal to the sum of integrals of $\exp\left[-\frac{1}{2}Az^2\right]$ of the segments

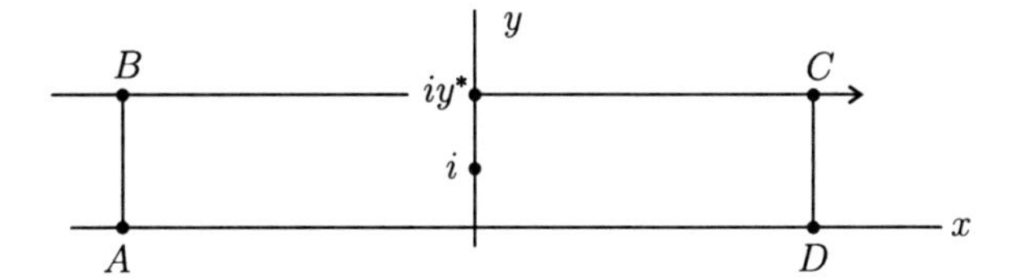

Fig. 5.31 To the proof of (5.306) for complex l

AB and CD. The latter tend to zero as these segments are moved to infinity because $\left|e^{-\frac{1}{2}Az^2}\right|$ on these segments tends to zero. Hence, the right hand side of (5.307) is equal to $\exp\left[\frac{1}{2A}l^2\right]$, and (5.306) holds true.

Two forms of (5.306) will be needed further: one is obtained by replacing l by il (with real l):

$$\sqrt{A}\int_{-\infty}^{+\infty} e^{-\frac{1}{2}Ax^2+ilx}\frac{dx}{\sqrt{2\pi}} = e^{-\frac{1}{2A}l^2} \tag{5.308}$$

and another by replacing l by $i\sqrt{z}l$ with real l and any complex number z:

$$\sqrt{A}\int_{-\infty}^{+\infty} e^{-\frac{1}{2}Ax^2+i\sqrt{z}lx}\frac{dx}{\sqrt{2\pi}} = e^{-z\frac{1}{2A}l^2}, \tag{5.309}$$

In this and all further formulas the argument of a complex number, $z = |z|\,e^{i\theta}$, is constrained by the condition $-\pi \leq \theta \leq \pi$, and the square root of a complex number is understood as $\sqrt{z} = |z|^{1/2}\,e^{i\theta/2}$. In particular, if z lies in the right half-plane, $\operatorname{Re} z \geq 0$, then $\operatorname{Re}\sqrt{z} \geq 0$.

A useful form of (5.304) is obtained if we change the variable of integration, $x \to v : v = ix$ and integrate over the imaginary axes, $[-i\infty, +i\infty]$:

$$\sqrt{A}\int_{-i\infty}^{+i\infty} e^{\frac{1}{2}Av^2+\sqrt{z}lv}\frac{dv}{\sqrt{2\pi i}} = e^{-z\frac{1}{2A}l^2}. \tag{5.310}$$

Changing l by $-l$, one can also write (5.310) in the form

$$\sqrt{A}\int_{-i\infty}^{+i\infty} e^{\frac{1}{2}Av^2-\sqrt{z}lv}\frac{dv}{\sqrt{2\pi i}} = e^{-z\frac{1}{2A}l^2}. \tag{5.311}$$

For $\operatorname{Re} z > 0$, one can put (5.311) in the form

$$\sqrt{Az} \int_{-i\infty}^{+i\infty} e^{z\left(\frac{1}{2}Av^2 - lv\right)} \frac{dv}{\sqrt{2\pi i}} = e^{-z\frac{1}{2A}l^2}. \tag{5.312}$$

To prove (5.312) we get rid of the linear term in the exponent by making the change, $v \to v + \frac{l}{A}$. The line of integration can be moved back to imaginary axes similarly to the move of the line of integration in proving (5.306). It remains to show that

$$\sqrt{Az} \int_{-i\infty}^{+i\infty} e^{\frac{1}{2}zAv^2} \frac{dv}{\sqrt{2\pi i}} = 1 \tag{5.313}$$

for complex z, $\operatorname{Re} z > 0$. Denote $\operatorname{Arg} z$ by θ ; $|\theta| < \pi/2$ since $\operatorname{Re} z > 0$. Let us make the change of variables, $v \to w = \xi + i\eta : w = v\sqrt{z}$. The line of integration in w-plane is the line passing through the points A, B shown in Fig. 5.32; denote it by L. We have

$$\sqrt{Az} \int_{-i\infty}^{+i\infty} e^{\frac{1}{2}zAv^2} \frac{dv}{\sqrt{2\pi i}} = \sqrt{A} \int_L e^{\frac{1}{2}Aw^2} \frac{dw}{\sqrt{2\pi i}}. \tag{5.314}$$

The integral over the segment CB is estimated as

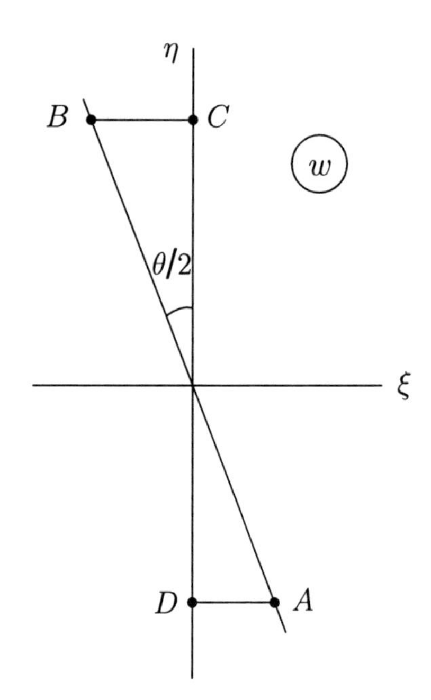

Fig. 5.32 To the proof (5.313)

$$\left| \int_{CB} e^{\frac{1}{2}A(\xi^2-\eta^2+2i\xi\eta)} d\xi \right| \le e^{-\frac{1}{2}A\eta^2} \int_0^{\eta\tan(\theta/2)} e^{\frac{1}{2}A\xi^2} d\eta \le$$
$$\le \eta e^{-\frac{1}{2}A(\eta^2-\eta^2\tan^2(\theta/2))}. \tag{5.315}$$

Since $|\tan(\theta/2)| < 1$, the right hand side of (5.315) tends to zero as the segment CB moves to infinity. Similarly, the integral over AD tends to zero as well. Thus, integral over L in (5.314) can be replaced by integral over the imaginary axis, and we arrive at (5.313).

It is essential for what follows that z in (5.312) can also be taken as pure imaginary: $\mathrm{Re}\, z = 0, z \neq 0$. For pure imaginary z the integral (5.312) does not converge absolutely and needs to be defined. We define it as the following limit:

$$\sqrt{Az} \int_{-i\infty}^{+i\infty} e^{z\left(\frac{1}{2}Av^2-lv\right)} \frac{dv}{\sqrt{2\pi i}} = \lim_{\varepsilon\to+0} \sqrt{A(z+\varepsilon)} \int_{-i\infty}^{+i\infty} e^{(z+\varepsilon)\left(\frac{1}{2}Av^2-lv\right)} \frac{dv}{\sqrt{2\pi i}}.$$

Since

$$\lim_{\varepsilon\to+0} \sqrt{A(z+\varepsilon)} \int_{-i\infty}^{+i\infty} e^{(z+\varepsilon)\left(\frac{1}{2}Av^2-lv\right)} \frac{dv}{\sqrt{2\pi i}} = \lim_{\varepsilon\to+0} e^{-(z+\varepsilon)\frac{1}{2A}l^2} = e^{-z\frac{1}{2A}l^2},$$

we see that (5.312) holds for pure imaginary z as well.[23]

If $\mathrm{Re}\, z < 0$, then (5.312) can be replaced by the equation

$$\sqrt{-Az} \int_{-\infty}^{+\infty} e^{z\left(\frac{1}{2}Au^2-lu\right)} \frac{du}{\sqrt{2\pi}} = e^{-z\frac{1}{2A}l^2},$$

where integration is conducted over real u. This equation follows from (5.312) by changing the integration variable $v \to u : v = iu$, and replacing $l \to il$, $z \to -z$.

---

[23] Putting in (5.312) $A = 1, l = 0, z = i, v = ix$ we obtain

$$\sqrt{i} \int_{-\infty}^{+\infty} e^{-i\frac{1}{2}x^2} \frac{dx}{\sqrt{2\pi}} = 1,$$

which yields the well-known relations

$$\int_{-\infty}^{+\infty} \cos\frac{1}{2}x^2 dx = \sqrt{\pi}, \qquad \int_{-\infty}^{+\infty} \sin\frac{1}{2}x^2 dx = \sqrt{\pi}.$$

Another useful form of (5.312) is obtained if we replace z by $1/z$ ($\mathrm{Re}\, z^{-1} \geq 0$ if $\mathrm{Re}\, z \geq 0$, so such a replacement is legitimate) and l by lz :

$$\sqrt{\frac{1}{z}A} \int_{-i\infty}^{+i\infty} e^{\frac{1}{2z}Av^2 - lv} \frac{dv}{\sqrt{2\pi i}} = e^{-z\frac{1}{2A}l^2}. \tag{5.316}$$

The finite-dimensional forms of (5.308), (5.309), (5.310), (5.311), (5.312) and (5.316) are:

$$\sqrt{\det A} \int_{R_m} e^{-\frac{1}{2}(Au,u)+i(l,u)} du = e^{-\frac{1}{2}(A^{-1}l,l)}, \tag{5.317}$$

$$\sqrt{\det A} \int_{R_m} e^{-\frac{1}{2}(Au,u)+i\sqrt{z}(l,u)} du = e^{-\frac{1}{2}z(A^{-1}l,l)}, \tag{5.318}$$

$$\sqrt{\det A} \int_{-i\infty}^{+i\infty} e^{\frac{1}{2}(Av,v)+\sqrt{z}(l,v)} dv = e^{-\frac{1}{2}z(A^{-1}l,l)}, \tag{5.319}$$

$$\sqrt{\det A} \int_{-i\infty}^{+i\infty} e^{\frac{1}{2}(Av,v)-\sqrt{z}(l,v)} dv = e^{-\frac{1}{2}z(A^{-1}l,l)}, \tag{5.320}$$

$$\sqrt{z^m \det A} \int_{-i\infty}^{+i\infty} e^{z\left(\frac{1}{2}(Av,v)-(l,v)\right)} dv = e^{-\frac{1}{2}z(A^{-1}l,l)}, \tag{5.321}$$

$$\sqrt{\frac{1}{z^m} \det A} \int_{-i\infty}^{+i\infty} e^{\frac{1}{2z}(Av,v)-(l,v)} dv = e^{-\frac{1}{2}z(A^{-1}l,l)}. \tag{5.322}$$

For brevity, the imaginary unity is included in the "volume element":

$$dv = \frac{dv_1}{\sqrt{2\pi i}} \frac{dv_2}{\sqrt{2\pi i}} \cdots \frac{dv_m}{\sqrt{2\pi i}}.$$

These relations are proved by transforming the coordinates to the principal coordinates of the tensor A_{ij}.

For $\mathrm{Re}\, z < 0$, the similar relation is

$$\sqrt{(-z)^m \det A} \int_{\infty}^{\infty} e^{z\left(\frac{1}{2}(Av,v)-(l,v)\right)} dv = e^{-\frac{1}{2}z(A^{-1}l,l)}. \tag{5.323}$$

Tending the dimension to infinity, we find the following formulas linking the minimum values of quadratic functionals with integration in functional spaces: for arbitrary z,

$$e^{\min_u I(u)} = \int e^{-\frac{1}{2}(Au,u)+i(l,u)} \mathcal{D}_A u, \tag{5.324}$$

$$e^{z \min_u I(u)} = \int e^{-\frac{1}{2}(Au,u)+i\sqrt{z}(l,u)} \mathcal{D}_A u, \tag{5.325}$$

$$e^{z \min_u I(u)} = \int_{-i\infty}^{+i\infty} e^{\frac{1}{2}(Av,v)+\sqrt{z}(l,v)} \mathcal{D}_A v, \tag{5.326}$$

for $\operatorname{Re} z > 0$,

$$e^{z \min_u I(u)} = \int_{-i\infty}^{+i\infty} e^{z[\frac{1}{2}(Av,v)-(l,v)]} \mathcal{D}_{zA} v = \int_{-i\infty}^{+i\infty} e^{zI(v)} \mathcal{D}_{zA} v, \tag{5.327}$$

$$e^{z \min_u I(u)} = \int_{-i\infty}^{+i\infty} e^{\frac{1}{2z}(Av,v)-(l,v)} \mathcal{D}_{\frac{1}{z}A} v, \tag{5.328}$$

for $\operatorname{Re} z < 0$,

$$e^{z \min_u I(u)} = \int_{-\infty}^{\infty} e^{z[\frac{1}{2}(Au,u)-(l,v)]} \mathcal{D}_{-zA} u = \int_{-\infty}^{\infty} e^{zI(u)} \mathcal{D}_{-zA} u. \tag{5.329}$$

In (5.327), (5.328) and (5.329) the parameter, z, is also included in the volume element: for $m-$dimensional truncation,

$$\mathcal{D}_{zA} v = \sqrt{z^m \det A} dv, \mathcal{D}_{\frac{1}{z}A} v = \sqrt{z^{-m} \det A} dv, \mathcal{D}_{-zA} u = \sqrt{(-z)^m \det A} du.$$

In the right hand sides of (5.324)-(5.328) l can be replaced by $-l$ because the left hand sides remain the same for such change.

Let now the functional (Au, u) be non-convex, i.e. it can take negative values for some u. Formula (5.327) still holds if we assume that (Au, u) is a non-degenerated functional (i.e. $(Au, u) \neq 0$ for any $u \neq 0$), and, thus, the functional $I(u) = \frac{1}{2}(Au, u) - (l, u)$ has a unique stationary point. Then, denoting by S.V.$I(u)$[24] the

[24] S.V. stands for stationary value.

value of $I(u)$ at the stationary point and admitting in (5.327) only pure imaginary z we get

$$e^{z\mathrm{S.V.}I(u)} = \int_{-i\infty}^{+i\infty} e^{zI(v)} \mathcal{D}_{zA} v. \tag{5.330}$$

The usefulness of such relations will be illustrated in Vol. 2.

5.13 Miscellaneous

In this section we collect some simple facts which help in working with variational problems.

Euler equations in curvilinear coordinates. In physical problems, the action functional is invariant with respect to the choice of coordinate system. This allows one to write easily the equations for the minimizer in curvilinear coordinates. We illustrate that by the following example. Let x, y be Cartesian coordinates. Consider the functional of functions of two variables, $u(x, y)$,

$$I(u) = \frac{1}{2} \int_V \left(\left(\frac{\partial u}{\partial x} \right)^2 + \left(\frac{\partial u}{\partial y} \right)^2 \right) dxdy. \tag{5.331}$$

The Euler equation of this functional is Laplace's equation,

$$\Delta\, u = 0.$$

We wish to write this equation in polar coordinates, r, θ. To this end, we rewrite the functional in polar coordinates. Since

$$\left(\frac{\partial u}{\partial x} \right)^2 + \left(\frac{\partial u}{\partial y} \right)^2 = \left(\frac{\partial u}{\partial r} \right)^2 + \left(\frac{1}{r} \frac{\partial u}{\partial \theta} \right)^2, \qquad dxdy = rdrd\theta,$$

we have

$$I(u) = \frac{1}{2} \int_V \left(r \left(\frac{\partial u}{\partial r} \right)^2 + \frac{1}{r} \left(\frac{\partial u}{\partial \theta} \right)^2 \right) drd\theta.$$

This functional must be stationary with respect to $u(r, \theta)$ at the minimizer. Thus,

$$\frac{\partial}{\partial r} r \frac{\partial u}{\partial r} + \frac{\partial}{\partial \theta} \frac{1}{r} \frac{\partial u}{\partial \theta} = 0.$$

This is the equation required. Derivation in other cases is similar.

Evenness and oddness of the minimizer. Let the admissible functions of a variational problem be defined in a symmetric domain V, i.e. the domain, which contains with each point x the point $-x$. Then one can introduce the notion of evenness and oddness: function $u(x)$ is even if

$$u(-x) = u(x),$$

and odd if

$$u(-x) = -u(x).$$

Each function may be uniquely presented as a sum of an even and an odd functions:

$$u(x) = u'(x) + u''(x), \qquad u'(-x) = -u'(x), u''(-x) = u''(x). \tag{5.332}$$

Indeed, $u'(x)$ and $u''(x)$ can be found uniquely in terms of $u(x)$ from (5.332):

$$u'(x) = \frac{1}{2}(u(x) - u(-x)), \qquad u''(x) = \frac{1}{2}(u(x) + u(x)).$$

If the functional $I(u)$ splits in the sum of functionals $I(u') + I(u'')$ and the constraints on admissible functions can be formulated as constraints to their odd and even parts, then the entire variational problem splits into two independent simpler problems for $u'(x)$ and $u''(x)$. For example, if the region, V, in (5.331) is symmetric, then

$$I(u) = I(u') + I(u''),$$

because $\nabla u'$ and $\nabla u''$ are even and odd vector fields, respectively,[25] therefore $\int \nabla u' \nabla u'' dxdy = 0$. If the boundary values of u are prescribed, and they are even, then the boundary values of u' are zero, and minimization over u' yields $u' = 0$. So, it is enough to perform minimization only over even functions. Similar arguments hold if the region is symmetric only with respect to one variable.

Divergence terms. If Lagrangian contains a divergence term, e.g.,

$$L\left(u, \frac{\partial u}{\partial x^i}\right) = L_0\left(u, \frac{\partial u}{\partial x^i}\right) + \frac{\partial}{\partial x^i} P^i\left(u, \frac{\partial u}{\partial x^i}\right),$$

then this term does not contribute to Euler equation. That follows from the divergence theorem,

[25] I.e. $\nabla u'\big|_{-x} = \nabla u'\big|_x$ and $\nabla u''\big|_{-x} = -\nabla u''\big|_x$.

$$\int_V L\left(u, \frac{\partial u}{\partial x^i}\right) dV = \int_V L_0\left(u, \frac{\partial u}{\partial x^i}\right) dV + \int_{\partial V} P^i\left(u, \frac{\partial u}{\partial x^i}\right) n_i dA.$$

We see that the divergence term contributes only to the boundary conditions.

Derivative of minimum value with respect to parameter. Consider a variational problem, the functional of which depends on a parameter, r :

$$\check{I} = \min_u I(u, r).$$

The minimizer, $\check{u}$, depends on the parameter, $\check{u} = \check{u}(r)$, and so does the minimum value: $\check{I} = I(\check{u}(r), r)$. The following formula holds true:

$$\frac{d\check{I}}{dr} = \left.\frac{\partial I(u, r)}{\partial r}\right|_{u=\check{u}(r)}. \tag{5.333}$$

The proof is simple:

$$\frac{dI(\check{u}(r), r)}{dr} = \left(\left.\frac{\delta I}{\delta u}\right|_{u=\check{u}(r)}, \frac{d\check{u}(r)}{dr}\right) + \left.\frac{\partial I(u, r)}{\partial r}\right|_{u=\check{u}(r)}.$$

The first term in the right hand side is zero, because $\check{u}(r)$ is the minimizer.

A modification of variational problems. Let the functional $I(u, v)$ have a stationary point, u^*, v^*, and v^* can be explicitly computed in terms of u^* : $v^* = \Phi(u^*)$, where $\Phi(u^*)$ is a function or an operator. Then u^* is a stationary point of the functional

$$\tilde{I}(u) = I(u, \Phi(u)). \tag{5.334}$$

Indeed,

$$\delta\tilde{I}(u) = \left(\left.\frac{\delta I(u, v)}{\delta u}\right|_{v=\Phi(u)}, \delta u\right) + \left(\left.\frac{\delta I(u, v)}{\delta v}\right|_{v=\Phi(u)}, \frac{\delta \Phi(u)}{\delta u} \cdot \delta u\right).$$

After substitution $u = u^*$ in the left hand side both terms vanish, thus $\delta\tilde{I}(u) = 0$ as claimed. Note that the substitution may result in appearance of additional stationary points, for which $\delta I(u, v)/\delta v|_{v=\Phi(u)} \neq 0$. We will use this point in transformations of variational principles of fluid mechanics.

Stationary points of complex-valued functionals and Cauchy-Riemann equations. Let $S(v)$ be a complex-valued functional of complex-valued functions, v. We denote their real and imaginary parts by S_1, S_2 and v_1, v_2 :

$$S = S_1(v_1, v_2) + iS_2(v_1, v_2), \qquad v = v_1 + iv_2.$$

Suppose that the functional $S(v)$ is smooth, i.e. its variation, δS, is a linear functional of δv, which we denote by

$$\left(\frac{\delta S}{\delta v}, \delta v\right).$$

The scalar product $(A, \delta v)$ is supposed to obey the two conditions: first, it is linear with respect to both arguments on the set of complex numbers, i.e. for any complex numbers, c_1 and c_2,

$$\begin{aligned}(c_1 A_1 + c_2 A_2, \delta v) &= c_1(A_1, \delta v) + c_2(A_2, \delta v),\\ (A, c_1\delta v_1 + c_2\delta v_2) &= c_1(A, \delta v_1) + c_2(A, \delta v_2);\end{aligned}$$

second, if $(A, \delta v) = 0$ for any δv, then $A = 0$. Therefore, if the equation

$$\left(\frac{\delta S}{\delta v}, \delta v\right) = (A, \delta v)$$

holds for any δv, then

$$\frac{\delta S}{\delta v} = A.$$

Under these assumptions the following generalized Cauchy-Riemann equations hold:

$$\frac{\delta S_1}{\delta v_1} = \frac{\delta S_2}{\delta v_2}, \qquad \frac{\delta S_2}{\delta v_1} = -\frac{\delta S_1}{\delta v_2}. \tag{5.335}$$

Indeed, let us vary v_1. Then variation of the functional $S(v)$ is

$$\left(\frac{\delta S_1(v_1, v_2)}{\delta v_1}, \delta v_1\right) + \left(i\frac{\delta S_2(v_1, v_2)}{\delta v_1}, \delta v_1\right).$$

On the other hand, since S is a functional of v, the same variation is

$$\left(\frac{\delta S}{\delta v}, \delta v\right) = (\xi + i\eta, \delta v_1), \qquad \frac{\delta S}{\delta v} \equiv \xi + i\eta,$$

where ξ and η are some real-valued operators. Equating both expressions and using arbitrariness of variations, we obtain the relations,

$$\frac{\delta S_1}{\delta v_1} = \xi, \qquad \frac{\delta S_2}{\delta v_1} = \eta.$$

Similarly, if we vary v_2, we have for the variation of $S(v)$

$$\left(\frac{\delta S_1(v_1, v_2)}{\delta v_2}, \delta v_2\right) + \left(i\frac{\delta S_2(v_1, v_2)}{\delta v_2}, \delta v_2\right).$$

This must be equal to

$$\left(\frac{\delta S}{\delta v}, \delta v\right) = (\xi + i\eta, i\delta v_2).$$

Hence,

$$\frac{\delta S_1}{\delta v_2} = -\eta, \qquad \frac{\delta S_2}{\delta v_2} = \xi.$$

So we arrived at (5.335).

If $S = S_1 + iS_2$ is a complex-valued functional of complex-valued functions, $v = v_1 + iv_2$, then, varying v_1 and v_2, we obtain the system of Euler equations for the stationary points:

$$\frac{\delta S_1(v_1, v_2)}{\delta v_1} = 0, \qquad \frac{\delta S_2(v_1, v_2)}{\delta v_1} = 0, \qquad \frac{\delta S_1(v_1, v_2)}{\delta v_2} = 0, \qquad \frac{\delta S_2(v_1, v_2)}{\delta v_2} = 0.$$

At first glance we have four equations for two functions, v_1 and v_2. In fact, however, due to (5.335), only two of these equations are independent. As such, we can take the equations, obtained by varying only the real part of the functional, $S_1(v_1, v_2)$:

$$\frac{\delta S_1(v_1, v_2)}{\delta v_1} = 0, \qquad \frac{\delta S_1(v_1, v_2)}{\delta v_2} = 0. \tag{5.336}$$

Part II
Variational Features of Classical Continuum Models

In continuum mechanics, many seemingly unrelated variational principles have been invented. What follows is an attempt to present these variational principles systematically, point out their interrelations, and fill in some gaps. Usually, the variational principles have several equivalent forms; the form depends on the choice of arguments of the action functional. The initial formulation of the variational principles, which directly follows from the principle of least action, employs the particle trajectories $x(X, t)$ as arguments of the action functional. Such a formulation is usually convenient in mechanics of solids. As a rule, the problems of fluid mechanics are better suited for being stated and investigated when the arguments of the action functional are the functions of Eulerian coordinates. Such variational principles are derived from initial "Lagrangian coordinate" formulations by the corresponding change of sought functions. Further transformations are based on the idea of duality. The chapters related to the mechanics of solids and to fluid mechanics can be read independently.

Chapter 6
Statics of a Geometrically Linear Elastic Body

6.1 Gibbs Principle

Gibbs principle. Equilibrium of elastic bodies is governed by Gibbs variational principles. We will use as a starting point the second Gibbs principle. We assume that the positions of the boundary particles are given at some part of the boundary, $\partial \mathring{V}_u$,

$$x(X) = x_{(b)}(X) \quad \text{for } X \in \partial \mathring{V}_u, \tag{6.1}$$

while on the remaining part of the boundary, $\partial \mathring{V}_f$, the surface forces, f_i, are known.
Gibbs variational principle. *Equilibrium states of elastic body correspond to the minimum of the functional*

$$I(x(X), S(X)) = \int_{\mathring{V}} \rho_0 U\left(x_a^i, S\right) d\mathring{V} - \int_{\partial \mathring{V}_f} f_i x^i (X) d\mathring{A} \tag{6.2}$$

on the set of all functions $x^i(X^a)$ obeying the boundary conditions (6.1) and functions $S(X^a)$ subject to the constraint

$$\int_{\mathring{V}} \rho_0 S d\mathring{V} = S_0. \tag{6.3}$$

Strictly speaking, Gibbs variational principle selects the stable equilibrium states. There might be unstable equilibrium states which correspond to the stationary points of the functional (6.2). Functional (6.2) is called energy functional.

Minimization with respect to entropy can be carried out explicitly by introducing the Lagrange multiplier, T, for the constraint (6.3). One has to minimize over S the functional

$$\int_{\mathring{V}} \rho_0 U\left(x_a^i, S\right) d\mathring{V} - T\left(\int_{\mathring{V}} \rho_0 S d\mathring{V} - S_0\right) - \int_{\partial \mathring{V}_f} f_i x^i (X) d\mathring{A}.$$

V.L. Berdichevsky, *Variational Principles of Continuum Mechanics*,
Interaction of Mechanics and Mathematics, DOI 10.1007/978-3-540-88467-5_6,

Introducing free energy,

$$F\left(x_a^i, T\right) = \min_S \left[U\left(x_a^i, S\right) - TS\right], \tag{6.4}$$

we have

$$\min_S I\left(x\left(X\right), S\left(X\right)\right) = \int_{\mathring{V}} \rho_0 F\left(x_a^i, T\right) d\mathring{V} - \int_{\partial \mathring{V}_f} f_i x^i\left(X\right) d\mathring{A} + TS_0. \tag{6.5}$$

The parameter T has the meaning of absolute temperature at which the equilibrium is achieved.

The functional (6.5) in which the additive constant TS_0 is dropped will be denoted by $I\left(x\left(X\right)\right)$. We arrived at the following version of Gibbs principle.

Gibbs variational principle. *Equilibrium states of an elastic body correspond to the minimum of the energy functional*

$$I\left(x\left(X\right)\right) = \int_{\mathring{V}} \rho_0 F\left(x_a^i, T\right) d\mathring{V} - \int_{\partial \mathring{V}_f} f_i x^i\left(X\right) d\mathring{A} \tag{6.6}$$

on the set of all functions $x^i\left(X^a\right)$ satisfying the boundary conditions (6.1).

Note that for inhomogeneous elastic bodies, internal energy and, correspondingly, free energy depend explicitly on Lagrangian coordinates but, for brevity, they are not mentioned explicitly among the arguments.

Free energy density $F\left(x_a^i, T\right)$ is invariant with respect to rigid rotation. Therefore, the distortion enters into F only in combinations, $\varepsilon_{ab} = \frac{1}{2}\left(g_{ij} x_a^i x_b^j - \mathring{g}_{ab}\right)$:

$$F = F\left(\varepsilon_{ab}, T\right).$$

If the elastic material is also subject to the body forces with the force per unit mass g_i, then an additional term must be included in the energy functional

$$I\left(x\left(X\right)\right) = \int_{\mathring{V}} \rho_0 F\left(\varepsilon_{ab}, T\right) d\mathring{V} - \int_{\mathring{V}} \rho_0 g_i x^i\left(X\right) d\mathring{V} - \int_{\partial \mathring{V}_f} f_i x^i\left(X\right) d\mathring{A}. \tag{6.7}$$

Geometrically linear deformation. By geometrically linear deformation one means the case when the displacements

$$u^i = x^i\left(X^a\right) - \mathring{x}^i\left(X^a\right)$$

and their gradients are small. The gradients of displacements are dimensionless, denote their order by ε. In geometrically linear theory one neglects the terms on the order of ε in comparison with unity. This yields several simplifications. First, the strain tensor becomes a linear function of the displacement gradients:

$$\varepsilon_{ab} = \frac{1}{2}\left(g_{ab} - \overset{\circ}{g}_{ab}\right) = \frac{1}{2}\left(g_{ij}\left(\frac{\partial \overset{\circ}{x}^i}{\partial X^a} + \frac{\partial u^i}{\partial X^a}\right)\left(\frac{\partial \overset{\circ}{x}^j}{\partial X^b} + \frac{\partial u^j}{\partial X^b}\right) - \overset{\circ}{g}_{ab}\right) =$$
$$= \frac{1}{2} g_{ij}\left(\overset{\circ}{x}{}^i_a \frac{\partial u^j}{\partial X^b} + \overset{\circ}{x}{}^j_b \frac{\partial u^i}{\partial X^a}\right). \tag{6.8}$$

For Cartesian Eulerian coordinates coinciding with the Lagrangian coordinates in the initial state, $g_{ij} = \delta_{ij}$, $\overset{\circ}{x}{}^i_a = \delta^i_a$, and (6.8) becomes

$$\varepsilon_{ab} = \frac{1}{2}\left(\frac{\partial u_a}{\partial X^b} + \frac{\partial u_b}{\partial X^a}\right). \tag{6.9}$$

Second, the derivatives with respect to Eulerian coordinates and Lagrangian coordinates coincide:

$$\frac{\partial}{\partial X^a} = x^i_a \frac{\partial}{\partial x^i} = \left(\overset{\circ}{x}{}^i_a + \frac{\partial u^i}{\partial X^a}\right)\frac{\partial}{\partial x^i} = \overset{\circ}{x}{}^i_a \frac{\partial}{\partial x^i} = \frac{\partial}{\partial x^a}.$$

Third, the components of tensors in Eulerian coordinates and the Lagrangian coordinates coincide within the accuracy accepted because

$$x^i_a = \overset{\circ}{x}{}^i_a + \frac{\partial u^i}{\partial X^a} \approx \overset{\circ}{x}{}^i_a = \delta^i_a.$$

Hence, only Eulerian indices can be used. In particularly, (6.9) will be written as

$$\varepsilon_{ij} = \frac{1}{2}\left(\frac{\partial u_i}{\partial x^j} + \frac{\partial u_j}{\partial x^i}\right) = u_{(i,j)}. \tag{6.10}$$

Fourth, due to smallness of displacements, one does not need to distinguish the initial region occupied by the body, $\overset{\circ}{V}$, and the final region, V. Accordingly, the regions of integration in the energy functional (6.7) can be taken coinciding with the corresponding regions in the deformed state. Therefore, one can drop the index $_0$ in the region notation in (6.7). Fifth, the densities in the deformed and undeformed states differ by terms of the order ε. Densities enter in all relations as factors at small terms. Thus, distinguishing the initial and final densities would yield only small corrections. Therefore, in linear theory densities in the initial and the deformed states can be identified. Further we drop index $_0$ in ρ_0 in (6.7) and assume that ρ is the given initial density. We will include the factor ρ in free energy density, so in what follows F means the free energy per unit volume. Besides, we will not mention explicitly temperature as an argument of F. Finally, replacing in (6.7) the particle positions, $x^i(X^a)$, by the displacement vector, $u^i(x)$, and dropping the additive constants we obtain for the energy functional of elastic body in case of geometrically linear deformation the following formula:

$$I(u) = \int_V F(\varepsilon_{ij})\, dV - l(u), \tag{6.11}$$

$$l(u) = \int_V \rho g_i u^i dV + \int_{\partial V_f} f_i u^i dA.$$

Minimum of the energy functional is sought on the set of displacements which have the prescribed values on ∂V_u :

$$u_i(x) = u_i^{(b)} \quad \text{on } \partial V_u. \tag{6.12}$$

Obviously, the minimum is achieved at the stress field,

$$\sigma^{ij} = \frac{\partial F(\varepsilon_{ij})}{\partial \varepsilon_{ij}},$$

satisfying the equations,

$$\frac{\partial \sigma^{ij}}{\partial x^j} + \rho g^i = 0 \text{ in } V, \qquad \sigma^{ij} n_j = f^i \text{ on } \partial V_f.$$

Without loss of generality, the energy density F may be assumed to possess the properties:

$$F|_{\varepsilon_{ij}=0} = 0, \qquad \left.\frac{\partial F}{\partial \varepsilon_{ij}}\right|_{\varepsilon_{ij}=0} = 0. \tag{6.13}$$

If (6.13) were not held, one can replace F by $\tilde{F} = F - F_0 - \sigma_0^{ij} \varepsilon_{ij}$, where $F_0 = F|_{\varepsilon_{ij}=0}$, $\sigma_0^{ij} = \partial F / \partial \varepsilon_{ij}|_{\varepsilon_{ij}=0}$, and replace also g_i and f_i by $g_i = g_i + \sigma_{0,j}^{ij}$, $\tilde{f}_i = f_i - \sigma_0^{ij} n_j$. However, such a replacement would change the physical meaning of g_i and f_i as densities of the external body and surface forces. In what follows, we prefer to keep this original meaning. A continuum model with the energy functional (6.11) is called physically nonlinear and geometrically linear elastic body. Physically and geometrically linear elastic body corresponds to quadratic function $F(\varepsilon_{ij})$.

Physical meaning of energy functional. If an elastic body is deformed by the prescribed boundary displasments, then the energy functional coincides with the free energy of the body, and its minimum value is the free energy at equilibrium. If the prescribed boundary displacements are zero, and the body is deformed by an external force, then, in linear theory, according to the Clapeyron theorem (5.46), the minimum value of the functional is equal to negative energy. If the body is deformed by both the prescribed non-zero displacements and external forces, then such simple interpretations of the minimum value of the energy functional are lost.

The question arises: What is the physical meaning of the energy functional to be minimized to determine the equilibrium states? The answer can be sought by studying a microscopical system that is described at the macro-level by the model of an elastic body. One can expect that the energy functional is the average of the Hamiltonian of the microsystem. The first term of the energy functional, energy, is the result of averaging of the energy of the microsystem. To interpret the second term, a linear functional, we recall that the action of an external force, f, on a particle with a generalized coordinate, q, is described by the term in Hamiltonian, $-fq$. Therefore, the linear functional $l(u)$ may be thought of as the average value of all terms in the micro-Hamiltonian of the form fq.

The issue on the physical meaning of the energy functional becomes especially important, when one complicates the elasticity model by taking into account additional phenomena, like cracks, dislocations or an interaction with electro-magnetic field. The guiding principle in a proper construction of the energy functional is that the energy functional has the meaning of the averaged micro-Hamiltonian of the system.

Consider now the conditions under which the minimization problem for the energy functional (6.11) is well posed.

6.2 Boundedness from Below

Uniqueness. First let no kinematic constraints be imposed (i.e. the surface tractions are prescribed everywhere on ∂V, and $\partial V = \partial V_f$). Then the elastic energy has a kernel: it becomes zero on the infinitesimally small rigid motions:

$$u_i = c_i + e_{ijk}\varphi^j x^k. \tag{6.14}$$

According to the necessary condition for the boundedness from below (5.30), the linear part of the functional $I(u)$ must vanish on the fields (6.14):

$$\int_V \rho g_i dV + \int_{\partial V} f_i dA = 0, \quad e^{ijk}\left(\int_V \rho g_j x_k dV + \int_{\partial V} f_i x_k dA\right) = 0. \tag{6.15}$$

Equations (6.15) mean that the resultant and the total moment of the external forces acting on the body are equal to zero.

Let us assume that the conditions (6.15) are satisfied. Then the energy functional is invariant with respect to rigid motions (6.14), and the minimizing element is not uniquely determined. To single out a unique solution, one has to impose additional constraints which exclude rigid motions. As such one can take the condition of zero average displacement,

$$\int_V u_i dV = 0 \tag{6.16}$$

and zero average rotation

$$e^{ijk} \int_V u_i x_j dV = 0. \tag{6.17}$$

Let the elastic properties be non-degenerated in the sense that, for some positive constant $\underline{\mu}$ and for any ε_{ij},

$$\underline{\mu} \varepsilon_{ij} \varepsilon^{ij} \leq F\left(\varepsilon_{ij}\right). \tag{6.18}$$

Then the conditions (6.15) are also sufficient conditions for the boundedness from below of the energy functional. In order to see that, we need the inequality

$$l(u) \leq c \sqrt{\int_V F\left(u_{(i,j)}\right) dV}. \tag{6.19}$$

If the inequality (6.19) holds, the boundedness from below is obtained in the same way as in Example 3 in Sect. 5.1.

The proof of the inequality (6.19) is based on one of the key inequalities of the elasticity theory, the Korn inequality:

$$\int_V u_{i,j} u^{i,j} dV \leq K \int_V \varepsilon_{ij} \varepsilon^{ij} dV. \tag{6.20}$$

For the Korn inequality to be valid the additional conditions, excluding the rotations of the body, must be imposed; otherwise, for the rotations, the left hand side of (6.19) is positive while the right hand side is equal to zero. Usually, either the condition (6.17) or the condition

$$\int_V (u_{i,j} - u_{j,i}) dV = 0$$

is used.

The Korn inequality can be written in a more impressive form by splitting the displacement gradient into the sum of the deformation and the rotation parts:

$$u_{i,j} = \varepsilon_{ij} + e_{ijk} \varphi^k, \quad \varphi^k = \frac{1}{2} e^{ijk} u_{i,j}.$$

In terms of ε_{ij} and φ_i the Korn inequality takes the form

$$\int\limits_V \varphi_i \varphi^i dV \leq \frac{K-1}{2} \int\limits_V \varepsilon_{ij} \varepsilon^{ij} dV.$$

This means that for any deformation of the body, the average value of the squared angle of rotation never exceeds the average value of the squared strains times a universal coefficient, $\frac{K-1}{2}$, which depends only on the geometry of the region. The proof of the Korn inequality can be found in the works cited in the bibliographic comments.

To prove (6.19) we also need the inequalities

$$\lambda^2 \int\limits_{\partial V} u_i u^i dA \leq \int\limits_V u_{i,j} u^{i,j} dV, \quad \mu^2 \int\limits_V u_i u^i dV \leq \int\limits_V u_{i,j} u^{i,j} dV. \tag{6.21}$$

These inequalities are obtained by summing the inequalities (5.38) written for each component of the displacement vector. Inequalities (6.21) are valid if the translational motions are excluded by, for example, conditions (6.16).

Let us derive (6.19). From (6.18), (6.20), (6.21), and the Cauchy inequality, we have

$$|l(u)| \leq \sqrt{\int\limits_V \rho g_i g^i dV} \sqrt{\int\limits_V u_i u^i dV} + \sqrt{\int\limits_{\partial V} f_i f^i dA \int\limits_{\partial V} u_i u^i dA} \leq$$

$$\leq c \sqrt{\int\limits_V u_{i,j} u^{i,j} dV} \leq c \sqrt{K \int\limits_V \varepsilon_{ij} \varepsilon^{ij} dV} \leq c \sqrt{\frac{K}{\underline{\mu}}} \sqrt{\int\limits_V F(\varepsilon_{ij})\, dV}.$$

Here,

$$c = \frac{1}{\mu} \sqrt{\int\limits_V \rho g_i g^i dV} + \frac{1}{\lambda} \sqrt{\int\limits_{\partial V} f_i f^i dA}. \tag{6.22}$$

So, the energy functional is bounded from below if the squared surface forces are integrable on the surface while the body forces are integrable in the volume.

In the case when the displacements are given on a part of the boundary, the energy functional is bounded from below for any functions, g_i and f_i, including those which have non-zero total force and non-zero total moment. Let us outline the proof when the displacements are given on the entire boundary ∂V.

Consider the integral

$$2 \int\limits_V \varepsilon_{ij} \varepsilon^{ij} dV = \int\limits_V \left(u_{i,j} u^{i,j} + u_{i,j} u^{j,i}\right) dV.$$

Integrating the second term by parts twice, we get

$$2\int_V \varepsilon_{ij}\varepsilon^{ij}dV =$$

$$= \int_V \left(u_{i,j}u^{i,j} + \left(u^i_{,i}\right)^2\right)dV + \int_{\partial V} \left(u_i u^{j,i} n_j - u^i n_i u^k_{,k}\right)dA. \tag{6.23}$$

The surface integral depends only on the displacements at the boundary. Indeed, expanding the derivative along the normal and tangent directions (see Sect. 14.1), we have

$$u_i u^{j,i} n_j - u^i n_i u^k_{,k} = u_i r^{i\alpha} u^j_{,\alpha} n_j - u^i n_i u^k_{,\alpha} r^\alpha_k$$

Hence, the surface integral is known due to the boundary conditions; we will denoted it by B. Dropping in the identity (6.23) the integral of $\left(u^i_{,i}\right)^2$ which is non-negative, we obtain the inequality

$$\int_V u_{i,j}u^{i,j}dV \leq 2\int_V \varepsilon_{ij}\varepsilon^{ij}dV - B,$$

which is a version of the Korn inequality. Analogous changes appear in the inequalities (6.21). Further steps are the same as in the case of given surface forces.

The boundedness from below of the energy functional shows that Gibbs principle is well-posed for the energy functional. The existence of the minimizing element in the energy space is established following the general scheme of Sect. 5.3.

Let the free energy density be a strictly convex function of the components of the strain tensor. The physical interpretation of this assumption can be seen from the inequality (5.89). Writing this inequality for function F, we have

$$0 < \Delta\sigma^{ij}\Delta\varepsilon_{ij}. \tag{6.24}$$

Here,

$$\sigma^{ij} = \frac{\partial F}{\partial \varepsilon_{ij}}$$

are the components of the stress tensor, $\Delta\varepsilon_{ij}$ is the difference between any two values of the strain tensor, and $\Delta\sigma^{ij}$ is the difference between the corresponding values of the stress tensor. Inequality (6.24) means that there is a one-to-one correspondence between the values of the stress tensor and the strain tensor. Besides, the stresses monotonously increase with increasing strains in the following sense. In the principal coordinate system of the tensor $\Delta\varepsilon_{ij}$ the inequality (6.24) becomes

$$0 < \Delta\sigma^{11}\Delta\varepsilon_{11} + \Delta\sigma^{22}\Delta\varepsilon_{22} + \Delta\sigma^{33}\Delta\varepsilon_{33}.$$

Consequently, the stress increment $\Delta\sigma^{11}$ (for $\varepsilon_{22} = \varepsilon_{33} = 0$) has the same sign as the strain increase ε_{11} for any (not necessarily small) $\Delta\varepsilon_{11}$.

For strictly convex function, $F\left(\varepsilon_{ij}\right)$, the energy functional is a strictly convex functional on the set of displacements with excluded rigid motions. As shown in Sect. 5.4, it has the only minimizing element.

6.3 Complementary Energy

Consider the Young-Fenchel transformation of function $F\left(\varepsilon_{ij}\right)$ with respect to ε_{ij}:

$$F^*\left(\sigma^{ij}\right) = \max_{\varepsilon_{ij}}\left[\sigma^{ij}\varepsilon_{ij} - F\left(\varepsilon_{ij}\right)\right]. \tag{6.25}$$

Here, σij are the components of a symmetric tensor. The function $F^*\left(\sigma^{ij}\right)$ is called the complementary energy. This term was motivated by the following geometrical interpretation of F^*. Consider the constitutive equation of a one-dimensional elastic body, $\sigma = \sigma(\varepsilon)$. The dependence $\sigma = \sigma(\varepsilon)$ is shown in Fig. 6.1. Since $\sigma = \partial F/\partial\varepsilon$,

$$F(\varepsilon) = \int_0^\varepsilon \sigma(\tilde{\varepsilon})\, d\tilde{\varepsilon}.$$

Hence, the energy $F(\varepsilon)$ is equal to the area beneath the curve $\sigma = \sigma(\varepsilon)$ in Fig. 6.1; this area is vertically shaded. The number $F^*(\sigma) = \sigma\varepsilon - F(\varepsilon)$ is equal to the area of the figure horizontally shaded. This area "complements" $F(\varepsilon)$ to the area of the rectangle, $\sigma\varepsilon$. In a multi-dimensional case, this simple geometric interpretation is lost.

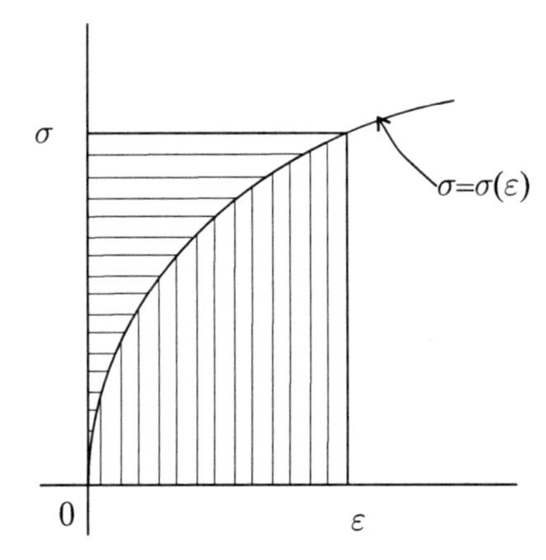

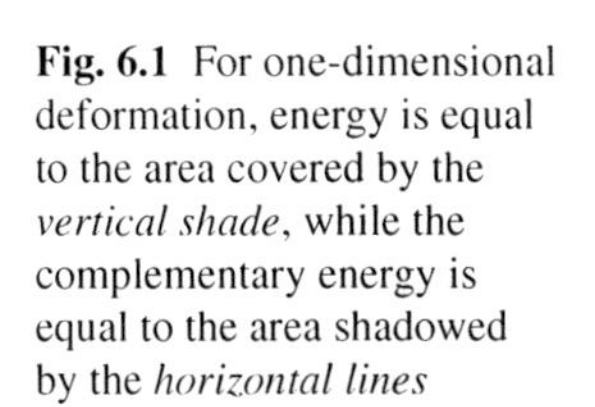

Fig. 6.1 For one-dimensional deformation, energy is equal to the area covered by the *vertical shade*, while the complementary energy is equal to the area shadowed by the *horizontal lines*

6.4 Reissner Variational Principle

Let us represent the free energy density by means of its Young-Fenchel transformation:

$$F\left(\varepsilon_{ij}\right) = \max_{\sigma^{ij}} \left[\sigma^{ij}\varepsilon_{ij} - F^*\left(\sigma^{ij}\right)\right]. \tag{6.26}$$

Reissner variational principle. *The true state of an elastic body is a stationary point of the functional*

$$I\left(\sigma, u\right) = \int_V \left(\sigma^{ij}u_{i,j} - F^*\left(\sigma^{ij}\right)\right)dV - \int_V \rho g_i u^i dV - \int_{\partial V_f} f_i u^i dA \tag{6.27}$$

on the set of all symmetric tensor fields, $\sigma^{ij}(x)$, and the fields,$u_i(x)$, obeying the constraints

$$u_i\left(x\right) = u_i^{(b)} \quad \text{on} \quad \partial V_u.$$

This statement follows from the general duality theory because

$$\min_{u\in(6.12)} I\left(u\right) = \min_{u\in(6.12)} \max_{\sigma} I\left(\sigma, u\right). \tag{6.28}$$

The solution of the minimax problem for the functional (6.27) is a saddle point, i.e. at this point $I\left(\sigma, u\right)$ has maximum with respect to functions σ^{ij} and the minimum with respect to functions u^i. Sometimes this variational principle is also called the mixed variational principle.

6.5 Physically Linear Elastic Body

Assume that there is a stress-free state of the body. We choose the initial state of the body coinciding with the stress-free state. Then,

$$\sigma^{ij} = \frac{\partial F(\varepsilon_{ij})}{\partial \varepsilon_{ij}} = 0 \qquad \text{at} \quad \varepsilon_{ij} = 0.$$

Expanding $F(\varepsilon_{ij})$ over ε_{ij} near the point $\varepsilon_{ij} = 0$ and keeping only the leading terms, we have

$$F(\varepsilon_{ij}) = F(0) + \frac{1}{2}C^{ijkl}\varepsilon_{ij}\varepsilon_{kl}. \tag{6.29}$$

Elastic material with energy (6.29) is called physically linear. For physically linear elastic material, the stresses σ^{ij} depend linearly on strains:

$$\sigma^{ij} = C^{ijkl}\varepsilon_{kl}. \tag{6.30}$$

Since ε_{ij} is a symmetric tensor, the antisymmetric parts of C^{ijkl} with respect to indices ij, kl and with respect to the transposition of the index couples (i, j) and (k, l) do not contribute to energy. Therefore, without loss of generality, one can accept the symmetry properties for the elastic moduli:

$$C^{ijkl} = C^{jikl}, \quad C^{ijkl} = C^{ijlk}, \quad C^{ijkl} = C^{klij}. \tag{6.31}$$

The inverse tensor, i.e. the tensor defined by the equations

$$C^{(-1)}_{ijkl} C^{klmn} = \delta_i^{(m} \delta_j^{n)}, \tag{6.32}$$

is called the tensor of elastic compliances. This tensor, by definition, possesses the same symmetry properties as C^{ijkl} :

$$C^{(-1)}_{ijkl} = C_{jikl}, \quad C^{(-1)}_{ijkl} = C^{(-1)}_{ijlk}, \quad C^{(-1)}_{ijkl} = C^{(-1)}_{klij}. \tag{6.33}$$

Contracting (6.30) with the tensor of elastic compliances, one find strains in terms of stresses:

$$\varepsilon_{ij} = C^{(-1)}_{ijkl} \sigma^{kl}. \tag{6.34}$$

For physically linear material the complementary energy coincides with energy expressed in terms of stresses:

$$F^* \left(\sigma_{ij} \right) = \frac{1}{2} C^{(-1)}_{ijkl} \sigma^{ij} \sigma^{kl}.$$

In the one-dimensional case, this has a simple geometric interpretation: for physically linear material the stress-strain curve is the straight line dividing the rectangle in Fig. 6.1 into two parts with equal areas.

In general, the tensor of elastic moduli depends on the tensors characterizing material symmetry.[1] In the particular case of isotropic material, the only tensor which is invariant with respect to all rotations is the metric tensor g^{ij}. Thus, C^{ijkl} are functions of g^{ij}. The only independent tensors of fourth order which can be constructed from g^{ij} are $g^{ij} g^{kl}$, $g^{ik} g^{jl}$, $g^{il} g^{kj}$ (the index i in g^{ij} may appear only in combination with either with j or k or l). Therefore, the general form of elastic moduli in the isotropic case is

$$C^{ijkl} = \lambda g^{ij} g^{kl} + \mu_1 g^{ik} g^{jl} + \mu_2 g^{il} g^{kj} \tag{6.35}$$

where λ, μ_1 and μ_2 are some scalars. Tensor (6.35) must satisfy the symmetry requirements (6.31). This yields $\mu_1 = \mu_2 \equiv \mu$. Finally,

---

[1] The corresponding theory can be found in [269].

$$C^{ijkl} = \lambda g^{ij} g^{kl} + \mu \left(g^{ik} g^{jl} + g^{il} g^{jk} \right). \tag{6.36}$$

Accordingly,

$$F = \frac{1}{2} \lambda \left(\varepsilon_i^i \right)^2 + \mu \varepsilon_{ij} \varepsilon^{ij}, \tag{6.37}$$

and

$$\sigma_{ij} = \lambda \varepsilon_k^k g_{ij} + 2\mu \varepsilon_{ij}. \tag{6.38}$$

The constants, λ and μ, are called Lame's constants. Resolving (6.38) with respect to ε_{ij}, we have

$$\sigma_k^k = (3\lambda + 2\mu) \varepsilon_k^k, \quad 2\mu \varepsilon_{ij} = \sigma_{ij} - \frac{\lambda}{3\lambda + 2\mu} \sigma_k^k g_{ij}. \tag{6.39}$$

The constant that appeared here, $3\lambda + 2\mu$, with the factor $\frac{1}{3}$ is called bulk modulus, K:

$$K = \frac{1}{3} (3\lambda + 2\mu).$$

Using (6.39), we find the complementary energy

$$\begin{aligned} F^* \left(\sigma^{ij} \right) &= \frac{1}{2} \lambda \frac{\left(\sigma_k^k \right)^2}{(3\lambda + 2\mu)^2} + \frac{1}{4\mu} \left(\sigma_{ij} - \frac{\lambda}{3\lambda + 2\mu} g_{ij} \sigma_k^k \right) \left(\sigma^{ij} - \frac{\lambda}{3\lambda + 2\mu} g^{ij} \sigma_k^k \right) = \\ &= \frac{1}{4\mu} \sigma_{ij} \sigma^{ij} - \frac{\lambda}{4\mu (3\lambda + 2\mu)} \left(\sigma_k^k \right)^2 \end{aligned} \tag{6.40}$$

or, in terms of the stress deviator, $\sigma'_{ij} = \sigma_{ij} - \frac{1}{3} g_{ij} \sigma_k^k$, and the constants K and μ,

$$F^* \left(g^{ij} \right) = \frac{1}{4\mu} \sigma'_{ij} \sigma'^{ij} + \frac{1}{2K} \left(\frac{1}{3} \sigma_k^k \right)^2. \tag{6.41}$$

As a check, one can differentiate F^* with respect to σ_{ij} to obtain the second equation (6.39):

$$\varepsilon_{ij} = \frac{\partial F^* \left(\sigma^{ij} \right)}{\partial \sigma^{ij}}.$$

The condition of positiveness of energy is conveniently expressed in terms of bulk modulus and shear modulus:

$$K > 0, \quad \mu > 0. \tag{6.42}$$

To derive (6.42) we note that the components of strain deviator

$$\varepsilon'_{ij} = \varepsilon_{ij} - \frac{1}{3} g_{ij} \varepsilon^k_k, \quad \left(\varepsilon'^k_k = 0\right)$$

and the strain trace, ε^k_k, can be changed independently. Replacing ε_{ij} in (6.37) by the sum, $\varepsilon'_{ij} + \frac{1}{3}\varepsilon^k_k g_{ij}$, we have

$$F = \frac{1}{2} K \left(\varepsilon^i_i\right)^2 + \mu \varepsilon'_{ij} \varepsilon'^{ij}.$$

Obviously, $F > 0$ for $\varepsilon_{ij} \neq 0$ if and only if (6.42) holds.

Instead of K, λ and μ, one often uses other two characteristics of isotropic elastic bodies, Young' modulus, E, and Poisson's coefficient, ν:

$$E \equiv \frac{\mu (3\lambda + 2\mu)}{\lambda + \mu} = 2(1+\nu)\mu, \quad \nu \equiv \frac{\lambda}{2(\lambda + \mu)}.$$

In terms of these characteristics the complementary energy is

$$F^* = \frac{1+\nu}{2E} \left[\sigma_{ij} \sigma^{ij} - \frac{\nu}{1+\nu} \left(\sigma^k_k\right)^2 \right].$$

Accordingly,

$$\varepsilon_{ij} = \frac{\partial F^* \left(\sigma^{ij}\right)}{\partial \sigma^{ij}} = \frac{1+\nu}{E} \left(\sigma_{ij} - \frac{\nu}{1+\nu} \sigma^k_k g_{ij} \right).$$

The constant $\lambda + \mu$ is always positive:

$$\lambda + \mu = \frac{3\lambda + 3\mu}{3} = \frac{1}{3} K + \mu > 0.$$

Therefore, $E > 0$. Poisson's coefficient, ν, may be negative, but since $E = 2(1+\nu)\mu > 0$, it is bounded from below:

$$\nu > -1.$$

As follows from its definition, Poisson's coefficient does not exceed $\frac{1}{2}$ and approaches $\frac{1}{2}$ when $\lambda \gg \mu$.

6.6 Castigliano Variational Principle

Castigliano principle. In this section, we construct a variational principle dual to Gibbs principle. The dual variational principle can be derived directly from the general formulation given for the integral functionals (5.142), however we first repeat the general scheme for constructing the dual principles taking into account the distinctive feature of the energy functional, its invariance with respect to rigid motions. This feature has already been incorporated implicitly in the Young-Fenchel transformation (6.26) which used only symmetric tensors σ^{ij}.

As in the general case, commuting the order of calculation of maximum and minimum in (6.28), we obtain

$$\min_{u\in(6.12)} I(u) = \max_{\sigma}\left[J(\sigma) + \min_{u\in(6.12)} \Phi(\sigma, u)\right],$$

$$J(\sigma) = \int_{\partial V_u} \sigma_i^j n_j u^i_{(b)} d\sigma - \int_V F^*\left(\sigma^{ij}\right) dV,$$

$$\Phi(\sigma, u) = -\int_V \left(\sigma^j_{i,j} + \rho g_i\right) u^i dV - \int_{\partial V_f} \left(f_i - \sigma_i^j n_j\right) u^i dA. \tag{6.43}$$

If the tensor field σ^{ij} satisfies the equations

$$\sigma^j_{i,j} + \rho g_i = 0 \quad \text{in } V, \tag{6.44}$$

$$\sigma_i^j n_j = f_i \quad \text{on } \partial V_f, \tag{6.45}$$

then $\Phi(\sigma, u) = 0$. If σ^{ij} does not satisfy these equations, $\min_u \Phi(\sigma, u) = -\infty$. Therefore,

$$\min_{u\in(6.12)} I(u) = \max_{\sigma\in(6.44),(6.45)} J(\sigma).$$

In the general case, one can show that the maximizing field σ^{ij} coincides with the stresses computed for the minimizing displacements. We arrive at the following **Castigliano variational principle.** *The true stresses maximize the functional $J(\sigma)$ on the set of all stress fields σ^{ij} obeying the constraints (6.44) and (6.45).*

Now, let us show how to derive the Castigliano principle from the dual variational principle for the general integral functional (5.142). It follows from the general theory that the dual functional $J(\sigma)$ has the form (6.43); however F^* in (6.43) represents not the function (6.25) but the Young-Fenchel transformation of F with respect to displacements gradients $u_{i,j}$:

$$F_1^* = \max_{u_{i,j}} \left(\sigma^{ij} u_{i,j} - F\left(\varepsilon_{ij}\right)\right). \tag{6.46}$$

Accordingly, in (6.46), and also in (6.44) and (6.45), the tensor σ^{ij} is not necessarily symmetric.

Let us rewrite (6.46), separating symmetric and antisymmetric components of the stresses and displacement gradients:

$$F_1^* = \max_{u_{i,j}} \left[\sigma^{(ij)} \varepsilon_{ij} + \sigma^{[ij]} u_{[i,j]} - F\left(\varepsilon_{ij}\right) \right].$$

Since $F\left(\varepsilon_{ij}\right)$ does not depend on the antisymmetric components of the displacement gradients $u_{[i,j]}$, then $F_1^* = +\infty$ for $\sigma^{[ij]} \neq 0$, and, consequently, if for some subregion of region V, $\sigma^{[ij]} \neq 0$, then the dual functional becomes equal to $-\infty$. Therefore, all non-symmetric stress tensors should be excluded. For $\sigma^{[ij]} = 0$ the function F_1^* coincides with the function (6.25), and the variational principle is equivalent to the general dual variational principle constructed for the functional (5.142).

Castigliano principle for stress functions. The general solution of the equilibrium (6.44) can easily be found. For simplicity, we consider only the case of zero external volume forces and simply-connected region V. We will show that the general solution of the equilibrium equations is

$$\sigma^{ij} = e^{ikl} e^{jmn} \psi_{km,ln}, \tag{6.47}$$

where ψ_{km} is an arbitrary symmetric tensor field.

Indeed, the general solution of the equation

$$A^i_{,i} = 0 \tag{6.48}$$

is

$$A^i = e^{ijk} B_{j,k},$$

where B_k are arbitrary functions of x^i. Functions B_k are not uniquely determined by A^i: adding to B_k an arbitrary potential vector field $\psi_{,k}$ does not change A^i.

Equilibrium equations for an elastic body have the divergence form (6.48), and, therefore their solution is

$$\sigma^{ij} = e^{jkl} B^i_{k,l}. \tag{6.49}$$

Functions $\psi^i_{,k}$ can be added to B^i_k without changing the components of the stress tensor. In particular, one can choose ψ^i to make the trace of tensor B^i_k equal to zero:

$$B^i_i = 0. \tag{6.50}$$

In order for the function (6.49) to be the solution to the equilibrium equations, the symmetry conditions for stresses has to be satisfied as well. Contracting (6.49) with e_{ijs} and using (3.19), we get

$$\sigma^{ij} e_{ijs} = B^i_{s,i} = 0. \tag{6.51}$$

Since (6.51) again has the form (6.48), the tensor B^i_s can be written as

$$B^i_s = e^{imn} \psi_{sm,n}.$$

The gradient of an arbitrary vector φ_s can be added to the tensor ψ_{sm}. Let us choose φ_s in such a way that ψ_{sm} be symmetric, i.e. if ψ_{sm} were not symmetric we add the gradient of the vector field ψ_s to obtain

$$e^{ism} \left(\psi_{sm} + \psi_{s,m} \right) = 0. \tag{6.52}$$

Denote by $\boldsymbol{\psi}$ and $\mathbf{f}$ the vector fields with the components ψ_s and $e^{ism}\psi_{sm}$. Then, in direct vector notation, (6.52) reads

$$\text{curl}\, \boldsymbol{\psi} = -\mathbf{f} \tag{6.53}$$

The vector field $\mathbf{f}$ has zero divergence: from (6.52) and (6.50)

$$\text{div}\, \mathbf{f} = \left(e^{ism} \psi_{sm} \right)_{,i} = e^{ism} \psi_{sm,i} = B^s_s = 0$$

As known from calculus, for any vector field $\mathbf{f}$ with zero divergence, (6.53) is solvable. Hence, one can always choose the tensor ψ_{mn} to be symmetric.

Components of the symmetric tensor ψ_{ij} are called the stress functions.

It is clear that (6.47) has some extra degrees of freedom: on the left hand side of (6.47) there are six functions connected by the three equilibrium equations, i.e. there are three functional degrees of freedom; on the right hand side of (6.47) there are six functional degrees of freedom. Thus, three more constraints can be imposed on ψ_{km}. Of course, these additional constraints should not prevent the functions (6.47) from covering all possible solutions of the equilibrium equations. It is most convenient to select the additional restrictions on the stress functions for each particular problem, using the simplifications related to the specifics of the problem.

Castigliano variational principle for stress functions. *The true stress functions deliver the maximum value to the functional*

$$J(\psi) = \int\limits_{\partial V_u} e^{ikl} e^{jmn} \psi_{km,ln} u_i^{(b)} n_j dA - \\ - \int\limits_V F^* \left(e^{ikl} e^{jmn} \psi_{km,ln} \right) dV \tag{6.54}$$

on the set of all stress functions obeying the boundary conditions

$$e^{jkl} \psi_{km,ls} e^{jms} n_j = f^i \quad \text{on } \partial V_f. \tag{6.55}$$

This variational principle does not determine the stress functions uniquely, but this does not affect the uniqueness of the stress state.

One may wonder how to simplify the energy expression, and, thus, the governing equations by imposing the constraints on the stress functions. Consider an isotropic physically linear homogeneous elastic body with the complementary energy (6.40). According to (3.19),

$$\begin{aligned}
\sigma^{ij}\sigma_{ij} &= e^{ikl}e^{jmn}\psi_{km,ln}e_{ik'l'}e_{jm'n'}\psi^{k'm',l'n'} = \\
&= \left(\delta^k_{k'}\delta^l_{l'} - \delta^k_{l'}\delta^l_{k'}\right)\left(\delta^m_{m'}\delta^n_{n'} - \delta^m_{n'}\delta^n_{m'}\right)\psi_{km,ln}\psi^{k'm',l'n'} = \\
&= \psi_{km,ln}\psi^{km,ln} - 2\psi_{km,ln}\psi^{lm,kn} + \psi_{km,ln}\psi^{ln,km}, \\
\sigma^k_k &= \left(g^{km}g^{ln} - g^{lm}g^{kn}\right)\psi_{km,ln} = \Delta\psi^k_k - \psi^{km}_{\ \ ,km}.
\end{aligned} \tag{6.56}$$

It is seen from (6.56) that by setting the three conditions

$$\psi^{ij}_{\ \ ,j} = 0, \tag{6.57}$$

we make vanish the last term in the second equation (6.56) and put the last two terms in the first equation (6.56) into divergence form:

$$\begin{aligned}
&-2\psi_{km,ln}\psi^{lm,kn} + \psi_{km,ln}\psi^{ln,km} = \\
&= -2\frac{\partial}{\partial x^l}\left(\psi_{km,n}\psi^{lm,kn}\right) + \frac{\partial}{\partial x^n}\left(\psi_{km,l}\psi^{ln,km}\right)
\end{aligned}$$

Thus, they do not affect the equations for ψ^{ij}.

Up to the divergence terms, the free energy becomes

$$\int_V F^* dV = \int_V \frac{1}{4\mu}\left(\psi_{km,ln}\psi^{km,ln} - \frac{\nu}{1+\nu}\left(\Delta\psi^k_k\right)^2\right)dV,$$

with Δ being Laplace's operator. Hence, Euler equations for the stress functions are

$$\Delta^2\psi^{ij} - \frac{\nu}{1+\nu}\Delta^2\psi^k_k g^{ij} = 0.$$

Note that the divergence terms may contribute to the boundary conditions.

Castigliano variational principle and compatibility of strains. Euler equations for the functional (6.54) read: in region V,

$$e^{ikl}e^{jmn}\partial_l\partial_n\frac{\partial F^*}{\partial\sigma^{ij}} = 0. \tag{6.58}$$

Since the derivatives of F^* are the components of strain tensor

$$\frac{\partial F^*}{\partial \sigma^{ij}} = \varepsilon_{ij},$$

(6.58) can be also written as

$$e^{ikl} e^{jmn} \partial_l \partial_n \varepsilon_{ij} = 0. \tag{6.59}$$

It is proved in differential geometry that these equations are the necessary and sufficient conditions for the existence of a vector field, $u_i(x)$, such that

$$\varepsilon_{ij} = \frac{1}{2}(u_{i,j} + u_{j,i}). \tag{6.60}$$

In other words, (6.59) gives the conditions of the existence of a displacement vector field such that the strain in the deformed state is generated by displacement from some undeformed state. One says that such strain is compatible.

On the other hand, stresses can be viewed as reactions to the compatibility constraints (6.60). Indeed, one can consider the strain in Gibbs variational principle as an arbitrary field such that (6.60) holds. Then the governing equations can be obtained by introducing Lagrange multipliers for the constraints (6.60), σ^{ij}, i.e. by adding to the energy functional the integral

$$\int_V \sigma^{ij} \left(\frac{1}{2}(u_{i,j} + u_{j,i}) - \varepsilon_{ij} \right) dV.$$

The reader can check that the Lagrange multipliers σ^{ij} have the meaning of stress components. If, instead of (6.60), the compatibility conditions are taken in the form (6.59), then the corresponding Lagrange multipliers have the meaning of stress functions.

Two-dimensional plane problems. Castigliano principle for stress functions takes a considerably simpler form in two-dimensional plane problems because in this case only one stress function is needed.

Let Ω be a bounded simply connected region in a two-dimensional plane, and x, y be some Cartesian coordinates in the plane. The boundary $\partial\Omega$ of the region Ω consists of two parts, $\partial\Omega_u$ and $\partial\Omega_f$; the displacement components, u_x and u_y, are prescribed on $\partial\Omega_u$:

$$u_x = u_x^{(b)}(s), \qquad u_y = u_y^{(b)}(s) \qquad \text{on} \qquad \partial\Omega_u,$$

s being the arc length along $\partial\Omega$, while at $\partial\Omega_f$ the external force $\{f_x(s), f_y(s)\}$ is known.

Then Castigliano principle states:

Castigliano variational principle. *The true stresses give the maximum value to the functional,*

$$J(\sigma) = \int_{\partial\Omega_f} \left[u_x^{(b)}(s)\left(\sigma_{xx}n_x + \sigma_{xy}n_y\right) + u_y^{(b)}(s)\left(\sigma_{xy}n_x + \sigma_{yy}n_y\right)\right] ds - \\ - \int_{\Omega} F^*\left(\sigma_{xx}, \sigma_{xy}, \sigma_{yy}\right) dxdy, \tag{6.61}$$

on the set of all functions σ_{xx}, $\sigma_{xy} = \sigma_{yx}$, *and* σ_{yy}, *which satisfy the constraints*

$$\frac{\partial \sigma_{xx}}{\partial x} + \frac{\partial \sigma_{xy}}{\partial y} = 0, \quad \frac{\partial \sigma_{yx}}{\partial x} + \frac{\partial \sigma_{yy}}{\partial y} = 0 \quad \text{in } \Omega, \tag{6.62}$$

$$\sigma_{xx}n_x + \sigma_{xy}n_y = f_x, \quad \sigma_{yx}n_x + \sigma_{yy}n_y = f_y \quad \text{on } \partial\Omega_f. \tag{6.63}$$

Equations (6.62) mean that there exist functions Φ_1 and Φ_2 such that

$$\sigma_{xx} = \frac{\partial \Phi_1}{\partial x}, \quad \sigma_{xy} = -\frac{\partial \Phi_1}{\partial x}, \quad \sigma_{yx} = \frac{\partial \Phi_2}{\partial y}, \quad \sigma_{yy} = -\frac{\partial \Phi_2}{\partial x}.$$

Since $\sigma_{xy} = \sigma_{yx}$,

$$\frac{\partial \Phi_1}{\partial x} + \frac{\partial \Phi_2}{\partial y} = 0.$$

Therefore, there exists a function ψ such that

$$\Phi_1 = \frac{\partial \psi}{\partial y}, \quad \Phi_2 = -\frac{\partial \psi}{\partial x}.$$

Thus, the general solution of the equilibrium equations is

$$\sigma_{xx} = \frac{\partial^2 \psi}{\partial y^2}, \quad \sigma_{xy} = -\frac{\partial^2 \psi}{\partial x \partial y}, \quad \sigma_{yy} = \frac{\partial^2 \psi}{\partial x^2}. \tag{6.64}$$

The function ψ is called the Airy function. If the components of the stress tensor are known, function ψ can be determined up to a linear function of x and y. The arbitrariness can be removed by setting

$$\psi = \frac{\partial \psi}{\partial x} = \frac{\partial \psi}{\partial y} = 0 \tag{6.65}$$

at some point C on $\partial\Omega_f$. It is convenient to measure the arc length on $\partial\Omega$ starting from the point C. The arc length is assumed to increase in a counter-clockwise direction.

The boundary conditions (6.63) take a simple form in terms of the Airy function. They are obtained by substitution (6.64) in (6.63) and taking into account (5.128) and (5.129):

$$\sigma_{xx}n_x + \sigma_{xy}n_y = \tau_y \frac{\partial^2 \psi}{\partial y^2} + \tau_x \frac{\partial^2 \psi}{\partial x \partial y} = \frac{d}{ds}\frac{\partial \psi}{\partial y} = f_x(s),$$

$$\sigma_{yx}n_x + \sigma_{yy}n_y = -\tau_y \frac{\partial^2 \psi}{\partial x \partial y} - \tau_x \frac{\partial^2 \psi}{\partial x^2} = -\frac{d}{ds}\frac{\partial \psi}{\partial x} = f_y(s). \tag{6.66}$$

Consequently, the boundary conditions define the value of derivatives of the stress functions on the contour $\partial\Omega_f$:

$$\frac{\partial \psi}{\partial y} = \int_0^s f_x(s)ds, \quad \frac{\partial \psi}{\partial x} = -\int_0^s f_y(s)ds. \tag{6.67}$$

The boundary conditions (6.67) are equivalent to prescribing at $\partial\Omega$ the function ψ and its normal derivative:

$$\frac{\partial \psi}{\partial n} \equiv n_x \frac{\partial \psi}{\partial x} + n_y \frac{\partial \psi}{\partial y}.$$

If the surface forces are prescribed on the entire boundary, then the boundary conditions (6.67) are also reduced to prescribing at the boundary function ψ and its normal derivative. To justify this statement we have to check that the boundary conditions (6.67) written for the entire boundary define a single-valued function ψ. Indeed, the necessary condition for the correctness of the variational problem in this case is vanishing of the total force and the total moment:

$$\int_0^l f_x(s)ds = 0, \quad \int_0^l f_y(s)ds = 0, \quad \int_0^l \left(xf_y(s) - yf_x(s)\right)ds = 0. \tag{6.68}$$

Here, l is the length of the contour $\partial\Omega$. The first two equations (6.68) mean that formulas (6.67) define on the contour $\partial\Omega$ two single-valued functions $\partial\psi/\partial x$ and $\partial\psi/\partial y$. Consequently, $\partial\psi/\partial n$ and $\partial\psi/\partial s \equiv \tau_x \partial\psi/\partial x + \tau_y \partial\psi/\partial y$ are also uniquely defined. Let us find the boundary value of the Airy function, $\psi(s)$. We have

$$\psi(s) = \int_0^s \frac{d\psi}{ds}ds = \int_0^s \left(\frac{\partial \psi}{\partial x}\frac{dx}{ds} + \frac{\partial \psi}{\partial y}\frac{dy}{ds}\right)ds = \int_0^s \left(\frac{\partial \psi}{\partial x}dx + \frac{\partial \psi}{\partial y}dy\right) =$$

$$= \left(x\frac{\partial \psi}{\partial x} + y\frac{\partial \psi}{\partial y}\right)\Bigg|_0^s - \int_0^s \left(x\frac{d}{ds}\frac{\partial \psi}{\partial x} + y\frac{d}{ds}\frac{\partial \psi}{\partial y}\right)ds.$$

Using (6.66), we obtain

$$\psi(s) = \left(x\frac{\partial \psi}{\partial x} + y\frac{\partial \psi}{\partial y} \right) \Big|_s + \int_0^s (xf_y - yf_x)ds. \tag{6.69}$$

Here, we take into account that ψ and derivatives of ψ are zero at $s = 0$. Due to (6.68), formula (6.69) defines a single-valued function $\psi(s)$ on $\partial\Omega$.

Substituting (6.64) into the functional (6.61), we get the following **Castigliano variational principle for stress functions.** *The maximum of the functional*

$$\int_{\partial\Omega_u} \left(u_x^{(b)}(s)\frac{d}{ds}\frac{\partial \psi}{\partial y} - u_y^{(b)}(s)\frac{d}{ds}\frac{\partial \psi}{\partial x} \right) ds - \int_{\Omega} F^* \left(\frac{\partial^2 \psi}{\partial y^2}, -\frac{\partial^2 \psi}{\partial x \partial y}, \frac{\partial^2 \psi}{\partial x^2} \right) dxdy$$

on the set of all functions ψ, satisfying the boundary conditions

$$\psi = f_1(s), \quad \frac{d\psi}{dn} = f_2(s) \quad \text{on } \partial\Omega_f$$

is achieved at the true stress function.

Functions $f_1(s)$ and $f_2(s)$ are calculated from $f_x(s)$ and $f_y(s)$ by means of (6.67) and (6.69).

Anti-plane problems. The problems in which the component of displacements in a Cartesian coordinates, u^3, is a function of $x^1 = x$ and $x^2 = y$, while two other components are zero, are called anti-plane problems. In this case, the body is a cylinder with a cross-section Ω, $\{x, y\} \in \Omega$. In anti-plane deformation, only two components of strain are non-zero:

$$2\varepsilon_{13} = \frac{\partial u}{\partial x}, \qquad 2\varepsilon_{23} = \frac{\partial u}{\partial y}, \qquad u^3 \equiv u(x, y).$$

Gibbs variational principle. *The true displacement field provides minimum to the functional*

$$I(u) = \int_{\Omega} F\left(\frac{\partial u}{\partial x}, \frac{\partial u}{\partial y} \right) dxdy - \int_{\partial\Omega_f} fu\, ds$$

on the set of all $u(x, y)$ obeying the condition

$$u = u^{(b)}(s) \qquad \text{at } \partial\Omega_u.$$

Following the general scheme, one obtains the dual variational principle.

Dual variational principle. *The true stresses provide minimum to the functional*

$$\int_{\Omega} F^*(\sigma_x, \sigma_y) dxdy - \int_{\partial\Omega_u} (\sigma_x n_x + \sigma_y n_y) u^{(b)}(s) ds$$

on the set of all fields, σ_x, σ_y, obeying the constraints

$$\frac{\partial \sigma_x}{\partial x} + \frac{\partial \sigma_y}{\partial y} = 0 \qquad \text{in } \Omega, \qquad \sigma_x n_x + \sigma_y n_y = f(s) \qquad \text{on } \partial\Omega_f.$$

In anti-plane problems, the stress function, ψ, is introduced by the relations

$$\sigma_x = \frac{\partial \psi}{\partial y}, \qquad \sigma_y = -\frac{\partial \psi}{\partial x}.$$

The dual variational principle in terms of the stress function takes the form:

Dual variational principle. *The true stress function provides minimum to the functional*

$$\int_{\Omega} F^* \left(\frac{\partial \psi}{\partial y}, -\frac{\partial \psi}{\partial x} \right) dxdy - \int_{\partial\Omega_u} \frac{d\psi}{ds} u^{(b)}(s) ds$$

on the set of all functions, $\psi(x, y)$, obeying the boundary condition

$$\frac{d\psi}{ds} = f(s) \qquad \text{on } \partial\Omega_f.$$

In the case when the displacements are given at all points of the boundary, minimum is taken over all functions, $\psi(x, y)$.

In elasticity theory, there are other boundary value problems of interest besides those already considered. For example, on a part Σ of the boundary, only one component of displacement, u^1 can be prescribed and the "complementary" components of external forces, f_2 and f_3, are also given. Then the energy functional of Gibbs principle will include the integral over Σ of $f_2 u_2 + f_3 u_3$. The corresponding changes in the Castigliano and the Reissner principles are straightforward.

6.7 Hashin-Strikman Variational Principle

In this section we formulate Hashin-Strikman variational principle for a linear inhomogeneous body. The free energy density of the body is

$$F(\varepsilon_{ij}) = \frac{1}{2} C^{ijkl}(x) \varepsilon_{ij} \varepsilon_{kl}.$$

If the displacements are prescribed on the boundary

$$u^i = u^i_{(b)} \quad \text{on } \partial V, \tag{6.70}$$

then the true displacement field minimizes the functional

$$I(u) = \int_V \frac{1}{2} C^{ijkl} \varepsilon_{ij} \varepsilon_{kl} dV \tag{6.71}$$

on the set of displacements with the prescribed boundary values (6.70). We select a homogeneous elastic body with the free energy

$$F_0 = \frac{1}{2} C_0^{ijkl} \varepsilon_{ij} \varepsilon_{kl}. \tag{6.72}$$

It is convenient to choose this body isotropic, i.e.

$$C_0^{ijkl} = \lambda_0 g^{ij} g^{kl} + \mu_0 \left(g^{ik} g^{jl} + g^{il} g^{jk} \right). \tag{6.73}$$

Suppose that the quadratic form, $F - F_0$, is positive definite:

$$F - F_0 = \frac{1}{2} \left(C^{ijkl} - \frac{1}{2} C_0^{ijkl} \right) \varepsilon_{ij} \varepsilon_{kl} > 0 \ \text{ for } \varepsilon_{ij} \neq 0.$$

Then $F - F_0$ can be presented by means of Young-Fenchel transformation:

$$F - F_0 = \max_{p^{ij}} \left[p^{ij} \varepsilon_{ij} - \frac{1}{2} H_{ijkl} p^{ij} p^{kl} \right], \tag{6.74}$$

where H_{ijkl} is the inverse tensor:

$$H_{ijkl} \left(C^{klmn} - C_0^{klmn} \right) = \delta_i^{(m} \delta_j^{n)}. \tag{6.75}$$

Repeating the line of reasoning of Sect. 5.9, we have:

$$\begin{aligned} I(u) &= \int_V \left[\frac{1}{2} C_0^{ijkl} \varepsilon_{ij} \varepsilon_{kl} + \frac{1}{2} \left(C^{ijkl} - C_0^{ijkl} \right) \varepsilon_{ij} \varepsilon_{kl} \right] dV = \\ &= \max_{p^{ij}} \int_V \left[\frac{1}{2} C_0^{ijkl} \varepsilon_{ij} \varepsilon_{kl} + p^{ij} \varepsilon_{ij} - \frac{1}{2} H_{ijkl} p^{ij} p^{kl} \right] dV. \end{aligned}$$

Therefore,

$$\begin{aligned} \check{I} = \min_{u \in (6.70)} I(u) = \min_{u \in (6.70)} \max_{p^{ij}(x)} \int_V \Big[\frac{1}{2} C_0^{ijkl} \varepsilon_{ij} \varepsilon_{kl} + p^{ij} \varepsilon_{ij} - \\ - \frac{1}{2} H_{ijkl}(x) p^{ij} p^{kl} \Big] dV. \end{aligned}$$

Changing the order of minimum and maximum, we have

$$\check{I} = \max_{p^{ij}(x)} \left[\tilde{J}(p) - \int_V \frac{1}{2} H_{ijkl}(x) p^{ij} p^{kl} dV \right] \tag{6.76}$$

where

$$\tilde{J}(p) = \min_{u \in (6.70)} \int_V \left(\frac{1}{2} C_0^{ijkl} \varepsilon_{ij} \varepsilon_{kl} + p^{ij}(x) \varepsilon_{ij} \right) dV. \tag{6.77}$$

Let $\mathring{u}^i$ be the minimizer in the following variational problem for the homogeneous body:

$$E_0 = \min_{u \in (6.70)} \int_V \frac{1}{2} C_0^{ijkl} \varepsilon_{ij} \varepsilon_{kl} dV.$$

We present displacements in (6.77) as the sum

$$u^i = \mathring{u}^i + u'^i, \ u'^i \mid_{\partial V} = 0. \tag{6.78}$$

Then

$$\tilde{J}(p) = E_0 + \int_V p^{ij} \mathring{\varepsilon}_{ij} dV + J(p), \tag{6.79}$$

$$J(p) = \min_{u' \in (6.78)} \int_V \left(\frac{1}{2} C_0^{ijkl} \varepsilon'_{ij} \varepsilon'_{kl} + p^{ij} \varepsilon'_{ij} \right) dV, \quad \varepsilon'_{ij} \equiv u'_{(i,j)}. \tag{6.80}$$

The term

$$\int C_0^{ijkl} \mathring{\varepsilon}_{ij} \varepsilon'_{ij} dV,$$

vanished due to the Euler equations for the minimizer, $\mathring{u}_i$, and the zero boundary conditions for u'_i. Plugging (6.79) in (6.76), we obtain the following **Hashin-Strikman variational principle.** *The true energy of the body, $\check{I}$, can be computed from the variational problem*

$$\check{I} - E_0 = \max_{p^{ij}(x)} \left[\int_V \left(p^{ij}(x) \mathring{\varepsilon}_{ij} - \frac{1}{2} H_{ijkl}(x) p^{ij}(x) p^{kl}(x) \right) dV + J(p) \right]. \tag{6.81}$$

The rest of this section is concerned with determining an explicit form of $J(p)$. To find an explicit dependence of $J(p)$ on $p^{ij}(x)$ one has to solve the boundary value problem[2]

[2] Writing (6.82) we used the symmetry of C_0^{mnrs} with respect to indices r, s, and therefore $C_0^{mnrs} \varepsilon_{rs} = C_0^{mnrs} \partial_r u_s$.

$$\partial_n \left(C_0^{mnrs} \partial_r u_s \right) = -\partial_n p^{mn}(x), \quad u_s \mid_{\partial V} = 0. \tag{6.82}$$

Here $p^{mn}(x)$ are assumed to be some smooth functions. If $p^{mn}(x)$ are piece-wise smooth and have a jump on some surface, S, while $u_s(x)$ remain continuous on S, then additionally,

$$\left[C_0^{mkrs} u_{r,s} + p^{mk} \right] n_k = 0 \quad \text{on } S. \tag{6.83}$$

As usual, $[\varphi]$ denotes the jump of φ on S. To get the solution of the boundary value problem (6.82) and (6.83) in an explicit form, we have to employ, as in the scalar case considered in Sect. 5.9, Green's function, which, for elasticity problems, is a tensor.

Green's tensor. Consider the variational problem

$$\int_V \left(\frac{1}{2} C_0^{mnrs} \partial_m u_n \partial_r u_s - u_s f^s(x) \right) dV \to \min_{u_s : u_s = 0 \text{ at } \partial V}. \tag{6.84}$$

Its minimizer obeys the boundary value problem

$$\partial_n C_0^{mnrs} \partial_r \check{u}_s = -f^m(x), \quad \check{u}_s \mid_{\partial V} = 0. \tag{6.85}$$

Let $f^m(x)$ be sufficiently smooth, e.g., piece-wise continuous. If we discretize (6.85), they become a system of linear algebraic equations. Solving this system, we find the value of $\check{u}_s$ at each point, x, as a linear function of the values of f^m at all points of the discretization grid. In continuum limit, $\check{u}_s$ becomes a linear functional of f^m of the form

$$\check{u}_s(x) = \int_V G_{sm}(x, x') f^m(x') dV'. \tag{6.86}$$

The kernel, $G_{sm}(x, x')$, is called Green's tensor. The operator of the linear problem (6.85) is symmetric, therefore its inversion, the operator (6.86), is also symmetric, i.e.

$$G_{sm}(x, x') = G_{ms}(x', x). \tag{6.87}$$

As follows from (6.85) and (6.86), Green's tensor is a solution of the boundary value problem,

$$\partial_n C_0^{mnrs} \partial_r G_{st}(x, x') = -\delta_t^m \delta(x - x'), \tag{6.88}$$

$$G_{st}(x, x') = 0 \text{ if } x \in \partial V. \tag{6.89}$$

By $\delta(x - x')$ we mean the delta-function in three-dimensional space:

$$\delta(x - x') = \delta_1(x_1 - x_1')\delta_1(x_2 - x_2')\delta_1(x_3 - x_3'),$$

where $\delta_1(\xi)$ is the usual "one-dimensional" delta-function of one variable, and x_i, x_i' the components of the vectors, x and x', respectively.

The δ-function in the right hand side of (6.88) causes Green's tensor to have a singularity at $x = x'$. To find the character of this singularity we solve (6.88) in the case of unbounded region.

Green's tensor for unbounded region. We are going to show that Green's tensor for unbounded region is

$$G_{st}(x, x') = \frac{1}{16\pi\mu_0(1-\nu_0)\,|x-x'|}\left((3-4\nu_0)\,g_{st} + n_s n_t\right) \tag{6.90}$$

where μ_0 and ν_0 are the shear modulus and Poisson's coefficient of the homogeneous elastic body, and n_s the unit vector:

$$n_s = \frac{x_s - x_s'}{|x-x'|}. \tag{6.91}$$

For a bounded region, Green's tensor is a sum of the tensor (6.90) and a tensor that is bounded as $x \to x'$.

Green's tensor for unbounded region can be found by applying to (6.88) Fourier transformation. First, we recall the basic features of Fourier transformations.

Fourier transformations. By Fourier transformations of a function, $u(x)$, one means the function of a "dual variable," k, defined by the relation

$$u(k) = \int u(x) e^{-ik\cdot x} d^3x. \tag{6.92}$$

Here $k \cdot x$ means the scalar product of two three-dimensional vectors, k and x, and the integration is conducted over the entire three-dimensional space. As is often done in physical literature, we keep for Fourier transformation of function, $u(x)$, the same notation, u, but with the other argument, k. This emphasizes that $u(x)$ and $u(k)$ are different presentations of the same function. Both $u(x)$ and $u(k)$ can be complex-valued. Integral (6.92) is converging if $u(x)$ decays at infinity fast enough. In this case, it can be shown that formula (6.92) can be inverted and yields

$$u(x) = \frac{1}{(2\pi)^3}\int u(k) e^{ik\cdot x} d^3k. \tag{6.93}$$

Besides, for any two functionals, $u(x)$ and $v(x)$, which decay fast enough at infinity, an important Parseval's equation holds:

$$\int u(x) v^*(x) d^3x = \frac{1}{(2\pi)^3}\int u(k) v^*(k) d^3k. \tag{6.94}$$

We are going to apply Fourier transformation not only to fast decaying functions but also to the generalized functions, like $\delta(x-x')$, in (6.88). This is not an elementary step, and we just mention here that Fourier transformation can be extended to the generalized functions. In particular, for the delta-function one has to use the following formal rules:

$$\delta(k) = \int \delta(x) e^{-ik\cdot x} d^3x = 1, \tag{6.95}$$

$$\delta(x) = \frac{1}{(2\pi)^3} \int e^{ik\cdot x} d^3k. \tag{6.96}$$

Parseval's equality (6.94) holds not only for fast decaying functions, but also for functions u and v, one of which is generalized, and another one is "usual" (fast decaying). Note that (6.93) and (6.94) can easily be obtained using (6.95) and (6.96). For example, multiplying (6.92) by $e^{ik\cdot x}$ and integrating over k we have

$$\begin{aligned}\int u(k) e^{ik\cdot x} d^3k &= \int u(x') e^{-ik\cdot x'} d^3x' e^{ik\cdot x} d^3k = \int u(x') \int e^{ik\cdot(x-x')} d^3k d^3x' \\ &= \int u(x')(2\pi)^3 \delta(x-x') d^3x' = (2\pi)^3 u(x).\end{aligned}$$

For a fast decaying function, $u(x)$, Fourier transformation of its derivatives

$$u_s(k) = \int \frac{\partial u(x)}{\partial x^s} e^{-ik\cdot x} d^3x$$

is computed in terms of $u(k)$ by integration by parts:

$$u_s(k) = -\int u(x) \frac{\partial}{\partial x^s} e^{-ik\cdot x} d^3x = ik_s \int u(x) e^{-ik\cdot x} d^3x = ik_s u(k).$$

The same formula remains true for generalized functions.

Derivation of (6.90). Denote by $G_{st}(k, k')$ the Fourier transformation of Green's tensor:

$$G_{st}(k, k') = \int G_{st}(x, x') e^{-i(k\cdot x - k'\cdot x')} d^3x d^3x'. \tag{6.97}$$

The sign at k' is chosen for convenience. Due to the symmetry of Green's tensor (6.87), its Fourier transformation possesses a symmetry:

$$\begin{aligned}G_{st}(k, k') &= \int G_{st}(x, x') e^{-i(k\cdot x' - k'\cdot x)} d^3x d^3x' \\ &= \int G_{ts}(x, x') e^{-i(k'\cdot x - k\cdot x')} d^3x d^3x' = G_{ts}(-k', -k). \end{aligned} \tag{6.98}$$

Multiplying (6.97) by $e^{i(k\cdot x-k'\cdot x')}$, integrating over k and k' and using (6.96) we obtain the inversion of (6.97):

$$\int G_{st}(k,k')e^{i(k\cdot x-k'\cdot x')}d^3kd^3k' =$$

$$= \int G_{st}(\tilde{x},\tilde{x}')e^{-i(k\cdot\tilde{x}-k'\cdot\tilde{x}')}d^3\tilde{x}d^3\tilde{x}'e^{i(k\cdot x-k'\cdot x')}d^3kd^3k' =$$

$$= \int G_{st}(\tilde{x},\tilde{x}')e^{ik\cdot(x-\tilde{x})+ik'\cdot(\tilde{x}'-x')}d^3kd^3k'd^3\tilde{x}d^3\tilde{x}' =$$

$$(2\pi)^6\int G_{st}(\tilde{x},\tilde{x}')\delta(x-\tilde{x})\delta(x'-\tilde{x}')d^3\tilde{x}d^3\tilde{x}' = (2\pi)^6\, G_{st}(x,x') \tag{6.99}$$

To find $G_{st}(k,k')$ we multiply both sides of (6.88) by $e^{i(k\cdot x-k'\cdot x')}$ and integrate over x and x'. Since

$$\int \delta(x'-\tilde{x}')e^{i(k\cdot x-k'\cdot x')}d^3xd^3x' = \int e^{-i(k-k')\cdot x}d^3x = (2\pi)^3\,\delta(k-k')$$

and

$$\int \partial_n\partial_r G_{st}\left(x,x'\right)e^{-i(k\cdot x-k'\cdot x')}d^3xd^3x' =$$

$$= \int G_{st}\left(x,x'\right)\partial_n\partial_r e^{-i(k\cdot x-k'\cdot x')}d^3xd^3x' = -k_nk_rG_{st}\left(k,k'\right),$$

we obtain for $G_{st}\left(k,k'\right)$ a system of linear equations:

$$C_0^{mnrs}k_rk_nG_{st}\left(k,k'\right) = (2\pi)^3\,\delta_t^m\delta\left(k-k'\right). \tag{6.100}$$

In the case of an isotropic body, when C_0^{mnrs} are given by (6.73), these equations take the form

$$(\lambda_0+\mu_0)\,k^mk^sG_{st}\left(k,k'\right)+\mu_0\,|k|^2\,G_t^m(k,k') = (2\pi)^3\,\delta_t^m\delta\left(k-k'\right). \tag{6.101}$$

Here

$$|k|^2 \equiv k^sk_s.$$

Contracting (6.101) with k_m, we get

$$(\lambda_0+2\mu_0)\,k^sG_{st}\left(k,k'\right) = (2\pi)^3\,\frac{k_t}{|k|^2}\delta\left(k-k'\right).$$

Plugging this result back in (6.100) we finally obtain

$$G_t^m(k,k') = \frac{(2\pi)^3}{\mu_0 |k|^2}\left(\delta_t^s - \frac{\lambda_0+\mu_0}{\lambda_0+2\mu_0}\frac{k^s k_t}{|k|^2}\right)\delta\left(k-k'\right). \tag{6.102}$$

Obviously this tensor obeys the symmetry condition (6.98). Green's tensor in x-coordinates is obtained from (6.99):

$$G_{st}\left(x,x'\right) = \frac{1}{(2\pi)^3}\int \frac{1}{\mu_0 |k|^2}\left(g_{st} - \frac{\lambda_0+\mu_0}{\lambda_0+2\mu_0}\frac{k^s k_t}{|k|^2}\right) e^{ik\cdot(x-x')}d^3k. \tag{6.103}$$

The integral here can be computed explicitly. We have

$$\int \frac{1}{|k|^2}e^{ik\cdot\tau}d^3k = \int_0^\infty\int_S e^{i(\nu\cdot n)\rho|\tau|}d\rho d^2\nu, \quad \rho\equiv|k|.$$

By ν and n we denote the unit vectors: $\nu = k/|k|$, $n = \tau/|\tau|$, and by $d^2\nu$ the area element on the unit sphere, S. Changing ρ by $\rho/|\tau|$ we obtain

$$\int \frac{1}{|k|^2}e^{ik\cdot r}d^3k = \frac{c}{|\tau|}$$

where c is a constant,

$$c = \int_0^\infty\int_S e^{i(\nu\cdot n)\rho}d\rho d^2\nu.$$

Choosing n to be the north pole of the unit sphere, S, and θ, φ to be the spherical coordinates, we can write

$$\int e^{i(\nu\cdot n)\rho}d^2\nu = \int_0^\pi\int_0^{2\pi} e^{i\rho\cos\theta}d\theta d\varphi = -2\pi\int_0^\pi e^{i\rho\cos\theta}d\cos\theta = 4\pi\frac{\sin\rho}{\rho}.$$

Since[3]

$$\int_0^\infty \frac{\sin\rho t}{\rho}d\rho = \frac{\pi}{2}\operatorname{sgn} t, \tag{6.104}$$

we obtain

$$c = \int_0^\infty\int_S e^{i(\nu\cdot n)\rho}d\rho d^2\nu = 2\pi^2. \tag{6.105}$$

So,

[3] $\operatorname{sgn} t = 1$ for $t > 0$, $\operatorname{sgn} t = 0$ for $t = 0$, $\operatorname{sgn} t = -1$, for $t < 0$.

$$\int \frac{1}{|k|^2} e^{ik\cdot\tau} d^3k = \frac{2\pi^2}{|\tau|}. \tag{6.106}$$

Comparing (6.106) with (6.93), we conclude that the Fourier transformation of $1/4\pi\ |x|$ is $1/\ |k|^2$.

To find the integral

$$\int \frac{1}{|k|^4} k_s k_t e^{ik\cdot\tau} d^3k = \int_0^\infty \int_S \nu_s \nu_t e^{i(\nu\cdot n)\rho|\tau|} d\rho d^2\nu,$$

we change the variable, $\rho \to \rho/\ |\tau|$, to obtain

$$\int \frac{1}{|k|^4} k_s k_t e^{ik\cdot\tau} d^3k = \frac{1}{|\tau|} N_{st} \tag{6.107}$$

where the notation is introduced

$$N_{st} = \int_0^\infty \int_S \nu_s \nu_t e^{i(\nu\cdot n)\rho} d\rho d^2\nu.$$

The tensor, N_{st}, is a function of the unit vector, n_s. A general form of such dependence is

$$N_{st} = \alpha g_{st} + \beta n_s n_t. \tag{6.108}$$

To determine the values of the constants α and β we contract (6.108) with g^{st} and $n^s n^t$ to obtain a system of two linear equations with respect to α and β:

$$3\alpha + \beta = N_s^s = \int_0^\infty \int_S e^{i(\nu\cdot n)\rho} d\rho d^2\nu \tag{6.109}$$

$$\alpha + \beta = N_{st} n^s n^t = \int_0^\infty \int_S (\nu\cdot n)^2 e^{i(\nu\cdot n)\rho} d\rho d^2\nu. \tag{6.110}$$

According to (6.105), the right hand side in (6.109) is $2\pi^2$. The right hand side in (6.110) is zero. Indeed,

$$\int_0^\infty \int_S (\nu\cdot n)^2 e^{i(\nu\cdot n)\rho} d\rho d^2\nu = 2\pi \int_0^\infty \int_0^\pi \cos^2\theta e^{i\rho\cos\theta} \sin\theta d\theta d\rho. \tag{6.111}$$

Changing the variable of integration, $\theta \to t$, $t = \cos\theta$, we have for the integral over θ,

$$\int_0^\pi \cos^2\theta e^{i\rho\cos\theta}\sin\theta d\theta = \int_{-1}^1 t^2 e^{i\rho t}dt =$$

$$\int_{-1}^1 t^2\cos\rho t dt = \frac{1}{\rho}\int_{-1}^1 t^2 d\sin\rho t = 2\frac{\sin\rho}{\rho} - 2\int_{-1}^1 \frac{\sin\rho t}{\rho}t dt. \tag{6.112}$$

Integrals over ρ of both terms in the right hand side of (6.112) are equal due to (6.104). Therefore, the integral (6.111) is zero. Then, the solution of (6.109) and (6.110) is $\alpha = -\beta = \pi^2$, and

$$N_{st} = \pi^2(g_{st} - n_s n_t). \tag{6.113}$$

Collecting the results, (6.106), (6.107) and (6.113), and using the relation

$$\frac{\lambda_0+\mu_0}{\lambda_0+2\mu_0} = \frac{1}{2(1-\nu_0)}, \tag{6.114}$$

we arrive at (6.90).

Explicit form of $J(p)$. We are going to show that the functional $J(p)$ (6.80) can be written in terms of Green's tensor, as

$$J(p) = \frac{1}{4}\int_V\int_V \frac{\partial^2 G_{ij}(x,x')}{\partial x^k \partial x'^m}\left(p^{im}(x)p^{jk}(x) + p^{im}(x')p^{jk}(x') - \right.$$
$$\left. -2p^{im}(x)p^{jk}(x')\right) dV dV'. \tag{6.115}$$

The derivation of (6.115) is similar to that of the scalar case given in Sect. 5.9. First, we consider smooth functions, $p^{ij}(x)$. Then, from Clapeyron's theorem (5.47),

$$J(p) = \frac{1}{2}\int_V u_{i,j}p^{ij}dV, \tag{6.116}$$

where u_i is the solution of the boundary value problem (6.82). According to (6.85), this solution can be written as

$$u_i(x) = \int_V G_{ij}(x,x')\frac{\partial p^{jk}(x')}{\partial x'^k}dV'. \tag{6.117}$$

Green's tensor has an integrable singularity, $1/|x-x'|$. Replacing $\partial p^{jk}(x')/\partial x'^k$ by $\partial(p^{jk}(x') - p^{jk}(x))/\partial x'^k$ and integrating by part, we have

$$u_i(x) = -\int_V \frac{\partial G_{ij}(x,x')}{\partial x'^k}\left(p^{jk}(x') - p^{jk}(x)\right)dV'. \tag{6.118}$$

If we differentiate (6.118) over x^m; then the singularity is still absolutely integrable; therefore, the differentiation is possible. We get

$$\frac{\partial u_i(x)}{\partial x^m} = -\int_V \frac{\partial^2 G_{ij}(x,x')}{\partial x^m \partial x'^k} \left(p^{jk}(x') - p^{jk}(x)\right) dV'. \tag{6.119}$$

Derivative with respect to x^m of the last term in (6.118),

$$\int_V \frac{\partial G_{ij}(x,x')}{\partial x'^j} dV' \frac{\partial p^{jk}(x)}{\partial x^m} = \int_{\partial V} G_{ij}(x,x') n_j(x') dA \frac{\partial p^{jk}(x)}{\partial x^m},$$

is equal to zero because $G_{ij}(x,x') = 0$ for $x' \epsilon \partial V$.

From (6.116) and (6.119),

$$J(p) = -\frac{1}{2} \int \int \frac{\partial^2 G_{ij}(x,x')}{\partial x^m \partial x'^k} p^{im}(x) \left(p^{jk}(x') - p^{jk}(x)\right) dV dV'. \tag{6.120}$$

The symmetry of Green's tensor yields a symmetry of its derivatives:

$$\frac{\partial^2 G_{ij}(x,x')}{\partial x^m \partial x'^k} = \frac{\partial^2 G^{ji}(x,x')}{\partial x'^k \partial x^m}. \tag{6.121}$$

Let us change in (6.120) $i \leftrightarrow j$, $m \leftrightarrow k$, $x \leftrightarrow x'$. We obtain

$$J(p) = -\frac{1}{2} \int \int \frac{\partial^2 G_{ji}(x',x)}{\partial x'^k \partial x^m} p^{jk}(x') \left(p^{im}(x) - p^{im}(x')\right) dV dV'. \tag{6.122}$$

Summing up (6.120) and (6.122) and using (6.121) we arrive at (6.115).

Now let $p^{ij}(x)$ have a discontinuity on some surface, S. As in the scalar case of Sect. 5.9, one can show that the minimizer of the variational problem,

$$\int_V \frac{1}{2} C_0^{ijkl} \partial_j u_i \partial_k u_l dV - \int_S \sigma_i u^i dA \rightarrow \min_{u:u=0 \text{ at } \partial V}, \tag{6.123}$$

where minimum is sought over all functions $u^i(x)$ vanishing on ∂V, is given by the formula

$$u_i(x) = \int_S G_{ij}(x,x') \sigma^j(x') dA'.$$

Therefore, the solution of the boundary value problem (6.82) and (6.83) is

$$u_i(x) = \int_V G_{ij}(x,x') \frac{\partial p^{jk}(x')}{\partial x'^k} dV' + \int_S G_{ij}(x,x') \left[p^{jk}\right] n_k dA'.$$

That yields, after integration by parts, (6.118). For any point, x, which does not belong to S, we may differentiate (6.118) to obtain (6.119). After that the derivation of (6.115) proceeds in the same way as for smooth $p^{ij}(x)$.

The case of negative $F - F_0$. If the constants, λ_0 and μ_0, are so large that

$$F_0 - F = \frac{1}{2}\left(C_0^{ijkl} - C^{ijkl}\right)\varepsilon_{ij}\varepsilon_{kl} > 0 \quad \text{for } \varepsilon_{ij} \neq 0,$$

then, denoting by H_{ijkl} the inverse tensor for $C_0^{ijkl} - C^{ijkl}$,

$$H_{ijkl}\left(C_0^{ijkl} - C^{ijkl}\right) = \delta_i^{(m}\delta_j^{n)},$$

we can write

$$F_0 - F = \max_{p^{ij}}\left[p^{ij}\varepsilon_{ij} - \frac{1}{2}H_{ijkl}p^{ij}p^{kl}\right]$$

or

$$F - F_0 = \min_{p^{ij}}\left(\frac{1}{2}H_{ijkl}p^{ij}p^{kl} - p^{ij}\varepsilon_{ij}\right).$$

Then

$$I(u) = \min_{p^{ij}}\int_V\left[\frac{1}{2}C_0^{ijkl}\varepsilon_{ij}\varepsilon_{kl} + \frac{1}{2}H_{ijkl}p^{ij}p^{kl} - p^{ij}\varepsilon_{ij}\right]dV$$

and

$$\check{I} = \min_{u\epsilon(6.70)} I(u) = \min_{p^{ij}}\left[\int_V \frac{1}{2}H_{ijkl}p^{ij}p^{kl} + \tilde{J}(p)\right]$$

where

$$\begin{aligned}\tilde{J}(p) &= \min_{u\epsilon(6.70)}\int_V\left(\frac{1}{2}C_0^{ijkl}\varepsilon_{ij}\varepsilon_{kl} - p^{ij}\varepsilon_{ij}\right)dV = \\ &= E_0 - \int_V p^{ij}\mathring{\varepsilon}_{ij}dV + J(p),\\ J(p) &= \min_{u\epsilon(6.78)}\int_V\left(\frac{1}{2}C_0^{ijkl}\varepsilon'_{ij}\varepsilon'_{kl} - p^{ij}\varepsilon'_{ij}\right)dV.\end{aligned}$$

Formula for $J(p)$ (6.115) remains unchanged, since it is invariant with respect to the substitution, $p^{ij} \to -p^{ij}$. We arrive at the following

Hashin-Strikman variational principle. *The true energy of the body, $\check{I}$, can be computed from the variational problem*

$$\check{I} - E_0 = \min_{p^{ij}(x)} \left[\int_V \left(\frac{1}{2} H_{ijkl} p^{ij} p^{kl} - p^{ij} \overset{\circ}{\varepsilon}_{ij} \right) dV + J(p) \right].$$

The variational principles formulated hold for any choice of the parameters, C_0^{ijkl}.

6.8 Internal Stresses

Elasticity theory considered in the previous sections was based on the assumption that there exists a global stress-free state of the body. The displacements in that theory are the displacements from the stress-free state to the deformed state. There are situations for which such an assumption is not physically adequate, though "elasticity," the way in which energy depends on strains, still holds. Consider, for example, a polycrystal. In general, even for zero tractions at the boundary of the polycrystal, the stresses are not zero: if we cut off a grain and set it free, it deforms into a free-stress state. Moreover, if we cut the polycrystal over all grain boundaries and let each grain to deform to a stress-free state, then, in general, the unstressed grains cannot be put together to form a continuous body without gaps and overlapping. This motivates a modification of the previous theory, which we consider first within a general nonlinear framework and then make simplifications specific to geometrical linearity. We begin with an example.

Consider a stress-free material (matrix) with a cavity (Fig. 6.2a) and a piece of other material (inclusion). The inclusion is shown in its stress-free state in Fig. 6.2b. After some deformation of the inclusion it fits the cavity. We put the inclusion into the cavity and glue it to the cavity surface (Fig. 6.2d). We obtain a material with nonzero stresses: the matrix is unstressed, while the inclusion, being deformed, has nonzero stresses. To keep such a system in equilibrium, some forces must be applied at the interface surface. If the external forces are removed, then the system comes to a new equilibrium state (Fig. 6.2e). The stresses in this state are nonzero. They are usually referred to as internal stresses. The example considered provides a physically adequate model for various defects in solids: precipitates and inclusions of various nature, voids, vacancies and interstitials in crystal lattices.

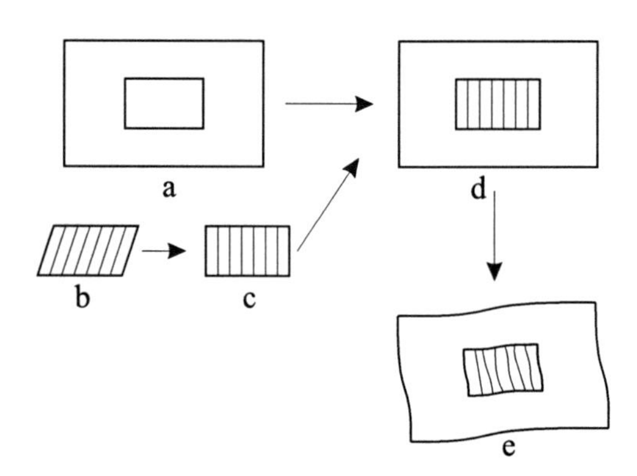

Fig. 6.2 Internal stresses caused by inclusion

Fitting this example to the general continuum mechanics scheme, we can identify the current state with the state in Fig. 6.2e and the initial state with the state in Fig. 6.2d. There is a displacement field which transforms the initial state into the current state. The initial state is not, however, stress-free. To have a stress-free state, we have to cut off the inclusion. That yields the stress-free states of the two pieces (Fig. 6.2a and b). These two pieces cannot be put together without overlapping. To characterize the transition from the initial state to a stress-free state quantitatively, first we introduce a Lagrangian smooth coordinate system, X^a, in the initial state. It can always be chosen Cartesian. The metric tensor in the initial state we denote as before by $\mathring{g}_{ab}$. Cutting off a small piece of material in the vicinity of point X and its unloading lead to deformation to a stress-free state. The metric tensor in the stress-free state differs from $\mathring{g}_{ab}$, and we denote it by g^*_{ab}. In our example, g^*_{ab} coincides with $\mathring{g}_{ab}$ inside the matrix, and differs from $\mathring{g}_{ab}$ in the inclusion. The deformation occurring in the transition from the initial to the stress-free state is characterized by the tensor

$$\varepsilon^*_{ab} = \frac{1}{2}\left(g^*_{ab} - \mathring{g}_{ab}\right).$$

It is called eigen-strain tensor, or just eigen-strain. In general, the eigen-strain, ε^*_{ab}, cannot be obtained by a smooth displacement field from the initial state. One says that the eigen-strain is incompatible.

Since g^*_{ab} corresponds to a stress-free state, the elastic strain tensor is

$$\varepsilon^{(e)}_{ab} = \frac{1}{2}\left(g_{ab} - g^*_{ab}\right).$$

Here, g_{ab} is the metric tensor in the current state,

$$g_{ab} = g_{ij} x^i_a x^j_b, \qquad x^i_a = \frac{\partial x^i(X)}{\partial X^a},$$

$x^i(X)$ being the position of the particle X in the current state. As before, the total strain, ε_{ab}, is, by definition, the strain associated with the transition from the initial to the current state:

$$\varepsilon_{ab} = \frac{1}{2}\left(g_{ab} - \mathring{g}_{ab}\right).$$

Obviously,

$$\varepsilon_{ab} = \varepsilon^{(e)}_{ab} + \varepsilon^*_{ab}.$$

The total strain is compatible, the elastic strain and the eigen-strain are not.

A typical problem of the internal stress theory is to find the stress field if the eigen-strain field is known. The internal stresses can be found from the following

Gibbs variational principle. *The true internal stress field provides minimum to the functional*

$$I = \int_V F(\varepsilon_{ab}^{(e)})dV,$$

over all particle positions in the current state.

In geometrically linear theory,

$$\varepsilon_{ij}^{(e)} = \frac{1}{2}\left(\partial_i u_j + \partial_j u_i - g_{ij}^*\right).$$

Therefore the functional simplifies to

$$I = \int_V F\left(\frac{1}{2}\left(\partial_i u_j + \partial_j u_i - g_{ij}^*\right)\right)dV.$$

The tensor g_{ij}^* is compatible if there exists a vector field, $u_i^*(x)$, such that

$$g_{ij}^* = \partial_i u_j^* + \partial_j u_i^*. \tag{6.124}$$

The vector field, u_i^*, can be interpreted as a displacement field from the initial to the stress-free state. If $F(\varepsilon_{ab}^{(e)})$ has the only minimum for zero $\varepsilon_{ab}^{(e)}$, and stresses

$$\sigma^{ij} = \frac{\partial F(\varepsilon_{ij}^{(e)})}{\partial \varepsilon_{ij}^{(e)}},$$

are zero for $\varepsilon_{ij}^{(e)} = 0$, then the minimum is achieved for $u_i = u_i^*$, i.e. in the stress-free state. Internal stresses appear only for incompatible g_{ij}^*.

In physically and geometrically linear theory,

$$F(\varepsilon_{ij}^{(e)}) = \frac{1}{2}C^{ijkl}\varepsilon_{ij}^{(e)}\varepsilon_{kl}^{(e)}.$$

To obtain the dual variational principle in this case, we present $F(\varepsilon_{ij}^{(e)})$ as

$$F(\varepsilon_{ij}^{(e)}) = \max_{\sigma^{ij}}\left(\sigma^{ij}\frac{1}{2}\left(\partial_i u_j + \partial_j u_i - g_{ij}^*\right) - \frac{1}{2}C_{ijkl}^{(-1)}\sigma^{ij}\sigma^{kl}\right).$$

Following the general scheme we arrive at

Dual variational principle. *The true stress state of the body provides minimum to the functional*

$$J = \int_V \frac{1}{2}C_{ijkl}^{(-1)}\sigma^{ij}\sigma^{kl}dV + l(\sigma), \qquad l(\sigma) = \frac{1}{2}\int_V \sigma^{ij}g_{ij}^*dV, \tag{6.125}$$

on the set of all stress fields obeying the constraints

$$\frac{\partial \sigma^{ij}}{\partial x^j} = 0 \qquad \text{in } V, \qquad \sigma^{ij} n_j = 0 \qquad \text{on } \partial V. \tag{6.126}$$

If g_{ij}^* are compatible, i.e. (6.124) holds, then $l(\sigma) = 0$ due to (6.126), and the minimizing stress is zero.

As we discussed in Sect. 6.6, in order to be compatible, g_{ij}^* must satisfy (6.59):

$$e^{ikl} e^{jmn} \partial_l \partial_n g_{ij}^* = 0.$$

Therefore, the tensor

$$\eta^{ij} = \frac{1}{2} e^{ikl} e^{jmn} \partial_l \partial_n g_{ij}^*$$

is a measure of the incompatibility of the eigen-strain.

The linear functional, $l(\sigma)$, can be expressed in terms of η^{ij}, if we eliminate stresses by introducing the stress functions, ψ_{ij}. Plugging in $l(\sigma)$ (6.125) the expression of stresses in terms of stress functions (6.47), integrating by parts and assuming, for simplicity, that g_{ij}^* and their derivatives vanish at the boundary, we have

$$l(\psi) = \frac{1}{2} \int_V \sigma^{ij} g_{ij}^* dV = \frac{1}{2} \int_V e^{ikl} e^{jms} \psi_{km,ls} g_{ij}^* dV = \int_V \psi_{km} \eta^{km} dV.$$

Denote by $F(\psi_{km,ls})$ the function which is obtained from $\frac{1}{2} C_{ijkl}^{(-1)} \sigma^{ij} \sigma^{kl}$ when one replaces the stress tensor σ^{ij} by its expression in terms of stress functions, $e^{ikl} e^{jms} \psi_{km,ls}$. We obtain

Dual variational principle for stress functions. *The true stress functions provide minimum to the functional*

$$J(\psi) = \int_V F(\psi_{km,ls}) dV + \int_V \psi_{km} \eta^{km} dV$$

on the set of all stress functions obeying the boundary condition

$$e^{ikl} e^{jms} \psi_{km,ls} n_j = 0 \qquad \text{on } \partial V.$$

Here the incompatibility measure, η^{km}, is assumed to be known.

6.9 Thermoelasticity

In the previous consideration temperature was assumed to be the same in the deformed and undeformed state. If the temperature changes from the initial value, T_0, in the stress-free state to a value, T, in the deformed state, then some additional stresses develop caused by the temperature difference, $T - T_0$. Accordingly, an interaction term between the strains and the temperature difference appears in the free energy. In linear theory, expanding $F\left(\varepsilon_{ij}, T\right)$ in Taylor series with respect to small ε_{ij} and $T - T_0$, we get

$$F\left(\varepsilon_{ij}, T\right) = F_0(T) - \beta^{ij}\varepsilon_{ij}\left(T - T_0\right) + \frac{1}{2}C^{ijkl}\varepsilon_{ij}\varepsilon_{kl}$$

where $F_0(T)$ is the free energy of the undeformed material. Then the stresses are

$$\sigma^{ij} = \frac{\partial F}{\partial \varepsilon_{ij}} = C^{ijkl}\varepsilon_{kl} - \beta^{ij}\left(T - T_0\right). \tag{6.127}$$

The meaning of the coefficients β^{ij} is simple: $-\beta^{ij}\left(T - T_0\right)$ are the stresses developed in a homogeneous deformation of the material clamped at the boundary. For isotropic materials, $-\beta^{ij} = \beta g^{ij}$. Usually, $\beta > 0$, i.e. the material is under compression, when temperature raises, and under tension when temperature dropes.

One can introduce the temperature expansion coefficients, α_{ij}, by the relation:

$$\varepsilon_{ij} = \alpha_{ij}\left(T - T_0\right) \qquad \text{when } \sigma_{ij} = 0.$$

From (6.127)

$$\alpha_{ij} = C^{-1}_{ijkl}\beta^{kl}.$$

In case of isotropic body,

$$\sigma^{ij} = \lambda\varepsilon^k_k g^{ij} + 2\mu\varepsilon^{ij} - \beta g^{ij}\left(T - T_0\right).$$

Therefore,

$$\alpha_{ij} = \alpha g_{ij}, \qquad \alpha = \frac{\beta}{3\lambda + 2\mu},$$

and the free energy of isotropic thermoelastic body has the form

$$F\left(\varepsilon_{ij}, T\right) = F_0(T) + \frac{1}{2}\lambda\left(\varepsilon^i_i\right)^2 + \mu\varepsilon_{ij}\varepsilon^{ij} - (3\lambda + 2\mu)\alpha\left(T - T_0\right)\varepsilon^k_k.$$

In general case of anisotropic body, the free energy of the thermoelastic body can be written as

$$F\left(\varepsilon_{ij}, T\right) = \frac{1}{2}C^{ijkl}\left(\varepsilon_{ij} - \alpha_{ij}\left(T - T_0\right)\right)\left(\varepsilon_{kl} - \alpha_{kl}\left(T - T_0\right)\right) + F_1(T), \tag{6.128}$$

where

$$F_1(T) = F_0(T) - \frac{1}{2} C^{ijkl} \alpha_{ij} \alpha_{kl} (T - T_0)^2 .$$

Comparing the free energy of thermoelastic bodies (6.128) with the free energy of elastic bodies with internal stresses (Sect. 6.8), we see that the thermoelastic body is a special case of elastic bodies with internal stresses when the eigenstrains are equal to $\alpha_{ij} (T - T_0)$.

6.10 Dislocations

Elasticity theory gives an adequate description of internal stresses caused by defects of perfect crystal lattices. In such modeling of a discrete system, the displacement field of a continuum is viewed as a smooth extrapolation of the displacements of the nodes of the crystal lattice. In this section we consider variational principles for the internal stresses caused by one type of the crystal defects, dislocations.

The mechanism of plastic deformation in crystals (at not very high temperatures) is as follows. Let a shear force is applied to a crystal (Fig. 6.3a).

If the force is small, the crystal lattice is just slightly deformed. If the force is large enough, it can cause an irreversible deformation of the crystal which remains after unloading (Fig. 6.3b). To move a crystal to state *b* from state *a* would require a very large force. However, such transformation can be achieved by a smaller force if a defect is introduced in the crystal, a dislocation, shown in Fig. 6.3c. When this defect passes the crystal, the crystal is transformed from state *a* to state *b*. The crystal lattice is deformed in a vicinity of dislocation. Thus, there are internal stresses in the body. We are going to formulate a mathematical problem describing these stresses.

As the initial stress-free state we take the perfect lattice. It is assumed to be modeled by some elastic continuum. It is shown in Fig. 6.4a. There is another stress-free state obtained as a result of plastic deformation (Fig. 6.4b). The total displacement, u_i, of material particles from state *a* to state *b* has a discontinuity on a slip plane Ω. The jump of displacements, $b_i = [u_i]$ is constant on Ω. It is called Burgers vector. The Burgers vector is tangential to the slip plane Ω. The magnitude of the Burgers

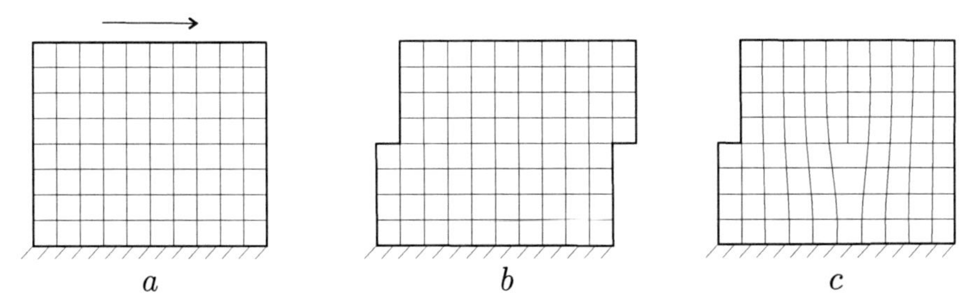

Fig. 6.3 A mechanism of plastic deformation in crystals. The grid nodes corresponds to positions of atoms

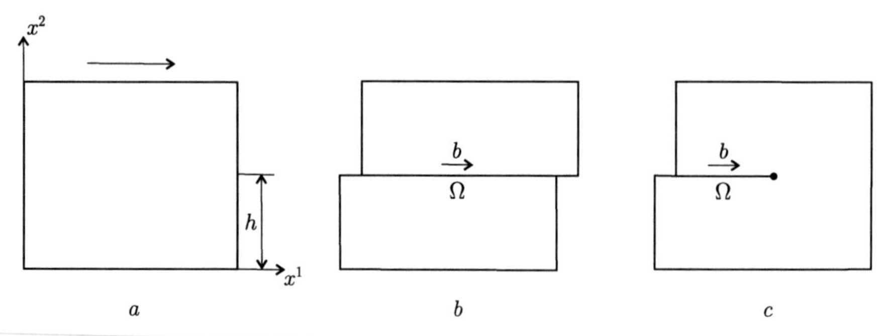

Fig. 6.4 Continuum model of a dislocation

vector is equal to the interatomic distance. In the intermediate state from a to b, the state with a dislocation, the displacements have a jump on the slip plane Ω shown in Fig. 6.4b.

The transition from state a to state b results in some plastic deformation. This plastic deformation is compatible: there is a displacement vector, $u_i^{(p)}$, from state a to state b. Assuming the geometric linearity, we can write

$$\varepsilon_{ij}^{(p)} = \frac{1}{2}\left(\frac{\partial u_i^{(p)}}{\partial x^j} + \frac{\partial u_j^{(p)}}{\partial x^i}\right). \tag{6.129}$$

The plastic displacement vector, $u_i^{(p)}$, is discontinuous. For example, for the case shown in Fig. 6.4b,

$$u_1^{(p)} = b\theta\left(x_2 - h\right), \quad u_2^{(p)} = 0$$

where $\theta\left(x_2\right)$ is the step function. Thus

$$\varepsilon_{12}^{(p)} = \frac{1}{2} b\delta\left(x_2 - h\right), \tag{6.130}$$

while all other components of the plastic strain tensor are zero. As we will see, for a state with dislocations one can introduce plastic strains, but they will not be compatible, i.e. there are no plastic displacement fields satisfying (6.129). The total displacements, i.e. the displacements from the state a to the state c do exist and have a jump on Ω.

Generalizing this picture, we consider an elastic body which contains a surface Ω, on which displacements have a given jump,

$$b_i = u_i^+ - u_i^- . \tag{6.131}$$

The signs, $\pm$, mark the values at the two sides of Ω. The Burgers vector, b_i, is tangential to Ω and its magnitude is equal to the interatomic distance.[4] The crystal lattice remains perfect inside Ω; however at the boundary of Ω, Γ, the positions of the atoms are far from the position of the perfect lattice. The distorted region can be viewed as a thin rod with the central line Γ. The material inside the rod is in a state similar to the state of a melted crystal. Γ is called the dislocation line, or just dislocation.

Modeling the slip plane by a mathematical surface, Ω, and the dislocation by a curve, Γ, make sense, because we are going to consider the stresses far away from the dislocation. The numerical simulations of crystal lattices have shown that such an approximation fails to predict the correct stress field only in a small vicinity of the dislocation line of the size of a few interatomic distances.

So, the only difference from the usual elasticity theory is to allow the displacements to have a given jump, b_i, at some surface, Ω, with the vector, b_i, tangent to that surface. To accord this kinematic picture with the theory of the previous sections, we note that the initial state is identified here with the stress-free state. Usually, Ω is a plane or piece-wise plane surface, and the vector, b_i, is constant on Ω. However, in further consideration this is not essential, and the surface Ω can be viewed as an arbitrary smooth surface. The constancy of b_i would cause a stress state with singularities at Γ. In order to avoid singularities, the jump of displacements could be smoothed from a constant value, b_i, inside Ω to zero at Γ. Smoothing should be made only in a small vicinity of Γ on the order of the interatomic distance. This region can be viewed as corresponding to the dislocation core.

Note that the "true" difference of the displacements on the two sides of the slip surface "in a crystal" is equal to the Burgers vector plus a small vector on the order of elastic strain. Such correction is negligibly small and will be ignored.

In physically and geometrically linear theory the internal stresses caused by a dislocation is determined from

Gibbs variational principle. *The true displacement field of an unloaded crystal provides the minimum value to the functional*

$$I(u) = \int_V \frac{1}{2} C^{ijkl} \varepsilon_{ij} \varepsilon_{kl} dV, \qquad \varepsilon_{ij} \equiv \frac{1}{2}\left(\partial_i u_j + \partial_j u_i\right),$$

on the set of all displacement fields with a prescribed jump on Ω (6.131).

Since, according to (6.131), the variations of displacements are continuous on Ω,

$$[\delta u_i] = 0,$$

(remember that $[\varphi]$ denotes the difference of the boundary values of function, φ, at the two sides of discontinuity surface, $[\varphi] = \varphi^+ - \varphi^-$) the minimum is achieved at the continuous surface forces at Ω :

[4] There are crystal defects, of which the magnitude of Burger's vector is not equal to the interatomic distance. They are called partial dislocations and not considered here. Partial dislocations possess an additional energy distributed over the slip surface.

$$[\sigma^{ij}]n_j = 0.$$

For definiteness, the normal vector on Ω is chosen directed from the side "−" to the side "+"; in addition, the positive direction on the contour Γ is taken in such a way that moving along Γ in positive direction one sees the surface Ω on the left (Fig. 6.5).

If some external surface forces, f_i, act at the boundary of the crystal, the functional to be minimized changes to

$$I(u) = \int_V \frac{1}{2} C^{ijkl} \varepsilon_{ij} \varepsilon_{kl} dV - \int_{\partial V} f_i u^i dA. \tag{6.132}$$

Let us construct the dual variational principle. Following the general scheme,

$$\begin{aligned}
&\min_{u \in (6.131)} I(u) = \\
&= \min_{u \in (6.131)} \max_{\sigma^{ij}} \left[\int_V \left(\sigma^{ij} u_{i,j} - \frac{1}{2} C^{(-1)}_{ijkl} \sigma^{ij} \sigma^{kl} \right) dV - \int_{\partial V} f_i u^i dA \right] \\
&= \max_{\sigma^{ij}} \min_{u \in (6.131)} \left[\int_{\partial V} (\sigma^{ij} n_j - f^i) u_i dA + \int_\Omega (\sigma^{ij}_- n_j u_i^- - \sigma^{ij}_+ n_j u_i^+) dA \right. \\
&\qquad \left. - \int_V (\partial_j \sigma^{ij}) u_i dV - \int_V \frac{1}{2} C^{(-1)}_{ijkl} \sigma^{ij} \sigma^{kl} dV \right].
\end{aligned} \tag{6.133}$$

Changing the order of minimum and maximum in (6.133), replacing u_i^+ by $u_i^- + b_i$ and using the arbitrariness of u_i in V and on ∂V and u_i^- on Ω, we obtain

Dual variational principle. *The true stress state of the body with a dislocation provides minimum to the functional*

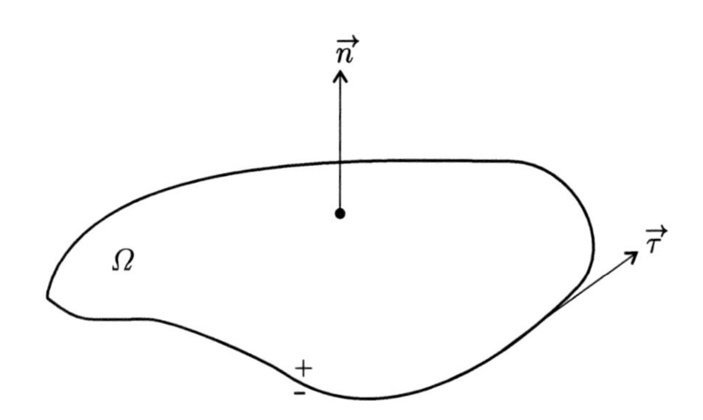

Fig. 6.5 Notations for a single dislocation

$$J(\sigma) = \int_V \frac{1}{2} C^{(-1)}_{ijkl} \sigma^{ij} \sigma^{kl} dV + l(\sigma), \tag{6.134}$$

$$l(\sigma) = \int_\Omega \sigma^{ij} n_j b_i dA \tag{6.135}$$

on the set of all stress fields obeying the constraints

$$\frac{\partial \sigma^{ij}}{\partial x^j} = 0 \quad \text{in } V, \quad [\sigma^{ij}] n_j = 0 \quad \text{on } \Omega, \quad \sigma^{ij} n_j = f^i \quad \text{on } \partial V. \tag{6.136}$$

The boundary values of σ^{ij} are, in principle, different on the two sides of Ω, but it does not matter which boundary values are used in (6.135) because $\sigma^{ij} n_j$ are continuous.

As we discussed in Section 6.1, the energy functional $I(u)$ has the meaning of averaged micro-Hamiltonian. Its minimum value, denote it by H, depends on the position of dislocation. Derivative of H with respect to that position is called the force acting on dislocation. If H takes the minimum value for some dislocation position, the force is zero, and the system is in equilibrium. Let us find the force. According to (6.133),

$$H = - \min_{\sigma \in (6.136)} J(\sigma). \tag{6.137}$$

To find the variation of H caused by an infinitesimally small variation of the dislocation line, we note that, though the stresses depend on the position of the dislocation, and the stress variations are not zero, these variations do not contribute in δH because the variation of $J(\sigma)$ with respect to stresses is zero. This is quite similar to the differentiation of the minimum value of a functional with respect to parameters (Section 5.13). Thus, to find the variation of H, one has to find the variation of the linear functional, $l(\sigma)$, caused by the variation of Γ. We consider infinitely small displacements of Γ which lie in the tangent plane to Ω : for such displacements no change of volume occurs during plastic deformation. In general, dislocations can move in the direction normal to the slip plane as well, such motion is called climbing. However, for such motion the atomic half plane, bounded by the dislocation line, shrinks or grows and, accordingly, consumes or emits vacancies and/or interstitials. The proper energy balance for such case should include the contribution from additional fields describing vacancies and interstitial, and we do not dwell on this issue here.

Let ν^i be the unit vector orthogonal to Γ and to the normal vector of the slip plane, n^i. Vector ν^i looks outside Ω. Denoting by $\delta\nu$ the displacement of Γ in the direction ν^i, we have from (6.135) and (6.137),

$$\delta H = - \int_\Gamma \sigma^{ij} n_j b_i \delta \nu ds. \tag{6.138}$$

The expression

$$f = -\sigma^{ij} n_j b_i$$

may be interpreted as the component of the force acting on the dislocation in the direction of ν^i.

An additional analysis shows that out of slip plane displacements at some point s do not involve the volume change (or, in physical terms, the flux of vacancies and/or interstitials), if the Burgers vector of the dislocation is tangent to Γ at this point. In this case, n_j in (6.138) is the normal vector to the increment of the slip surface. Thus, if δx^i is the virtual displacement of the dislocation line with the magnitude $\delta \nu$, then $n^i = e^{ijk} \delta x_j \tau_k / \delta \nu$, τ_k being the tangent vector to Γ. Hence, the force is

$$f_m = -\sigma^{ik} b_i e_{kms} \tau^s.$$

This formula makes sense only if $b_i = b\tau_i$. Therefore, in particular, the spherical part of the stress tensor, pressure does not contribute to the force.

The relations obtained hold for a set of dislocations as well: this case corresponds to the slip surface, Ω, consisting of several disjoined pieces.

6.11 Continuously Distributed Dislocations

Crystals contain a huge number of dislocations. A typical total length of dislocation lines is on a cosmic scale: about 10^{14} m in 1 m^3. Therefore, it makes sense to consider a continuum theory which mimics some features of dislocation networks. Remarkably, it can be done in such a way that the continuum theory transforms in the theory of single dislocations presented in the previous section by concentrating the continuum characteristics of dislocation networks on dislocation lines. In this section we describe this continuum theory and the corresponding variational principles.

Plastic strains. Let us introduce the δ-functions associated with the slip surface, Ω. We will need the usual three-dimensional δ-function, which we denote here by $\delta_3(x)$:

$$\delta_3(x) = \delta(x_1)\delta(x_2)\delta(x_3),$$

the δ-function of Ω

$$\delta(\Omega) = \int_{\Omega} \delta_3(x - x_{\Omega}) dA,$$

and δ-function of Γ

$$\delta(\Gamma) = \int_\Gamma \delta_3(x - x_\Gamma) ds.$$

Here x_Ω and x_Γ denote the points on Ω and Γ, respectively.

These δ – functions are linked by Kunin's identity:[5]

$$e^{jkl}\partial_k(n_l\delta(\Omega)) = \tau^j\delta(\Gamma). \tag{6.139}$$

Each dislocation causes some plastic deformation of the crystal. We are going to motivate the following formula: a plastic strain, associated with a dislocation, is[6]

$$\varepsilon_{ij}^{(p)} = b_{(i}n_{j)}\delta(\Omega). \tag{6.140}$$

Indeed, by plastic deformation one usually means the residual deformation, which remains in material after unloading. The residual deformation is measured using the values of displacements at the boundary of the material. Let loading of the material from a stress-free state create a dislocation inside the material. After unloading the displacements at the boundary are not zero due to the presence of the dislocation. By displacements we mean here the displacements from the state that corresponds to the perfect lattice. We define the plastic strain of a specimen in terms of the boundary values of displacements as

$$\varepsilon_{ij}^{(p)} = \frac{1}{|V|}\int_{\partial V} u_{(i}n_{j)} dA. \tag{6.141}$$

For example, if there is just shear of the two sides of the boundary, S_+ and S_-, like the one shown in Fig. 6.6, then the only non-zero component of the displacement vector at the boundary is u_1, and (6.141) yields a meaningful result: all the components of plastic strain are zero except $\varepsilon_{12}^{(p)}$:

$$\varepsilon_{12}^{(p)} = \frac{1}{2}\frac{1}{|V|}\left(u_1^+ S - u_1^- S\right) = \frac{1}{2}\frac{u_1^+ - u_1^-}{h}.$$

---

[5] Indeed, for any smooth function, $\varphi(x)$,

$$\int \varphi(x) e^{jkl}\partial_k(n_l\delta(\Omega))d^3x = -\int\int_\Omega \partial_k\varphi(x)e^{jkl}n_l\delta(x - x_\Omega)d^3x dA$$
$$= -\int_\Omega \partial_k\varphi(x_\Omega)e^{jkl}n_l dA = -\int_\Omega \partial_k\varphi(x_\Omega)\varepsilon^{\alpha\beta}r_\alpha^j r_\beta^k dA$$
$$= -\int_\Omega \frac{\partial\varphi}{\partial\zeta^\beta}\varepsilon^{\alpha\beta}r_\alpha^j dA = -\int_\Omega \nabla_\beta(\varphi\varepsilon^{\alpha\beta}r_\alpha^j)dA = -\int_\Gamma \varphi\varepsilon^{\alpha\beta}\nu_\beta r_\alpha^j ds = \int_\Gamma \varphi\tau^j ds.$$

Here we used (14.10), (14.13), (14.37), (14.28) and (14.21).

[6] This formula transforms into (6.130) for the special case of Fig. 6.3b.

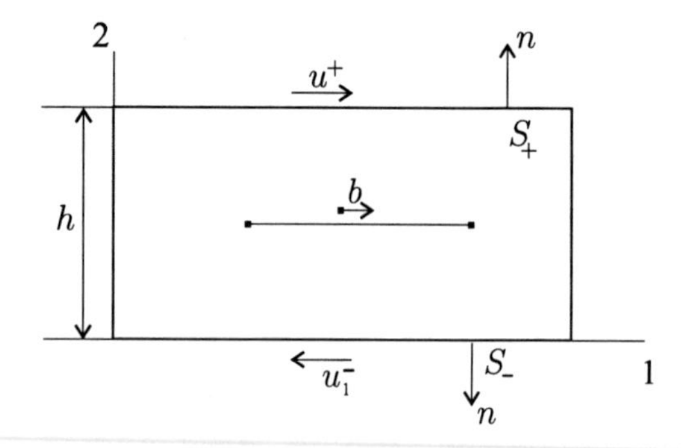

Fig. 6.6 Notation to the motivation of formula (6.141)

Another point in favor of the definition (6.141) is that, for a homogeneous deformation inside the specimen, $u_i = a_{ij}x^j$, the integral in the right hand side of (6.141) is equal to $a_{(ij)}$ (due to the divergence theorem (3.89)), as it should be.

Let the specimen contain a dislocation with a slip surface, Ω. The strains inside the specimen, by assumption, can be computed within the framework of linear elasticity. The elastic moduli are assumed to be constant over the specimen. We are going to show that the plastic strain of the specimen (6.141) is equal to

$$\varepsilon_{ij}^{(p)} = \frac{1}{|V|}\int_{\Omega} b_{(i}n_{j)}dA. \tag{6.142}$$

We have from (6.141):

$$\begin{aligned}\varepsilon_{ij}^{(p)} &= \frac{1}{|V|}\left(\int_{\partial V} u_{(i}n_{j)}dA - \int_{\Omega} b_{(i}n_{j)}dA\right) + \frac{1}{|V|}\int_{\Omega} b_{(i}n_{j)}dA \\ &= \frac{1}{|V|}\int_V u_{(i,j)}dV + \frac{1}{|V|}\int_{\Omega} b_{(i}n_{j)}dA.\end{aligned} \tag{6.143}$$

The first integral in (6.143) is equal to

$$\frac{1}{|V|}\int_V u_{(i,j)}dV = \frac{1}{|V|}C_{ijkl}^{(-1)}\int_V \sigma^{kl}dV. \tag{6.144}$$

On the other hand,

$$\int_V \sigma^{kl}dV = \int_{\partial V} \sigma^{ks}n_s x^l dA,$$

because, due to momentum equations,

$$\int_{\partial V} \sigma^{ks}n_s x^l dA = \int_V \frac{\partial(\sigma^{ks}x^l)}{\partial x^s}dV = \int_V \left(\frac{\partial \sigma^{ks}}{\partial x^s}x^l + \sigma^{ks}\frac{\partial x^l}{\partial x^s}\right)dV = \int_V \sigma^{kl}dV.$$

The presence of a surface of discontinuity, Ω, does not change the result because $\sigma^{ks} n_s$ are continuous on Ω. Since at the boundary the surface force is zero, so does the volume average of the stress. Thus, the integral (6.144) is zero, and (6.142) holds true.

So the plastic strain of a body with a dislocation is given by (6.142). This formula can be written as the volume average of functions (6.140):

$$\varepsilon_{ij}^{(p)} = \frac{1}{|V|} \int_V b_{(i} n_{j)} \delta(\Omega) dV. \tag{6.145}$$

This relation holds true when Ω consists of several disjointed pieces, i.e. for a set of dislocations. So, if we associate with each dislocation the plastic strain (6.140), then the total plastic deformation of the specimen is the volume average of "local" plastic deformations (6.140). That shows a feasibility of (6.140).

Plastic distortion and dislocation density tensor. The volume average of the characteristics of dislocation networks (6.140) is the macroscopic plastic strain. If we construct a theory of internal stresses in which $\varepsilon_{ij}^{(p)}$ can be smooth functions, admitting a limit transition to the δ-function (6.140), we embed the case of discrete dislocation in a continuum theory. At first glance, we can take the smooth functions, $\varepsilon_{ij}^{(p)}$, as the primary characteristics of dislocation networks. Here, however, we face a difficulty. Plastic strain (6.140) depends on the slip surface, Ω. The slip surface is determined by the history of dislocation motion. Thus, the plastic strain depends on the history of motion. The physical state of the material is affected only by the current positions of the dislocations (if dislocations move slow). Therefore, there should be no dependence of thermodynamic functionals on plastic strains. The first gradients of plastic strains do not help: they are history-dependent as well. The resolution of this difficulty is the following: instead of plastic strain, one introduces the plastic distortion as the primary characteristics of plastic deformation

$$\beta_{ij} = b_i n_j \delta(\Omega). \tag{6.146}$$

Plastic strain is the symmetric part of the plastic distortion,

$$\varepsilon_{ij}^{(p)} = \beta_{(ij)}.$$

Plastic distortion contains three additional degrees of freedom, the plastic rotation, $\beta_{[ij]}$. Now one can form the characteristics which are history-independent and associated only with the current position of the dislocation line:

$$\alpha_i^j = e^{jkl} \partial_k \beta_{il}. \tag{6.147}$$

Indeed, due to the identity (6.139),

$$\alpha_i^j = b_i \tau^j \delta(\Gamma). \tag{6.148}$$

Tensor α_j^i is called the dislocation density tensor.[7]

The plastic distortion and the dislocation density tensor can now be taken as continuous. In the special case (6.146), we return to the kinematic relation for discrete dislocations (6.148).

The dislocation density tensor obeys the identity

$$\partial_j \alpha_i^j = 0.$$

It follows from (6.147).

The dislocation density tensor can be interpreted as a measure of incompatibility of plastic deformation: according to (6.147), $\alpha_i^j = 0$ if and only if there exists a smooth field of plastic displacements, $u_i^{(p)}$, such that the plastic distortion is the gradient of plastic displacements:

$$\beta_{ij} = \partial_j u_i^{(p)}.$$

In case of dislocations, $\alpha_i^j \neq 0$, and the plastic displacements, $u_i^{(p)}$, do not exist.

The dislocation density tensor being determined by only the current positions of dislocations is a proper argument of thermodynamic functions.

Elastic distortion and lattice rotation. Along with the plastic distortion, β_{ij}, one can introduce the elastic distortion, $\beta_{ij}^{(e)}$, by the relation

$$\beta_{ij} + \beta_{ij}^{(e)} = \frac{\partial u_i}{\partial x^j}$$

where u_i is the displacement from the state corresponding to the perfect lattice. As follows from this definition,

$$e^{jkl} \partial_k \beta_{il} + e^{jkl} \partial_k \beta_{il}^{(e)} = 0.$$

Therefore, the dislocation density tensor can be also written as

$$\alpha_i^j = -e^{jkl} \partial_k \beta_{il}^{(e)}. \tag{6.149}$$

The symmetric part of the elastic distortion, $\beta_{ij}^{(e)}$, is the elastic strain, $\varepsilon_{ij}^{(e)}$; the anti-symmetric part, ω_{ij}, describes the rotation of the crystal lattice. Typically, for

[7] Historically, the dislocation density tensor was introduced from other reasoning. It was first used as a measure of the lattice curvature by Nye [234]. Later Bilby and Kröner constructed its continuous version that involves all nine degrees of freedom of plastic distortions.

metals $\varepsilon_{ij}^{(e)} \sim 10^{-4}$ and $\omega_{ij} \sim 10^{-2}$; therefore $\varepsilon_{ij}^{(e)}$ can be dropped in (6.149), and α_i^j is determined by the lattice rotation only:

$$\alpha_i^j = -e^{jkl}\partial_k\omega_{il}. \tag{6.150}$$

The lattice rotations can be measured experimentally, and formula (6.150) is used to estimate the values of the dislocation density tensor.

Functional $l(\sigma)$**.** To specify a dislocation network, one can prescribe the tensor β_{ij}. In terms of this tensor, the linear functional (6.135) can be written as

$$l(\sigma) = \int_V \sigma^{ij}\beta_{ij}dV.$$

The symmetry of the stress tensor allows us to write $l(\sigma)$ in terms of plastic deformation only:

$$l(\sigma) = \int_V \sigma^{ij}\varepsilon_{ij}^{(p)}dV. \tag{6.151}$$

Formula (6.151) is exact in the sense that it remains true if the smooth functions $\varepsilon_{ij}^{(p)}$ are replaced by the singular ones, (6.140).

In fact, $l(\sigma)$ depends only on the derivatives of $\varepsilon_{ij}^{(p)}$ (or β_{ij}), because in (6.151) σ^{ij} is not an arbitrary tensor but a tensor satisfying the equilibrium equations. This can be seen explicitely if we replace the stress tensor by the stress functions from (6.47):

$$l(\sigma) = \int_V e^{ikl}e^{jms}\psi_{km,ls}\beta_{ij}dV.$$

Here ψ_{km} is a symmetric tensor field. Assume, for simplicity, that β_{ij} are zero in some vicinity of the boundary. Integrating by parts we obtain the linear functional in terms of dislocation density tensor:

$$l(\sigma) = \int_V e^{ikl}\psi_{km,l}\alpha_i^m dV. \tag{6.152}$$

Formula (6.152) shows that the functional $l(\sigma)$, in fact, does not depend on the history of dislocation motion. Accordingly, energy of the body does not depend on the history of dislocation motion as well.

Further integration by parts in (6.152) yields the equation

$$l(\sigma) = \int_V \psi_{km}\eta^{km}dV,$$

where η^{km} is the incompatibility measure

$$\eta^{km} = \frac{1}{2}\left(e^{ilk}e^{jsm} + e^{ilm}e^{jsk}\right)\beta_{ij,ls} = \frac{1}{2}\left(e^{kil}\alpha^{m}_{i,l} + e^{mil}\alpha^{k}_{i,l}\right). \tag{6.153}$$

The incompatibility η^{km} is taken symmetric over k, m because ψ_{km} is symmetric. Interestingly, the incompatibility depends only on the symmetric part of plastic distortion, the plastic strain:[8]

$$\eta^{km} = e^{ilk}e^{jsm}\varepsilon^{(p)}_{ij,ls}. \tag{6.154}$$

On the other hand, according to (6.152) the incompatibility can be expressed in terms of the dislocation density tensor and, therefore, is history-independent. Hence, the combinations of second derivatives of plastic strain (6.154) are also history-independent and can serve as the characteristics of the current physical state of the crystal.

Dislocation networks. In crystals, dislocations form a random network.[9] It can be characterized by random fields $\beta_{ij}(x,\omega)$, $\varepsilon^{(p)}_{ij}(x,\omega)$, $\alpha_{ij}(x,\omega)$ or $\eta_{ij}(x,\omega)$. Denote by bar the mathematical expectations of these fields, e. g.,

$$\bar{\varepsilon}^{(p)}_{ij}(x) = M\varepsilon^{(p)}_{ij}, \quad \bar{\beta}_{ij}(x) = M\beta_{ij}, \quad \bar{\alpha}_{ij}(x) = M\alpha_{ij},$$

and by prime the fluctuations:

$$\varepsilon'^{(p)}_{ij} = \varepsilon^{(p)}_{ij} - \bar{\varepsilon}^{(p)}_{ij}, \quad \beta'_{ij} = \beta_{ij} - \bar{\beta}_{ij}, \quad \alpha'_{ij}(x) = \alpha_{ij} - \bar{\alpha}_{ij}.$$

We assume that $\bar{\beta}_{ij}$ and $\bar{\alpha}_{ij}$ are some smooth fields, while β'_{ij} and α'_{ij} may be singular. Since the operations of mathematical expectation and differentiation commute,

$$\bar{\alpha}^{j}_{i} = e^{jkl}\partial_k\bar{\beta}_{il}, \quad \alpha'^{j}_{i} = e^{jkl}\partial_k\beta'_{il}.$$

We are going to split the variational problem for the energy functional into two variational problems, one is for the averaged characteristics, and another one for fluctuations.

Let $\bar{\sigma}^{ij}$ be the minimizer of the functional (6.134) in which $l(\sigma)$ is computed on the averaged plastic strains,

---

[8] Indeed, changing the dummy indices, $i \leftrightarrow j$, $l \leftrightarrow s$, we can write

$$e^{ilm}e^{jsk}\beta_{[ij],ls} = e^{jsm}e^{ilk}\beta_{[ji],sl} = -e^{ilk}e^{jsm}\beta_{[ij],ls}.$$

Thus, the antisymmetric part of the plastic distortion disappears in (6.153).

[9] This part of the Section uses the notion of random fields introduced further in Chapter 16. The readers, who are not familiar with this notion, are advised to look at the definitions and the corresponding notations in Section 16.1.

$$l(\sigma) = \int_V \sigma^{ij} \bar{\varepsilon}_{ij}^{(p)} dV.$$

Our goal is to prove the following

Variational principle. *Let $\bar{I}(u)$ be the energy functional of an elastic body with the eigen-strains $\bar{\varepsilon}_{ij}^{(p)}$,*

$$\bar{I}(u) = \int_V \frac{1}{2} C^{ijkl} \left(u_{(i,j)} - \bar{\varepsilon}_{ij}^{(p)} \right) \left(u_{(k,l)} - \bar{\varepsilon}_{kl}^{(p)} \right) dV - \int_{\partial V} f_i u^i dA,$$

and $\bar{\sigma}^{ij}$ the stress field of the minimizer of $\bar{I}(u)$. Then

$$\min_u I(u) = \min_u \bar{I}(u) + \mathcal{E}\left(\alpha'\right) - \int_V \bar{\sigma}^{ij} \varepsilon_{ij}^{\prime(p)} dV, \tag{6.155}$$

where $\mathcal{E}\left(\alpha'\right)$ is the elastic energy of the dislocation network with the dislocation density tensor α'_{ij} and zero tractions at the boundary.

The proof proceeds as follows. We present the plastic strains in (6.151) as the sum of averaged plastic strains and fluctuations,

$$l(\sigma) = \int_V \sigma^{ij} \bar{\varepsilon}_{ij}^{(p)} dV + \int_V \sigma^{ij} \varepsilon_{ij}^{\prime(p)} dV.$$

Then the functional (6.134) takes the form

$$J(\sigma) = \int_V \frac{1}{2} C_{ijkl}^{-1} \sigma^{ij} \sigma^{kl} dV + \int_V \sigma^{ij} \bar{\varepsilon}_{ij}^{(p)} dV + \int_V \sigma^{ij} \varepsilon_{ij}^{\prime(p)} dV. \tag{6.156}$$

If we drop the last term and minimize over admissible σ^{ij}, we get

$$J(\bar{\sigma}) = \min_{\sigma^{ij} \in (6.136)} \left[\int_V \frac{1}{2} C_{ijkl}^{(-1)} \sigma^{ij} \sigma^{kl} dV + \int_V \sigma^{ij} \bar{\varepsilon}_{ij}^{(p)} dV \right]. \tag{6.157}$$

The constraints (6.136) can be written in a weak form: for smooth functions, u_i,

$$\int_V \sigma^{ij} u_{i,j} dV = \int_{\partial V} f_i u^i dA. \tag{6.158}$$

Interpreting u_i as the Lagrange multipliers, we can write (6.157) as

$$J(\bar{\sigma}) = \min_{\sigma^{ij}} \max_{u_i} \left[\int_V \left(\frac{1}{2} C_{ijkl}^{-1} \sigma^{ij} \sigma^{kl} + \sigma^{ij} \bar{\varepsilon}_{ij}^{(p)} - \sigma^{ij} u_{(i,j)} \right) dV + \int_{\partial V} f_i u^i dA \right].$$

Changing the order of maximization and minimization and computing minimum over symmetric tensors σ^{ij}, we get

$$J(\bar{\sigma}) = \max_{u_i} \left[\int_{\partial V} f_i u^i dA - \frac{1}{2} C^{ijkl} \left(u_{(i.j)} - \bar{\varepsilon}_{ij}^{(p)} \right) \left(u_{(k,l)} - \bar{\varepsilon}_{kl}^{(p)} \right) dV \right]$$

$$= - \min_{u_i} \left[\int_V \frac{1}{2} C^{ijkl} \left(u_{(i.j)} - \bar{\varepsilon}_{ij}^{(p)} \right) \left(u_{(k,l)} - \bar{\varepsilon}_{kl}^{(p)} \right) dV - \int_{\partial V} f_i u^i dA \right]. \quad (6.159)$$

Let us present any admissible stress field in the minimization problem for the functional $J(\sigma)$ (6.156) as a sum

$$\sigma^{ij} = \bar{\sigma}^{ij} + \sigma'^{ij}. \quad (6.160)$$

The field σ'^{ij} must obey the constraints that follow from (6.136):

$$\frac{\partial \sigma'^{ij}}{\partial x^j} = 0 \text{ in } V, \quad \left[\sigma'^{ij} \right] n_j = 0 \text{ on } \Omega, \quad \sigma'^{ij} n_j = 0 \text{ on } \partial V. \quad (6.161)$$

Recall that Ω denotes the set of slip planes of all dislocations in the body. Plugging (6.160) in (6.155) we obtain

$$J(\sigma) = \int_V \frac{1}{2} C_{ijkl}^{(-1)} \bar{\sigma}^{ij} \bar{\sigma}^{kl} dV + \int_V \frac{1}{2} C_{ijkl}^{(-1)} \sigma'^{ij} \sigma'^{kl} dV + \int_V \frac{1}{2} C_{ijkl}^{(-1)} \bar{\sigma}^{ij} \sigma'^{kl} dV$$

$$+ \int_V \bar{\sigma}^{ij} \bar{\varepsilon}_{ij}^{(p)} dV + \int_V \sigma'^{ij} \bar{\varepsilon}_{ij}^{(p)} dV + \int_V \bar{\sigma}^{ij} \varepsilon_{ij}'^{(p)} dV + \int_V \sigma'^{ij} \varepsilon_{ij}'^{(p)} dV. \quad (6.162)$$

Note that for any tensor field σ'^{ij}, obeying the constraints (6.161),

$$\int_V \left(C_{ijkl}^{(-1)} \bar{\sigma}^{ij} + \bar{\varepsilon}_{kl}^{(p)} \right) \sigma'^{kl} dV = 0.$$

This is Euler's equation for the variational problem (6.157). Therefore, the sum of the third and the fifth terms in (6.162) vanishes. The sum of the first and the fourth terms in (6.162) is $J(\bar{\sigma})$. Thus,

$$J(\sigma) = J(\bar{\sigma}) + \int_V \frac{1}{2} C_{ijkl}^{(-1)} \sigma'^{ij} \sigma'^{kl} dV + \int_V \sigma'^{ij} \varepsilon_{ij}'^{(p)} dV + \int_V \bar{\sigma}^{ij} \varepsilon_{ij}'^{(p)} dV. \quad (6.163)$$

Let ψ_{km} be the stress functions for the stress tensor σ'^{ij},

$$\sigma'^{ij} = e^{ikl} e^{jmn} \psi_{km,ln}.$$

According to (6.161), ψ_{km} obey the conditions,

$$\left[e^{ikl} e^{jmn} \psi_{km,ln} \right] n_j = 0 \text{ on } \Omega, \quad e^{ikl} e^{jmn} \psi_{km,ln} n_j = 0 \text{ on } \partial V. \quad (6.164)$$

Denote by $U(\psi)$ the elastic energy,

$$\int_V \frac{1}{2} C^{(-1)}_{ijkl} \sigma'^{ij} \sigma'^{kl} dV,$$

computed as a functional of ψ_{km}. Then, using (6.152), we have

$$\min J(\sigma) = J(\bar{\sigma}) + \min_{\psi_{km} \in (6.164)} \left[U(\psi) + \int_V e^{ikl} \psi_{km,l} \alpha_i'^m dV \right] + \int_V \bar{\sigma}^{ij} \varepsilon_{ij}'^{(p)} dV. \tag{6.165}$$

The second term in (6.165) is negative energy of the dislocation network with the dislocation density tensor $\alpha_i'^j$ and zero surface forces, $-\mathcal{E}(\alpha')$. Recalling that

$$\min I(u) = -\min J(\sigma),$$

we obtain from (6.165) formula (6.155).

The case of prescribed displacements at the boundary. If the displacements are given at some part ∂V_u of the boundary of the solid,

$$u^i = u^i_{(b)} \text{ on } \partial V_u, \tag{6.166}$$

then the previous formulas change in the following way: the functional $l(\sigma)$ gets an additional term,

$$l(\sigma) = \int_V \sigma^{ij} \varepsilon^{(p)}_{ij} dV - \int_{\partial V_u} \sigma_i^j n_j u^i_{(b)} dA;$$

tensor $\bar{\sigma}^{ij}$ becomes the minimizer of the variational problem

$$J(\bar{\sigma}) = \min_{\sigma^{ij} \in (6.136)} \left[\left[\int_V \frac{1}{2} C^{-1}_{ijkl} \sigma^{ij} \sigma^{kl} dV + \int_V \sigma^{ij} \bar{\varepsilon}^{(p)}_{ij} dV - \int_{\partial V_u} \sigma_i^j n_j u^i_{(b)} dA \right] \right];$$

the weak form of the constraints for σ^{ij} becomes

$$\int_V \sigma^{ij} u_{i,j} dV = \int_{\partial V_f} f_i u^i dA + \int_{\partial V_u} \sigma_i^j n_j u^i_{(b)} dA,$$

where u^i are smooth functions taking at ∂V_u the boundary values $u^i_{(b)}$, equation (6.159) is replaced by the equation

$$J(\bar{\sigma}) = -\min_{u \in (6.166)} \left[\left[\int_V \frac{1}{2} C^{ijkl} \left(u_{(i,j)} - \bar{\varepsilon}^{(p)}_{ij} \right) \left(u_{(k,l)} - \bar{\varepsilon}^{(p)}_{kl} \right) dV - \int_{\partial V_f} f_i u^i dA \right] \right];$$

the last of the constraints (6.161) is replaced by

$$\sigma'^{ij} n_j = 0 \text{ on } \partial V_f;$$

the constraints (6.164) are replaced by

$$\left[e^{ikl} e^{jmn} \psi_{km,ln}\right] n_j = 0 \text{ on } \partial V, \tag{6.167}$$

$$e^{ikl} e^{jmn} \psi_{km,ln} n_j = 0 \text{ on } \partial V_f.$$

Variational principle. *The minimum value of the energy functional can be presented in terms of solutions of two variational problems:*

$$\min_{u \in (6.166)} I(u) = \min_{u \in (6.166)} \bar{I}(u) + \mathcal{E}(\alpha') - \int_V \bar{\sigma}^{ij} \varepsilon_{ij}^{\prime(p)} dV, \tag{6.168}$$

$$-\mathcal{E}(\alpha') = \min_{\psi \in (6.167)} \left[U(\psi) + \int_V e^{ikl} \psi_{km,l} \alpha_i'^m dV \right].$$

The functional $\mathcal{E}(\alpha')$ has the meaning of elastic energy of the dislocation network with zero tractions at ∂V_f and zero displacements at ∂V_u.

Formula for energy of a crystal with dislocations. The variational principle formulated yields important consequence, a formula for energy of a crystal with dislocations. For simplicity, we consider the case of zero tractions at ∂V_f. Then $\min I(u)$ has the meaning of total energy of the crystal. Functional $I(u)$ is the elastic energy of the crystal, a quadratic functional of elastic strains, $\varepsilon_{ij}^{(e)} = u_{(i,j)} - \bar{\varepsilon}_{ij}^{(p)}$,

$$\int_V \frac{1}{2} C^{ijkl} \varepsilon_{ij}^{(e)} \varepsilon_{kl}^{(e)} dV.$$

Let us assume that the random fields $\varepsilon_{ij}^{\prime(p)}$ and α'_{ij} have a correlation radius, a, which is much smaller than the characteristic size of region V. Assume also that the field $\varepsilon_{ij}^{\prime(p)}$ is ergodic, i.e. its volume integral coincides with mathematical expectation and, thus, equal to zero (the size of the volume must be much bigger than a). Then

$$\int_V \bar{\sigma}^{ij} \varepsilon_{ij}^{\prime(p)} dV \simeq 0.$$

Let us divide V in a large number of boxes, $B_1, \ldots, B_N$, the size of which is much bigger than a. Then $\alpha_i'^m$ can be considered statistically independent in different boxes. The linear functional,

$$\int_V e^{ikl} \psi_{km,l} \alpha_i'^m dV,$$

becomes a sum of statistically independent linear functionals with zero mean value,

$$\int_V e^{ikl}\psi_{km,l}\alpha_i^{\prime m}dV = \sum_{a=1}^{N}\int_{B_a} e^{ikl}\psi_{km,l}\alpha_i^{\prime m}dV.$$

Then the mathematical expectation of energy is equal to the sum of mathematical expectations of energies of each box (see footnote in Section 16.6). Hence,

$$M\mathcal{E} = \int_V U_m dV$$

where $U_m\Delta V$ is the energy of the box. To find U_m, one has to remove all dislocations outside the box and find energy of the system assuming that the displacements at ∂V_u are zero. Since $\mathcal{E}$ is a sum of a large number of energies of boxes, it coincides with $M\mathcal{E}$ as $N \to \infty$, and we arrive at the following formula for the total energy of the crystal, E,

$$E = \min_{u\in(6.166)} \int \frac{1}{2}C^{ijkl}\left(u_{(i,j)} - \bar{\varepsilon}_{ij}^{(p)}\right)\left(u_{(k,l)} - \bar{\varepsilon}_{kl}^{(p)}\right)dV + \int_V U_m dV. \qquad (6.169)$$

Here an addition to elastic energy appears, the energy of microstructure, U_m, the energy of crystal defects. It is essential that for calculation U_m one has to use not the true dislocation density, α_i^j, but its fluctuations, $\alpha_i^{\prime j}$. Note that our derivation holds true, if the tensor of elastic moduli depends on coordinates. Therefore, (6.169) holds for polycrystals as well.

Chapter 7
Statics of a Geometrically Nonlinear Elastic Body

In the geometrically nonlinear case, displacements and their gradients are not small. The major new feature which that brings into the theory is non-convexity of the energy functional.

7.1 Energy Functional

Free energy. In elasticity theory, the free energy density F is a known function of distortion x_a^i and temperature. Due to invariance of energy with respect to rotations, distortion can enter in F in combinations like g_{ab}, $|x|_{ab}$, γ_{ab}, or ε_{ab} (see (3.33), (3.34), (3.35), (3.36) and (3.37)). The function F also depends on the physical constants – the components of some fixed tensors with Lagrangian indices, which do not change in the process of deformation and characterize the elastic properties. Physical constants of inhomogeneous bodies are some functions of material points, and through such functions the free energy explicitly depends on Lagrangian coordinates. This is not emphasized in notation, and only significant (i.e. depending on motion) variables are written as the arguments of F. In the first three sections of this chapter we mean by $F\left(x_a^i\right)$ the free energy per unit volume of the undeformed state, so the total free energy of the body is

$$\int_{\mathring{V}} F d\mathring{V}.$$

We take zero as the "reference point" of free energy and assume that there are no stresses in the initial state. Then F satisfies the conditions

$$F = 0, \qquad \frac{\partial F}{\partial \varepsilon_{ab}} = 0 \quad \text{for}\, \varepsilon_{ab} = 0. \tag{7.1}$$

There are two significantly different cases in modeling of elastic bodies: the case of small deformations and the case of finite deformations. Deformations are called small if they are negligible in comparison with unity. Otherwise, the deformations

V.L. Berdichevsky, *Variational Principles of Continuum Mechanics*,
Interaction of Mechanics and Mathematics, DOI 10.1007/978-3-540-88467-5_7,

are called finite. In the theory of small deformations, either the tensor ε_{ab} or the tensor γ_{ab} can be used as the measure of deformation, depending on convenience. According to formula (3.37), for small deformations these two tensors coincide.

In the theory of small deformations, the function F can be taken as a quadratic form with respect to the strain tensor:

$$F = \frac{1}{2}C^{abcd}\varepsilon_{ab}\varepsilon_{cd}. \tag{7.2}$$

We consider here only isothermal equilibrium processes, and therefore the terms depending on temperature are dropped.

Since any elastic deformation of the body involves an accumulation of energy, the quadratic form (7.2) must be positive:

$$\frac{1}{2}\underline{\mu}\varepsilon_{ab}\varepsilon^{ab} \leq F, \quad \underline{\mu} = const > 0. \tag{7.3}$$

Remember that the juggling of indices is done by means of the metric of the undeformed state $\mathring{g}_{ab}$.

Formula (7.2) defines a physically linear material. The only difference from the physically linear material considered in the previous section is that distortion x_a^i may differ considerably from $\mathring{x}_a^i$.

In the case of an isotropic material (i.e. a material for which $F(\varepsilon_{ab}) = F\left(\varepsilon_{cd}\alpha_a^c\alpha_b^d\right)$ for any orthogonal matrix with the components α_a^c), there are two material characteristics, λ and μ, and

$$F = \frac{1}{2}\lambda\left(\varepsilon_a^a\right)^2 + \mu\varepsilon_{ab}\varepsilon^{ab}. \tag{7.4}$$

The quadratic form (7.2) can be considered as the first term of Taylor's expansion of the free energy with respect to ε_{ab}. The next terms are cubic in strains:

$$F = \frac{1}{2}C^{abcd}\varepsilon_{ab}\varepsilon_{cd} + \frac{1}{3!}C^{aba'b'cd}\varepsilon_{ab}\varepsilon_{a'b'}\varepsilon_{cd}.$$

If the components of tensors C^{abcd} and $C^{aba'b'cd}$ are of the same order of magnitude and the deformations are small, then in statics the cubic form can be ignored. There are, however, the materials for which the components of the tensor $C^{aba'b'cd}$ are much greater than the components of the tensor of elastic moduli. For such materials, the cubic terms may be of the same order as the quadratic ones and must be taken into account. In dynamics, the cubic terms, even being small, bring qualitatively new effects.

In what follows, by the theory of small elastic deformations we mean the theory in which F is given by the formula (7.2) or by an equivalent formula

$$F = \frac{1}{2} C^{abcd} \gamma_{ab} \gamma_{cd}. \tag{7.5}$$

In the range of finite deformations, expressions (7.2) and (7.5) are not equivalent. The model with the energy density (7.5) is called the semi-linear elastic body. The name is due to the fact that the tensor $|x|_{ab}$ used to construct the deformation measure γ_{ab} depends semi-linearly on distortion (see Sect. 3.1).

Formulas (7.2) and (7.5) can be used in the case of finite but not very large deformations. Indeed, consider, for example, a homogeneous collapse of a body into a point. In this case, the stress tensors ε_{ab} and γ_{ab} change from zero to $-\frac{1}{2}\mathring{g}_{ab}$ and from zero to $-\mathring{g}_{ab}$, respectively. If the energy is defined by formulas (7.2) or (7.5) for any finite deformations, then the energy change for the collapse is finite. Consequently, to collapse a body into a point one has to perform finite work. Besides, the external forces are finite because the stresses are on the order of μ at the collapse. Experiments show that solids cannot be compressed to a point by finite forces.

For finite strains, it is natural to use the components of tensor g_{ab} or tensor $|x|_{ab}$ as the strain measures, since their deviations from initial values are not small and the role of ε_{ab} and γ_{ab} as the measures of deviation from the initial values is lost.

In the construction of the elastic energy in the case of finite deformations, the fact that the energy has to tend to infinity when the continuum is being compressed to a point or infinitely expanded has to be taken into account. When an element of the continuum is compressed to a point, the contravariant components of the metric tensor, g^{ab}, tend to infinity; for an infinite expansion, the covariant components of the metric tensor, g_{ab}, tend to infinity. For an isotropic material the simplest expression satisfying the condition of infinite energy in the cases of infinite compression or infinite expansion can be constructed from the contractions $\mathring{g}^{ab} g_{ab}$ and $\mathring{g}_{ab} g^{ab}$:

$$F = c_1 \mathring{g}^{ab} g_{ab} + c_2 \mathring{g}_{ab} g^{ab}. \tag{7.6}$$

The function (7.6) is to be corrected to satisfy the conditions (7.1), and to coincide with the function (7.4) for small deformations. After addition of the necessary additive constant and inclusion of another possible term, $\hat{g}/\mathring{g}$, one obtains

$$F = \frac{\mu}{4} \left[(1+c) \left(\mathring{g}^{ab} g_{ab} - 3 \right) + (1-c) \left(\mathring{g}_{ab} g^{ab} - 3 \right) \right] - \frac{\mu c}{2} \left(\frac{\hat{g}}{\mathring{g}} - 1 \right). \tag{7.7}$$

where μ is the shear modulus for small deformations, and c is the second material constant.

In the theory of finite deformations of incompressible materials, the last term in (7.7) is zero and the free energy density becomes

$$F = \frac{\mu}{4} \left[(1+c) \left(\mathring{g}^{ab} g_{ab} - 3 \right) + (1-c) \left(\mathring{g}_{ab} g^{ab} - 3 \right) \right]. \tag{7.8}$$

Expression (7.8) was suggested by Moony. The Moony material has one constant c experimentally determined for large deformations (the coefficient μ is the shear modulus for small deformations).

For $c = 1$, the Moony formula reduces to the formula derived by Treloar for rubbers from statistical reasoning,

$$F = \frac{1}{2}\mu\left(\mathring{g}^{ab} g_{ab} - 3\right). \tag{7.9}$$

Instead of contractions $\mathring{g}^{ab} g_{ab}$ and $\mathring{g}_{ab} g^{ab}$, the contractions $\mathring{g}^{ab} |x|_{ab}$ and $\mathring{g}_{ab}$ $|x|^{(-1)ab}$ can be used in constructing the free energy ($|x|^{(-1)ab}$ is the inverse tensor of tensor $|x|_{ab}$). Then the Moony and Treloar formulas are replaced by the expressions

$$F = \frac{\mu}{4}\left[(1+c)\left(\mathring{g}^{ab}|x|_{ab} - 3\right) + (1-c)\left(\mathring{g}_{ab}\,|x|^{(-1)ab} - 3\right)\right],$$

$$F = \frac{1}{2}\mu\left(\mathring{g}^{ab}\,|x|_{ab} - 3\right).$$

There exists a number of other suggestions on the dependence of the free energy on the deformation measures; we restrict the exposition by the examples given.

External forces. In principle, the body and the surface external forces may depend on the displacements and their gradients, and the work of external forces

$$\delta A = \delta A_{\text{body}} + \delta A_{\text{surf}},$$

$$\delta A_{\text{body}} = \int\limits_{\mathring{V}} \rho_0 g_i \delta x^i d\mathring{V}, \quad \delta A_{\text{surf}} = \int\limits_{\partial \mathring{V}_f} f_i \delta x^i d\mathring{A}, \tag{7.10}$$

may be non-holonomic, i.e. not-admitting the functional, $l\,(x\,(X))$, such that δA is equal to variation of $l\,(x\,(X))$. In this case, the variational equation of the first law of thermodynamics,

$$\delta F = \delta A,$$

is not reduced to a variational principle. The conditions on the external forces for which a functional $l\,(x\,(X))$ exists can be obtained by the approach outlined in Appendix C (vol. 2). Here we summarize the results. The work of the body forces is holonomic if and only if, a function Φ of particle positions exists, such that

$$g_i = -\frac{\partial \Phi\,(x\,(X))}{\partial x^i}.$$

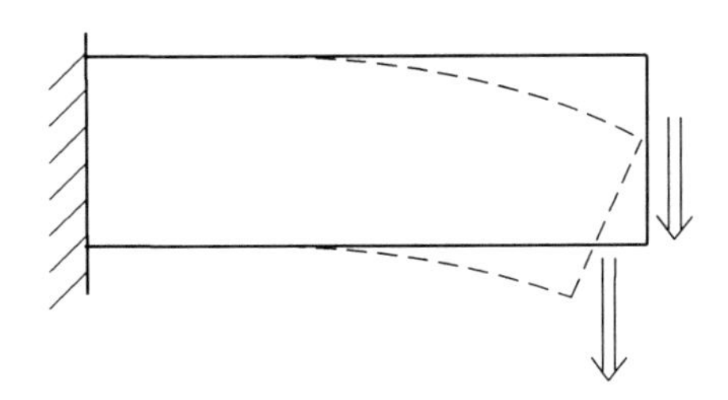

Fig. 7.1 A "dead" surface force

Then δA_{body} is a variation of the following functional:

$$\delta A_{\text{body}} = -\delta \int_{\mathring{V}} \rho_0 \Phi d\mathring{V} = -\delta \int_{V} \rho \Phi dV. \tag{7.11}$$

An important example is the gravity force:

$$\Phi = -g_i x^i \left(X^a\right),$$

where $g_i \equiv \mathit{const}$ is the gravity acceleration.

A surface load is called "dead" if the surface force per unit area in the undeformed state depends only on Lagrangian coordinates, i.e. in (7.10) $f_i = f_i\left(X^a\right)$. In the process of deformation, the "dead" loads conserve their magnitude and the direction in space at every fixed point of the body's boundary (Fig. 7.1). As a rule, the loads caused by gravity are "dead."

In the case of "dead" loads, the work of the surface forces has the potential

$$\int_{\partial \mathring{V}_f} f_i(X) x^i (X) d\mathring{A}.$$

Another case of holonomic work of surface forces is the hydrostatic load. It appears when elastic bodies interact with ideal fluid. Then the surface forces per unit area in the deformed state, f_i, are directed along the normal vector to the deformed surface,

$$f_i = p n_i, \tag{7.12}$$

p being fluid pressure. Pressure may be considered as a known function of Eulerian coordinates, and, therefore, of $x\,(X)$,

$$p = p\,(x\,(X)).$$

Then the work of surface forces has the following potential:

$$\delta A_{\text{surf}} = \delta \left[\int_{V} p\,(x) dV. \right] \tag{7.13}$$

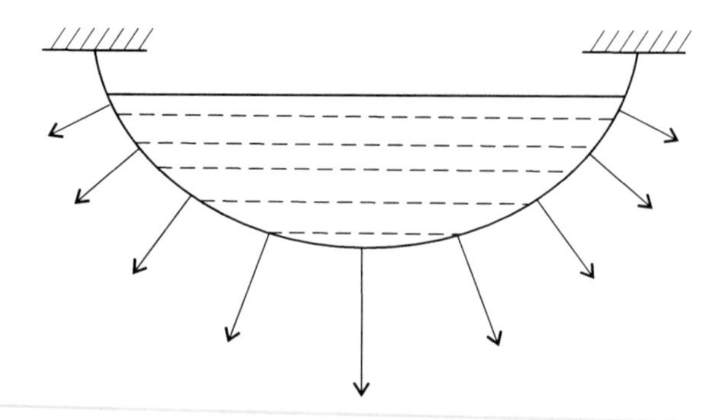

Fig. 7.2 Hydrostatic load acting on an elastic shell

In particular, for a body under constant external hydrostatic pressure,

$$\delta \mathrm{A}_{\mathrm{surf}} = p\delta\left(|V|\right).$$

Inhomogeneous hydrostatic load appears, for example, in the problems on the deformation of elastic shells containing fluid (see Fig. 7.2).

If at every boundary point of the body the direction of the vector f_i is linked to the direction of the normal vector of the surface, the load is called following. The simplest example of the following load is a surface force which, at every boundary point, is normal to the surface, i.e. which is defined by the formula (7.12). The difference from the hydrostatic load is that p is not a universal function of Eulerian coordinates, but can also depend on the Lagrangian coordinates of the surface points. Such cases are encountered in some problems on deformation of shells and beams. For example, for the load shown in Fig. 7.3, $p = const \neq 0$ at the end of the beam and $p = 0$ on the lateral surface of the beam. The work of following loads is usually non-holonomic.

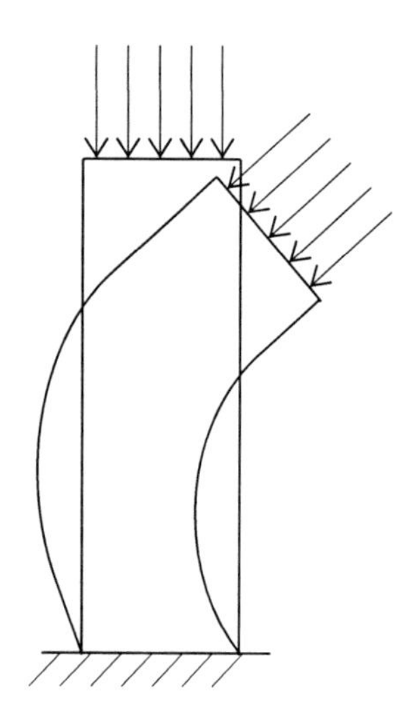

Fig. 7.3 An example of following load

The issue of the holonomicity of the work of external forces is of importance in the problems of vibrations of elastic bodies. For holonomic external loads, their total work per one cycle of vibrations is equal to zero, and it can be expected that the amplitude of vibrations will not increase. If the work of external forces is not holonomic, then the total work during one cycle depends on the path. There might exist such cycles on which the total work is positive, and the amplitude of vibrations grows.

Non-convexity of free energy. The geometrically nonlinear theory of elasticity could be studied by means of general methods of variational calculus set forth in Chap. 5 if the free energy density F were a convex function of the distortion x_a^i. However, the invariance of F with respect to rigid motion practically excludes such a possibility. Let us discuss this issue in more detail.

Consider the quadratic form

$$\Delta F = \frac{1}{2} \frac{\partial^2 F}{\partial x_a^i \partial x_b^j} \bar{x}_a^i \bar{x}_b^j.$$

In order for F to be convex, it is necessary and sufficient for the quadratic form ΔF to be positively definite (see (5.90)). Let us take $g_{ab} = x_a^i x_{ib}$ as the arguments of F. Since

$$\frac{\partial g_{dc}}{\partial x_a^i} = x_{ic} \delta_d^a + x_{id} \delta_c^a,$$

we have

$$\frac{\partial F}{\partial x_a^i} = 2 \frac{\partial F}{\partial g_{ac}} x_{ic}, \qquad \frac{\partial^2 F}{\partial x_a^i \partial x_b^j} = 4 \frac{\partial^2 F}{\partial g_{ac} \partial g_{bd}} x_{ic} x_{jd} + \frac{\partial F}{\partial g_{ab}} g_{ij}. \tag{7.14}$$

Denote $\bar{x}_a^i x_{ib}$ by $\bar{x}_{ab}$. The quadratic form ΔF becomes

$$\Delta F = 2 \frac{\partial^2 F}{\partial g_{ac} \partial g_{bd}} \bar{x}_{ac} \bar{x}_{bd} + \frac{1}{2} \frac{\partial F}{\partial g_{ab}} g^{cd} \bar{x}_{ac} \bar{x}_{bd}. \tag{7.15}$$

The tensor $\bar{x}_{ac}$ can be presented as the sum of its symmetric and antisymmetric parts

$$\bar{x}_{ac} = \bar{\varepsilon}_{ac} + e_{ach} \bar{\omega}^h. \tag{7.16}$$

Here,

$$\bar{\omega}^h = \frac{1}{2} e^{ach} \bar{x}_{ac}, \; \bar{\varepsilon}_{ab} = \bar{x}_{(ab)}.$$

Substituting (7.16) into the expression for ΔF and using the notation for the Lagrangian components of stress tensor,[1]

$$\sigma^{ab} = \frac{\rho}{\rho_0}\frac{\partial F}{\partial \varepsilon_{ab}} = 2\frac{\rho}{\rho_0}\frac{\partial F}{\partial g_{ab}},$$

we obtain

$$\Delta F = \frac{1}{2}C'^{abcd}\bar{\varepsilon}_{ab}\bar{\varepsilon}_{cd} + \frac{\rho_0}{\rho}\sigma^{ab}g^{cd}\bar{\varepsilon}_{ac}e_{bdh}\bar{\omega}^h$$
$$+\frac{\rho_0}{2\rho}\sigma^{ab}g^{cd}e_{ach'}e_{bdh}\bar{\omega}^h\bar{\omega}^{h'}. \tag{7.17}$$

Here,

$$C^{abcd} = 4\frac{\partial^2 F}{\partial g_{ab}\partial g_{cd}}$$

is the tensor of the instantaneous elastic moduli (for a physically linear material they coincide with the similarly denoted quantities in (7.2)), and

$$C'^{abcd} = C^{abcd} + \frac{\rho_0}{2\rho}\sigma^{ac}g^{bd} + \frac{\rho_0}{2\rho}\sigma^{ad}g^{bc}. \tag{7.18}$$

If we set $\bar{\omega}^h = 0$, then the convexity condition becomes

$$\Delta F = \frac{1}{2}C'^{abcd}\bar{\varepsilon}_{ab}\bar{\varepsilon}_{cd} \geq 0. \tag{7.19}$$

For small deformations, $C^{abcd}\mu, \quad \sigma^{ab}\mu\varepsilon, \quad \varepsilon$ being the magnitude of strain, the last two terms in (7.18) are much smaller than the first one and can be neglected. Then (7.19) is warranted by the validity of (7.3).

Let us now set $\bar{\varepsilon}_{ab} = 0$. It follows from the formula (7.17) that for the convexity of F it is necessary that

$$\sigma^{ab}g^{cd}e_{ach}e_{bdh'}\bar{\omega}^h\bar{\omega}^{h'} \geq 0 \quad \text{for any } \bar{\omega}^h. \tag{7.20}$$

Since, according to (3.19),

$$g^{cd}e_{ach}e_{bdh'} = \frac{1}{\hat{g}}g^{cd}\varepsilon_{ach}\varepsilon_{bdh'} = \frac{1}{\hat{g}}\left(g_{ab}g_{hh'} - g_{ah'}g_{bh}\right),$$

the inequality (7.20) can be written as

[1] Recall that by F we denoted in this section free energy per unit initial volume.

$$\left(\sigma_c^c g_{ab} - \sigma_{ab}\right) \bar{\omega}^a \bar{\omega}^b \geq 0. \tag{7.21}$$

Here, the indices are juggled by means of the metric g_{ab}.

Consider the Lagrangian coordinate system for which the metric tensor, g_{ab}, coincides with δ_{ab}, while the stress tensor is diagonal. The inequality (7.21) means that in this coordinate system, the sum of any two diagonal elements of the stress tensor must be non-negative:

$$\sigma_1^1 + \sigma_2^2 \geq 0, \quad \sigma_1^1 + \sigma_3^3 \geq 0, \quad \sigma_2^2 + \sigma_3^3 \geq 0. \tag{7.22}$$

The inequalities (7.22) are necessary for convexity of energy. It is clear that the stress state for which the inequalities (7.22) are not satisfied can always be realized (for example, hydrostatic pressure). Hence, the elastic energy is not a convex function of distortion.

The non-convexity of F can also be made obvious by a concrete example of a physically linear material: we set $x^1 = x(X)$, $X = X^1$, $x^2 = X^2$, $x^3 = X^3$, $\mathring{g}_{11} = 1$, and take

$$F = \frac{1}{2} C \varepsilon^2, \quad \varepsilon = \frac{1}{2}\left(\left(\frac{dx}{dX}\right)^2 - 1\right).$$

The graph of the function $F(dx/dX)$ is shown in Fig. 7.4. The non-convexity of F is obvious.

In this example, F is convex in a neighborhood of the undeformed state, and becomes non-convex for finite deformations. It turns out that for three-dimensional deformations the situation is more complex: the function F is non-convex in any small vicinity of the undeformed state. Indeed, consider the nine-dimensional space of variables x_a^i. The undeformed state corresponds to the point $x_a^i = \mathring{x}_a^i$. Since in the undeformed state $\sigma^{ab} = 0$, the quadratic form ΔF (7.17) computed at the point $\mathring{x}_a^i$ is

$$\frac{1}{2} \mathring{C}^{abcd} \bar{\varepsilon}_{ab} \bar{\varepsilon}_{cd} \geq 0, \quad \mathring{C}^{abcd} = C^{abcd}\Big|_{\varepsilon_{ab}=0}.$$

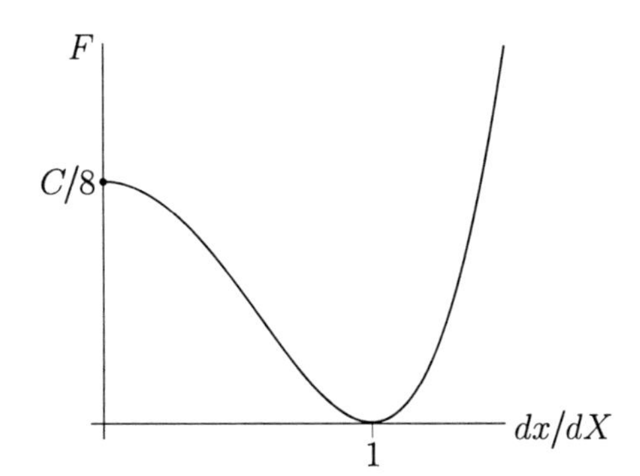

Fig. 7.4 Free energy as a function of distortion for one-dimensional deformation

Therefore, the convexity condition at the point $\mathring{x}_a^i$ coincides with the condition (7.3): energy is accumulated for any small deformations. The condition of the strict convexity of F is not satisfied: $\Delta F = 0$ for any $\bar{\omega}^h$. If at the point $\mathring{x}_a^i$ the condition of strict convexity were held, then there would be a neighborhood of the point $\mathring{x}_a^i$ for which the condition of strict convexity would be satisfied, and the convexity of F could be guaranteed in this neighborhood. However, for any infinitesimally small shift from the point $\mathring{x}_a^i$ the quadratic form ΔF ceases to be positive. In order to see that, consider the expression (7.17) for small deformations. The components of the tensor C'^{abcd} differ by quantities on the order of ε from the components of the tensor $\mathring{C}^{abcd}$. Since the quadratic form $\mathring{C}^{abcd}\bar{\varepsilon}_{ab}\bar{\varepsilon}_{cd}$ is positive definite, the replacement of the first term in (7.17) by $\frac{1}{2}\mathring{C}^{abcd}\bar{\varepsilon}_{ab}\bar{\varepsilon}_{cd}$ cannot result in the change of the sign of ΔF. Similarly, g^{cd} can be replaced by $\mathring{g}^{cd}$, and ρ_0/ρ by 1 in the second and the third terms of (7.17). The expression for ΔF becomes

$$\Delta F = \frac{1}{2}\mathring{C}^{abcd}\bar{\varepsilon}_{ab}\bar{\varepsilon}_{cd} + \sigma^{ab}\bar{\varepsilon}_a^d e_{bdh}\bar{\omega}^h + \frac{1}{2}\left(\sigma_c^c \mathring{g}_{ab} - \sigma_{ab}\right)\bar{\omega}^a\bar{\omega}^b. \tag{7.23}$$

The quadratic form ΔF has the following structure:

$$\Delta F = a^{ij}x_i x_j + 2b^{i\alpha}x_i y_\alpha + c^{\alpha\beta}y_\alpha y_\beta,$$

where the variables x_i and y_α represent $\bar{\varepsilon}_{ab}$ and $\bar{\omega}^a$, respectively, the coefficients a^{ij} are on the order of μ, and the coefficients $b^{i\alpha}$ and $c^{\alpha\beta}$ are small quantities on the order of $\mu\varepsilon$. In such a quadratic form the interaction term, $2b^{i\alpha}x_i y_\alpha$, in the first approximation, does not affect the positiveness of the form. Indeed, the form can be written as

$$\Delta F = a^{ij}\left(x_i + a_{ik}^{(-1)}b^{k\alpha}y_\alpha\right)(i \to j) + c^{\alpha\beta}y_\alpha y_\beta - a_{ij}^{(-1)}b^{i\alpha}b^{j\beta}y_\alpha y_\beta.$$

The last term is small compared to the previous one, and in the first approximation can be ignored. Therefore the quadratic form ΔF is positive definite if and only if $a^{ij}x_i x_j \geq 0$ and $c^{\alpha\beta}y_\alpha y_\beta \geq 0$. The first of these inequalities is the condition of energy accumulation, while the second is the inequality (7.21). In any infinitesimally small neighborhood of the undeformed state there are points for which the inequality (7.21) is not satisfied. So, energy of the elastic body is not convex.

7.2 Gibbs Principle

In what follows, we assume the external forces to be "dead" and given on a part of the boundary, $\partial\mathring{V}_f$, while the position of the other part of the boundary, ∂V_u, is given:

$$x^i(X) = x_{(b)}^i(X) \quad \text{on } \partial V_u. \tag{7.24}$$

Gibbs principle. *Equilibrium states are the stationary points of the functional,*

$$I(x(X)) = \int_{\mathring{V}} F\left(x_a^i\right) d\mathring{V} - l,$$

$$l = \int_{\mathring{V}} \rho_0 g_i(X) x^i(X) d\mathring{V} + \int_{\partial\mathring{V}_f} f_i(X) x^i(X) d\mathring{A}, \tag{7.25}$$

on the set of functions $x(X)$ satisfying the conditions (7.24).

The stationary points of the functional $I(x(X))$ are the solutions of the equilibrium equations

$$\mathring{\nabla}_a p_i^a + \rho_0 g_i = 0, \quad p_i^a \equiv \frac{\partial F}{\partial x_a^i},$$

with the boundary conditions on ∂V_f,

$$p_i^a \mathring{n}_a = f_i,$$

and the boundary conditions (7.24) on $\partial\mathring{V}_u$.

The constitutive equations. The form of constitutive equations depends on the choice of the arguments of F. If F is a function ε_{ab} or g_{ab}, then the Piola-Kirchhoff tensor is linked to the strain by the equations

$$p_i^a = 2\frac{\partial F}{\partial g_{ab}} x_{ib} = \frac{\partial F}{\partial \varepsilon_{ab}} x_{ib}.$$

Here we used (7.14). We see that Piola-Kirchhoff tensor depends not only on strain but also on rotation of material elements. To derive the constitutive equations for the case when F is a function of γ_{ab} or $|x|_{ab}$, we need to find the derivatives $\partial\, |x|_{bc} / \partial x_a^i$. By definition, $|x|_{bc}$ are determined by the conditions that the distortion, x_a^i, can be presented as the product of a symmetric non-negative matrix $|x|_{bc}$ and an orthogonal matrix λ^{ic} : $x_b^i = |x|_{bc}\, \lambda^{ic}$. Recall that the orthogonality conditions are

$$g_{ij}\lambda^{ia}\lambda^{jb} = \mathring{g}^{ab}, \quad \mathring{g}_{ab}\lambda^{ia}\lambda^{jb} = g^{ij}. \tag{7.26}$$

Juggling of Eulerian and Lagrangian indices is done by means of the metrics g_{ij} and $\mathring{g}_{ab}$, respectively. Therefore, the orthogonality conditions can also be written as

$$\lambda^{ia}\lambda_{ib} = \delta_b^a, \quad \lambda^{ia}\lambda_{ja} = \delta_j^i.$$

Let δx_b^i be an infinitesimally small increment of distortion. Varying the equality $x_b^i = |x|_{bc}\, \lambda^{ic}$, we have

$$\lambda^{ic}\delta\,|x|_{bc} + |x|_{bc}\,\delta\lambda^{ic} = \delta x^i_b. \tag{7.27}$$

Contracting (7.27) with a non-degenerated matrix, λ_{id}, we obtain an equivalent system of equations:

$$\delta\,|x|_{bd} + |x|^c_b\,\lambda_{id}\delta\lambda^i_c = \lambda_{id}\delta x^i_b. \tag{7.28}$$

The tensor $\lambda_{id}\delta\lambda^i_c$ is antisymmetric with respect to d, c. This can be seen from varying the orthogonality condition (7.26): $\lambda_{id}\delta\lambda^i_c + \lambda_{ic}\delta\lambda^i_d = 0$. Therefore, the tensor $\lambda_{id}\delta\lambda^i_c$ is in one-to-one correspondence with a three-dimensional vector $\delta\lambda^b$,

$$\lambda_{id}\delta\lambda^i_c = e_{dch}\delta\lambda^h, \quad \delta\lambda^h \equiv \frac{1}{2}e^{dch}\lambda_{id}\delta\lambda^i_c. \tag{7.29}$$

Equation (7.28) contains nine equalities. First we write down the three of them which are obtained by contracting (7.28) with e^{bda}. Since, according to (3.19),

$$|x|^c_b\,\lambda_{id}\delta\lambda^i_c e^{bda} = |x|^c_b\,e_{dch}\delta\lambda^h e^{bda} = |x|^c_b\left(\delta^a_c\delta^b_h - \delta^b_c\delta^a_h\right)\delta\lambda^h,$$

such contraction yields a system of three equations with respect to vector $\delta\lambda^h$:

$$\left(|x|^b_b\,\delta^a_h - |x|^a_h\right)\delta\lambda^h = \lambda_{id}\delta x^i_b e^{dba}. \tag{7.30}$$

In the principal coordinate system of the tensor $|x|_{ab}$, the tensor $|x|^b_b\,\delta^a_h - |x|^a_h$ is diagonal and its diagonal components are equal to $|x|^2_2 + |x|^3_3$, $|x|^3_3 + |x|^1_1$, $|x|^1_1 + |x|^2_2$. The numbers, $|x|^1_1$, $|x|^2_2$, and $|x|^3_3$ are positive. Thus, the inverse tensor y^h_c can be introduced by the equation

$$y^h_c\left(|x|^b_b\,\delta^a_h - |x|^a_h\right) = \delta^a_c.$$

In the principal coordinate system of the tensor $|x|_{ab}$, the components of the tensor y^h_c are

$$y^1_1 = \left(|x|^2_2 + |x|^3_3\right)^{-1},\ y^2_2 = \left(|x|^3_3 + |x|^1_1\right)^{-1},\ y^3_3 = \left(|x|^1_1 + |x|^2_2\right)^{-1}.$$

The solution to (7.30) is given by the formula

$$\delta\lambda^h = y^h_c e^{dbc}\lambda_{id}\delta x^i_b. \tag{7.31}$$

From the first equality (7.29) and (7.31) we have,[2]

$$\lambda_{id}\delta\lambda^i_c = e_{dch}y^h_{c'}e^{d'bc'}\lambda_{id'}\delta x^i_b. \tag{7.32}$$

---

[2] In substituting (7.31) into (7.29), one has to re-denote the dummy summation indices d and c. Since, by our convention, only the first few letters of the Latin alphabet are reserved for the Lagrangian indices, we increase the number of "admissible letters" by using the letters with a prime.

Substitution of (7.32) into (7.28) yields an expression for $\delta\,|x|_{bd}$ in terms of δx_a^i:

$$\delta\,|x|_{bd} = \lambda_{id}\delta x_b^i - |x|_b^c\, e_{dch} y_{c'}^h e^{d'b'c'} \lambda_{id'} \delta x_{b'}^i. \tag{7.33}$$

Each of the two terms on the right hand side of (7.33) is non-symmetric with respect to b, d; however, it is easy to verify that their difference is symmetric. From (7.33) we find that

$$\frac{\partial\,|x|_{bd}}{\partial x_a^i} = \lambda_{id}\delta_b^a - |x|_b^c\, e_{dch} y_{c'}^h e^{d'ac'} \lambda_{id'}. \tag{7.34}$$

Therefore, if F is considered as a function of $|x|_{bd}$ or γ_{bd}, the expression for the Piola-Kirchhoff tensor is

$$\begin{aligned} p_i^a &= \frac{\partial F}{\partial\,|x|_{ad}}\lambda_{id} - \frac{\partial F}{\partial\,|x|_{bd}}\,|x|_b^c\, e_{dch} y_{c'}^h e^{d'ac'} \lambda_{id'} = \\ &= \frac{\partial F}{\partial \gamma_{ad}}\lambda_{id} - \frac{\partial F}{\partial \gamma_{bd}}\,|x|_b^c\, e_{dch} y_{c'}^h e^{d'ac'} \lambda_{id'}. \end{aligned} \tag{7.35}$$

The condition of local minimum. The functional $I\,(x\,(X))$ is not convex because F is a non-convex function of x_a^i. Physically, this is natural: if the functional $I\,(x\,(X))$ were convex, then, according to the uniqueness theorem (Sect. 5.4) it would have only one stationary point, an apparent contradiction to the experimentally observed instability of elastic bodies, i.e. to the existence of several equilibrium states for a given load.

An important characteristic of the stationary point $x\,(X)$ is the second variation of $I\,(x\,(X))$:

$$\Delta I = \lim_{\sigma\to 0} \frac{I\,(x\,(X) + \sigma \bar{x}\,(X)) - I\,(x\,(X))}{\sigma^2}.$$

Recall that the first variation,

$$\lim_{\sigma\to 0} \frac{I\,(x\,(X) + \sigma \bar{x}\,(X)) - I\,(x\,(X))}{\sigma},$$

computed at a stationary point is equal to zero for all admissible $\bar{x}\,(X)$.

The functional ΔI is a quadratic functional with respect to $\bar{x}\,(X)$. If the functional ΔI is positive, then small disturbances of external forces will result in small changes of the minimizer.

Let us obtain the formula for the second variation. It is obvious that

$$\Delta I = \int_V \Delta F dV.$$

where ΔF is given by (7.17), with $\bar{x}_{ac} = x_{ic}\partial\bar{x}^i/\partial X^a$.

The expression for ΔF admits significant simplifications if the deformations are small enough to omit not only the terms on the order of deformations, ε, but also the terms on the order of $\sqrt{\varepsilon}$. Indeed, consider the second term in (7.23). According to the inequality $\left|x_i y^i\right| \le \frac{1}{2}\left(\lambda^{-1} x_i x^i + \lambda y_i y^i\right)$, which holds for any x_i, y^i, $(i = 1, \ldots, r)$, and $\lambda > 0$, we have

$$\left|2\sigma^{ab} e_{bdh}\bar{\omega}^h \bar{\varepsilon}_a^d\right| \le \frac{1}{\lambda}\sigma^{ab} e_{bdh}\bar{\omega}^h \sigma_a^{b'} e_{b'h}^d \bar{\omega}^h + \lambda \bar{\varepsilon}_{ab}\bar{\varepsilon}^{ab}.$$

Stresses are on the order $\mu\varepsilon$. If we choose $\lambda = \mu\sqrt{\varepsilon}$, then the above expression is on the order $\mu\sqrt{\varepsilon}\left(\bar{\varepsilon}_{ab}\bar{\varepsilon}^{ab} + \bar{\omega}_h \bar{\omega}^h\right)$. So, neglecting the terms on the order of $\sqrt{\varepsilon}$ compared to unity, we obtain

$$\Delta I = \frac{1}{2}\int\limits_{\mathring{V}} \left(\mathring{C}^{abcd}\bar{\varepsilon}_{ab}\bar{\varepsilon}_{cd} + \left(\sigma_c^c \mathring{g}_{ab} - \sigma_{ab}\right)\bar{\omega}^a \bar{\omega}^b\right) d\mathring{V}.$$

Consider the condition of local convexity, $\Delta I \ge 0$. Choose the Lagrangian coordinate system in such a way that $x_a^i = \delta_a^i$. Then, denoting $\bar{x}_a = x_{ia}\bar{x}^i(X)$, we get

$$\bar{\varepsilon}_{ab} = \frac{1}{2}\left(\frac{\partial \bar{x}_a}{\partial X^b} + \frac{\partial \bar{x}_b}{\partial X^a}\right), \quad \bar{\omega}^a = \frac{1}{2}e^{abc}\left(\frac{\partial \bar{x}_b}{\partial X^c} - \frac{\partial \bar{x}_c}{\partial X^b}\right). \tag{7.36}$$

The functions $\bar{x}_a$ are equal to zero on $\partial \mathring{V}_u$. Thus, the equilibrium state is the point of local minimum of the functional $I(x(X))$, if for any functions $\bar{x}_a$ equal to zero on $\partial \mathring{V}_u$, the inequality holds:

$$\frac{1}{2}\int\limits_{\mathring{V}} \mathring{C}^{abcd}\bar{\varepsilon}_{ab}\bar{\varepsilon}_{cd} d\mathring{V} \ge -\frac{1}{2}\int\limits_{\mathring{V}} \left(\sigma_c^c \delta_{ab} - \sigma_{ab}\right)\bar{\omega}^a \bar{\omega}^b d\mathring{V}, \tag{7.37}$$

where $\bar{\varepsilon}_{ab}$ and $\bar{\omega}^a$ are expressed through $\bar{x}_a$ by means of (7.36). Note that for an isotropic body, the tensor $\mathring{C}^{abcd}$ does not "feel" the transition to the coordinate system chosen above within the accepted accuracy. For an isotropic body, the local convexity condition has the form

$$\frac{1}{2}\int\limits_{\mathring{V}} \left(\lambda\left(\bar{\varepsilon}_a^a\right)^2 + 2\mu\bar{\varepsilon}_{ab}\bar{\varepsilon}^{ab}\right) d\mathring{V} \ge -\frac{1}{2}\int\limits_{\mathring{V}} \left(\sigma_c^c \delta_{ab} - \sigma_{ab}\right)\bar{\omega}^a \bar{\omega}^b d\mathring{V}. \tag{7.38}$$

If the tensor $\sigma_c^c \delta_{ab} - \sigma_{ab}$ is positive at every point of the body (i.e. the quadratic form $\left(\sigma_c^c \delta_{ab} - \sigma_{ab}\right)\bar{\omega}^a \bar{\omega}^b$ is positive for $\bar{\omega}^a \neq 0$), then the inequality (7.37) holds true, and the equilibrium state is the point of local minimum of the functional $I(x(X))$. If the tensor $\sigma_c^c \delta_{ab} - \sigma_{ab}$ is negative, or if in some parts of the body the tensor $\sigma_c^c \delta_{ab} - \sigma_{ab}$ is positive while in others it is negative, then the investigation of the inequality (7.37) is a difficult problem.

The inequality (7.37) is similar in structure to the Korn inequality, and simple sufficient conditions can be given in terms of the Korn constant K for satisfying the inequality (7.37).

For generic boundary conditions non-convexity yields non-uniqueness of the stationary points. There is a special case, so-called "hard device" boundary conditions, when the particle positions are prescribed everywhere at the boundary. In this case, the minimum value may be unique even for non-convex energy. There is a vast literature on this subject, some papers are cited in bibliographic comments.

7.3 Dual Variational Principle

In this section a generalization of the Castigliano principle for the geometrically nonlinear case is given.

In constructing the Castigliano principle in geometrically linear theory, we presented F in terms of its Young-Fenchel transformation

$$F\left(\varepsilon_{ij}\right) = \max_{p^{ij}} \left(p^{ij}\varepsilon_{ij} - F^*\left(\varepsilon_{ij}\right)\right). \tag{7.39}$$

In the geometrically nonlinear theory the formula analogous to (7.39) would be

$$F\left(x_a^i\right) = \max_{p_i^a} \left[p_i^a x_a^i - F^*\left(p_i^a\right)\right]. \tag{7.40}$$

Equation (7.40), however, does not hold since the function F is not convex, and it can only be asserted that

$$F\left(x_a^i\right) \geq \max_{p_i^a} \left[p_i^a x_a^i - F^*\left(p_i^a\right)\right] \tag{7.41}$$

where $F^*\left(p_i^a\right)$ is the Young-Fenchel transformation of the function $F\left(x_a^i\right)$:

$$F^*\left(p_i^a\right) = \max_{x_a^i} \left[p_i^a x_a^i - F\left(x_a^i\right)\right].$$

Therefore, we will try to construct the Legendre transformation of the function $F\left(x_a^i\right)$, the function $F^\times\left(p_i^a\right)$, which is not necessarily single-valued everywhere, but is such that the function $p_i^a x_a^i - F^\times\left(p_i^a\right)$ at its stationary points over p_i^a coincides with the function $F\left(x_a^i\right)$. We assume that free energy F is a convex function of $|x|_{ab}$. Its Young-Fenchel transformation will be denoted by $G\left(n^{ab}\right)$:

$$G\left(n^{ab}\right) = \max_{|x|_{ab}} \left(n^{ab}|x|_{ab} - F\left(|x|_{ab}\right)\right).$$

Here, the maximum is computed over all symmetric positive tensors $|x|_{ab}$, and achieved at such $|x|_{ab}$ that $n^{ab} = \partial F / \partial |x|_{ab}$. Note that for small deformations $n^{ab} = 2\sigma^{ab}$. For finite deformations this is not true; the corresponding relation can be obtained using (3.38) or from further relations of this section.

Let us show that for an isotropic material:

1. The function $F^{\times}$ is given by the equation

$$F^{\times}\left(p_i^a\right) = G\left(|p|^{ab} s_b^c\right), \tag{7.42}$$

where $|p|^{ab}$ is the "modulus" of the Piola-Kirchhoff tensor defined by the polar expansion: $p_i^a = |p|^{ab} \mu_{ib}$ (μ_{ib} are the components of an orthogonal matrix). The function $F^{\times}$ has several branches, each of which is defined by the choice of the matrix s_b^c. In the principal coordinate system of the tensor $|p|^{ab}$, the matrix s_b^c has eight possible values:

$$\begin{Vmatrix} 1 & 0 & 0 \\ 0 & 1 & 0 \\ 0 & 0 & 1 \end{Vmatrix}, \quad \begin{Vmatrix} 1 & 0 & 0 \\ 0 & -1 & 0 \\ 0 & 0 & -1 \end{Vmatrix}, \quad \begin{Vmatrix} -1 & 0 & 0 \\ 0 & 1 & 0 \\ 0 & 0 & -1 \end{Vmatrix}, \quad \begin{Vmatrix} -1 & 0 & 0 \\ 0 & -1 & 0 \\ 0 & 0 & 1 \end{Vmatrix}, \tag{7.43}$$

$$\begin{Vmatrix} -1 & 0 & 0 \\ 0 & 1 & 0 \\ 0 & 0 & 1 \end{Vmatrix}, \quad \begin{Vmatrix} 1 & 0 & 0 \\ 0 & -1 & 0 \\ 0 & 0 & 1 \end{Vmatrix}, \quad \begin{Vmatrix} 1 & 0 & 0 \\ 0 & 1 & 0 \\ 0 & 0 & -1 \end{Vmatrix}, \quad \begin{Vmatrix} -1 & 0 & 0 \\ 0 & -1 & 0 \\ 0 & 0 & -1 \end{Vmatrix}. \tag{7.44}$$

Matrix $\|s_b^c\|$ is one of the matrices (7.43) if $\det \|p_i^a\| > 0$, and one of the matrices (7.44) if $\det \|p_i^a\| < 0$.

2. For given x_a^i, the point at which the function $\Phi\left(p_i^a, x_a^i\right) = p_i^a x_a^i - F^{\times}\left(p_i^a\right)$ is stationary with respect to p_i^a is uniquely defined; moreover, the branch of the function $F^{\times}$, for which $\Phi\left(p_i^a, x_a^i\right)$ has a stationary point, is determined by the values of x_a^i. At the stationary point, the following equality holds:

$$p_i^a x_a^i - F^{\times}\left(p_i^a\right) = F\left(x_a^i\right). \tag{7.45}$$

To prove the first assertion, we need to calculate $F^{\times} = p_i^a x_a^i - F\left(x_a^i\right)$, where x_a^i must be found by given p_i^a from the equation

$$p_i^a = \frac{\partial F\left(x_a^i\right)}{\partial x_a^i}. \tag{7.46}$$

Denote by $\mathcal{X}$ the set in the space of variables x_a^i defined by the condition $\det \|x_a^i\| > 0$, and by $\mathcal{P}$ the set run by $p_i^a = \partial F / \partial x_a^i$ when x_a^i take on the values in the set $\mathcal{X}$.

Note that, in general, p_i^a cannot take on any prescribed values. This can be seen, for example, from Fig. 7.4 for $dx/dX > 0$: the stress,

$$p = \frac{1}{2} C \frac{dx}{dX}\left(\left(\frac{dx}{dX}\right)^2 - 1\right),$$

can take on any positive values, while its possible negative values are bounded from below. In the one-dimensional case, the minimum value of stresses has a simple interpretation for semi-linear material. In this case, $p_1^1 = C(|x|_{11} - 1)$, and, since $|x|_{11} \geq 0$, $p_1^1 \geq -C$. At the value $p_1^1 = -C$ the material element collapses into a point.

Considering p_i^a, a known element of the set $\mathcal{P}$, we seek the solutions of the system of equations (7.46) from the set $\mathcal{X}$. According to (7.35), these equations can be written as

$$p_i^a = \frac{\partial F}{\partial |x|_{ad}} \lambda_{id} - \frac{\partial F}{\partial |x|_{bd}} |x|_b^c e_{dch} y_{c'}^h e^{d'ac'} \lambda_{id'}. \tag{7.47}$$

For an isotropic material, the tensors $|x|_{ab}$ and $\partial F / \partial |x|_{ab}$ are coaxial, and therefore the tensor $|x|_b^c \, \partial F / \partial |x|_{bd}$ is symmetric with respect to c, d. In the last term of (7.47) this tensor is contracted with an antisymmetric object e_{dch}. Therefore, this term vanishes, and the system of equations (7.47) becomes

$$p_i^a = n^{ab} \lambda_{ib}, \quad n^{ab} = \frac{\partial F}{\partial |x|_{ab}}. \tag{7.48}$$

Let us introduce the polar decomposition of the tensors p_i^a and n^{ab}:

$$p_i^a = |p|^{ab} \mu_{ib}, \quad n^{ab} = |n|^{ab'} s_{b'}^b.$$

In the principle coordinate system of the tensor n^{ab}, the non-diagonal elements of the matrix s_b^a are equal to zero, while the diagonal elements are equal to either $+1$ or -1, depending on whether the eigenvalue of the tensor n^{ab} is positive or negative.[3] Hence, in the principal coordinate system of the tensor $|n|^{ab}$, $\left\| s_b^a \right\|$ is one of the matrices (7.43) and (7.44).

In terms of the polar decomposition, the system of equations (7.48) can be written as

$$|p|^{ab} \mu_{ib} = |n|^{ab} s_b^d \lambda_{id}.$$

Due to the uniqueness of the polar decomposition,

$$|p|^{ab} = |n|^{ab}, \quad \mu_{ib} = \lambda_{ib'} s_b^{b'}. \tag{7.49}$$

The relations (7.49) determine how the solution of (7.46) should be obtained. First, the polar decomposition of the tensor p_i^a is constructed. That gives us the tensors $|p|^{ab}$ and μ_{ib}. Since the tensors $|p|^{ab}$ and $|n|^{ab}$ are equal, the matrix $\left\| s_b^a \right\|$ has the

[3] The case being considered is the generic one with $\det \left\| |n|^{ab} \right\| \neq 0$. The case when some of the eigenvalues of $|n|^{ab}$ are equal to zero is obtained by the limit transition.

form (7.43) or (7.44) in the principal coordinate system of the tensor $|p|^{ab}$. If we fix the matrix $\left\|s_b^a\right\|$, then $n^{ab} = |p|^{ab'} s_{b'}^b$ are specified. For known n^{ab}, we find $|x|_{ab}$ from the equation

$$n^{ab} = \frac{\partial F}{\partial |x|_{ab}}.$$

Finally, the "orthogonal part" of distortion is found from (7.49): $\lambda_{ib} = \mu_{ib'} s_b^{b'}$. So, each solution corresponds to some choice of the matrix $\left\|s_b^a\right\|$.

It follows from (7.49) and the condition $\det\left\|\lambda_d^i\right\| = +1$ that $\det\left\|\mu_a^i\right\| = \det\left\|s_b^c\right\|$. Therefore, for $\det\left\|\mu_a^i\right\| = +1$, s_b^c are the components of one of the matrices (7.43) and for $\det\left\|\mu_a^i\right\| = -1$, and s_b^c are the components of one of the matrices (7.44).

Since

$$|x|_{ab} = \frac{\partial G}{\partial n^{ab}}, \quad x_a^i = |x|_{ab}\,\lambda^{ib}, \tag{7.50}$$

the distortion is reconstructed using the components of the Piola-Kirchhoff tensor by the formula

$$x_a^i = \left.\frac{\partial G}{\partial n^{ab}}\right|_{n^{ab}=|p|^{ab'} s_{b'}^b} \mu^{ib'} s_{b'}^b. \tag{7.51}$$

For every Piola-Kirchhoff tensor p_i^a, we have eight solutions corresponding to eight different choices of the matrices $\left\|s_b^a\right\|$.

Note that the formula for the second variation F (7.23) shows that $\det\left\|\frac{\partial^2 F}{\partial x_a^i \partial x_b^j}\right\|$ is zero in a small vicinity of the planes $p_1^1 + p_2^2 = 0$, $p_2^2 + p_3^3 = 0$, $p_3^3 + p_1^1 = 0$, and for other values of x_a^i, this determinant is nonzero. Therefore, the bifurcations of the solutions of (7.46) are possible only in a small vicinity of these planes.

The branch corresponding to the first matrix (7.43) can be obtained by the following invariant condition:

$$\mathring{x}_a^i \lambda_i^a > 1. \tag{7.52}$$

The other branches do not satisfy the condition (7.52). Indeed, let λ_b^a be the projection of λ_i^a on the initial basis. The matrix, $\lambda_b^a = \mathring{x}_b^i \lambda_i^a = \mathring{x}_b^i \mu_i^c s_c^a$, is orthogonal:

$$\mathring{g}_{ab} \lambda_{a'}^a \lambda_{b'}^b = \mathring{g}_{ab} \mathring{x}_{a'}^i \lambda_i^a \mathring{x}_{b'}^j \lambda_j^b = g_{ij} \mathring{x}_{a'}^i \mathring{x}_{b'}^j = \mathring{g}_{a'b'}.$$

Denote by λ_{1b}^a the rotational part of distortion corresponding to the first matrix $\left\|s_b^a\right\|$ (7.43), and by λ_{2b}^a the rotational part of distortion corresponding to any other choice of the matrices (7.43) . Since $\lambda_b^a = \mathring{x}_b^i \mu_i^c s_c^a$, and $\mathring{x}_b^i \mu_i^c$ is fixed by the choice of the Piola-Kirchhoff tensor, the sum $\lambda_{1a}^a + \lambda_{2a}^a$, according to the construction of

s_c^a is equal to double the value of one of the diagonal elements of the matrix λ_{1b}^a in the principal coordinate system of the tensor $|p|^{ab}$. Since the components of the orthogonal matrix are less than or equal to 1,

$$\lambda_{1a}^a + \lambda_{2a}^a \leq 2.$$

Consequently, if λ_{1a}^a satisfies the condition (7.52), then λ_{1a}^a does not.

It can be proved analogously that the branch corresponding to the fourth matrix (7.44) is selected by the invariant condition

$$\mathring{x}_a^i \lambda_i^a < -1.$$

The condition (7.52) has a simple geometric interpretation. Let us use the representation of orthogonal matrices (3.40). It follows from (3.41) that $\lambda_a^a = 1 + 2\cos\theta$, where θ is the angle of rotation around the rotation axis. According to (7.52), $\cos\theta > 0$, and the inequality (7.52) means that the unique solution of (7.46) can be selected by the condition that the material filaments do not rotate for angles greater than $\pi/2$.

Let us calculate the function $p_i^a x_a^i - F\left(x_a^i\right)$ at its stationary points. From (7.51) we have

$$F^{\times}\left(p_i^a\right) = p_i^a x_a^i - F\left(x_a^i\right) = p_i^a \left.\frac{\partial G}{\partial n^{ab}}\right|_{n^{ab}=|p|^{ab'} s_{b'}^b} \mu^{ib'} s_{b'}^b - F =$$

$$= |p|^{ab'} s_{b'}^b \left.\frac{\partial G}{\partial n^{ab}}\right|_{n^{ab}=|p|^{ab'} s_{b'}^b} - F = \left.\left(n^{ab}\frac{\partial G}{\partial n^{ab}}\right)\right|_{n^{ab}=|p|^{ab'} s_{b'}^b} - F = G\left(|p|^{ab'} s_{b'}^b\right). \tag{7.53}$$

The function $F^{\times}$ has several possible values at every point p_i^a, depending on the choice of the matrix $\left\|s_b^a\right\|$.

Let us prove the second assertion. Consider the expression

$$\Phi\left(p_i^a, x_a^i\right) = p_i^a x_a^i - F^{\times}\left(p_i^a\right) = p_i^a x_a^i - G\left(|p|^{ab} s_b^d\right)$$

as a function of p_i^a; we seek its stationary points with respect to p_i^a when x_a^i are fixed. The function Φ has several branches, each of which is determined by the choice of the matrix $\left\|s_b^a\right\|$. Having chosen some matrix $\left\|s_b^a\right\|$, let us calculate the derivatives of $\Phi\left(p_i^a, x_a^i\right)$ with respect to p_i^a. Using the formula (7.34), written for the derivatives of $|p|_{ab}$ with respect to p_i^a, as the stationarity condition we get the system of equations for p_i^a:

$$x_a^i - \frac{\partial F^{\times}}{\partial p_i^a} = x_a^i - \left.\frac{\partial G}{\partial n^{ab}}\right|_{n^{ab}=|p|^{ab'} s_{b'}^b} s_{b'}^b \mu^{ib'} = 0. \tag{7.54}$$

It is taken into account here that, due to material isotropy, the tensor $s_{b'}^{b} \partial G / \partial n^{ab}$ is coaxial with the tensor n^{ab}; therefore, its contractions with the terms in the expression $\partial \, |p|^{bd} / \partial p_i^a$, analogous to the second term in (7.34), are zero. Introducing the polar decomposition for x_a^i into (7.54), we get

$$|x|_{ad} \, \lambda^{id} = \left. \frac{\partial G}{\partial n^{ab}} \right|_{n^{ab} = |p|^{ab'} s_{b'}^{b}} s_{b'}^{b} \mu^{ib'}.$$

The tensor $\partial G / \partial n^{ab}$ is symmetric. Besides, it is positive for every n^{ab} as follows from (7.50). Consequently, due to the uniqueness of the polar decomposition,

$$|x|_{ab} = \left. \frac{\partial G}{\partial n^{ab}} \right|_{n^{ab} = |p|^{ab'} s_{b'}^{b}}, \qquad \lambda^{id} = \mu^{ib} s_b^d.$$

Hence,

$$|p|^{ab'} s_{b'}^{b} = n^{ab}, \quad \mu^{ib} = \lambda^{id} s_d^b, \tag{7.55}$$

where n^{ab} is calculated by $|x|_{ab}$ from the formula $n^{ab} = \partial F / \partial \, |x|_{ab}$.

Consider the first equation (7.55) in the principal coordinate system of the tensor $|x|_{ab}$ (and n^{ab}). This equation suggests that the tensor $|p|^{ab}$ is diagonal and

$$|p|_1 \, s_1 = n_1, \quad |p|_2 \, s_2 = n_2, \quad |p|_3 \, s_3 = n_3. \tag{7.56}$$

Here and further, $|p|_1, |p|_2, |p|_3, s_1, s_2, s_3$, and n_1, n_2, n_3 are the diagonal components of the tensors $|p|_{ab}, s_b^d$ and n^{ab} in the principal coordinate system. Since $|p|_1, |p|_2$ and $|p|_3$ are positive, (7.56) with known n_1, n_2 and n_3 uniquely define $|p|_1 = |n_1|$, $|n|_2 = |n_2|$, $|p|_3 = |n_3|$, and

$$s_1 = \operatorname{sgn} n_1, \quad s_2 = \operatorname{sgn} n_2, \quad s_3 = \operatorname{sgn} n_3. \tag{7.57}$$

Subsequently, μ^{ia} are calculated from the second equation (7.55).

Equation (7.57) uniquely define the matrix $\left\| s_b^a \right\|$ by matrix $\left\| x_a^i \right\|$ (in the principal coordinate system of the tensor $|x|_{ab}$, s_1, s_2, s_3 are given by the formulas (7.57)). Hence, (7.54) have solutions with respect to p_i^a only in the case when the choice of x_a^i and s_b^d is coordinated in such a way as to satisfy the equalities (7.57). Consequently, despite the function $\Phi \left(p_i^a, x_a^i \right)$ being non-single-valued, its branch which has a stationary point with respect to p_i^a is picked out uniquely. At this stationary point, the function $\Phi \left(p_i^a, x_a^i \right)$ has the value

$$p_i^a x_a^i - F^{\times} \left(p_i^a \right) = \frac{\partial F}{\partial \, |x|_{ab}} \, |x|_{ab} - G(n)|_{n = \partial F / \partial |x|} = F \left(|x|_{ab} \right).$$

Based on the equality (7.45), Gibbs principle can be rewritten as

$$\delta_x \delta_p J = 0,$$

$$J = \int\limits_{\mathring{V}} \left(p_i^a x_a^i - F^\times \left(p_i^a\right)\right) d\mathring{V} - \int\limits_{\mathring{V}} \rho_0 g_i x^i d\mathring{V} - \int\limits_{\partial \mathring{V}_f} f_i x^i d\mathring{A}. \tag{7.58}$$

The symbols δ_x and δ_p mean variations with respect to $x^i\ (X^a)$ and p_i^a, respectively.

Let us find first the conditions for the functional to be stationary with respect to $x^i\ (X^a)$. It is easy to see that these are the equilibrium equations and the boundary conditions on $\partial \mathring{V}_f$:

$$\mathring{\nabla}_a p_i^a + \rho_0 g_i = 0 \quad \text{in } \mathring{V}, \qquad p_i^a \mathring{n}_a = f_i \quad \text{on } \partial \mathring{V}_f. \tag{7.59}$$

Now, instead of taking the variation with respect to all p_i^a, let us take the variation only with respect to those functions p_i^a which satisfy (7.59). For such functions p_i^a, the functional J will take the form

$$J\,(p) = \int\limits_{\partial \mathring{V}_u} p_i^a \mathring{n}_a x_{(b)}^i d\mathring{A} - \int\limits_{\mathring{V}} F^\times \left(p_i^a\right) d\mathring{V}.$$

We obtain

Principle of complementary work. *The true stress state of the isotropic elastic body is the stationary point of the functional $J\,(p)$ on the set of all functions p_i^a which are the solutions of the equilibrium equations (7.59).*

Small deformations of anisotropic bodies. Constructing the principle of complementary work in geometrically nonlinear theory, we used isotropy of the body only once–in replacing (7.46) by (7.48). Such a replacement was possible because the second term of the formula (7.35) for the Piola-Kirchhoff tensor is zero for isotropic bodies. For anisotropic bodies, this term is not equal to zero; however, in the case of small deformations it is on the order of ε compared to the first term:

$$\frac{\partial F}{\partial\, |x|_{bd}} |x|_b^c\, e_{dch} y_c^h e^{d'ac'} \lambda_{id'} = \frac{\partial F}{\partial\, |x|_{bd}} \gamma_b^c e_{dch} y_c^h e^{d'ac'} \lambda_{id'} \sim \mu \varepsilon^2.$$

Therefore, in the theory of small deformations, the considered construction remains valid for anisotropic bodies as well, and the principle of complementary work formulated above is still applicable.

The extended dual variational principle. It is possible to avoid the difficulties associated with using a multi-valued functional if, along with the variation of the Piola-Kirchhoff tensor, we retain the variation of the orthogonal part of distortion, λ_a^i. The corresponding expanded principle of complementary work is constructed in the following way.

Gibbs principle can be reformulated as a statement on stationarity of the functional

$$\int_{\mathring{V}} F\left(|x|_{ab}\right) d\mathring{V} - \int_{\mathring{V}} \rho_0 g_i x^i \left(X^a\right) d\mathring{V} - \int_{\partial \mathring{V}_f} f_i x^i \left(X^a\right) d\mathring{A}, \tag{7.60}$$

over all positive symmetric tensors $|x|_{ab}$ and the functions $x^i\left(X^a\right)$ satisfying the boundary conditions $x^i = x^i_{(b)}$ on $\partial \mathring{V}_u$ and compatibility conditions: for each $|x|_{ab}$ and $x^i\left(X^a\right)$ there exists an orthogonal matrix, λ^{ib}, such that the constraints hold:

$$\frac{\partial x^i}{\partial X_a} - |x|_{ab}\, \lambda^{ib} = 0. \tag{7.61}$$

The energy density is considered to be strictly convex function of $|x|_{ab}$; the elastic body can be anisotropic.

Let us introduce the Lagrange multipliers, p_i^a, for the constraints (7.61) and rewrite the functional (7.60) as

$$\int_{\mathring{V}} \left(p_i^a \left(\frac{\partial x^i}{\partial X_a} - |x|_{ab}\, \lambda^{ib} \right) + F\left(|x|_{ab}\right) \right) d\mathring{V} - \int_{\mathring{V}} \rho_0 g_i x^i \left(X^a\right) d\mathring{V} - \int_{\partial \mathring{V}_f} f_i x^i \left(X^a\right) d\mathring{A}. \tag{7.62}$$

Functions p_i^a, $|x|_{ab}$, $x^i\left(X^a\right)$, and λ^{ib} are varied independently in the functional (7.62). Varying $x^i\left(X^a\right)$, we get (7.59) for p_i^a. The functional (7.62) is strictly convex with respect to $|x|_{ab}$. Therefore, varying $|x|_{ab}$ can be replaced by minimization over $|x|_{ab}$. As a result, taking into account (7.59), for the functional (7.62) we get

$$J\left(p, \lambda\right) = \int_{\partial \mathring{V}_u} p_i^a \mathring{n}_a x^i_{(b)} d\mathring{A} - \int_{\mathring{V}} G\left(p_i^a \lambda^{ib}\right) d\mathring{V}.$$

Here, as before, $G\left(n^{ab}\right)$ is the Young-Fenchel transformation of $F\left(|x|_{ab}\right)$ with respect to $|x|_{ab}$.

The true stress states are the stationary points of the functional $J\left(p, \lambda\right)$ on the set of all p_i^a satisfying the equilibrium equations and the boundary conditions (7.59) and on the set of all orthogonal matrices λ^{ia}.

The stationarity condition for $J\left(p, \lambda\right)$ over λ_a^i have the form

$$\left.\frac{\partial G}{\partial n^{ab}}\right|_{n^{ab}=p_i^a \lambda^{ib}} p_i^a \lambda_j^b = \left.\frac{\partial G}{\partial n^{ab}}\right|_{n^{ab}=p_i^a \lambda^{ib}} p_j^a \lambda_i^b. \tag{7.63}$$

Equation (7.63) has a simple meanings. Since $\partial G / \partial n^{ab} = |x|_{ab}$, while $|x|_{ab}\, p_i^a \lambda_j^b = p_{ij}$ are the components of the Cauchy stress tensor, (7.63) are the symmetry conditions for the Cauchy stress tensor and, thus, are the angular momentum equations.

The minimization problem. Since F is bounded from below, it is natural to pose the minimization problem for the functional $I(x(X))$. If the positions of the particles are given at the entire boundary, and the body forces are absent, then the functional $I(x(X))$ is bounded from below by zero, and posing the minimization problem is possible. Let us consider the other extreme case when the kinematic constraints are absent while, the functional $I(x(X))$ has the form (7.25). The energy has a kernel as it is invariant with respect to the rigid motions:

$$F\left(x'(X)\right) = F\left(x(X)\right), \quad x'^{i}\left(X^{a}\right) = c^{i} + \alpha^{i}_{j} x^{j}\left(X^{a}\right). \tag{7.64}$$

Here, $x'(X)$ is the position of the particles after translation of the deformed state $x(X)$ for a constant vector, c^i, and rotation with an orthogonal matrix, α^i_j. The translations form a cone in the functional space, and according to (5.30) for the boundedness from below of $I(x(X))$ it is necessary that the work of external forces be equal to zero at any translation. This condition is equivalent to vanishing the resultant of external forces:

$$\int\limits_{\mathring{V}} \rho_0 g_i d\mathring{V} + \int\limits_{\partial\mathring{V}} f_i d\mathring{A} = 0. \tag{7.65}$$

The work of external forces on rotations is

$$\alpha_{ij}\left(\int\limits_{\mathring{V}} \rho_0 g_i x^j d\mathring{V} + \int\limits_{\partial\mathring{V}} f_i x^j d\mathring{A}\right). \tag{7.66}$$

Unlike the geometrically linear case, this work is not necessarily equal to zero because the orthogonal transformations do not form a cone. Hence, the total momentum of the prescribed external forces does not need to be equal to zero for the existence of a minimizer. The mechanical interpretation of this fact is simple. Consider an example shown in Fig. 7.5. For an arbitrary chosen position of a solid the "dead" forces applied are not in equilibrium (Fig. 7.5a). Nevertheless, the body does have the equilibrium states; they are shown in Fig. 7.5b,c.

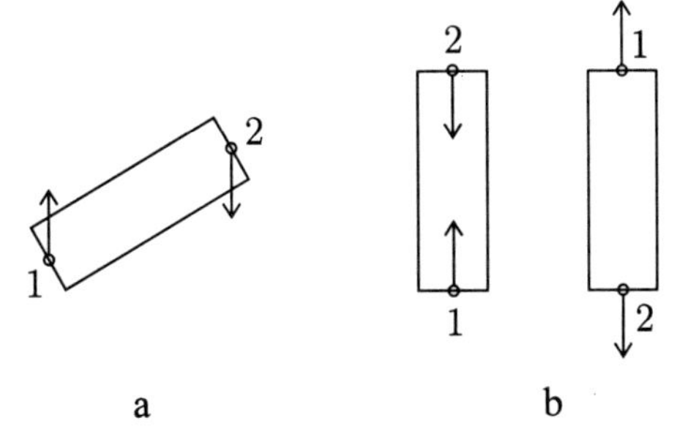

Fig. 7.5 "Dead" load can be given in such a way that the total moment is not zero (**a**); nevertheless, the equilibrium positions exist (**b**)

The functional (7.66) is bounded from below due to boundedness of the components of orthogonal matrices. If $x^i(X^a)$ is a stationary point, then variation of (7.66) with respect to rotations yields vanishing of the total moment computed at the stationary point:

$$e_{ijk}\left(\int_{\mathring{V}} \rho_0 g^j x^k(X^a)\, d\mathring{V} + \int_{\partial\mathring{V}} f^j x^k(X^a)\, d\mathring{A}\right) = 0. \tag{7.67}$$

Suppose that external forces acting on the body are such that there exists a solution f_i^a of the equilibrium equations,

$$\mathring{\nabla}_a f_i^a + \rho_0 g_i = 0 \text{ in } \mathring{V}, \quad f_i^a \mathring{n}_a = f_i \text{ on } \partial\mathring{V}, \tag{7.68}$$

with a finite integral,

$$\|f\|_{L_2}^2 \equiv \int_{\mathring{V}} f_i^a f_a^i d\mathring{V} < +\infty.$$

Assume also that free energy is bounded from below by a quadratic form

$$\frac{1}{2}\mu\gamma_{ab}\gamma^{ab} \leq F.$$

Then, the functional $I(x(X))$ is also bounded from below. Indeed,

$$l = \int_{\mathring{V}} f_i^a x_a^i d\mathring{V} = \int_{\mathring{V}} f_i^a \lambda^{ib} |x|_{ab} d\mathring{V} = \int_{\mathring{V}} f_i^a \lambda_a^i d\mathring{V} + \int_{\mathring{V}} f_i^a \lambda^{ib} \gamma_{ab} d\mathring{V}.$$

Using the Cauchy inequality, we get

$$|l| \leq \|f\|_{L_2}\sqrt{3\left|\mathring{V}\right|} + \|f\|_{L_2}\sqrt{\int_{\mathring{V}} \gamma_{ab}\gamma^{ab} d\mathring{V}}.$$

The remaining part of the proof is similar to that considered in Sect. 5.1.

Analogously investigated is the case of mixed boundary conditions. So, the functional $I(x(X))$ is bounded from below and posing the minimization problem for the energy functional is meaningful.

Let us construct the corresponding dual variational problem. Due to the non-convexity of $I(x(X))$ it is difficult to presuppose the existence of a functional I^*, such that $\max I^* = \min I$. However, it is possible to construct the functional I^* for which

$$\max I^* \leq \min I. \tag{7.69}$$

The construction of the functional I^* is based on the calculation of the Young-Fenchel transformation of F.

The Young-Fenchel transformation of free energy. Let us denote by $\bar{F}\left(x_a^i\right)$ the function

$$\bar{F}\left(x_a^i\right) = \begin{cases} F\left(x_a^i\right), & \det \left\|x_a^i\right\| > 0 \\ +\infty, & \det \left\|x_a^i\right\| \leq 0 \end{cases}$$

and by $F^*\left(p_i^a\right)$ – its Young-Fenchel transformation,

$$F^*\left(p_i^a\right) = \max\left[p_i^a x_a^i - \bar{F}\left(x_a^i\right)\right] = \max_{x_a^i, \det\left\|x_a^i\right\|>0}\left[p_i^a x_a^i - F\left(x_a^i\right)\right].$$

In order to calculate $F^*\left(p_i^a\right)$ we will need the following two assertions.

1. Let q^{ab} be an arbitrary tensor and let μ_{ab} be the components of an orthogonal matrix with positive determinant:

$$\mathring{g}^{ab}\mu_{ac}\mu_{bd} = \mathring{g}_{cd}, \quad \det\|\mu_{ab}\| = +1.$$

Then

$$\max_{\mu_{ab}} q^{ab}\mu_{ab} = |q|_1 + |q|_2 + |q|_3 \operatorname{sgn}\det\left\|q^{ab}\right\|. \tag{7.70}$$

Here, $|q|_1, |q|_2, |q|_3$ are the eigenvalues of the tensor $|q|^{ab}$ arranged in decreasing order.

To prove (7.70) consider first the case when $\det\left\|q^{ab}\right\| > 0$. Let us present q^{ab} in terms of its polar decomposition: $q^{ab} = |q|^{ac}\lambda_c^b$, λ_c^b being the components of an orthogonal matrix. Since $\det\|q\|^{ab} > 0$, we have $\det\left\|\lambda_c^b\right\| = +1$. The matrix with the components $\lambda_c^b\mu_{ab}$ belongs to the set of the orthogonal matrices with positive determinant; therefore,

$$\max_{\mu_{ab}} q^{ab}\mu_{ab} = \max_{\mu_{ab}} |q|^{ab}\mu_{ab}.$$

Taking as a trial matrix $\|\mu_{ab}\|$ the unit matrix, we get the low bound

$$|q|_1 + |q|_2 + |q|_3 \leq \max_{\mu_{ab}} q^{ab}\mu_{ab}. \tag{7.71}$$

In the principal coordinate system of the tensor $|q|^{ab}$, the contraction $|q|^{ab}\mu_{ab}$ has the form $|q|_1\mu_{11} + |q|_2\mu_{22} + |q|_3\mu_{33}$. Since the components of the orthogonal matrix do not exceed unity,

$$\max_{\mu_{ab}} |q|^{ab} \mu_{ab} \leq |q|_1 + |q|_2 + |q|_3 . \tag{7.72}$$

Low and upper bounds coincide and yield (7.70).

Consider now the case $\det \left\| q^{ab} \right\| < 0$. In the polar decomposition of the tensor q^{ab}, the matrix $\left\| \lambda_c^b \right\|$ has a determinant equal to -1; therefore,

$$\max_{\substack{\mu_{ab} \\ \det\|\mu_{ab}\|=+1}} q^{ab} \mu_{ab} = \max_{\substack{\mu_{ab} \\ \det\|\mu_{ab}\|=-1}} |q|^{ab} \mu_{ab}, \tag{7.73}$$

and we have to prove that this maximum (7.70) is equal to $|q|_1 + |q|_2 - |q|_3$.

Let us use the representation of the orthogonal matrices (3.42). In the principal coordinate system of the tensor $|q|^{ab}$, we have

$$\begin{aligned} |q|^{ab} \mu_{ab} = &\cos\theta \left[|q|_1 \left(1 - c_1^2\right) + |q|_2 \left(1 - c_2^2\right) + |q|_3 \left(1 - c_3^2\right) \right] \\ &- |q|_1 c_1^2 - |q|_2 c_2^2 - |q|_3 c_3^2. \end{aligned}$$

The expression in square brackets is non-negative because $c_1^2 + c_2^2 + c_3^2 = 1$. Therefore, the maximum is reached at $\theta = 0$. After substituting c_3^2 by $1 - c_1^2 - c_2^2$, we need to seek the maximum over c_1, c_2 $(c_1^2 + c_2^2 \leq 1)$ of the expression

$$|q|_1 + |q|_2 - |q|_3 - 2\left(|q|_1 - |q|_3\right) c_1^2 - 2\left(|q|_2 - |q|_3\right) c_2^2.$$

It is equal to $|q|_1 + |q|_2 - |q|_3$ which is what was claimed. If $\det \left\| q^{ab} \right\| = 0$, then $|q|_3 = 0$, and, as easy to see, (7.70) still holds.

Formula (7.70) can be presented in a form which does not employ the ordering of the eigenvalues:

$$\max_{\mu_{ab}} q^{ab} \mu_{ab} = \max_{s_{ab}} \left\{ |q|^{ab} s_{ab} \right\},$$

where the maximum is calculated over the tensors s_{ab}, which have the form (7.43) in the principal coordinate system of the tensor $|q|^{ab}$ if $\det \left\| q^{ab} \right\| > 0$ and the form (7.44) if $\det \left\| q^{ab} \right\| < 0$.

2. For an isotropic material, the point at which the function $p_i^a x_a^i - F\left(x_a^i\right)$ reaches its maximum value over x_a^i is the point at which the tensor $|x|_{ab}$ is coaxial with the tensor $|p|^{ab}$.

Indeed, if the maximizing element of the function $p_i^a x_a^i - F\left(x_a^i\right)$ is an internal point of the set $\mathcal{X}$, then this statement follows from the formula (7.49). If the maximizing element is on the boundary of the set $\mathcal{X}$, i.e. $\det \left\| x_a^i \right\| = 0$, then, x_a^i is the solution of the system of equations

$$p_i^a = \frac{\partial F}{\partial x_a^i} + \varkappa \frac{\partial}{\partial x_a^i} \det \left\| x_a^i \right\| = \frac{\partial F}{\partial |x|_{ab}} \lambda_{ib} + \varkappa e_{a'b'c'} \lambda_i^{a'} |x|_b^{b'} |x|_c^{c'} e^{abc},$$

where $\varkappa$ is the Lagrange multiplier for the constraint $\det \left\| x_a^i \right\| = 0$. Hence,

$$|p|^{ac}\,\mu_{ic}\lambda^{id} = \frac{\partial F}{\partial\,|x|_{ad}} + \varkappa e^{d}_{\cdot b'c'}\,|x|^{b'}_{b}\,|x|^{c'}_{c}\,e^{abc}. \tag{7.74}$$

The two terms on the right side of (7.74) are coaxial to the tensor $|x|_{ab}$ and are symmetric with respect to a, d. Therefore, they are coaxial with the tensor $|p|^{ac}$, and the tensors $|p|^{ab}$ and $|x|_{ab}$ are coaxial.

Now everything is prepared to derive the following formula: for an isotropic elastic material, in the framework of the theory of finite deformations:

$$F^*\left(p^a_i\right) = \max_{x^i_a}\left(p^a_i x^i_a - F\left(x^i_a\right)\right) = \begin{cases} G\left(|p|^{ab}\right), & \det\left\|p^a_i\right\| > 0 \\ \max\limits_{s^c_b}\left\{G\left(|p|^{ab}\,s^c_b\right)\right\}, & \det\left\|p^a_i\right\| < 0 \end{cases} \tag{7.75}$$

where the maximum is calculated over the matrices s^c_b, which have the form (7.44) in the principal coordinate system of the tensor $|p|^{ab}$.

Indeed, in searching for the maximum, $\max\left(p^a_i x^i_a - F\left(x^i_a\right)\right)$, one can consider only such x^i_a for which the tensor $|x|_{ab}$ is coaxial to the tensor $|p|^{ab}$. Then, $p^a_i x^i_a = |p|^{ab}\,\mu_{ib}\,|x|_{ac}\,\lambda^{ic} = q^{ab}\nu_{ab}$, where the tensor $q^{ab} = |p|^{ac}\,|x|^b_c$ is symmetric because the tensors $|p|^{ac}$ and $|x|^b_c$ are coaxial, while ν_{ab} are the components of the orthogonal matrix with $\det\|\nu_{bc}\| = +1$ for $\det\left\|p^a_i\right\| = +1$, and $\det\|\nu_{bc}\| = -1$ for $\det\left\|p^a_i\right\| = -1$. One can seek for maximum over x^i_a successively, first finding the maximum over λ^i_a and then over $|x|_{ab}$. Finding the maximum over λ^i_a is equivalent to maximization of $q^{ab}\nu_{ab}$. According to (7.70), we have:

for $\det\left\|p^a_i\right\| > 0$,

$$\begin{aligned} F^*\left(p^a_i\right) &= \max_{\substack{\nu_{ab},\,|x|_{ab} \\ \det\|\nu_{ab}\|=+1}}\left[|p|^{ab}\,|x|^c_b\,\nu_{bc} - F\left(|x|_{ab}\right)\right] = \\ &= \max_{|x|_{ab}}\left[|p|^{ab}\,|x|_{ab} - F\left(|x|_{ab}\right)\right] = G\left(|p|^{ab}\right), \end{aligned}$$

for $\det\left\|p^a_i\right\| < 0$,

$$\begin{aligned} F^*\left(p^a_i\right) &= \max_{\substack{\nu_{ab},\,|x|_{ab} \\ \det\|\nu_{ab}\|=+1}}\left[|p|^{ab}\,|x|^c_b\,\nu_{bc} - F\left(|x|_{ab}\right)\right] = \\ &= \max_{|x|_{ab}}\left[\max\left\{|p|^{ab}\,|x|^c_b\,s_{ac}\right\} - F\left(|x|_{ab}\right)\right], \end{aligned}$$

where the inner maximum is taken over the matrices s_{bc}, which have the form (7.44) in the principal coordinate system of the tensor $|p|^{ab}$. Changing the order of the maximization over s_{ac} and $|x|_{ab}$, we obtain (7.75).

Note that in the case when the maximizing element of the function $p^a_i x^i_a - F\left(x^i_a\right)$ is an internal point of the set $\mathcal{X}$, the assertion of the (7.75) follows directly from the formula for the value of the function $F^{\times} = p^a_i x^i_a - F\left(x^i_a\right)$ at the points stationary with respect to x^i_a (7.53).

Semi-linear material. Let us find the Young-Fenchel transformation of the function F for the semi-linear isotropic material with $\lambda = 0$[4], in the principal coordinate system of the tensor n^{ab},

$$G(n) = \max_{|x|_i} \Big[n_1 |x|_1 + n_2 |x|_2 + n_3 |x|_3 - \mu \left[(|x|_1 - 1)^2 + (|x|_2 - 1)^2 + (|x|_3 - 1)^2 \right] \Big] = f(n_1) + f(n_2) + f(n_3),$$

where

$$f(n) = \max_{x \geq 0} \left[nx - \mu (x-1)^2 \right] = \begin{cases} n + \frac{n^2}{4\mu}, & n \geq -2\mu \\ -\mu & n \leq -2\mu \end{cases}$$

According to the formula (7.46), for $\det \| p_i^a \| > 0$,

$$F^* \left(p_i^a \right) = f \left(|p|_1 \right) + f \left(|p|_2 \right) + f \left(|p|_3 \right),$$

and for $\det \| p_i^a \| < 0$,

$$F^* \left(p_i^a \right) = \max_{\substack{s_1, s_2, s_3 = \pm 1 \\ s_1 s_2 s_3 = -1}} \left[f \left(|p|_1 s_1 \right) + f \left(|p|_2 s_2 \right) + f \left(|p|_3 s_3 \right) \right].$$

Let us number the eigenvalues of the tensor $|p|^{ab}$ in decreasing order. Since $f(p)$ is a non-decreasing function, we have for $\det \| p_i^a \| < 0$,

$$F^* \left(p_i^a \right) = f \left(|p|_1 \right) + f \left(|p|_2 \right) + f \left(-|p|_3 \right).$$

Dual variational principle. *The maximum value of the functional*

$$I^*(p) = \int_{\partial \mathring{V}_u} p_i^a \mathring{n}_a x^i_{(b)} d\mathring{A} - \int_{\mathring{V}} F^* \left(p_i^a \right) d\mathring{V}$$

on the set of all functions p_i^a satisfying the equilibrium equations (7.59) does not exceed the minimum value of energy functional.

If the minimizing element of the energy functional is in the convexity region, then maximum value of I^* coincides with the minimum value of I.

The dual variational problem is convex. To evaluate the errors which the convexification may cause, it is worth finding the convex function $F^{**} \left(x_a^i \right)$, the Young-Fenchel transformation of the function $F^* \left(p_i^a \right)$. Due to the properties of the

[4] The expression for F^* for $\lambda \neq 0$ is more complex and can be found in [66].

Young-Fenchel transformation, $F^{**} \leq F$, and, if F^{**} does not differ considerably from F, it can be expected that the solutions of the dual convex problem and the initial non-convex problem are close, at least in energy norm. The calculations for the two-dimensional case, not reproduced here, result in

$$F^{**} = \begin{cases} \mu\left(\gamma_1^2 + \gamma_2^2\right) & \text{for } \gamma_1 + \gamma_2 \geq 0 \\ \frac{1}{2}\mu\left(\gamma_1 - \gamma_2\right)^2 & \text{for } \gamma_1 + \gamma_2 \leq 0 \end{cases}$$

where γ_1, γ_2 are the eigenvalues of the tensor γ_{ab}. Consequently, if the true strains $\gamma_1 = |x|_1 - 1$, and $\gamma_2 = |x|_2 - 1$ along the axes X^1, X^2 are such that $\gamma_1 + \gamma_2 \geq 0$, then $F = F^{**}$, and the solutions of the initial and the dual problems coincide. If the true strain happens to be in the region $\gamma_1 + \gamma_2 < 0$, then, generally speaking, the solutions of the initial and the dual problems differ. The energy measure of the errors is controlled by the difference

$$F\left(x_a^i\right) - F^{**}\left(x_a^i\right) = \begin{cases} 0 & \text{for } \gamma_1 + \gamma_2 \geq 0 \\ \frac{1}{2}\mu\left(\gamma_1 - \gamma_2\right)^2 & \text{for } \gamma_1 + \gamma_2 \leq 0 \end{cases}$$

7.4 Phase Equilibrium of Elastic Bodies

The variational approach considered so far dealt with only mechanic deformations of an elastic body. Gibbs variational principles are deeper; in particular, they also control the thermodynamic equilibrium of two-phase elastic bodies in which the phases can transform one into another (for example, the equilibrium of a metal with its melt). In this section we derive the conditions of the phase equilibrium from the second Gibbs principle.

Let two phases of a solid occupy some region V. Subregions of V occupied by each phase are denoted by V_1 and V_2 and their common boundary by Σ (Fig. 7.6). Let $U_1\left(x_a^i, S\right)$ and $U_2\left(x_a^i, S\right)$ be the internal energy densities of two phases per unit mass. The quantities corresponding to the two phases are supplied by the indices 1 and 2.

The motion of the surface Σ over the particles corresponds to transformation of one phase into another. For definiteness, on the boundary of region V, the particle positions are assumed to be given while the system is adiabatically isolated.

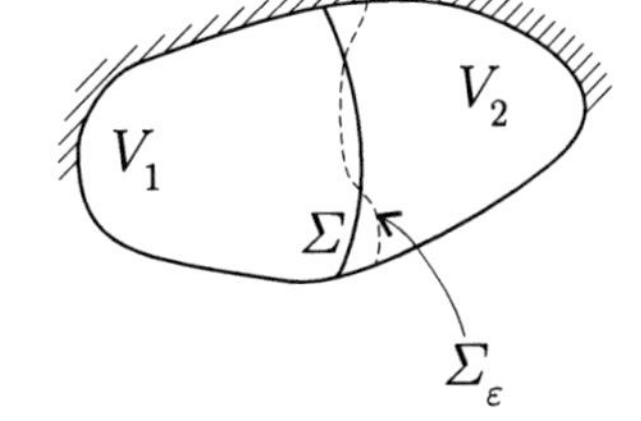

Fig. 7.6 Notation in consideration of phase equilibrium

According to the second Gibbs principle, in thermodynamical equilibrium the functional

$$\mathcal{U} = \int_{V_1} \rho U_1 \left(x_a^i, S\right) dV + \int_{V_2} \rho U_2 \left(x_a^i, S\right) dV \tag{7.76}$$

takes the minimum value on the set of all functions $x^i(X^a)$, $S(X^a)$ and surfaces Σ that satisfy the conditions

$$\int_{V_1} \rho S dV + \int_{V_2} \rho S dV = \mathcal{S}_0 = const \tag{7.77}$$

$$x^i\left(X^a\right) = x^i_{(b)}\left(X^a\right) \quad \text{on } \partial V, \tag{7.78}$$

and an additional condition on continuity of particle positions on Σ

$$\left[x^i\left(X^a\right)\right] = 0 \quad \text{on } \Sigma. \tag{7.79}$$

As before, $[\varphi]$ denotes the difference of the limit values of function φ on the two sides of the surface Σ: $[\varphi] = \varphi_1 - \varphi_2$. Equation (7.79) excludes appearance of voids and slipping of the two sides of Σ. The derivatives of the functions $x^i(X^a)$ and the entropy $S(X^a)$ can be discontinuous on Σ. To include into consideration the inhomogeneous solids, we have to take into account that energy density may depend on Lagrangian coordinates, X^a, through the dependence on X^a the physical characteristics, some tensors K^B with a set of Lagrangian indices denoted by B.

Denote by p_i^a and μ_b^a the tensors

$$p_i^a = \rho_0 \frac{\partial U}{\partial x_a^i}, \qquad \mu_b^a = -\rho_0 \frac{\partial U}{\partial x_a^i} x_b^i + \rho_0 (U - TS)\, \delta_b^a.$$

We are going to show that the conditions of thermodynamical equilibrium are

$$\mathring{\nabla}_a p_i^a = 0, \qquad \frac{\partial U}{\partial S} = T = const \quad \text{in } V, \tag{7.80}$$

$$\left[p_i^a\right] \mathring{n}_a = 0 \quad \text{on } \Sigma, \tag{7.81}$$

$$\left[\mu_b^a\right] \mathring{n}_a = 0 \quad \text{on } \Sigma. \tag{7.82}$$

where $\mathring{\nabla}_a$ is the covariant derivative in the undeformed state with respect to Lagrangian coordinates; without loss of generality, it may be considered coinciding with the partial derivatives, $\partial/\partial X^a$. The first two equations are the conditions of the mechanical equilibrium. The third equation is the condition of the phase (chemical) equilibrium.

Indeed, let us find the variation of the functional $\mathcal{U}$. Let $x^i(X^a)$ be a stationary point of $\mathcal{U}$, and let $x^i(X^a, \varepsilon)$ be close particle positions. The space derivatives of the particle positions, $x^i\left(X^a,\varepsilon\right)$, may be discontinuous on some surface Σ_ε which is close to Σ. As in the derivation of the Lagrange equations (see Appendix A), it is convenient to introduce the mapping $V\to V$: $X'^a=X'^a\left(X^b,\varepsilon\right)$ for which the surface Σ transforms into the surface Σ_ε. Due to the proximity of Σ and Σ_ε, we can write $X'^a=X^a+\delta X^a$, where δX^a are some smooth functions in V. The functions δX^a are zero on ∂V.

Let us denote by $\delta_\Pi x^i$ and $\delta_\Pi S$ the full variations of x^i and S, i.e.

$$\delta_\Pi x^i=x^i\left(X'^a,\varepsilon\right)-x^i\left(X^a\right),\quad \delta_\Pi S=S\left(X'^a,\varepsilon\right)-S\left(X^a\right).$$

The full variations $\delta_\Pi x^i$, according to (7.78) and (7.79), satisfy the constraints

$$\delta_\Pi x^i=0\quad \text{on}\,\partial V,\quad \left[\delta_\Pi x^i\right]=0\quad \text{on}\,\Sigma. \tag{7.83}$$

Let us show that the constraint on the variations $\delta_\Pi S$ which follow from (7.77) have the form

$$\int\limits_{\mathring{V}}\left[\rho_0\delta_\Pi S+S\mathring{\nabla}_a\left(\rho_0\delta X^a\right)\right]d\mathring{V}=0. \tag{7.84}$$

Indeed, keeping the small terms of the first order with respect to $d\varepsilon$, we have for the volume element[5]

$$\begin{aligned}\rho_0\left(X'\right)\sqrt{\mathring{g}\left(X'\right)}d^3X' &= \left[\rho_0\left(X\right)\sqrt{\mathring{g}\left(X\right)}+\frac{\partial\rho_0\sqrt{\mathring{g}}}{\partial X^a}\delta X^a\right]\left|\frac{\partial X'}{\partial X}\right|d^3X=\\ &=\rho_0\sqrt{\mathring{g}}\left[1+\frac{1}{\rho_0\sqrt{\mathring{g}}}\frac{\partial\rho_0\sqrt{\mathring{g}}}{\partial X^a}\delta X^a\right]\left[1+\frac{\partial\delta X^a}{\partial X^a}\right]d^3X=\\ &=\rho_0\sqrt{\mathring{g}}\left[1+\frac{1}{\rho_0\sqrt{\mathring{g}}}\frac{\partial\rho_0\sqrt{\mathring{g}}\delta X^a}{\partial X^a}\right]d^3X\\ &=\rho_0\left[1+\frac{1}{\rho_0}\mathring{\nabla}_a\left(\rho_0\delta X^a\right)\right]d\mathring{V}.\end{aligned} \tag{7.85}$$

Therefore,

[5] In (7.85), the formulas were used,

$$\mathring{\nabla}_cA^c=\frac{1}{\sqrt{\mathring{g}}}\frac{\partial\sqrt{\mathring{g}}A^c}{\partial X^c},\quad \rho dV=\rho_0d\mathring{V}.$$

$$\delta \int\limits_{V_1} \rho S dV = \int\limits_{V_1'} \rho_0 \left(X'\right) S \left(X'\right) \sqrt{\mathring{g}\left(X'\right)} d^3 X' -$$

$$- \int\limits_{V_1} \rho_0 \left(X\right) S \left(X\right) \sqrt{\mathring{g}\left(X\right)} d^3 X = \int\limits_{V_1} \left[\rho_0 \delta_{\Pi} S + S \mathring{\nabla}_a \left(\rho_0 \delta X^a\right)\right] d\mathring{V}.$$

Formula (7.84) follows from the last equation and a similar equation written for the region V_2.

Analogously to (7.84), we get the relation

$$\delta \mathcal{U} = \int\limits_{V_1} \left[\rho_0 \delta_{\Pi} U_1 + U_1 \mathring{\nabla}_a \left(\rho_0 \delta X^a\right)\right] d\mathring{V} + \int\limits_{V_2} \left[\rho_0 \delta_{\Pi} U_2 + U_2 \mathring{\nabla}_a \left(\rho_0 \delta X^a\right)\right] d\mathring{V}.$$

Consider the quantity $\delta_{\Pi} U$. In calculating $\delta_{\Pi} U$, the operator δ_{Π} should be applied to the tensor with the Lagrangian indices – the distortion x_a^i and the characteristics of the medium, K^B. We will assume that the operator δ_{Π} includes the parallel transport over the Lagrangian indices in the initial state. Then, $\delta_{\Pi} K^B = \delta X^a \mathring{\nabla}_a K^B$. For $\delta_{\Pi} x_a^i$, we have

$$\delta_{\Pi} x_a^i = \frac{\partial x'^i \left(X'^b, \varepsilon\right)}{\partial X'^a} - \mathring{\Gamma}_{ac}^b \delta X^c x_b^i - \frac{\partial x^i \left(X^b\right)}{\partial X'^a} = \frac{\partial \delta_{\Pi} x^i}{\partial X^a} - x_b^i \mathring{\nabla}_a \delta X^b.$$

After integration by parts, the equation $\delta \mathcal{U} = 0$ becomes

$$\int\limits_V \left[-\delta_{\Pi} x^i \mathring{\nabla} p_i^a + \rho_0 \delta_{\Pi} S \left(\frac{\partial U}{\partial S} - T\right) + \right.$$

$$\left. + \delta X^a \left(-\mathring{\nabla}_b \mu_a^b + (U - TS) \mathring{\nabla}_a \rho_0 + \rho_0 \frac{\partial U}{\partial K^B} \mathring{\nabla}_a K^B\right)\right] d\mathring{V} +$$

$$+ \int\limits_{\Sigma} \left(\left[p_i^a\right] \mathring{n}_a \delta_{\Pi} x^i + \left[\mu_b^a\right] \mathring{n}_a \delta X^b\right) d\sigma = 0. \tag{7.86}$$

Here, T is the Lagrange multiplier for the constraint (7.84). The equilibrium conditions (7.80), (7.81) and (7.82) follow from (7.86).

There is an additional equation, which must be satisfied in equilibrium:

$$-\mathring{\nabla}_b \mu_a^b + (U - TS) \mathring{\nabla}_a \rho_0 + \rho_0 \frac{\partial U}{\partial K^B} \mathring{\nabla}_a K^B = 0 \quad \text{in } V, \tag{7.87}$$

However, one can check that (7.87) is a consequence of (7.80).

Equations (7.80) are the conditions of thermodynamic equilibrium obtained in Sect. 7.2.

Equations (7.81) shows that the necessary condition of thermodynamic equilibrium is the equality of forces acting on the two sides of the surface Σ.

Of the three equations (7.82), only one is independent. This follows from (7.81) and the compatibility condition on the surface of the discontinuity[6]

$$\left[x_a^i\right] = \lambda^i \mathring{n}_a. \tag{7.89}$$

Indeed,

$$\left[\mu_b^a\right]\mathring{n}_a = \left[-\rho_0 \frac{\partial U}{\partial x_a^i} x_b^i + \rho_0 (U - TS)\,\delta_b^a\right]\mathring{n}_a =$$
$$= -p_i^a \mathring{n}_a \left[x_b^i\right] + \left[\rho_0 (U - TS)\right]\mathring{n}_b = \left(-p_i^a \mathring{n}_a \lambda^i + \left[\rho_0 (U - TS)\right]\right)\mathring{n}_b.$$

In these relations, by $p_i^a \mathring{n}_a$ one can mean the value of $p_i^a \mathring{n}_a$ on either side of the discontinuity surface since $\left[p_i^a\right]\mathring{n}_a = 0$.

The independent equation in (7.82) can be singled out by contracting (7.82) with the normal vector $\mathring{n}_b$:

$$\left[\mu_b^a\right]\mathring{n}_a \mathring{n}_b = 0. \tag{7.90}$$

Unlike the "force" conditions which are not difficult to write from other reasoning, (7.90) is not trivial. It appears by taking the variations of the surface Σ over the particles, i.e. by allowing the phase transformations in the system to occur.

The condition of phase equilibrium (7.90) has one distinguishing characteristic: unlike the other relations of mechanics, it contains the internal energy itself, not its derivatives. Adding of constants to the energies of the phases, U_1 and U_2, changes the phase equilibrium conditions. Due to that, it becomes important how the initial state is chosen and how U and S are measured. In the theory of phase transitions it is assumed that for all phases which can transform to each other, the same initial state may be chosen from which energy and entropy of different phases are measured.

---

[6] Let $\varphi(X^a)$ be a continuous function on the surface Σ, but its derivatives may be discontinuous on Σ. Then the discontinuities of the derivatives of φ on Σ are not arbitrary and satisfy the compatibility conditions

$$\left[\frac{\partial \varphi}{\partial X^a}\right] = \lambda \mathring{n}_a. \tag{7.88}$$

To prove (7.88) we introduce the parametric equations of surface Σ, $X^a = \mathring{r}^a(\zeta^\alpha)$, $\alpha = 1, 2$ and use the decomposition of the derivative of the type (14.17)

$$\frac{\partial}{\partial X^a} = \left(\mathring{r}_a^\alpha \mathring{r}_\alpha^b + \mathring{n}_a \mathring{n}^b\right)\frac{\partial}{\partial X^b} =$$
$$= \mathring{r}_a^\alpha \frac{\partial}{\partial \zeta^\alpha} + \mathring{n}_a \mathring{n}^b \frac{\partial}{\partial X^b}.$$

Hence,

$$\left[\frac{\partial \varphi}{\partial X^a}\right] = \mathring{r}_a^\alpha \frac{\partial [\varphi]}{\partial \zeta^\alpha} + \mathring{n}_a \mathring{n}^b \left[\frac{\partial \varphi}{\partial X^b}\right].$$

For functions φ that are continuous on $\Sigma([\varphi] = 0)$, this equation reduces to (7.88), where $\lambda = \mathring{n}_b \left[\partial \varphi / \partial X^b\right]$. Relations (7.89) follow from (7.88).

Due to this assumption, the initial values of energy, entropy, density and other phase characteristics turn out to be continuous on the phase boundary.

In order to clarify the meaning of the condition (7.90), let us consider a special case of elastic bodies, a liquid, when the internal energy depends on distortion only through the density $\rho = \rho_0 \sqrt{\mathring{g}} / \sqrt{g} \Delta$, $\Delta = \det \left\| x^i_a \right\|$. Since $\partial \rho / \partial x^i_a = -\rho X^a_i$, the tensor μ^a_b is spherical:

$$\mu^a_b = \rho_0 \mu \delta^a_b, \quad \mu = U - TS + \frac{p}{\rho}.$$

Here, $p = \rho^2 \partial U / \partial \rho$ is the pressure. Since $[\rho_0] = 0$, it follows from (7.90) that $[\mu] = 0$. The quantity μ is the chemical potential (see (5.104)), and the condition $[\mu] = 0$ is the necessary condition for the thermodynamic equilibrium of two liquid phases, established by Gibbs. Accordingly, the tensor μ^a_b can be called the chemical potential tensor of an elastic medium.

Chapter 8
Dynamics of Elastic Bodies

8.1 Least Action vs Stationary Action

The extrapolation to dynamics of the minimization principles formulated above encounters difficulties, the essence of which can be observed for systems with one degree of freedom.

Consider the harmonic oscillator – a material point on a spring. The deviation of the point from the equilibrium position is denoted by $x(t)$; the kinetic energy is equal to $\frac{1}{2}m\dot{x}^2$, and the energy of the spring is $\frac{1}{2}kx^2$. According to the Hamilton principle, the true trajectory is the stationary point of the functional

$$I = \frac{1}{2}\int_0^{\Delta t} \left(m\dot{x}^2 - kx^2\right)dt$$

on the set of functions, $x(t)$, which take at the initial and final moments the given values

$$x(0) = x_0, \quad x(\Delta t) = x_1.$$

The question is: does the true trajectory provide the minimum for the functional I? In order to investigate this question, it is convenient to use instead of functions $x(t)$ the functions $u(t)$ which are equal to zero in the initial and final moments:

$$x \to u : x(t) = vt + x_0 + u(t), \, v = (x_2 - x_1)/\Delta t = const, \; u(0) = u(\Delta t) = 0.$$

The constant, v, has the meaning of the average velocity. The functional $I(u)$ takes the form (additive constant is omitted)

$$I(u) = J(u) + l(u),$$

$$J(u) = \frac{1}{2}\int_0^{\Delta t} \left(m\dot{u}^2 - ku^2\right)\, dt, \quad l(u) = -k\int_0^{\Delta t} (vt + x_o)\, u dt.$$

V.L. Berdichevsky, *Variational Principles of Continuum Mechanics*,
Interaction of Mechanics and Mathematics, DOI 10.1007/978-3-540-88467-5_8,

Posing the minimization problem is possible if the functional $I(u)$ is bounded from below. For this to be true, it is necessary that the functional $J(u)$ be bounded from below.[1] For every function $u(x)$, the set of admissible functions contains the function $\lambda u(x)$ (with any λ); therefore, for the quadratic functional to be bounded from below it is necessary and sufficient that it is non-negative. Let us write down this condition. It is convenient to use a new argument $\tau = \frac{2\pi}{\Delta t} t$. From the inequality $J(u) \geq 0$, it follows that, for any functions $u(\tau)$ such that $u(0) = u(2\pi) = 0$, the inequality

$$\int_0^{2\pi} \left(\frac{du}{d\tau}\right)^2 d\tau \geq c \int_0^{2\pi} u^2 d\tau, \quad c = \frac{k(\Delta t)^2}{4\pi^2 m} \tag{8.1}$$

should hold. We arrive at the Wirtinger inequality (5.24). The inequality (8.1) holds for all $c \leq 1/4$. Moreover, for $c < 1/4$ the functional $J(u)$ will be strictly convex, as a quadratic positive functional. For $c > 1/4$, the functional $J(u)$ is not bounded from below. Indeed, $1/4$ is the best constant in the Wirtinger inequality for zero values of the function at the ends; therefore, for $c > 1/4$ there exists at least one function u_0 for which $J(u_0) < 0$, while $l(u_0)$ has a finite value. Thus, for the sequence $\{\lambda u_0\}$, $\lambda \to \infty$, $J(\lambda u_0) \to -\infty$.

So, posing the minimization problem is possible only for sufficiently small Δt, $\Delta t < \pi\sqrt{m/k}$.

For continuous media, the problem is more complicated. Let us consider, for example, the action functional of the wave equation

$$I(u) = \frac{1}{2} \int_0^{\Delta t} \int_{-\pi}^{\pi} \left(u_{,t}^2 - u_{,x}^2\right) dx dt$$

and set the kinematic boundary conditions $u(0, x) = u_0(x)$, $u(\Delta t, x) = u_1(x)$, $u(t, -\pi) = u(t, \pi) = 0$. Let the functions $u_0(x)$ and $u_1(x)$ be odd, so that we will only need odd admissible functions $u(x)$.[2] Functions $u(x)$ can be presented in the form of the Fourier series:

$$u(t, x) = \sum_{k=1}^{\infty} u_k(t) \sin kx. \tag{8.2}$$

Substituting (8.2) into $I(u)$, we get

$$I(u) = \sum_{k=1}^{\infty} I_k, \quad I_k = \frac{\pi}{2} \int_0^{\Delta t} \left(\dot{u}_k^2 - k^2 u_k^2\right) dt.$$

---

[1] Indeed, if there exists a sequence $\{u_n\}$ for which $J(u_n) \to -\infty$, while $l(u_n) \leq const$, then $I(u_n) \to -\infty$. If $l(u_n) \to +\infty$, then for the sequence $\{-u_n\}$, we have $I(u_n) \to -\infty$.

[2] See Sect. 5.13.

Since the functions $u_k(t)$ are independent, the boundedness from below of each functional I_k is necessary for the boundedness from below of the functional $I(u)$. For this, as shown above, it is necessary that for each k

$$k^2 \Delta t^2 \leq \pi^2.$$

It is impossible to satisfy these inequalities for all k. Therefore the functional $I(u)$ is not bounded from below, however small is Δt. Apparently, posing the minimization problem is possible for a quasi-continuum [164], for which short waves (with large k) are excluded.

The noted difficulty is related to the fact that the Hamilton principle in essence yields a Dirichlet-type problem, but this problem is ill-posed for hyperbolic equations.

The noted difficulty is closely related to ill-posedness of hyperbolic equations for the Dirichlet-type problems.

A well-posed dynamical problem is the Cauchy problem, when one prescribes the initial values of the required function and its time derivative. The Hamilton principle can be rendered in the form which corresponds to a problem with the prescribed initial value of the required function and the average value of its time derivative over some interval Δt. Indeed, instead of fixing $u(\Delta t, x)$, one can prescribe

$$\frac{u(\Delta t, x) - u(0, x)}{\Delta t} = \frac{1}{\Delta t} \int_{t}^{t+\Delta t} u_{,t} dt.$$

This assertion, however, has the same peculiarity as the Hamilton principle: the minimization problem is ill-posed unless the short waves are excluded.

8.2 Nonlinear Eigenvibrations

The above pertains to general dynamic problems for continua. However, there is an important class of dynamic problems for which the setting of the variational problems is very close to that for static problems. These are the so-called natural vibrations. By natural one means periodic vibrations which occur without action of external forces. On the boundary of the continuum, homogeneous boundary conditions are assumed (clamped boundaries, zero external surface forces, etc.). Natural vibrations may occur only at some special values of frequencies called eigenfrequencies. Let us construct the variational principle for determining the eigenfrequencies and the modes of vibration.

Let the true motion of a continuum be a stationary point of the functional

$$\int_{t_0}^{t_1}\int_V L\left(x, u^\varkappa, u^\varkappa_{,t}, u^\varkappa_{,i}\right) dV dt.$$

For definiteness, we will assume the clamped boundaries conditions

$$u^\varkappa = 0 \quad \text{on } \partial V. \tag{8.3}$$

Natural vibration with a frequency ω is a stationary point of the functional

$$\int_{-\frac{\pi}{\omega}}^{\frac{\pi}{\omega}}\int_V L\left(x, u^\varkappa, u^\varkappa_{,t}, u^\varkappa_{,i}\right) dV dt$$

on the set of functions $u^\varkappa$ satisfying the condition (8.3) and the periodicity condition

$$u^\varkappa\left(x, -\frac{\pi}{\omega}\right) = u^\varkappa\left(x, \frac{\pi}{\omega}\right).$$

In the variational problem, it is convenient to make a change of the independent variable $t \to \theta$: $\omega t = \theta$, $-\pi \leq \theta \leq \pi$. Then, the eigenvibrations are the stationary points of the functional

$$\int_{-\pi}^{\pi}\int_V L\left(x, u^\varkappa, \omega u^\varkappa_{,\theta}, u^\varkappa_{,i}\right) dV d\theta \tag{8.4}$$

on the set of functions $u^\varkappa$ satisfying the condition (8.3) and the periodicity condition

$$u^\varkappa(x, -\pi) = u^\varkappa(x, \pi). \tag{8.5}$$

One can assume that 2π is the smallest period of the functions $u^\varkappa(x, \theta)$.

The variational problem (8.3), (8.4) and (8.5) admits an interesting interpretation if the Lagrangian L is the difference of the kinetic energy and the free energy,

$$L = K\left(x, u^\varkappa, u^\varkappa_{,t}\right) - F\left(x, u^\varkappa, u^\varkappa_{,i}\right),$$

where K and F are nonnegative, and K is a homogeneous function of the second order with respect to $u^\varkappa_{,t}$, i.e. for any $\omega > 0$,

$$K\left(x, u^\varkappa, \omega u^\varkappa_{,t}\right) = \omega^2 K\left(x, u^\varkappa, u^\varkappa_{,t}\right).$$

In order to simplify further consideration, we will also assume that K and F are strictly convex and equal to zero for $u^{\varkappa} \equiv 0$.

Let us show that in this case the variational problem (8.3), (8.4) and (8.5) is equivalent to the following one: find stationary points of free energy functional

$$\int_{-\pi}^{\pi}\int_{V} F\left(x, u^{\varkappa}, u^{\varkappa}_{,i}\right) dV d\theta \tag{8.6}$$

on the set of function $u^{\varkappa}$, satisfying the constraints (8.3) and (8.5) and the condition

$$\int_{-\pi}^{\pi}\int_{V} K\left(x, u^{\varkappa}, u^{\varkappa}_{,\theta}\right) dV d\theta = A^2, \tag{8.7}$$

where A is some given constant.

Indeed, introducing the Lagrange multiplier for the constraint (8.7), the functional of the problem can be written as

$$\int_{-\pi}^{\pi}\int_{V} F dV d\theta - \lambda \left(\int_{-\pi}^{\pi}\int_{V} K dV d\theta - A^2\right). \tag{8.8}$$

The Lagrange multiplier λ is nonnegative: for $\lambda < 0$, the functional (8.8) is strictly convex and has the only stationary point, $u^{\varkappa} = 0$. Therefore, each stationary point of the variational problem (8.3), (8.5), (8.6) and (8.7) is the stationary point of the initial problem with $\omega^2 = \lambda$.

Conversely, each stationary point of the initial problem is the stationary point of the variational problem (8.3), (8.5), (8.6) and (8.7) for some A; it is sufficient to calculate the value of A, corresponding to the stationary point being considered, by means of formula (8.7).

In the variational problem (8.3), (8.5), (8.6) and (8.7), the Lagrange multiplier for the constraint (8.7) has the meaning of the squared frequency, and the constant A is a measure of the amplitude of the vibrations.

As a rule, for any value of A there exists a countable number of stationary points and the corresponding values of eigenfrequencies. They continuously change as the value of A is changed.

8.3 Linear Vibrations: The Rayleigh Principle

For $A \to 0$, the amplitude of vibrations tends to zero. If F and K are smooth functions and K becomes zero for $u^{\varkappa}_{,t} = 0$, then for infinitesimally small amplitudes, F can be replaced by a quadratic form with respect to $u^{\varkappa}$, $u^{\varkappa}_{,i}$, and K can be replaced by a quadratic form with respect to $u^\varkappa$, $u^\varkappa_{,i}$, and K can be replaced by a quadratic form with respect to $u^\varkappa_\theta$: $2K = \rho_{\varkappa\varkappa'} u^\varkappa_\theta u^{\varkappa'}_\theta$. It is easy to check that the stationary points have the form $u^\varkappa = v^\varkappa(x)\cos\theta$ or $u^\varkappa = v^\varkappa(x)\sin\theta$. After integrating over θ, the variational principle (8.3), (8.5)–(8.7), becomes

Rayleigh principle. *The modes of eigenvibrations are the stationary points of the free energy functional* [3]

$$\mathcal{F} = \int_V F\left(x, v^\varkappa, v^\varkappa_{,i}\right) dV, \tag{8.9}$$

on the set of functions $v^\varkappa$ vanishing on ∂V and having a given value of kinetic energy

$$\mathcal{K} = \int_V \frac{1}{2} \rho_{\varkappa\varkappa'} v^\varkappa v^{\varkappa'} dv = A^2. \tag{8.10}$$

The change of the constant A corresponds to multiplying $v^\varkappa$ by a constant; therefore, without loss of generality we can set $A = 1$ in (8.10).

The Rayleigh variational principle is equivalent to the assertion of the stationarity of the so-called Rayleigh quotient,

$$R = \frac{\mathcal{F}}{\mathcal{K}} = \frac{\int_V F\left(x, v^\varkappa, v^\varkappa_{,i}\right) dV}{\int_V \frac{1}{2} \rho_{\varkappa\varkappa'} v^\varkappa v^{\varkappa'} dV}.$$

Indeed, the Rayleigh quotient does not change if the function $v^\varkappa$ is multiplied by a constant; in particular, this constant can be chosen in such a way as to make $\mathcal{K} = 1$ and thus $R = \mathcal{F}$.

The non-convexity, which is characteristic for the variational problems of dynamics, is preserved for eigenvibrations as well: the constraint $\mathcal{K} = const$ extracts a non-convex set in the functional space – a sphere (in an appropriate norm).[4]

If the function F is not smooth at zero, then the variational principle (8.3), (8.5), (8.6) and (8.7), generally speaking, does not transform to the Rayleigh principle even in the case of infinitesimally small amplitudes. As an example, note the problem of eigenvibrations when continuum has different moduli for $u_{,\varkappa} > 0$ and $u_{,\varkappa} < 0$; e.g., in the one-dimensional case, $2F = \alpha u^2_{,x} + \beta u_{,x}\left|u_{,x}\right|$,

---

[3] It is easy to check that the stationary points of the functional (8.6) do not change if the same factor, k, is included in the left-hand sides of (8.6) and (8.7), and the constant A^2 is replaced by kA^2. In transition from (8.6), (8.7) to (8.9), (8.10) the coefficient $k = 1/\pi$ is introduced; the constant A^2 in (8.10) differs from the constant A^2 in (8.7) by the same factor.

[4] If the condition $\mathcal{K} = A^2$ could be changed by the condition $\mathcal{K} \le A^2$, then the problem would be convex. In the variational principle formulated, it is possible only to replace the condition $\mathcal{K} = A^2$ by the inequality $\mathcal{K} \ge A^2$ which selects a non-convex set.

$\alpha + \beta > 0, \alpha - \beta > 0$. For such elastic body, due to non-linearity, of the problem, the eigenfunctions cannot be presented as the product of a function depending on θ and a function depending on x.

8.4 The Principle of Least Action in Eulerian Coordinates

The principle of least action for an elastic body in Lagrangian coordinates was formulated in Sect. 4.3. Sometimes it is of interest to formulate it also in Eulerian coordinates. Although it involves only rewriting the action functional in different terms, there are a number of technical details deserving consideration.

In Eulerian coordinates, it is convenient to define the action functional on displacements, $u^i(x,t)$:

$$u^i(x,t) = x^i(X,t) - \overset{\circ}{x}{}^i(X)\big|_{X=X(x,t)} = x^i - \overset{\circ}{x}{}^i(X(x,t)).$$

The displacements are uniquely related to the particle trajectories. The domain of displacements, $u^i(x,t)$, is the region, $V(t)$, occupied by the continuum at the time t.

Assigning the particle positions in the initial and the final times corresponds to assigning the displacements,

$$u^i(x,t_0) = 0, \quad u^i(x,t_1) = u_1^i(x). \tag{8.11}$$

The functions $u_1^i(x)$ are defined in the region $\overset{1}{V} = V(t_1)$. The points with the coordinates $x^i - u_1^i(x)$ for $x \in \overset{1}{V}$ are in the region $\overset{\circ}{V} = V(t_0)$. The corresponding mapping of $\overset{1}{V}$ onto $\overset{\circ}{V}$ is one-to-one.

Let us express the arguments of the Lagrangian in terms of the displacements and their derivatives. We begin with the Lagrangian coordinates X^a. The relation between the Lagrangian and the Eulerian coordinates at the initial instant is

$$X^a = \overset{\circ}{X}{}^a(\overset{\circ}{x})$$

where $\overset{\circ}{x}{}^i$ are the Eulerian coordinates of the points in $\overset{\circ}{V}$. Since $\overset{\circ}{x}{}^i = x^i - u^i(x^k,t)$, the dependence of the Lagrangian coordinates on the Eulerian coordinates at any instant has the form

$$X^a = \overset{\circ}{X}{}^a\left(x^i - u^i(x,t)\right). \tag{8.12}$$

Here, $\overset{\circ}{X}{}^a(x^i)$ are the known functions, the inverse of the functions $\overset{\circ}{x}{}^i(X^a)$. In particular, if the Lagrangian and the Eulerian coordinates coincide at the initial time, then $X^i = \overset{\circ}{x}{}^i$ and $X^i(x^i,t) = x^i - u^i(x,t)$. However, as has been noted, a particular choice of Lagrangian coordinates at the stage of deriving the equations excludes the possibility to check the invariance of all relations with respect to transformations of the Lagrangian coordinates; therefore, we use the general formula (8.12).

Assigning the particle positions at the boundary of the body corresponds to assigning a part of the boundary $\Sigma(t)$ of the region $V(t)$ and the values of the displacement vector at this part:

$$u^i = u^i_{(b)}(x,t) \quad \text{on } \Sigma(t). \tag{8.13}$$

These values are not arbitrary. They satisfy the condition

$$\mathring{X}^a\left(x^i - u^i_{(b)}(x,t)\right) \in \mathring{\Sigma} \text{ for } x \in \Sigma.$$

($\mathring{\Sigma}$ is the part of the boundary of the body in the space of Lagrangian coordinates for the points of which the displacements are given).

Let us find the relation between the distortion and the displacement gradient. Differentiating (8.12) with respect to x^i we have[5]

$$\frac{\partial X^a}{\partial x^i} = \frac{\partial \mathring{X}^a}{\partial x^k}\left(\delta^k_i - u^k_i\right), \quad u^k_i = \frac{\partial u^k}{\partial x^i}.$$

Hence,

$$\det\left\|\frac{\partial X^a}{\partial x^i}\right\| = \det\left\|\frac{\partial \mathring{X}^a}{\partial x^i}\right\| \cdot \det\left\|\delta^k_i - u^k_i\right\|. \tag{8.14}$$

Equation (8.14) shows that the matrix $\left\|\delta^k_i - u^k_i\right\|$ is non-singular. Therefore, it is possible to introduce the tensor s^i_j, which is the inverse of the tensor $\delta^k_i - u^k_i$:

$$s^i_j\left(\delta^j_k - u^j_k\right) = \delta^i_k. \tag{8.15}$$

The distortion x^i_a is given by the equation

$$x^i_a = \mathring{x}^k_a s^i_k. \tag{8.16}$$

This can be checked by direct inspection:

$$x^i_a X^a_j = \mathring{x}^k_a s^i_k \mathring{X}^a_m\left(\delta^m_j - u^m_j\right) = \delta^i_j.$$

Here, the formulas (8.15) and (8.16) are used.

In terms of the displacement gradient, the density ρ, according to (8.14) is

[5] Remember that the observer's coordinate system is Cartesian.

$$\rho = \frac{\rho_o \sqrt{\mathring{g}}}{\sqrt{\hat{g}}} = \frac{\rho_o \det \left\| \frac{\partial \mathring{x}}{\partial X} \right\|}{\det \left\| \frac{\partial x}{\partial X} \right\|} = \frac{\rho_o \det \left\| \frac{\partial X}{\partial x} \right\|}{\det \left\| \frac{\partial \mathring{X}}{\partial x} \right\|} = \rho_o \det \left\| \delta^i_j - u^i_{;j} \right\| . \tag{8.17}$$

Let us express the velocity in terms of the derivatives of the displacements. Taking the time derivative of the equation

$$u^i = x^i (X, t) - \mathring{x}^i (X)$$

for constant Lagrangian coordinates, we get

$$\frac{\partial u^i}{\partial t} + v^k \frac{\partial u^i}{\partial x^k} = v^i . \tag{8.18}$$

The relations (8.18) can be considered as a system of three linear algebraic equations with respect to the particle velocity, v^i. According to (8.15), the solution of this system of equation is

$$v^i = s^i_k u^k_t . \tag{8.19}$$

Here, $u^k_t \equiv \partial u^k / \partial t$.

For simplicity, let us assume that the external forces on $\partial V_t - \Sigma_t$ are equal to zero. Then the action functional can be written as

$$I (u) = \int_{t_0}^{t_1} \int_{V(t)} L \left(x^i, u^i, u^i_k, u^i_t\right) dV dt, \tag{8.20}$$

$$L = \rho \left(\frac{1}{2} v^2 - F \left(x^i_a\right) - \Phi \left(x^i\right) \right), \quad v^2 \equiv v_i v^i , \tag{8.21}$$

$\Phi \left(x^i\right)$ being the potential of body forces. In calculating L, one should express x^i_a, v and ρ in terms of the displacement gradient by (8.16), (8.19) and (8.17).

Note that for inhomogeneous media characteristics of which depend on the Lagrangian coordinates, the function, L, depends explicitly on the space coordinates, x^i, and on the displacements, u^i, through the difference, $x^i - u^i$. Besides, L depends explicitly on the Eulerian coordinates due to the presence of the external body forces.

Least action principle. *The true displacements are the stationary points of the functional (8.20) on the set of displacements satisfying the conditions (8.11) and (8.13).*

Let us show that the formulated variational principle indeed yields the equations of elasticity theory. Let us first calculate the variation of the functional $I (u)$ (8.20), fixing the region $V (t) \times [t_0, t_1]$. Then

$$\delta I(u) = \int_{t_0}^{t_1} \int_{V(t)} \partial L dV dt =$$

$$= \int_{t_0}^{t_1} \int_{V(t)} \left(\frac{\partial L}{\partial u^i} \partial u^i + \frac{\partial L}{\partial u^i_k} \frac{\partial \partial u^i}{\partial x^k} + \frac{\partial L}{\partial u^i_t} \frac{\partial \partial u^i}{\partial t} \right) dV dt = 0. \tag{8.22}$$

As before, the symbol ∂ denotes the variation at constant Eulerian coordinates.

First, let $\partial u^i = 0$ on $V(t)$. Integrating by parts, we get

$$\delta I(u) = \int_{t_0}^{t_1} \int_{V(t)} \partial u^i \frac{\delta L}{\delta u^i} dV dt = 0,$$

$$\frac{\delta L}{\delta u^i} = \frac{\partial L}{\partial u^i} - \frac{\partial}{\partial x^k} \frac{\partial L}{\partial u^i_k} - \frac{\partial}{\partial t} \frac{\partial L}{\partial u^i_t}. \tag{8.23}$$

From (8.23) the equations follow

$$\frac{\delta L}{\delta u^i} = 0. \tag{8.24}$$

These equations are equivalent to the momentum equations of the elasticity theory. Indeed, due to the non-singularity of the matrix $\left\| \delta^k_i - u^k_i \right\|$, the system of equations (8.24) is equivalent to the system of equations

$$\frac{\delta L}{\delta u^k} \left(\delta^k_i - u^k_i \right) = 0. \tag{8.25}$$

Note the identity

$$-u^k_i \frac{\delta L}{\delta u^k} = \partial_i L - \frac{\partial L}{\partial x^i} + \frac{\partial}{\partial x^m} \left(\frac{\partial L}{\partial u^k_m} u^k_i \right) + \frac{\partial}{\partial t} \left(\frac{\partial L}{\partial u^k_t} u^k_i \right) \tag{8.26}$$

which can be checked by direct inspection. Here, the partial derivative of the function $L\left(x^i, u^i, u^i_k, u^i_t\right)$ with respect to x^i is denoted by $\partial_i L$, in contrast to $\partial L / \partial x^i$ which means "full" derivative with respect to x^i, i.e. the derivative taking into account the dependence of all the arguments of L on x^i.

From (8.25) and (8.26) we find that

$$\frac{\delta L}{\delta u^k}\left(\delta_i^k - u_i^k\right) = \frac{\delta L}{\delta u^i} - u_i^k \frac{\delta L}{\delta u^k} = \frac{\partial L}{\partial u^i} - \frac{\partial}{\partial x^m}\frac{\partial L}{\partial u_m^i} -$$

$$-\frac{\partial}{\partial t}\frac{\partial L}{\partial u_t^i} + \partial_i L - \frac{\partial L}{\partial x^i} + \frac{\partial}{\partial x^m}\left(\frac{\partial L}{\partial u_m^k} u_i^k\right) + \frac{\partial}{\partial t}\left(\frac{\partial L}{\partial u_t^k} u_i^k\right) =$$

$$= \partial_i L + \frac{\partial L}{\partial u^i} - \frac{\partial}{\partial x^m}\left[\frac{\partial L}{\partial u_m^k}\left(\delta_i^k - u_i^k\right) + L\delta_i^m\right] - \frac{\partial}{\partial t}\left[\frac{\partial L}{\partial u_t^k}\left(\delta_i^k - u_i^k\right)\right]. \quad (8.27)$$

Since the free energy depends on the Eulerian coordinates only through the difference $x^i - u^i$,

$$\partial_i L + \frac{\partial L}{\partial u^i} = -\rho \frac{\partial \Phi}{\partial x^i}. \quad (8.28)$$

In order to calculate the other terms in (8.27), we need the formulas

$$\left(\delta_i^k - u_i^k\right)\frac{\partial \rho}{\partial u_m^k} = -\rho\delta_i^m, \qquad \frac{\partial s_j^i}{\partial u_m^k} = s_k^i s_j^m. \quad (8.29)$$

The first follows from the equalities (3.20) and (8.17), the second is derived in the same way as (4.24). Due to (8.29), we have

$$\left(\delta_i^k - u_i^k\right)\frac{\partial}{\partial u_t^k}\left(\frac{1}{2}\rho v^2\right) = \rho v_i, \quad \left(\delta_i^k - u_i^k\right)\frac{\partial}{\partial u_m^k}\frac{v^2}{2} = v_i v^m. \quad (8.30)$$

Therefore,

$$\frac{\partial L}{\partial u_m^k}\left(\delta_i^k - u_i^k\right) + L\delta_i^m = \rho v^m v_i - \sigma_i^m. \quad (8.31)$$

Here, we introduced the notation

$$\sigma_i^j = \rho\left(\delta_i^k - u_i^k\right)\frac{\partial F}{\partial u_j^k} = \rho\frac{\partial F}{\partial u_j^i} - \rho\frac{\partial F}{\partial u_j^k}u_i^k. \quad (8.32)$$

Using (8.29) and (8.16), it is easy to see that

$$\sigma_i^j = \rho\frac{\partial F}{\partial x_a^i}x_a^j.$$

Therefore, σ_i^j has the meaning of the components of the Cauchy stress tensor.

From the relations (8.29), (8.30), (8.31) and (8.32) follow the known momentum equations

$$\frac{\partial \rho v_i}{\partial t} + \frac{\partial}{\partial x^m}\left(\rho v_i v^m - \sigma_i^m\right) + \rho\frac{\partial \Phi}{\partial x^i} = 0. \quad (8.33)$$

The constitutive equations (8.32) are sometimes written in the form which takes into account the fact that the free energy depends on the components of the strain tensor with the observer's indices $\varepsilon_{ij} = \varepsilon_{ab} X_i^a X_j^b$. Since $\varepsilon_{ab} = \frac{1}{2} g_{ij} \left(x_a^i x_b^j - \overset{\circ}{x}{}_a^i \overset{\circ}{x}{}_b^j \right)$, the quantities ε_{ij} are expressed through the displacement gradient as

$$\varepsilon_{ij} = \frac{1}{2} \left(g_{ij} - \overset{\circ}{x}{}_a^m \overset{\circ}{x}{}_b^m X_i^a X_j^b \right) = \frac{1}{2} \left(g_{ij} - \left(\delta_i^k - u_i^k \right) \left(g_{kj} - u_{kj} \right) \right) = u_{(ij)} - \frac{1}{2} u_k^k u_{kj}.$$

Consequently, for $F = F\left(\varepsilon_{ij}\right)$, from (8.32) we have

$$\sigma^{ij} = \rho \frac{\partial F}{\partial \varepsilon_{ij}} - 2\rho \frac{\partial F}{\partial \varepsilon_{ik}} \varepsilon_k^j.$$

This relation is called Murnagan's equation. Note that this formula, unlike formula (8.32), is valid only for isotropic media because the free energy of anisotropic body depends not only on the components of the strain tensor ε_{ij} but also on the displacement gradient. This can be checked by considering the example of a physically linear material with $2F = C^{abcd} \varepsilon_{ab} \varepsilon_{cd}$. Switching to the components with the observer's indices, we have $2F = C^{abcd} x_a^i x_b^j x_c^k x_d^l \varepsilon_{ij} \varepsilon_{kl}$. Hence, F depends not only on ε_{ij}, but also – through x_a^i – on u_j^i.

Now let the displacement variations and the variations of the part of the boundary $S(t) = \partial V(t) - \Sigma(t)$ be non-zero. The calculation of δI results in

$$\delta I = \int_{t_0}^{t_1} \int_{S(t)} \left[\partial u^i \left(\frac{\partial L}{\partial u_k^i} n_k - \frac{\partial L}{\partial u_t^i} c_x \right) + L \delta n \right] d\sigma dt. \tag{8.34}$$

Here, (3.92) is used, and it is taken into account that due to the momentum equations (8.24) the volume integral is equal to zero; c_x is the velocity of the surface $S(t)$ along its normal in the observer's coordinate system, δn is the virtual displacement of the surface $S(t)$ in the direction of the normal.

Let us rewrite the integrand in (8.34) using the relations (8.30) and (8.31):

$$\begin{aligned}
\partial u^i \left(\frac{\partial L}{\partial u_s^i} n_s - \frac{\partial L}{\partial u_t^i} c_x \right) + L\delta n &= \partial u^j s_j^k \left(\delta_k^i - u_k^i \right) \left(\frac{\partial L}{\partial u_s^i} n_s - \frac{\partial L}{\partial u_t^i} c_x \right) + L\delta n = \\
&= \partial u^j s_j^k \left[\left(\rho v^s v_k - \sigma_k^s - L \delta_k^s \right) n_s - \rho v_k c_x \right] + L\delta n = \\
&= \partial u^j s_j^k \rho v_k \left(v^s n_s - c_x \right) + L \left(\delta n - \partial u^j s_j^k n_k \right) - \partial u^j s_j^k \sigma_k^s n_s.
\end{aligned} \tag{8.35}$$

The first term in the right hand side of (8.35) is equal to zero: the difference $v^s n_s - c_x$ has the meaning of the surface velocity over the particles, and it is equal to zero since $S(t)$ does not move over the particles. The second term is also equal to zero. Indeed, the equality is valid:

$$\delta x^k = \delta u^k = s_j^k \partial u^j. \tag{8.36}$$

It follows from the relation (4.31), written for the displacements:

$$\delta u^j = \partial u^j + u^j_k \delta u^k. \tag{8.37}$$

The solution of (8.37), considered as a system of linear equations with respect to δu^k, is given by the formulas (8.36). In terms of the variations with constant Lagrangian coordinates, the condition that the boundary particles remain on the boundary yields

$$\delta n = \delta x^k n_k. \tag{8.38}$$

From (8.38) and (8.36),

$$\delta n - \partial u^j s^k_j n_k = 0,$$

thus the second term in the right hand side of (8.35) vanishes. We obtain for the variation of the action functional,

$$\delta I = -\int_{t_0}^{t_1} \int_{S(t)} \partial u^i s^k_j \sigma^s_k n_s d\sigma dt = 0.$$

Due to the arbitrariness of ∂u^i and the non-singularity of the matrix s^k_j, the natural condition on $S(t)$ are $\sigma^s_k n_s = 0$.

Chapter 9
Ideal Incompressible Fluid

9.1 Least Action Principle

Consider a continuum in some vessel, $\mathring{V}$. Continuum occupies the entire vessel at the initial instant, t_0. Let us prescribe the initial and the final positions of each point of the continuum,

$$x(t_0, X) = \mathring{x}(X), \quad x(t_1, X) = \overset{1}{x}(X), \tag{9.1}$$

and find the trajectories which minimize the action functional,

$$I(x(t, X)) = \int_{t_0}^{t_1} \int_{\mathring{V}} \frac{1}{2} \rho \frac{dx^i(t, X)}{dt} \frac{dx_i(t, X)}{dt} d\mathring{V} dt, \tag{9.2}$$

with ρ being a constant. Such a variational problem corresponds to motion of non-interacting particle. The particles are driven only by inertia. Obviously, the trajectory of each particle is a straight line in four-dimensional time-space connecting the initial and final positions (Fig. 9.1). Such motion of continuum is compressible. Besides, the mapping, $x(t, X)$, is not a one-to-one mapping because trajectories may intersect. Let us complicate this inertial motion by imposing the incompressibility condition

$$\det \left\| \frac{\partial x(t, X)}{\partial X} \right\| = \det \left\| \frac{\partial \mathring{x}(X)}{\partial X} \right\|. \tag{9.3}$$

Then mapping, $x(t, X)$, becomes a one-to-one mapping, at least locally.

Continuum model which incorporates two features, inertia of particles and incompressibility of motion, is called ideal incompressible fluid. Motion of ideal incompressible fluid is a stationary point in the following variational principle.

Least action principle. *The true motions of ideal incompressible fluid are the stationary points of the action functional (9.2) on the set of all motions extracted by the constraints (9.1) and (9.3).*

V.L. Berdichevsky, *Variational Principles of Continuum Mechanics*,
Interaction of Mechanics and Mathematics, DOI 10.1007/978-3-540-88467-5_9,

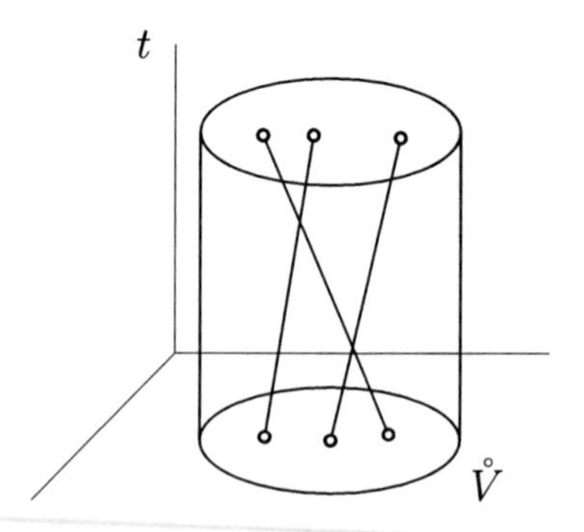

Fig. 9.1 Minimizer of the variational problem (9.1) and (9.2)

In this formulation particles are allowed to detach from the wall. One can also impose an additional condition that the boundary particles remain on the wall all the time:

$$x(t, X) \in \partial V(t) \quad \text{if } X \in \partial \mathring{V}, \tag{9.4}$$

where the current position of the vessel, $V(t)$, is prescribed and may not coincide with the initial position, $\mathring{V}$.

To derive the equations governing the motion of ideal incompressible fluid from the least action principle, we get rid of the constraint (9.3) by introducing a Lagrange multiplier, p. The expanded functional is

$$\int_{t_0}^{t_1} \int_{\mathring{V}} \left[\frac{1}{2} \rho \frac{dx^i}{dt} \frac{dx_i}{dt} + p \left(\frac{\det \left\| \frac{\partial x}{\partial X} \right\|}{\det \left\| \frac{\partial \mathring{x}}{\partial X} \right\|} - 1 \right) \right] d\mathring{V} dt,$$

and for its variation we obtain

$$\int_{t_0}^{t_1} \int_{\mathring{V}} \left[\rho v_i \frac{d\delta x^i}{dt} + p \frac{\partial \delta x^i}{\partial x^i} \right] d\mathring{V} dt =$$

$$= \left[\int_{\mathring{V}} \rho v_i \delta x^i d\mathring{V} \right]_{t_0}^{t_1} + \int_{t_0}^{t_1} \int_{\mathring{V}} p \delta x^i n_i dA dt - \int_{t_0}^{t_1} \int_{\mathring{V}} \delta x^i \left(\rho \frac{dv_i}{dt} + \frac{\partial p}{\partial x^i} \right) d\mathring{V} dt. \tag{9.5}$$

Here we used (4.19). The first term in (9.5) vanishes due to (9.1): $\delta x^i = 0$ at $t = t_0, t_1$. If particles do not detach from the wall, $\delta x^i n_i = 0$ at $\partial \mathring{V}$, and the second term vanishes as well. Therefore, equating (9.5) to zero we obtain the equations

$$\rho \frac{dv_i}{dt} = -\frac{\partial p}{\partial x^i}, \tag{9.6}$$

which are the momentum equations of ideal incompressible fluid. Pressure in the fluid is the Lagrange multiplier for the incompressibility condition. This was the way which was used by Lagrange to derive these equations [168]. Momentum equations must be complimented by the incompressibility condition which is usually taken in the differential form:

$$\frac{\partial v^i}{\partial x^i} = 0. \tag{9.7}$$

If fluid is allowed to detach from the wall in the course of motion, and some free surface forms, then, due to arbitrariness of $\delta x^i n_i$,

$$p = 0 \tag{9.8}$$

on the free surface.

To model the fluid evaporation or other physical mechanisms by which a non-zero pressure develops inside the cavity, formed by the detached fluid, one has to take into account the work of pressure on the free surface displacements. Then the action functional gains an additional term,[1]

$$\int_{V_c} p_0 dV dt,$$

where p_0 is the pressure in the cavity, V_c. Accordingly the boundary condition (9.8) is replaced by the condition

$$p = p_0 \tag{9.9}$$

at the free surface.

All the equations obtained remain valid if the fluid is inhomogeneous, i.e. its density depends on Lagrangian coordinates: $\rho = \rho(X)$. Another modification of the action functional appears if there are body forces with the potential $\Phi(x)$, acting on the fluid. The action functional becomes

$$\int_{t_0}^{t_1} \int_{\mathring{V}} \left[\frac{1}{2} \rho \frac{dx^i}{dt} \frac{dx_i}{dt} - \rho \Phi(x(t, X)) \right] d\mathring{V} dt. \tag{9.10}$$

If the fluid is homogeneous and does not detach from the wall, then the second term in (9.10) is not essential because it does not depend on fluid motion:

$$\int_{t_0}^{t_1} \int_{\mathring{V}} \rho \Phi(x(t, X)) d\mathring{V} dt = \int_{t_0}^{t_1} \int_{V(t)} \rho \Phi(x) dV dt$$

[1] See (7.13).

If the fluid is inhomogeneous and/or has free surfaces this term is essential. We will encounter its contributions later in Sect. 13.2.

9.2 General Features of Solutions of Momentum Equations

Momentum equations of ideal incompressible homogenous fluid (9.6) can be "integrated" in the following sense. Let us project these equations on the Lagrangian frame:

$$\rho x_a^i \frac{dv_i}{dt} = -x_a^i \frac{\partial p}{\partial x^i} = -\frac{\partial p}{\partial X^a}. \tag{9.11}$$

Denoting by v_a the Lagrangian components of velocity,

$$v_a = x_a^i v_i,$$

and using the fact that

$$\begin{aligned} x_a^i \frac{dv_i}{dt} &= \frac{dv_a}{dt} - v_i \frac{dx_a^i}{dt} = \frac{dv_a}{dt} - v_i \frac{\partial v^i}{\partial X^a} = \\ &= \frac{dv_a}{dt} - \frac{1}{2}\frac{\partial}{\partial X^a} v^2, \quad v^2 \equiv v_i v^i, \end{aligned} \tag{9.12}$$

we have

$$\frac{dv_a}{dt} = \frac{\partial}{\partial X^a}\left(-\frac{p}{\rho} + \frac{v^2}{2}\right).$$

Let us introduce a function, φ, by the equation

$$\frac{d\varphi}{dt} = -\frac{p}{\rho} + \frac{v^2}{2}. \tag{9.13}$$

Then the momentum equations can be written as

$$\frac{d}{dt}\left(v_a - \frac{\partial \varphi}{\partial X^a}\right) = 0.$$

Hence, the Lagrangian covariant components of velocity, v_a differ from $\partial\varphi/\partial X^a$ only by some field, $\mathring{v}_a$, which does not depend on time:

$$v_a = \frac{\partial \varphi}{\partial X^a} + \mathring{v}_a(X), \tag{9.14}$$

or, in Eulerian coordinates,

$$v_i = \frac{\partial \varphi}{\partial x^i} + X_i^a \mathring{v}_a(X). \tag{9.15}$$

Function φ can be chosen arbitrarily at the initial instant. If φ is set to be equal to zero initially, then $\mathring{v}_a(X)$ have the meaning of the initial Lagrangian velocity components.

Equation (9.14) determines the general structure of solutions of Euler equations: the Lagrangian components of velocity, v_a, may depend on time only through the potential part, $\partial\varphi/\partial X^a$.

There are other forms of this statement. As follows from (9.14) the antisymmetric part of the gradient of velocity does not depend on time:

$$\partial_{[a} v_{b]} = \partial_{[a} \mathring{v}_{b]}(X). \tag{9.16}$$

The antisymmetric tensor $\partial_{[a} v_{b]}$ is in one-to-one correspondence with the vector:

$$\omega^c \equiv \varepsilon^{abc} \partial_{[a} v_{b]} = \frac{1}{\sqrt{\hat{g}}} e^{abc} \partial_{[a} v_{b]}, \quad \partial_{[a} v_{b]} = \frac{1}{2}\sqrt{\hat{g}} e_{abc} \omega^c.$$

Due to the incompressibility condition, the determinant of the metric tensor, $\hat{g}$, can be set equal to unity.

The vector with the contravariant Lagrangian components, ω^a, is called the vorticity vector. In Eulerian coordinates, the vorticity vector has the components

$$\omega^i = e^{ijk} \frac{\partial v_k}{\partial x^j}, \tag{9.17}$$

and, due to the law of transformation of vector components,

$$\omega^i = x_a^i \omega^a. \tag{9.18}$$

Conversely,

$$\omega^a = X_i^a \omega^i.$$

According to (9.16), the contravariant Lagrangian components of vorticity, ω^c, do not change in time at each fluid particle:

$$\frac{d\omega^a}{dt} = 0. \tag{9.19}$$

Note that the covariant components of vorticity, $\omega_b = g_{ba}\omega^a$, may change in time.

Another form of (9.14) is obtained by differentiation of (9.18) with respect to time at constant Lagrangian coordinates. Using the independence of ω^a on time and formula the time defivative of distortion (3.43), we obtain the evolution equations for Eulerian components of vorticity,

$$\frac{d\omega^i}{dt} = \frac{\partial v^i}{\partial x^k}\omega^k. \tag{9.20}$$

This equation can also be written in a different form where summation over k is conducted with another index of the velocity gradient,

$$\frac{d\omega_i}{dt} = \frac{\partial v_k}{\partial x^i}\omega^k. \tag{9.21}$$

This follows from the identity

$$\left(\frac{\partial v_k}{\partial x^i} - \frac{\partial v_i}{\partial x^k}\right)\omega^k = 0. \tag{9.22}$$

The identity follows from the inversion of (9.17) (similarly to (3.53)):

$$\frac{\partial v_k}{\partial x^i} - \frac{\partial v_i}{\partial x^k} = e_{ikm}\omega^m,$$

and vanishing of the sum, $e_{ikm}\omega^k\omega^m = 0$.

Conservation of vorticity is often stated in the integral form: for any closed contour, Γ, the circulation of velocity over this contour, $\oint v_i dx^i$, does not change in time. This statement is equivalent to the conservation of the Lagrangian contravariant components of vorticity, ω^a. Indeed, a moving fluid contour Γ is the image of a contour, $\mathring{\Gamma}$, which is stationary in Lagrangian coordinates. Therefore, the circulation of velocity can be presented as the circulation of velocity over this stationary contour:

$$C \equiv \int\limits_{\Gamma} v_i dx^i = \int\limits_{\mathring{\Gamma}} v_i x^i_a dX^a = \int\limits_{\mathring{\Gamma}} v_a dX^a.$$

Transferring the integral over $\mathring{\Gamma}$ to the integral over a surface $\mathring{S}$ which has the boundary $\mathring{\Gamma}$, we have

$$C = \int\limits_{\mathring{\Gamma}} v_a dX^a = \int\limits_{\mathring{S}} \partial_{[a} v_{b]} dX^a \wedge dX^b.$$

Since $\partial_{[a} v_{b]}$ do not depend on time, C also does not depend on time. The arbitrariness of the contour $\mathring{\Gamma}$ yields the equivalence of the conservation of velocity circulations and the conservation of ω^a.

Let us return to (9.14). In this equation, one can set $\varphi = 0$ at the initial instant and consider $\mathring{v}_a(X)$ as the functions known from the initial conditions. Then the particle trajectories can be sought from a system of equations of the first order,

$$\frac{\partial x^i(t,X)}{\partial X^a}\frac{\partial x_i(t,X)}{\partial t}=\overset{\circ}{v}_a(X)+\frac{\partial\varphi(t,X)}{\partial X^a}. \tag{9.23}$$

The incompressibility condition,

$$\det\left\|\frac{\partial x}{\partial X}\right\|=\det\left\|\frac{\partial \overset{\circ}{x}}{\partial X}\right\|, \tag{9.24}$$

closes this system of equations for $x^i(t,X)$ and $\varphi(t,X)$. Pressure can be computed from (9.13) after the potential φ and velocity are found.

Equation (9.23) can be resolved with respect to time derivative, $\partial x_i(t,X)/\partial t$: using (3.23) we have

$$\frac{\partial x_i(t,X)}{\partial t}=X_i^a\left(\overset{\circ}{v}_a(X)+\frac{\partial\varphi(t,X)}{\partial X^a}\right)=\frac{1}{2}e_{ijk}x_a^j x_b^k e^{abc}\left(\overset{\circ}{v}_c(X)+\frac{\partial\varphi(t,X)}{\partial X^c}\right).$$

It is an attractive idea to eliminate φ and obtain a system of equations of the first order with respect to only the particle positions, $x^i(t,X)$. This is done further in a formulation of ideal fluid dynamics as the dynamics of vortex lines.

Note some other versions of (9.14). The three-dimensional vector, $\overset{\circ}{v}_a$, can be presented[2] (at least, locally) in terms of three potentials, $\alpha(X)$, $\beta(X)$ and $\varphi_0(X)$ as

$$\overset{\circ}{v}_a=\frac{\partial\varphi_0}{\partial X^a}+\alpha\frac{\partial\beta}{\partial X^a}.$$

It is convenient to redefine φ, denoting by φ the sum $\varphi+\varphi_0$. Equation (9.14) takes the form

$$v_a=\frac{\partial\varphi}{\partial X^a}+\alpha\frac{\partial\beta}{\partial X^a}. \tag{9.25}$$

The scalars, φ, α and β are called Clebsch's potentials. In Eulerian coordinates, (9.25) is

$$v_i=\frac{\partial\varphi}{\partial x^i}+\alpha\frac{\partial\beta}{\partial x^i}. \tag{9.26}$$

Note that Clebsch's potentials are not defined uniquely for a given velocity field. For example, for any constant k the change $\alpha\to k\alpha$, $\beta\to\beta/k$ does not alter the velocity field.

---

[2] This follows from the theorem on the canonical presentation of differential forms, $dA=A_1dx^1+\ldots+A_ndx^n$: for an even n, $n=2s$, there are independent functions, $p_1(x),\ldots,p_s(x),q^1(x),\ldots,q^s(x)$ such that $dA=p_1dq^1+\ldots+p_sdq^s$, while for an odd n, $n=2s+1$, there are independent functions, $H(x)$, $p_1(x),\ldots,p_s(x),q^1(x),\ldots,q^s(x)$, such that $dA=p_1dq^1+\ldots+p_sdq^s-dH$.

Equation (9.26) must be complemented by the conditions that the potentials α and β do not change along the particle trajectories:

$$\frac{d\alpha}{dt} = \frac{\partial \alpha}{\partial t} + v^i \frac{\partial \alpha}{\partial x^i} = 0, \qquad \frac{d\beta}{dt} = \frac{\partial \beta}{\partial t} + v^i \frac{\partial \beta}{\partial x^i} = 0. \tag{9.27}$$

Equation (9.13) for the potential φ in Eulerian coordinates is

$$\frac{\partial \varphi}{\partial t} + v^i \frac{\partial \varphi}{\partial x^i} - \frac{v^2}{2} + \frac{p}{\rho} = 0. \tag{9.28}$$

Using (9.26) and (9.27) we can also put it in the form

$$\frac{\partial \varphi}{\partial t} + \alpha \frac{\partial \beta}{\partial t} + \frac{v^2}{2} + \frac{p}{\rho} = 0. \tag{9.29}$$

For potential flows, either α or β is zero. Then (9.29) transforms into the Cauchy-Lagrange integral

$$\frac{\partial \varphi}{\partial t} + \frac{v^2}{2} + \frac{p}{\rho} = 0.$$

9.3 Variational Principles in Eulerian Coordinates

Instead of the particle trajectories $x\,(X, t)$, the action functional may be considered as a functional of functions of Eulerian coordinates: the Lagrangian coordinates $X\,(x, t)$, displacements, density and velocity, Clebsch potentials, etc. Doing so, a one-to-one correspondence of the new characteristics with the particle trajectories should be maintained; otherwise, some "degrees of freedom" may appear or disappear.

Consider the changes in the variational formulations caused by various choices of the action functional arguments.

Variation of the Lagrangian coordinates. Let $X\,(x, t)$ be the arguments of the action functional. The functions $X\,(x, t)$ are defined in the region $V\,(t)$ occupied by the fluid at the instant t.

Let us find Lagrangian L in terms of the functions $X\,(x, t)$. According to (3.58), velocity is function of time and space derivatives of $X\,(t, x)$:

$$v^k = -x_a^k \frac{\partial X^a}{\partial t}. \tag{9.30}$$

Here x_a^k are viewed as the components of the matrix which is inverse to the matrix $\left\| \partial X^a / \partial x^i \right\|$. Using (3.25), we can write that explicitly:

$$v^k = -\frac{1}{2\det\left\|X_i^a\right\|} e^{ijk} X_i^b X_j^c e_{abc} \frac{\partial X^a}{\partial t}. \tag{9.31}$$

So,

$$L = \frac{1}{2}\rho(X) g_{ij} x_a^i x_b^j \frac{\partial X^a}{\partial t}\frac{\partial X^b}{\partial t}. \tag{9.32}$$

Here we allow density to be a function of X. This function, $\rho(X)$, is assumed to be known.

The incompressibility condition is a constraint on the space derivatives of $X(x,t)$:

$$\det\left\|\frac{\partial X^a}{\partial x^i}\right\| = 1 \tag{9.33}$$

(for simplicity, we set $\left\|\mathring{X}_i^a\right\| = 1$). The values of the functions $X^a(x^i,t)$ at the initial instant are known:

$$X^a(x^i,t_0) = \mathring{X}^a(x^i). \tag{9.34}$$

Besides, the values of the function $X^a(x^i,t)$ at the finial instant are also given:

$$X^a(x^i,t_1) = \overset{1}{X}{}^a(x^i). \tag{9.35}$$

The functions $\mathring{X}^a(x^i)$ are defined in the region $\mathring{V} = V(t_0)$, and $\overset{1}{X}{}^a(x^i)$ in the region $\overset{1}{V} = V(t_1)$. It is assumed that there exists a mapping $\mathring{V} \to \overset{1}{V}$, for which the functions $\mathring{X}^a(x^i)$ and $\overset{1}{X}{}^a(x^i)$ coincide.

If the equation of the boundary of the region $\mathring{V}$ at the initial moment is

$$f(X) = 0,$$

then the non-detachment of the fluid particles from the boundary means that the equation,

$$f(X(x,t)) = 0, \tag{9.36}$$

is the equation of the boundary of the given region $V(t)$.

Consider the functional

$$I = \int_{t_0}^{t_1}\int_{V(t)} L dV dt \tag{9.37}$$

where L is the function (9.32).

Least action principle. *The true motion of the ideal incompressible fluid is a stationary point of the functional (9.37) with Lagrangian (9.32) on the set of all functions $X(x,t)$, satisfying the constraints (9.33), (9.34), (9.35) and (9.36).*

Let us show the validity of this principle by direct calculation. Indeed, denoting further by L the function (9.32) with the added Lagrange multiplier term,

$$-p\left(\det\left\|\frac{\partial X^a}{\partial x^i}\right\|-1\right),$$

we have

$$\delta I=\int_{t_0}^{t_1}\int_{V(t)}\frac{\delta L}{\delta X^a}\partial X^a dVdt+\int_{t_0}^{t_1}\int_{\partial V(t)}\partial X^a\left(\frac{\partial L}{\partial X_i^a}n_i-\frac{\partial L}{\partial X_t^a}c_x\right)dAdt=0. \quad (9.38)$$

Here $X_t^a\equiv\partial X^a/\partial t$. For $\partial X^a=0$ on $\partial V(t)$ it follows from (9.38) that

$$\frac{\delta L}{\delta X^a}=\frac{\partial L}{\partial X^a}-\frac{\partial}{\partial x^k}\frac{\partial L}{\partial X_k^a}-\frac{\partial}{\partial t}\frac{\partial L}{\partial X_t^a}=0. \quad (9.39)$$

Equation (9.39) can be transformed into the usual form of the momentum equations for ideal incompressible fluid. To do that we contract (9.39) with the components of the non-singular matrix X_i^a:

$$X_i^a\frac{\delta L}{\delta X^a}=X_i^a\frac{\partial L}{\partial X^a}-\frac{\partial}{\partial x^k}\left(X_i^a\frac{\partial L}{\partial X_k^a}\right)-\frac{\partial}{\partial t}\left(X_i^a\frac{\partial L}{\partial X_t^a}\right)+\frac{\partial L}{\partial X_k^a}\frac{\partial X_k^a}{\partial x^i}+\frac{\partial L}{\partial X_t^a}\frac{\partial X_t^a}{\partial x^i}=$$
$$=-\partial_i L-\frac{\partial}{\partial x^k}\left(X_i^a\frac{\partial L}{\partial X_k^a}-L\delta_i^k\right)-\frac{\partial}{\partial t}\left(X_i^a\frac{\partial L}{\partial X_t^a}\right)=0. \quad (9.40)$$

As before, $\partial_i L$ is the derivative of the function $L\left(x^i,X^a,X_i^a,X_t^a,p\right)$ with respect to x^i with X^a, X_i^a, X_t^a, p held constant. Note that $\partial L/\partial p=0$. Then,

$$\partial_i L=0,\quad X_i^a\frac{\partial L}{\partial X_k^a}-L\delta_i^k=-\rho v_i v^k-pX_i^a\frac{\partial}{\partial X_k^a}\det\left\|X_k^a\right\|-L\delta_i^k=$$
$$=-\rho v_i v^k-p\delta_i^k,\quad \frac{\partial L}{\partial X_t^a}X_i^a=-\rho v_i. \quad (9.41)$$

Substituting (9.41) into (9.40) yields momentum equations of ideal incompressible fluid (9.6).

The surface integral in (9.38) is equal to zero due to the non-detachment condition ($c_x-v^i n_i=0$) and the constraint on the variations

$$\frac{\partial f}{\partial X^a}\partial X^a=\frac{\partial f}{\partial x^i}x_a^i\partial X^a=0\ (\text{or } n_i x_a^i\partial X^a=0),$$

which is obtained by varying (9.36).

Variations of velocity. In hydrodynamics of ideal incompressible fluid, one usually uses a closed system of equations in terms of velocity. Therefore, the variational features of the velocity field are especially interesting. Is it possible to vary velocity independently? At first glance the answer is positive: the velocity is the first derivative of the particle trajectories $x^i(X^a, t)$, and there is a one-to-one correspondence between the functions $x^i(X^a, t)$ and $v^i = dx^i(X^a, t)/dt$ (with an additional initial condition, e.g., $x^i(X^a, t_0) = \overset{\circ}{x}{}^i(X^a)$). However, in such reasoning, one point is missed, which we explain by the following example.

Consider the minimization problem for the functional

$$\int_{t_0}^{t_1} \left(\frac{dx}{dt}\right)^2 dt \to \min \tag{9.42}$$

with the constraints

$$x(t_0) = x_0, \quad x(t_1) = x_1. \tag{9.43}$$

The function providing the minimum is obviously a linear function of t:

$$x(t) = x_0 + \frac{x_1 - x_0}{t_1 - t_0}(t - t_0).$$

Let us try to reformulate the variational problem (9.42) and (9.43) in terms of velocity $v(t) = dx(t)/dt$. Due to the end conditions (9.43), the function $v(t)$ satisfies the constraint

$$\int_{t_0}^{t_1} v(t)dt = x_1 - x_0. \tag{9.44}$$

It is easy to check that the minimization problem for the functional

$$\int_{t_0}^{t_1} v^2 dt \tag{9.45}$$

over all functions $v(t)$ satisfying the condition (9.44) is equivalent to the original one, and yields the same solution:

$$v(t) = \frac{x_1 - x_0}{t_1 - t_0} = const.$$

If in the minimization problem the value of $x(t)$ is given only at one end, then in the transformed problem the function $v(t)$, apparently, must not satisfy the constraint (9.44), and the functional (9.45) should be minimized over all functions $v(t)$. The minimum value of the functional is zero and achieved at $v(t) \equiv 0$, We obtain the same solution by minimizing the functional (9.42) with the one end constraint.

This example shows that velocity can be varied independently only in the absence of the constraint for the particle trajectories at the final instant, t_1. If the positions of the particles at $t = t_1$ are given, then the velocity must satisfy an additional constraint similar to (9.44).

The formulation of constraints for velocity in integral form, like (9.44), in case of continuum is difficult. However, such constraints can be given in the differential form.

Assigning the particle positions at the initial and final instants means that for any admissible velocity field there exist functions $X^a\left(x^i, t\right)$ (Lagrangian coordinates), which are the solution of the system of equations

$$\frac{\partial X^a}{\partial t} + v^k \frac{\partial X^a}{\partial x^k} = 0 \tag{9.46}$$

with the initial conditions and final conditions (9.34) and (9.35), and for all t and x the incompressibility condition (9.33) holds. The non-detachment of the flow at the boundary means that

$$v^i n_i = c_x \text{ on } \partial V(t). \tag{9.47}$$

Lin variational principle. *The stationary points of the functional*

$$\int_{t_0}^{t_1} \int_{V(t)} \frac{1}{2} \rho v_i v^i dV dt \tag{9.48}$$

on the set of functions $v^i\left(x^i, t\right)$ and $X^a\left(x^i, t\right)$, satisfying the conditions (9.33), (9.34), (9.35), (9.46) and (9.47) are the solutions of the momentum equations of ideal incompressible homogeneous fluid.

Here, the incompressibility constraint (9.33) can also be used in its differential form, $v^i_{,i} = 0$.

The differential constraints make this variational problem similar to some problems of the optimal control theory.

The Lin variational principle is in essence a reformulation of the least action principle for function $X^a\left(x^i, t\right)$, since velocity can be expressed only through the functions $X^a\left(x^i, t\right)$ from (9.46).

The expansion of the set of admissible functions. In the least action principle, it is necessary that the admissible functions $x^i\left(X^a, t\right)$ take on the assigned values at the initial and the final instants. Otherwise, varying the action functional, we obtain an additional term:

$$\left[\int_{V(t)} \rho v_i \delta x^i dV\right]_{t_0}^{t_1}.$$

For arbitrary δx^i at $t = t_0, t_1$, the least action principle yields a quite special flow with $v_i = 0$ for $t = t_0$ and $t = t_1$. However, we may weaken the conditions $\delta x^i = 0$ at $t = t_1, t_2$ in such a way that the least action principle yields some sensible classes of flows.

Suppose that for the initial and final instants, some system of hypersurfaces is fixed in regions $\overset{\circ}{V}$ and $\overset{1}{V}$. The hypersurfaces are the level surfaces of some function $\overset{0}{\beta}(x)$ in $\overset{\circ}{V}$ and function $\overset{1}{\beta}(x)$ in $\overset{1}{V}$. Now assume that for any admissible motion of the continuum, the hypersurfaces $\overset{0}{\beta}(x) = c$ become the hypersurfaces $\overset{1}{\beta}(x) = c$ (see Fig. 9.2). This means that there exists a function, $\beta(x, t)$, such that

$$\frac{d\beta}{dt} = \frac{\partial \beta}{\partial t} + v^k \frac{\partial \beta}{\partial x^k} = 0,$$
$$\beta(x, t_0) = \overset{0}{\beta}(x), \quad \beta(x, t_1) = \overset{1}{\beta}(x). \tag{9.49}$$

The motion of the fluid is incompressible:

$$\frac{\partial v^i}{\partial x^i} = 0, \tag{9.50}$$

and, since $V(t)$ is a given region, the normal component of velocity is a known function on $\partial V(t)$:

$$v^i n_i = c_x. \tag{9.51}$$

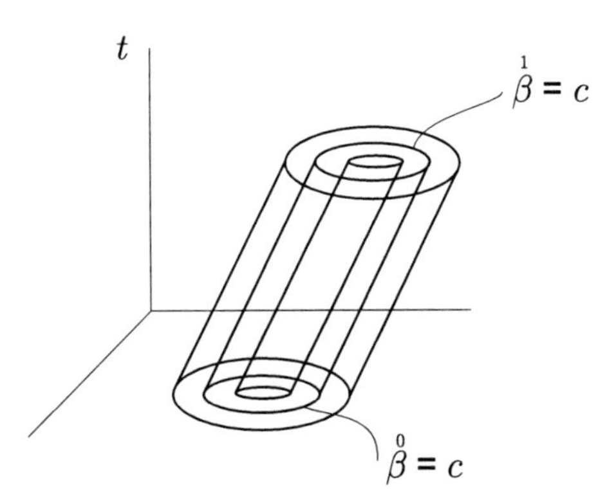

Fig. 9.2 Weakened constraints at the initial and final instants

Variational principle. *On the set of velocity fields, $v^i(x,t)$, which is selected by the constraints (9.49), (9.50) and (9.51), the minimizing element of the kinetic energy functional*

$$\int_{t_0}^{t_1}\int_{V(t)} \frac{1}{2}\rho v^i v_i dVdt$$

satisfies the momentum equations of the ideal incompressible homogeneous fluid.

Indeed, introducing the Lagrange multipliers $\rho\alpha$ and $\rho\varphi$ for the constrains (9.49) and (9.50), respectively, we obtain the functional

$$\rho\int_{t_0}^{t_1}\int_{V(t)}\left[\frac{v^2}{2}+\varphi v^i_{,i}-\alpha\left(\frac{\partial\beta}{\partial t}+v^i\frac{\partial\beta}{\partial x^i}\right)\right]dVdt.$$

Variation of the functional with respect to α gives (9.49); variation with respect to φ results in (9.50); variation with respect to v^i yields the velocity expression in terms of the Clebsch potentials,

$$v^i=\frac{\partial\varphi}{\partial x^i}+\alpha\frac{\partial\beta}{\partial x^i}; \tag{9.52}$$

finally, varying β we obtain the constancy of α for every particle:

$$\frac{d\alpha}{dt}=0. \tag{9.53}$$

The integral that appeared in the integration by parts is equal to zero due to the velocity constraint (9.51) and the constraints for β for (9.49).

Fluid motion with a free surface. Now, let the fluid have a free unknown surface $\Sigma(t)$, while the other part of the boundary, $S(t)$, is given and fluid does not detach from $S(t)$. Let the external body forces with the potential $\Phi(x)$ also be acting on the fluid. In this case, the following variational principle holds:

Variational principle. *The true motion of the ideal incompressible homogeneous fluid with a free surface is the stationary point of the functional*

$$\int_{t_0}^{t_1}\int_{V(t)}\rho\left(\frac{v^2}{2}-\Phi(x)\right)dV\,dt \tag{9.54}$$

on the set of all velocity fields which satisfy the constraints (9.49), (9.50), the conditions (9.51) at the wall, and on the set of regions $V(t)$ with the boundary $S(t)+\Sigma(t)$, which satisfy the conditions,

$$\text{volume of } V(t) = \text{volume of } \overset{\circ}{V}, \quad V(t_0) = \overset{\circ}{V}, \quad V(t_1) = \overset{1}{V}. \tag{9.55}$$

Let us introduce the Lagrange multipliers $\rho\alpha$ and $\rho\varphi$ for the constraints (9.49) and (9.50), respectively, and replace the functional (9.54) by the functional

$$\rho \int_{t_0}^{t_1} \int_{V(t)} \left[\frac{v^2}{2} - \Phi(x) + \varphi v^i_{,i} - \alpha \left(\frac{\partial \beta}{\partial t} + v^i \frac{\partial \beta}{\partial x^i} \right) \right] dV dt. \tag{9.56}$$

The variation of the functional (9.56) with respect to φ and α results in (9.49) and (9.50). The variation of v^i, β and the boundary yields the equation

$$\rho \int_{t_0}^{t_1} \int_{V(t)} \left[\partial v^i \left(v_i - \varphi_{,i} - \alpha \beta_{,i} \right) + \partial \beta \left(\alpha_{,i} + \partial_i \left(\alpha v^i \right) \right) \right] dV dt +$$
$$+ \rho \int_{t_0}^{t_1} \int_{\partial V(t)} \left(\alpha \partial \beta \left(c_x - v^i n_i \right) + \varphi \partial v^i n_i \right) dA dt + \rho \int_{t_0}^{t_1} \int_{\Sigma(t)} \left(\frac{v^2}{2} - \Phi \right) \delta n dA dt = 0. \tag{9.57}$$

Here, δn is the distance by which the surface $\Sigma(t)$ is displaced in the normal direction, when its position is varied.

First, let the variations ∂v^i, $\partial \beta$ and δn be equal to zero on $\partial V(t)$. Then (9.52) and (9.53) follow from (9.57). For nonzero values of ∂v^i, $\partial \beta$ and δn on $\partial V(t)$, (9.57) is reduced to the relation

$$\rho \int_{t_0}^{t_1} \int_{\Sigma(t)} \left[\alpha \partial \beta \left(c_x - v^i n_i \right) + \varphi \partial v^i n_i + \left(\frac{v^2}{2} - \Phi \right) \delta n \right] dA dt = 0. \tag{9.58}$$

Here, we used the fact that $c_x = v^i n_i$ and $\partial v^i n_i = 0$ on $\partial V(t) - \Sigma(t)$.

Let us first set ∂v^i and δn equal to zero on $\Sigma(t)$. There are two possible cases: $\overset{\circ}{\beta} = a = const$ on $\partial \overset{\circ}{V}$ and $\overset{\circ}{\beta} \neq const$ on $\partial \overset{\circ}{V}$. In the first case, due to the condition (9.49), the boundary equation of the region $V(t)$ is of the form $\beta(x,t) = a$. Consequently, the variation $\partial \beta$ on $\partial V(t)$ is equal to zero, and the equality (9.58) is satisfied identically for $\partial v^i = \delta n = 0$. In the second case,[3] the variations $\partial \beta$ are arbitrary on $\partial V(t)$ and (9.58) yields (9.51) on the free surface.

To derive the boundary conditions following from (9.58) for nonzero variations ∂v^i and δn, we need the following relation:

---

[3] For simplicity, we will assume that the surfaces $\beta(x,t) = const$ transversely cross the surface $\partial V(t)$.

$$\int_{t_0}^{t_1}\int_{\Sigma(t)} \varphi \partial v^i n_i dA dt = -\int_{t_0}^{t_1}\int_{\Sigma(t)} \frac{d\varphi}{dt}\delta n dA dt + \left[\int_{\Sigma(t)} \varphi \delta n dA\right]_{t_0}^{t_1}. \tag{9.59}$$

The proof of (9.59) proceeds as follows. Let $x^i = r^i\left(\zeta^1, \zeta^2, t\right)$ be the parametric equations of the surface $\Sigma(t)$, and let ζ^1, ζ^2 be the Lagrangian coordinates of the points on the surface $\Sigma(t)$ such that $v^i = r^i_{,t}$ on $\Sigma(t)$. Then, denoting the variation with ζ^α held constant by δ, we can write

$$\delta v^i = \partial v^i + \delta r^k v^i_{,k} = \delta r^i_{,t}.$$

Hence,

$$\partial v^i n_i = n_i \delta r^i_{,t} - \delta r^k n_i v^i_{,k}. \tag{9.60}$$

Note the formulas

$$\frac{dn^i}{dt} = -r^{i\alpha} n_k v^k_{,\alpha}, \qquad \frac{1}{\sqrt{a}}\frac{d\sqrt{a}}{dt} = r^\alpha_i v^i_{,\alpha}, \tag{9.61}$$

where a is the determinant of the metric tensor $a_{\alpha\beta}$ on the surface $\Sigma(t)$, $r^{i\alpha} = a^{\alpha\beta} r^i_\beta$, $r^i_\beta \equiv r^i_{,\beta}$. The first one is obtained from the formula for the variation of the normal vector (14.47), in which δn^i and δx^i should be replaced by dn^i and $v^i dt$, respectively. The second relation (9.61) follows from (3.20) and the definition of the surface metric tensor (14.3):

$$\frac{1}{\sqrt{a}}\frac{d\sqrt{a}}{dt} = \frac{1}{2a}\frac{\partial a}{\partial a_{\alpha\beta}}\frac{da_{\alpha\beta}}{dt} = \frac{1}{2}a^{\alpha\beta}\frac{d}{dt} r^i_\alpha r_{i\beta} = r^\alpha_i v^i_{,\alpha}.$$

By means of (9.60) and (9.61), we get

$$\begin{aligned} \varphi \partial v^i n_i dA &= \varphi n_i \frac{d\delta r^i}{dt} dA - \varphi \delta r^k n_i v^i_{,k} dA = \\ &= \frac{d}{dt}\left(\varphi n_i \delta r^i \sqrt{a}\right) d\zeta^1 d\zeta^2 - \frac{d\varphi}{dt} n_i \delta r^i dA + \varphi \delta r_i r^{i\alpha} n_k v^k_{,\alpha} dA - \\ &\quad -\varphi n_i \delta r^i r^\alpha_k v^k_{,\alpha} dA - \varphi \delta r^k n_i v^i_{,k} dA. \end{aligned} \tag{9.62}$$

Since $n_i \delta r^i = \delta n$, and according to the decomposition of Kronecker's delta (14.15), $\delta^j_i = n_i n^j + r^\alpha_i r^j_\alpha$, we have

$$\begin{aligned} \delta r^k \cdot n_i v^i_{,k} &= \delta r^k \cdot n_i v^i_{,j} \delta^j_k = \delta r^k \cdot n_i v^i_{,j}\left(n^j n_k + r^\alpha_k r^j_\alpha\right) \\ &= \left(\delta r^k \cdot n_k\right) n_i n^j v^i_{,j} + \delta r^k \cdot r^\alpha_k n_i v^i_{,\alpha}. \end{aligned}$$

The last three terms in (9.62) sum up to zero (it should be taken into account that $v^i_{,i} = v^i_{,j}\delta^j_i = v^i_{,j}\left(n^j n_i + r^j_\alpha r^\alpha_i\right) = n_i n^j v^i_{,j} + r^j_\alpha r^\alpha_i v^i_{,j} = 0$). So, the equality holds,

$$\varphi \partial v^i n_i dA = \frac{d}{dt}\left(\varphi n_i \delta r^i \cdot \sqrt{a}\right) d\zeta^1 d\zeta^2 - \frac{d\varphi}{dt}\delta n dA,$$

from which the formula (9.59) follows.

Based on (9.59), (9.58) can be rewritten as

$$\rho \int_{t_0}^{t_1}\int_{\Sigma(t)} \left(\frac{v^2}{2}c - \Phi - \frac{d\varphi}{dt}\right)\delta n dA dt + \rho \left[\int_{\Sigma(t)} \varphi \delta n dA\right]_{t_0}^{t_1} = 0. \tag{9.63}$$

The regions occupied by the fluid in the initial and the final instants are given, therefore $\delta n = 0$ for $t = t_0, t_1$, and, due to the constraint (9.55) and the condition $\delta n = 0$ on $\partial V(t) - \Sigma(t)$, also

$$\int_{\Sigma(t)} \delta n dA = 0. \tag{9.64}$$

Thus (9.63) yields the only condition

$$\frac{v^2}{2} - \Phi - \frac{d\varphi}{dt} = \frac{p_0}{\rho}. \tag{9.65}$$

Here, the constant p_0 is the Lagrange multiplier for the constraint (9.64).

One can also write (9.65) in a different way by adding the zero term $\alpha d\beta/dt$:

$$\frac{v^2}{2} - \Phi - \frac{d\varphi}{dt} - \alpha\frac{d\beta}{dt} = \frac{v^2}{2} - \Phi - \frac{\partial\varphi}{\partial t} - v^k\left(\varphi_{,k} + \alpha\beta_{,k}\right) - \alpha\frac{\partial\beta}{\partial t} = \frac{p_0}{\rho}.$$

Thus,

$$\frac{\partial\varphi}{\partial t} + \alpha\frac{\partial\beta}{\partial t} + \frac{v^2}{2} + \Phi = -\frac{p_0}{\rho}. \tag{9.66}$$

Therefore, the constant p_0 has the meaning of pressure, and the equality (9.66) shows that pressure is constant on the free surface. This assertion concludes the justification of the variational principle for the functional (9.54).

9.4 Potential Flows

After dropping the constraints on β (9.49), we get the following

Variational principle. *The stationary points of the functional (9.54) on the set of velocity fields satisfying the incompressibility condition (9.50) and the impenetrability of the walls condition (9.51), and on the set of regions $V(t)$ satisfying (9.55), are the potential flows of the ideal incompressible fluid.*

Let us construct the corresponding dual variational principle. Introducing the Lagrange multiplier for the constraint (9.50) and using the equality (9.51), we rewrite the functional (9.54) as

$$\rho \int_{t_0}^{t_1} \int_{V(t)} \left(\frac{v^2}{2} - \Phi(x) + \varphi \frac{\partial v^i}{\partial x^i} \right) dVdt =$$

$$\rho \int_{t_0}^{t_1} \int_{V(t)} \left(\frac{v^2}{2} - v^i \varphi_{,i} - \Phi(x) \right) dVdt + \rho \int_{t_0}^{t_1} \int_{\partial V(t)} \varphi c_x dAdt. \tag{9.67}$$

Finding the extremum of the functional (9.67) over v^i is equivalent to minimization of the integrand with respect to v^i. After calculating the minimum, we get the following

Dual variational principle. *The stationary points of the functional*

$$-\rho \int_{t_0}^{t_1} \int_{V(t)} \left(\frac{1}{2} \varphi_{,i} \varphi^{,i} + \Phi(x) \right) dVdt + \rho \int_{t_0}^{t_1} \int_{\partial V(t)} \varphi c_x dAdt \tag{9.68}$$

on all functions φ *and all regions* $V(t)$ *satisfying the condition (9.55) are the potential flows of the ideal incompressible fluid.*

The functional (9.68) can be written in a different way using the identity

$$\frac{d}{dt} \int_{V(t)} \varphi dV = \int_{V(t)} \frac{\partial \varphi}{\partial t} dV + \int_{\partial V(t)} \varphi c_x dA.$$

We have (up to a factor -1)

$$\int_{t_0}^{t_1} \int_{V(t)} \left(\varphi_{,t} + \frac{1}{2} \varphi_{,i} \varphi^{,i} + \Phi(x) \right) dVdt - \left[\int_{V(t)} \varphi dV \right]_{t_0}^{t_1}. \tag{9.69}$$

The integrand of the first integral in the formula (9.69) is (negative) pressure.

One can get rid of the last term in (9.69) by setting an additional constraint

$$\left[\int_{V(t)} \varphi dV \right]_{t_0}^{t_1} = 0 \tag{9.70}$$

on the admissible values of the potential φ. It is clear that this constraint does not affect the equations and the boundary conditions for φ. We get the following.

Variational principle. *The stationary points of the functional*

$$\rho \int_{t_0}^{t_1} \int_{V(t)} \left(\varphi_{,t} + \frac{1}{2} \varphi_{,i} \varphi^{,i} + \Phi(x) \right) d^3 x dt \tag{9.71}$$

on all functions φ and in all regions $V(t)$ satisfying the constraints (9.70) and (9.55), respectively, are the potential flows of ideal fluid.

Consider a particular case of motion of the ideal incompressible fluid over the plane Ω. Denote the hight of the fluid by $h(x^\alpha, t)$. The variational principle for the functional (9.71) becomes

Luke variational principle. *The potential flows of the ideal incompressible homogeneous fluid over a plane are the stationary points of the functional*

$$\rho \int_{t_0}^{t_1} \int_{\Omega} \int_{0}^{h(x^\alpha, t)} \left(\varphi_{,t} + \frac{1}{2} \varphi_{,i} \varphi^{,i} + \Phi(x) \right) dx^1 dx^2 dx^3 dt \tag{9.72}$$

on the set of all functions φ and h, satisfying the constraints

$$\int_{\Omega} \left[h(x^\alpha, t) - h_0(x^\alpha) \right] dx^1 dx^2 = 0, \quad \left[\int_{\Omega} \int_{0}^{h} \varphi dx^1 dx^2 dx^3 \right]_{t_0}^{t_1} = 0,$$

$$h(x^\alpha, t_0) = h_0(x^\alpha), \quad h(x^\alpha, t_1) = h_1(x^\alpha). \tag{9.73}$$

The first constraint (9.73) can be disregarded if we include the Lagrange multiplier for this constraint in the potential of the external body forces (or, after some redefinition of φ, in the potential, φ).

Consider now the flows without free surfaces.

Kelvin variational principle. *On the set $\mathcal{M}$ of velocity fields, selected by the constraints (9.50) and (9.51), the minimizing element of the kinetic energy functional*

$$I(v) = \int_{t_0}^{t_1} \int_{V(t)} \frac{1}{2} \rho v^i v_i dV dt$$

corresponds to the potential flow of ideal incompressible fluid.

The set $\mathcal{M}$ is convex. The kinetic energy, as a positive quadratic functional, is strictly convex. Therefore, the minimizing element of the kinetic energy functional is unique.

Constructing the dual variational principle, we get

Dirichlet principle. *On all functions φ, the maximizing element of the functional*

$$J = \int_{t_0}^{t_1} \rho \left[\int_{V(t)} c_x \varphi dA - \frac{1}{2} \int_{V(t)} \frac{\partial \varphi}{\partial x^i} \frac{\partial \varphi}{\partial x_i} dV \right] dt$$

the velocity potential of the flow of the ideal incompressible fluid which satisfies the boundary condition (9.51) and

$$\max_{\varphi} J = \min_{v} I. \tag{9.74}$$

Time is just a parameter in the Kelvin and Dirichlet principles. Therefore, the minimization problem for the functional I can be replaced by the minimization problem for the functional

$$\mathcal{K} = \int_{V(t)} \frac{1}{2} \rho v^2 dV \tag{9.75}$$

with the constraints (9.50) and (9.51), at each instant while the maximization problem for the functional J corresponds to maximization of the functional

$$\mathcal{K}^* = \rho \int_{\partial V(t)} c_x \varphi dA - \frac{1}{2} \rho \int_{\partial V(t)} \frac{\partial \varphi}{\partial x^i} \frac{\partial \varphi}{\partial x_i} dV \tag{9.76}$$

on all functions φ.

9.5 Variational Features of Kinetic Energy in Vortex Flows

There is a one-to-one correspondence between velocity and vorticity. Therefore kinetic energy can be considered as a functional of vorticity. It turns out that this functional is the minimum value in a variational problem to which we proceed.

First we have to show that velocity and vorticity are in one-to-one correspondence indeed. Consider a velocity field, v^i, in a closed bounded region V. The velocity field is incompressible:

$$\partial_i v^i = 0 \tag{9.77}$$

and satisfies the impermeability condition:

$$v^i n_i = 0 \quad \text{on } \partial V. \tag{9.78}$$

For a known velocity field, one can find the corresponding vorticity field:

$$\omega^i = e^{ijk}\partial_j v_k. \tag{9.79}$$

The vorticity field is divergence-free:

$$\partial_i \omega^i = 0. \tag{9.80}$$

Let now a divergence-free vorticity field be given. Then (9.77), (9.78) and (9.79) define a unique velocity field. Indeed, assume the opposite, that there are two solutions of (9.77), (9.78) and (9.79), v_1^i and v_2^i. Then the difference $v^i = v_1^i - v_2^i$ obeys (9.77) and (9.78) and the equation

$$e^{ijk}\partial_j v_k = 0.$$

The latter means that the velocity field is potential, $v_i = \partial_i \varphi$. According to (9.77) and (9.78) the potential is subject to the boundary value problem,

$$\Delta\varphi = 0, \qquad \left.\frac{\partial\varphi}{\partial n}\right|_{\partial V} = 0,$$

which has the solution, $\varphi = const$, and, hence, $v^i = 0$, $v_1^i = v_2^i$. The one-to-one correspondence, $\omega^i \Longleftrightarrow v^i$, holds true also if $v^i n_i$ at ∂V is given and not necessarily equal to zero.

If we do not impose on the velocity fields the constraints (9.77) and (9.78) then the one-to-one correspondence is lost: there are many velocity fields which satisfy (9.79) for a given ω^i. It turns out that (9.77) and (9.78) are Euler equations in the following

Variational principle. *The true dependence of kinetic energy of vorticity provides the minimum value to the functional*

$$\int_V \frac{1}{2}\rho v^i v_i dV$$

on the set of all velocity fields selected by the constraints (9.79).

To derive Euler equations, we introduce Lagrange multipliers for the constraints (9.79), $\rho\psi_i$. The functional to be minimized becomes

$$\rho\int_V \left[\frac{1}{2}v^i v_i - \psi_i\left(e^{ijk}\partial_j v_k - \omega^i\right)\right] dV \tag{9.81}$$

or, after integration by parts,

$$\rho \int_V \left[\frac{1}{2} v^i v_i + \partial_j \psi_i e^{ijk} v_k + \psi_i \omega^i \right] dV - \rho \int_{\partial V} \psi_i e^{ijk} n_j v_k dA. \tag{9.82}$$

Euler equations are

$$v^i = e^{ijk} \partial_j \psi_k \quad \text{in } V, \tag{9.83}$$

$$\psi_i e^{ijk} n_j = 0 \quad \text{on } \partial V. \tag{9.84}$$

Equation (9.77) follows from (9.83). Let us show that boundary condition (9.84) yields (9.78). Indeed, using Greek indices for vector projections on the tangent planes, we can write (9.84) as

$$\psi_\alpha = 0. \tag{9.85}$$

On the other hand, denoting by r^i (ξ^α) the position vector of ∂V, ξ^α being coordinates on ∂V, and using (14.7), we have

$$\begin{aligned} n_i v^i &= \frac{1}{\sqrt{a}} e_{imn} r_1^m r_2^n e^{ijk} \partial_j \psi_k = \frac{1}{\sqrt{a}} \left(\delta_m^j \delta_n^k - \delta_m^k \delta_n^j \right) r_1^m r_2^n e^{ijk} \partial_j \psi_k \\ &= \frac{1}{\sqrt{a}} \left(r_1^m r_2^n \partial_m \psi_n - r_1^n r_2^m \partial_m \psi_n \right) = \frac{1}{\sqrt{a}} \left(r_2^n \partial_1 \psi_n - r_1^n \partial_2 \psi_n \right) \\ &= \frac{1}{\sqrt{a}} \left(\partial_1 \psi_2 - \partial_2 \psi_1 \right) = \varepsilon^{\alpha\beta} \psi_{\beta|\alpha}, \qquad \psi_\alpha \equiv r_\alpha^m \psi_m. \end{aligned} \tag{9.86}$$

Here $r_\alpha^m \equiv \partial r^m / \partial \xi^\alpha$, the vertical bar in indices denotes the covariant surface derivative (see Sect. 14.1 for the definition). The boundary condition (9.78) follows from (9.86) and (9.85).

After minimization with respect to velocity fields the functional (9.81) is equal to

$$-\rho \int_V \left[\frac{1}{2} \left(curl \, \vec{\psi} \right)^2 - \vec{\psi} \cdot \vec{\omega} \right] dV. \tag{9.87}$$

According to the general scheme of construction the dual variational principles, to obtain kinetic energy of the flow one has to maximize the functional (9.87) with respect to all fields, $\vec{\psi}$, which satisfy the constraint (9.84). We arrive at the following

Dual variational principle. *The dependence of kinetic energy on vorticity can be found from the variational problem*

$$-\mathcal{K} = \min_{\vec{\psi} \in (9.84)} \rho \int_V \left[\frac{1}{2} \left(curl \, \vec{\psi} \right)^2 - \vec{\psi} \cdot \vec{\omega} \right] dV. \tag{9.88}$$

The vector field, $\overrightarrow{\psi}$, is called stream vector field; in case of two-dimensional flows $(v_1, v_2 \neq 0,\ v_3 = 0)$, it has the only non-zero component, ψ_3, which is called stream function.

Note that the functional in (9.88) is invariant with respect to shifts $\overrightarrow{\psi} \to \overrightarrow{\psi} + \nabla\varphi$ where the potential, φ, is constant at the boundary.[4] Therefore, its minimum value is achieved at many fields. To select a unique minimizing element, one can set an additional constraint

$$\partial_i \psi^i = 0 \tag{9.89}$$

which obviously, eliminates gradient invariance.

Euler equations in the variational problem (9.88) are

$$\Delta\psi_i - \partial_i \partial_j \psi^j = -\omega_i. \tag{9.90}$$

If the constraint (9.89) is set, Euler equations simplify to

$$\Delta\psi_i = -\omega_i. \tag{9.91}$$

The three equations (9.91) are not independent: the divergence of (9.91) is identically zero.

The variational principles formulated can be extended to the case of non-zero normal velocity at the boundary. To do that, consider the function, $v_n = v_i n^i$, at ∂V. We introduce a two-dimensional vector, χ^α, on ∂V such that its "surface divergence" is equal to v_n:

$$v_n = -\chi^\alpha_{|\alpha}. \tag{9.92}$$

Since the velocity field is divergence-free,

$$\int\limits_{\partial V} v_n dA = 0,$$

[4] For such a shift the functional gets an increment

$$-\rho \int\limits_V \omega^i \partial_i \varphi dV$$

which is equal, due to (9.80) to

$$-\rho \int\limits_{\partial V} \omega^i n_i \varphi dA.$$

Since φ is a constant, c, on ∂V,

$$-\rho \int\limits_{\partial V} \omega^i n_i \varphi dA = -\rho c \int\limits_{\partial V} \omega^i n_i dA = -\rho c \int\limits_V \partial_i \omega^i dV = 0.$$

the necessary condition for solvability of (9.92) is satisfied. For a given v_n, one can obtain a solution of (9.92) assuming that the field χ^α is potential, $\chi^\alpha = \chi^{|\alpha}$. So we assume that the vector field, χ^α, is known on ∂V.

Variational principle. *For a given vorticity, the true velocity field provides minimum to the functional*

$$\int_V \frac{1}{2} v^i v_i dV - \int_{\partial V} v_\alpha \chi^\alpha dA. \tag{9.93}$$

Here v_α are the tangent components of velocity on ∂V.

To obtain Euler equations for the functional (9.93) we introduce Lagrange multipliers for the constraint (9.79):

$$\int_V \left[\frac{1}{2} v^i v_i - \psi_i \left(e^{ijk} \partial_j v_k - \omega^i \right) \right] dV - \int_{\partial V} v_\alpha \chi^\alpha dA. \tag{9.94}$$

The only difference from (9.81) is that after integration by parts we have to vanish for all v^i the surface integral

$$\int_{\partial V} \left(-\psi_i e^{ijk} n_j v_k - v_\alpha \chi^\alpha \right) dA.$$

Since, due to (14.10),

$$\psi_i e^{ijk} n_j v_k = -\psi_i v_k \varepsilon^{\alpha\beta} r^i_\alpha r^k_\beta = -\varepsilon^{\alpha\beta} \psi_\alpha v_\beta ,$$

this surface integral is zero for

$$\varepsilon^{\alpha\beta} \psi_\alpha = \chi^\beta$$

or

$$\psi_\alpha = \varepsilon_{\alpha\beta} \chi^\beta. \tag{9.95}$$

Hence, inside the region V we find as before (9.83), while at the boundary we obtain using (9.86), (14.6) and (9.83),

$$v^i n_i = \varepsilon^{\alpha\beta} \psi_{\beta|\alpha} = \varepsilon^{\alpha\beta} \left(\varepsilon_{\beta\gamma} \chi^\gamma \right)_{|\alpha} = -\chi^\gamma_{|\gamma},$$

i.e. the prescribed boundary value of normal velocity.

Computation of the functional (9.94) on the minimizing velocity field yields the following

Dual variational principle. *The true stream vector field is the maximizer of the functional*

$$-\int_V \left[\frac{1}{2}\left(curl\,\vec{\psi}\right)^2 - \psi_i\omega^i\right] dV \tag{9.96}$$

on the set of all stream vector fields satisfying the boundary condition (9.95).

For $\chi^\alpha = 0$, the maximum value of the functional (9.96) is kinetic energy of the flow. The maximizer is a linear functional of ω^i:

$$\hat{\psi}_i = \int_V R_{ij}(x, x')\omega^j(x')d^3x'.$$

Accordingly, kinetic energy is a quadratic functional of vorticity:

$$\mathcal{K} = \frac{1}{2}\rho \int_V \int_V R_{ij}(x, x')\omega^i(x)\omega^j(x')d^3xd^3x'.$$

If vorticity depends on time, then the variational principles hold at each instant.

The variational principles formulated have pure mathematical origin as can be seen from the following reasoning.

Consider a divergence-free vector field $\vec{v}$ in a closed region V . Then, for a given $\vec{v}$,

$$\min_{\vec{\psi}} \frac{1}{2} \int_V \left|\vec{v} - curl\vec{\psi}\right|^2 dV = 0, \tag{9.97}$$

because there exists a vector field $\vec{\chi}$ such that $\vec{v} = curl\vec{\chi}$, and minimum is achieved on $\vec{\psi} = \vec{\chi} + \nabla\varphi$, φ is an arbitrary function.

Let additionally the normal component of $\vec{v}$ vanishes at the boundary. If $x^i = r^i(\xi^\alpha)$, $\alpha = 1, 2$, are the parametric equations of the boundary, then

$$v^i n_i = n_i e^{ijk}\partial_j\chi_k = r_1^j r_2^k(\partial_j\chi_k - \partial_k\chi_j) = \frac{\partial\chi_2}{\partial\xi^1} - \frac{\partial\chi_1}{\partial\xi^2} = 0, \tag{9.98}$$

where $\chi_\alpha = r_\alpha^i\chi_i$ are the tangent components of vector $\vec{\chi}$ on the boundary. According to (9.98), the surface vector, χ_α, is a potential vector: there is a function on the boundary, $\Phi(\xi^\alpha)$, such that $\chi_\alpha = \partial\Phi(\xi^\alpha)/\partial\xi^\alpha$. Taking any function, φ, which has the boundary value $\Phi(\xi^\alpha)$, and replacing $\vec{\chi}$ by $\vec{\chi} - \nabla\varphi$, we obtain the stream vector field with the zero tangent components at the boundary. Therefore, if we narrow the admissible vector fields $\vec{\psi}$ in (9.97) by the vector fields with the vanishing tangent components at the boundary,

$$\vec{\psi} \times \vec{n} = 0 \text{ on } \partial V, \tag{9.99}$$

minimum in (9.97) remains equal to zero:

$$\min_{\vec{\psi}\in(9.99)}\left[\frac{1}{2}\int_V v_i v^i dV - \int_V v_i e^{ijk}\partial_j\psi_k dV + \frac{1}{2}\int_V \left|curl\vec{\psi}\right|^2 dV\right] = 0. \quad (9.100)$$

The second term in (9.100) can be written as

$$-\int_V v_i e^{ijk}\partial_j\psi_k dV = -\int_{\partial V} v_i e^{ijk} n_j \psi_k dV + \int_V \psi_k e^{ijk}\partial_j v_i dV. \quad (9.101)$$

The boundary integral vanishes because of (9.99), while the volume integral is equal to

$$\int_V \psi_k e^{ijk}\partial_j v_i dV = -\int_V \psi_k \omega^k dV. \quad (9.102)$$

Since, for a given velocity field, vorticity is known, minimum over $\vec{\psi}$ in (9.100) yields the variational principle for the energy functional (9.88). According to (9.100), the minimum value of the energy functional is equal to negative kinetic energy of the flow.

Our derivation shows that variational principle for energy functional (9.88) also holds true for vorticity fields with non-zero normal component at the boundary.

Energy functional (9.88) is invariant with respect to transformations, $\vec{\psi} \to \vec{\psi} + \nabla\varphi$, where φ is constant at the boundary (see footnote in this section).
Therefore, an additional constraint can be set on $\vec{\psi}$ without changing the minimum value of the functional.

9.6 Dynamics of Vortex Lines

Setting the problem. The governing equations of ideal incompressible fluid possess an infinite number of integrals of motion, which are additional to energy, the circulations of velocity over closed fluid contours. As we have seen in Sect. 2.2, in statistical mechanics one usually deals with a Hamiltonian system which has the only integral, energy. Therefore to develop statistical mechanics of ideal fluid one needs to eliminate all degrees of freedom which are "driven" due to the integrals of motion and keep only the independent degrees of freedom. Conservation of velocity circulations reduces the number of independent degrees of freedom drastically. For example, ideal fluid in a container with an embedded rigid body is, in general, an infinite-dimensional system. However, if the circulations of velocity over any fluid contour is zero, i.e. the fluid flow is potential, then the number of independent degrees of freedom is only six: they are the degrees of freedom of the rigid body (see Sect. 13.2 for details). It turns out that the dynamics of only "essential degrees of freedom" is dynamics of vortex lines. It is the subject of this section.

Governing equations of vortex line dynamics. Consider a flow of ideal fluid in a bounded vessel, V, which is at rest in some inertial frame. The fluid neither penetrates through nor detaches from the wall, ∂V.

Conservation of velocity circulations over fluid contours is equivalent to conservation of the Lagrangian components of vorticity:

$$\omega^a = \mathring{\omega}^a \left(X^b \right). \tag{9.103}$$

The material lines tangent to $\mathring{\omega}^a$, i.e. the lines determined by the parametric equations,

$$\frac{dX^a(\sigma)}{d\sigma} = \lambda \omega^a (X^b),$$

are called vortex lines. Any vortex line consists of the same fluid particles because of the independence of ω^a on time. Two cases should be distinguished: $\omega^a \mathring{n}_a = 0$ at ∂V and $\omega^a \mathring{n}_a \neq 0$ at ∂V (recall that $\mathring{n}_a$ are the Lagrangian components of the unit normal vector at the boundary in the initial state). In the first case, vortex lines do not end at the boundary, in the second they do (an example of the second case is tornado). We consider in this section the first case. The necessary modifications for non-zero normal vorticity are made in the next section for quasi-two-dimensional flows. In three-dimensional space a vortex line that does not end at the boundary can be either closed or fill out some subregion of V. Strictly speaking, the term "dynamics of vortex lines" is meaningful for the case of closed vortex lines; for dense vortex lines it is more appropriate to deal with dynamics of "vortex particles". The latter is beyond the scope of our consideration. We focus on the case when a Lagrangian coordinate system can be attached to vortex lines. One of the coordinates, say, X^3, is directed along the vortex lines. We use for this coordinate the notation $\eta \equiv X^3$. The couple of other coordinates, $X^\mu, \mu = 1, 2$, marks different vortex lines; the couple $\left\{ X^1, X^2 \right\}$ is denoted in this section by X. In the coordinate system chosen only one component of vorticity, $\mathring{\omega}^3$, is not zero.

In Lagrangian coordinates, the condition that the vorticity field is divergence-free at the initial instant takes the form

$$\frac{\partial}{\partial X^a} \sqrt{\mathring{g}} \mathring{\omega}^a = 0, \quad \sqrt{\mathring{g}} \equiv \det \left\| \frac{\partial x^i \left(0, X^a\right)}{\partial X^a} \right\|. \tag{9.104}$$

Hence, in the vortex line coordinate system,

$$\frac{\partial}{\partial \eta} \sqrt{\mathring{g}} \mathring{\omega}^3 = 0,$$

and $\sqrt{\mathring{g}}\mathring{\omega}^3$ is constant along each vortex line but may change from one vortex line to another. We denote this function by $\mathring{\omega}$. It measures the intensity of vortex lines. Function $\mathring{\omega}(X^\mu)$ is assumed to be known from the initial conditions.

Due to incompressibility,

$$\Delta \equiv \det \left\| \frac{\partial x^i(t, X^a)}{\partial X^a} \right\| = \sqrt{\mathring{g}}, \tag{9.105}$$

and one can also put $\mathring{\omega} = \mathring{\omega}^3 \Delta$.

The position vector of the points of vortex lines will be denoted further by $r^i(t, \eta, X)$. Vorticity field is determined by $r^i(t, \eta, X)$ and $\mathring{\omega}(X)$: according to (9.12) and (9.103)

$$\omega^i = \frac{\partial r^i(t, \eta, X)}{\partial \eta} \mathring{\omega}^3 = \frac{\partial r^i(t, \eta, X)}{\partial \eta} \frac{\mathring{\omega}(X)}{\sqrt{\mathring{g}}}. \tag{9.106}$$

The independence of the Lagrangian component of vorticity, $\mathring{\omega}^3$, on time determines the dynamics of vortex lines. Indeed, let $\mathring{\omega}^3$ and the positions of vortex lines, $r^i(t, \eta, X)$, be known at some instant, t. Then we know at this instant the vorticity field from (9.106) and can find velocity, $v^i(t, x)$, solving the kinematic problem of the previous section:

$$\partial_i v^i = 0, \qquad e^{ijk} \partial_j v_k = \omega^i(t, x), \qquad v_i n^i \big|_{\partial V} = 0. \tag{9.107}$$

This problem determines velocity as a functional of the positions of vortex lines; denote it by $V^i(t, x | r(t, \eta, X))$. Then, we can find the positions of the vortex lines at the instant $t + \Delta t$ by putting

$$r^i(t + \Delta t, \eta, X) = r^i(t, \eta, X) + V^i(t, x | r(t, \eta, X))\big|_{x = r(t, \eta, X)} \Delta t. \tag{9.108}$$

Knowing the new positions of the vortex lines, we repeat the procedure, thus determining the dynamics of vortex lines.

Equation (9.108) means that $r^i(t, \eta, X)$ are sought from the system of equations

$$\frac{\partial r^i(t, \eta, X)}{\partial t} = V^i(t, x | r(t, \eta, X))\big|_{x = r(t, \eta, X)}. \tag{9.109}$$

We have to show that fluid dynamics so defined obeys momentum equations of ideal fluid. To this end we have to check the equality

$$curl \left(\frac{d\vec{v}}{dt} \right) = 0,$$

where velocity is the solution of (9.107). We have[5]

$$e^{ijk}\partial_j \frac{dv_k}{dt} = e^{ijk}\partial_j\left(\frac{\partial v_k}{\partial t} + v^m\partial_m v_k\right) = \\ = \frac{d}{dt}\omega^i + e^{ijk}\partial_j v^m\partial_m v_k.$$

Note that

$$e^{ijk}\partial_j v^m\partial_m v_k = e^{ijk}\left(\partial_j v_m - \partial_m v_j\right)\partial^m v_k \tag{9.110}$$

because $e^{ijk}\partial_m v_j\partial^m v_k \equiv 0$ due to symmetry of $\partial_m v_j\partial^m v_k$ over indices j, k. Besides, $\partial_j v_m - \partial_m v_j = e_{jms}\omega^s$, and, using (3.19) and (9.22) and incompressibility condition, $\partial_i v^i = 0$, we have

$$e^{ijk}\partial_j v^m\partial_m v_k = -\omega^k\partial_k v^i.$$

Therefore,

$$e^{ijk}\partial_j \frac{dv_k}{dt} = \frac{d\omega^i}{dt} - \omega^k\partial_k v^i. \tag{9.111}$$

From (9.106) $d\omega^i/dt = \omega^k\partial_k v^i$. Thus, the right hand side of (9.111) vanishes, and the particle trajectories, $r^i(t, \eta, X)$, obey the momentum equations of ideal incompressible fluid.

Setting (9.106), (9.107) and (9.109), we automatically obtain an incompressible motion of particles. Remarkably, we can reformulate the equations of vortex line dynamics as dynamics of geometrical lines when motion of particles over the vortex lines is ignored. This is done in the following way. Let us admit to consideration any motions, $r^i(t, \eta, X^\mu)$, including the compressible ones. Equation (9.106) is no longer true. For compressible motion, i.e. for $\Delta \neq \sqrt{\mathring{g}}$, we replace (9.106) with the equation

$$\omega^i = \frac{1}{\Delta}\frac{\partial r^i(t, \eta, X)}{\partial \eta}\mathring{\omega}(X). \tag{9.112}$$

---

[5] For any function, φ, we define $d\varphi/dt$ as

$$\frac{d\varphi}{dt} = \frac{\partial}{\partial t}\varphi(t, r(t, \eta, X))|_{\eta, X=const}.$$

Thus,

$$\frac{d\varphi}{dt} = \frac{\partial\varphi}{\partial t} + V^i(t, x|r(t, \eta, X))\partial_i\varphi.$$

That means that $\mathring{\omega}(X)$ is expressed in terms of the only non-zero Lagrangian component of vorticity, $\hat{\omega}^3$, as

$$\mathring{\omega}(X) = \hat{\omega}^3 \Delta.$$

The vortex intensity does not depend on η, and therefore the vector field (9.112) is divergence-free:

$$\partial_i \omega^i = \frac{1}{\Delta} \frac{\partial}{\partial X^a} \left(\Delta \frac{\partial X^a}{\partial x^i} \omega^i \right) = \frac{1}{\Delta} \frac{\partial}{\partial \eta} \left(\hat{\omega}^3 \Delta \right) = \frac{1}{\Delta} \frac{\partial}{\partial \eta} (\mathring{\omega}) = 0. \tag{9.113}$$

Besides, $\mathring{\omega}$ does not depend on time. For compressible motion, the only non-zero Lagrangian component of vorticity, $\hat{\omega}^3 = \mathring{\omega}/\Delta$, may depend on time.

Instead of (9.109), we set for $r^i(t, \eta, X)$ the differential equations

$$\frac{\partial r^i(t, \eta, X)}{\partial t} = V^i(t, x | r(t, \eta, X)) + \lambda \omega^i \big|_{x=r(t,\eta,X)}, \tag{9.114}$$

where V^i is determined by the vorticity field from (9.107) and (9.112), and λ is an arbitrarily prescribed function of coordinates and time. The parameter λ controls the particle velocity over vortex lines.

Equation (9.114) can also be written in the form

$$e_{ijk} \left(\frac{\partial r^i(t, \eta, X)}{\partial t} - V^i(t, x | r(t, \eta, X)) \big|_{x=r(t,\eta,X)} \right) \omega^k = 0 \tag{9.115}$$

which emphasizes that in (9.114) only the velocity normal to the vortex line is essential.

Both (9.109) and (9.115) give the same velocity normal to vortex lines. Therefore, (9.115) describe the dynamics of vortex lines correctly. We are going to show that (9.115) forms a Hamiltonian system of equations.

Variational principles. All possible motions can be split into "equivortical sheets"; each sheet is the set of motions $r(t, \eta, X)$ with the same vortex intensity $\mathring{\omega}$. Each sheet contains the true motions, i.e. motions obeying (9.115), and the motions that are not realized. For a given vortex intensity, the total kinetic energy of the flow becomes a functional of $r^i(t, \eta, X)$, because, as we have seen in the previous section, there is one-to-one correspondence between velocity and vorticity, and vorticity is expressed in terms of $r^i(t, \eta, X)$ by (9.112):

$$\mathcal{K} = \mathcal{K}\left(r^i(t, \eta, X)\right) = \frac{1}{2}\rho \int_V V_i(t, x | r(t, \eta, X)) V^i(t, x | r(t, \eta, X)) d^3x. \tag{9.116}$$

The dynamics of vortex lines is governed by the following

Variational principle. *The true motion of the vortex lines, $\check{r}(t,\eta,X)$, is a stationary point of the functional*

$$J = \int_{t_0}^{t_1} [\mathcal{A} - \mathcal{K}]\, dt, \tag{9.117}$$

$$\mathcal{A} = \frac{1}{3}\rho \int_{\mathring{V}} e_{ijk} r^j(t,\eta,X) \frac{\partial r^i(t,\eta,X)}{\partial t} \frac{\partial r^k(t,\eta,X)}{\partial \eta} \mathring{\omega}(X)\, d^2X d\eta, \tag{9.118}$$

on the set of all motions with the prescribed vortex intensity, which have the same initial and final positions:

$$r(t_0,\eta,X) = \check{r}(t_0,\eta,X), \qquad r(t_1,\eta,X) = \check{r}(t_1,\eta,X). \tag{9.119}$$

Here $d^2X \equiv dX^1 dX^2$, and $\mathring{V}$ is the region run by Lagrangian variables.

If vortex lines occupy some subregion, V', of V then the integrand in (9.118) must be integrated over this region, V' ($\mathring{\omega} = 0$ outside of V'). If $V' = V$, then

$$n_i \delta r^i = 0 \quad \text{at } \partial V, \tag{9.120}$$

δr^i being the variation of $r^i(t,\eta,X)$. If V' is strictly inside V, $n_i \delta r^i$ are arbitrary at $\partial V'$.

The functional does not depend on the choice of parameter along the vortex lines. This yields the above-mentioned peculiarity of the vortex line dynamics: this is dynamics of geometrical lines. That means that a vortex line has not three but two functional degrees of freedom. For example, if the vortex line crosses each plane $x^3 = const$ at one point, then its dynamics is completely determined by two functions $x^\alpha = r^\alpha(t, x^3)$, $\alpha = 1,2$.

Another feature of the variational principle is that, in contrast to Lagrange variational principle, the motion, $r^i(t,\eta,X)$, is compressible, while the motion of the ideal fluid under consideration is not.

The functional $\mathcal{A}$ is an analogue of the shortened action in classical mechanics.

We will derive the variational principle formulated from another one which is easier to justify.

First, we set up a one-to-one correspondence between v^i and ψ_k, imposing the following constraints on ψ_k :

$$\partial_k \psi^k = 0 \quad \text{in } V, \tag{9.121}$$

$$\overrightarrow{\psi} \times \overrightarrow{n} = 0 \quad \text{on } \partial V. \tag{9.122}$$

As we discussed in the previous section, the condition (9.122) means the impermeability of the boundary.

Consider a functional

$$I(\psi) = \frac{1}{2}(A\psi, \psi) - (l, \psi) \tag{9.123}$$

$$(A\psi, \psi) \equiv \rho \int_V \left(curl \vec{\psi}\right)^2 dV, \quad (l, \psi) = \rho \int_V \vec{\omega} \cdot \vec{\psi}\, dV.$$

For a given divergence-free vorticity field, we seek the minimum value of the functional $I(\psi)$ with respect to all vector fields $\vec{\psi}$ subject to the constraints (9.121) and (9.122).

Functional $I\left(\vec{\psi}\right)$ is a quadratic functional being minimized on a linear set. According to the dual variational principle of the previous section, its minimum value coincides with the negative kinetic energy of the flow:

$$\min_{\vec{\psi} \in (9.121),(9.122)} I\left(\vec{\psi}\right) = -\frac{1}{2}\rho_0 \int_V \vec{v} \cdot \vec{v}\, dV = -\frac{1}{2}\rho_0 \int_V \vec{\omega} \cdot \vec{\psi}\, dV, \quad \vec{v} = curl \vec{\psi}. \tag{9.124}$$

The minimization problem for the functional $I\left(\vec{\psi}\right)$ determines the kinetic energy as a functional of vorticity, and in accordance with (9.112) as a functional of the vortex line positions. Substituting (9.124) into (9.117) we arrive at the following **Variational principle.** *The true trajectory of vortex line dynamics, $\check{r}^i(t, \eta, X)$, is a stationary point of the functional*

$$\mathbb{I}(r(t, \eta, X), \psi(t, x)) = \int_{t_0}^{t_1} dt\rho \left[\frac{1}{3} \int_{\mathring{V}} e_{ijk} r^j \frac{\partial r^i}{\partial t} \frac{\partial r^k}{\partial \eta} \mathring{\omega} d^2 X d\eta + \frac{1}{2} \int_V \left(curl \vec{\psi}\right)^2 dV \right.$$
$$\left. - \int_{\mathring{V}} \psi_k(t, r) \frac{\partial r^k}{\partial \eta} \mathring{\omega} d^2 X d\eta \right] \tag{9.125}$$

on the set of all motions, $r^i(t, \eta, X)$, of vortex lines, having the same vortex intensity the same initial and final positions (9.119), and the set of all vector fields, $\psi^i\left(t, r^k\right)$, with vanishing tangent components on the boundary.

Motion of vortex lines, $r^i(t, \eta, X)$, in (9.125) is allowed to be compressible. The Jacobian Δ enters explicitly in the second integral (9.125) if we write the second integral as an integral over Lagrangian variables[6]:

$$\int_{\mathring{V}} \left(curl \vec{\psi}\right)^2 \Delta d^2 X d\eta.$$

[6] Recall that volume element, dV, is equal to $\Delta d^2 X d\eta$ in Lagrangian coordinates.

The third integral can also be written, according to (9.112), as

$$\int_{\mathring{V}} \psi_k(t,r)\frac{\partial r^k}{\partial \eta}\mathring{\omega} d^2 X d\eta = \int_V \psi_k \omega^k dV.$$

Then minimization over ψ_k is equivalent to the variational problem for the functional (9.123). One obtains the same equations for ψ_k when all integrals are written in Lagrangian variables.

When varying $r^i(t,\eta,X)$, the variation of the second integral is zero, because for fixed $\psi^i\left(t,r^k\right)$ the second integral does not depend on $r^i(t,\eta,X)$. So, varying the functional $\mathbb{I}$ with respect to $r^i(t,\eta,X)$, it is enough to vary only the first and the third integral. For the variation of the first integral we have

$$\begin{aligned}
&\delta\int_V \frac{1}{3}e_{ijk}r^j\frac{\partial r^i}{\partial t}\frac{\partial r^k}{\partial \eta}\mathring{\omega}(X)d^2Xd\eta\\
&=\int_V\left(\frac{1}{3}e_{ijk}\delta r^j\frac{\partial r^i}{\partial t}\frac{\partial r^k}{\partial \eta}+\frac{1}{3}e_{ijk}r^j\frac{\partial \delta r^i}{\partial t}\frac{\partial r^k}{\partial \eta}+\frac{1}{3}e_{ijk}r^j\frac{\partial r^i}{\partial t}\frac{\partial \delta r^k}{\partial \eta}\right)\mathring{\omega}(X)d^2Xd\eta\\
&=\int_V\left(\frac{1}{3}e_{ijk}\delta r^j\frac{\partial r^i}{\partial t}\frac{\partial r^k}{\partial \eta}-\frac{1}{3}e_{ijk}\delta r^i\frac{\partial}{\partial t}\left(r^j\frac{\partial r^k}{\partial \eta}\right)-\frac{1}{3}e_{ijk}\delta r^k\frac{\partial}{\partial \eta}\left(r^j\frac{\partial r^i}{\partial t}\right)\right)\mathring{\omega}(X)d^2Xd\eta\\
&\quad+\frac{\partial}{\partial t}\int_V\frac{1}{3}e_{ijk}r^j\delta r^i r^k_\eta\mathring{\omega}(X)d^2Xd\eta=\int_V e_{ijk}\delta r^j\frac{\partial r^i}{\partial t}\frac{\partial r^k}{\partial \eta}\mathring{\omega}(X)d^2Xd\eta\\
&\quad+\frac{\partial}{\partial t}\int_V\frac{1}{3}e_{ijk}r^j\delta r^i r^k_\eta\mathring{\omega}(X)d^2Xd\eta.
\end{aligned}\tag{9.126}$$

The divergence terms over η vanish due to closedness of the vortex lines. The variation of the third integral is

$$\begin{aligned}
\delta\int_V \psi_k(t,r)\frac{\partial r^k}{\partial \eta}\mathring{\omega}(X)d^2Xd\eta &= \int_V\left(\partial_i\psi_k\delta r^i\frac{\partial r^k}{\partial \eta}+\psi_k\frac{\partial \delta r^k}{\partial \eta}\right)\mathring{\omega}(X)d^2Xd\eta\\
&=\int_V\left(\partial_j\psi_k-\partial_k\psi_j\right)\delta r^j\frac{\partial r^k}{\partial \eta}\mathring{\omega}(X)d^2Xd\eta
\end{aligned}\tag{9.127}$$

or

$$\delta\mathcal{K}=\rho\int_V e_{ijk}\delta r^j\left(curl\vec{\psi}\right)^i\frac{\partial r^k}{\partial \eta}\mathring{\omega}(X)d^2Xd\eta=\rho\int_V e_{ijk}\delta r^j\left(curl\vec{\psi}\right)^i\omega^k dV.\tag{9.128}$$

So, for the variation of the action functional we obtain

$$\frac{1}{\rho}\delta\mathbb{I} = \int\limits_V e_{ijk}\delta r^j \left(\frac{\partial r^i}{\partial t} - \left(curl\vec{\psi}\right)^i\right) \frac{\partial r^k}{\partial\eta}\mathring{\omega}(X)\, d^2X d\eta + \\ + \frac{\partial}{\partial t}\int\limits_V \frac{1}{3} e_{ijk} r^j \delta r^i r^k_\eta \mathring{\omega}(X)\, d^2X d\eta. \quad (9.129)$$

The divergence term over time vanishes due to (9.119). Finally, one obtains the equations of vortex line dynamics:

$$e_{ijk}\left(\frac{\partial r^i}{\partial t} - \left(curl\vec{\psi}\right)^i\right)\omega^k = 0. \quad (9.130)$$

Here $curl\vec{\psi}$ is assumed to be presented in terms of vorticity, and vorticity is expressed in terms of vortex line positions, $r^i(t, \eta, X)$, according to (9.112). We arrive at (9.115).

Derivation from the least action principle. In Hamiltonian mechanics, elimination of the integrals of motion occurs by means of the following procedure: one makes a change of variables $(p, q) \to (p', q', I, \varphi)$, where $I_1(p, q), \ldots, I_s(p, q)$ is the set of integrals, $\varphi^1, \ldots, \varphi^s$ the dual variables, p', q' the remaining variables, in order to put the Lagrange function in the form: $p'\dot{q}' + I\dot{\varphi} - H(p', q', I)$.[7] Then conservation of $I_1, \ldots, I_s$ follows from the independence of Hamilton function on φ. It is not clear how to perform such a procedure for ideal fluid flow. Originally, the above-formulated variational principle with the eliminated integrals of motion was found as a remarkable feature of dynamics of vortex lines without reference to a general algorithm. The existence of such algorithm for fluid flow is still an open issue. It became clear, however, how to derive the variational principle of vortex line dynamics from the least action principle of Sect. 9.1. It is especially interesting, because one has to explain the appearance in the action functional "strange" compressibility of the originally incompressible flow and a "strange" factor 1/3. We conclude this section with that derivation.

Consider the action functional (9.2), which we write in the form

$$I(x(t, X)) = \rho \int\limits_{t_0}^{t_1}\int\limits_{\mathring{V}} \frac{1}{2} x^i_t x_{it} \Delta\, d^3X\, dt, \qquad x^i_t \equiv \frac{dx^i(t, X)}{dt}.$$

Here, for simplicity, we set $g_0 = 1$ and assume that fluid is homogeneous, $\rho_0 = const$. Besides, to reduce technicalities, we accept that the vessel is at rest

[7] This procedure is discussed in [4].

and fluid does not detach from the wall of the vessel. To enforce the incompressibility condition, we introduce in the action functional the corresponding Lagrange multiplier:

$$I\left(x\left(t,X\right),\varphi(t,X)\right)=\rho\int_{t_0}^{t_1}\int_{\mathring{V}}\left(\frac{1}{2}x_t^i x_{it}\Delta-\dot{\varphi}\Delta\right)d^3X\,dt,\qquad \dot{\varphi}\equiv\frac{d\varphi(t,X)}{dt}. \tag{9.131}$$

Here and in what follows we are concerned only with Lagrangian and ignore the boundary terms; otherwise we should add in (9.131) the integral

$$\left[\rho\int_{\mathring{V}}\varphi d^3X\right]_{t_0}^{t_1}.$$

One can check by direct inspection that the stationary points of the functional (9.131) are the solutions of equations of ideal incompressible fluid: variation over φ yields incompressibility condition:

$$\frac{d\Delta}{dt}=0,$$

while variation over $x^i\left(t,X\right)$ and use of (4.19) results in the equation

$$\frac{d}{dt}\rho\Delta x_{it}+\frac{\partial}{\partial x^i}\rho\Delta\left(\frac{1}{2}x_t^k x_{kt}-\dot{\varphi}\right)=0.$$

This is the momentum equation, in which pressure is

$$p=\rho\Delta\left(\frac{1}{2}x_t^k x_{kt}-\dot{\varphi}\right). \tag{9.132}$$

The stationary points of the functional (9.131) are also the stationary points of the functional $J\left(x\left(t,X\right),\varphi(t,X),v(t,X)\right)$:

$$J\left(x\left(t,X\right),\varphi(t,X),v(t,X)\right)= \tag{9.133}$$

$$\rho\int_{t_0}^{t_1}\int_{\mathring{V}}\left(v_i(t,X)x_t^i\Delta-\frac{1}{2}v_i(t,X)v^i(t,X)\Delta-\dot{\varphi}\Delta\right)d^3X\,dt.$$

At the stationary point, $v_i(t,X)$ coincides with velocity:

$$v^i(t,X)=x_t^i. \tag{9.134}$$

Comparison of (9.132) and (9.134) with (9.13) and (9.15) shows that at the stationary point,

$$v_i(t, X) = \frac{\partial \varphi}{\partial x^i} + X_i^a \mathring{v}_a(X). \tag{9.135}$$

Here $\mathring{v}_a(X)$ can be considered as functions known from the initial conditions. Now we employ a note on the functional modification (5.334) and plug (9.135) into the functional (9.133). We obtain the functional that depends on $x(t, X)$ and $\varphi(t, X)$ only:

$$\int_{t_0}^{t_1} \int_{\mathring{V}} \left(X_i^a \mathring{v}_a(X) x_t^i - \frac{1}{2} g^{ij} (\partial_i \varphi + X_i^a \mathring{v}_a(X))(\partial_j \varphi + X_j^a \mathring{v}_a(X)) - \frac{\partial \varphi(t, x)}{\partial t} \right) \Delta \, d^3 X \, dt. \tag{9.136}$$

Here we used the fact that $\dot{\varphi} = \partial \varphi / \partial t + \varphi_{,i} x_t^i$.

We are going to show that the functional (9.136) is identical to the functional of vortex line dynamics. The substitution (9.135) increased the set of stationary points, and this is how the incompressible fluid motions become compressible motions of vortex dynamics.

To transform the functional (9.136), we first note that the last term can be presented as a boundary integral:

$$\rho \int_{t_0}^{t_1} \int_{\mathring{V}} \frac{\partial \varphi(t, x)}{\partial t} \Delta \, d^3 X \, dt = \rho \int_{t_0}^{t_1} \int_{V} \frac{\partial \varphi(t, x)}{\partial t} \, d^3 x \, dt = \left[\rho \int_V \varphi \, d^3 x \right]_{t_0}^{t_1},$$

and, thus, does not affect the differential equations. Therefore, the search of the stationary point with respect to φ is reduced to minimization of the functional,

$$K(\varphi, x(t, X)) = \rho \int_{\mathring{V}} \frac{1}{2} g^{ij} (\partial_i \varphi + X_i^a \mathring{v}_a(X)) \, (\partial_j \varphi + X_j^a \mathring{v}_a(X)) \Delta \, d^3 X \,, \tag{9.137}$$

over φ at each instant. In this functional, $X_j^a \mathring{v}_a(X)$ are considered as known; denote them by u_i. Transforming (9.137) to Eulerian variables, we obtain the variational problem

$$\mathcal{K}(x(t, X)) = \min_{\varphi} \int_V \frac{1}{2} \rho (\partial_i \varphi + u_i)(\partial^i \varphi + u^i) \, d^3 x. \tag{9.138}$$

Any vector field, u_i, can be presented as a sum of a divergence-free vector field, u_i', obeying the boundary condition $u_i' n^i = 0$, and potential vector field, $\partial_i \chi$. Hence, the minimizing function, $\check{\varphi}$, is the solution of the boundary value problem

$$\Delta(\breve{\varphi}+\chi)=0 \qquad \text{in } V, \qquad \frac{\partial(\breve{\varphi}+\chi)}{\partial x^i}n^i=0 \qquad \text{on } \partial V.$$

Obviously, $\breve{\varphi}=-\chi+const$, and the minimum value in the variational problem (9.138) is

$$\int\limits_V \frac{1}{2}\rho u_i' u'^i \, d^3x.$$

As we discussed, u_i' are uniquely determined by vorticity,

$$\begin{aligned}\omega^i &= e^{ijk}\partial_j u_k' = e^{ijk}\partial_j v_k = e^{ijk}X_j^b X_k^a \partial_b \mathring{v}_a(X)\\ &= \frac{1}{\Delta}x_c^i e^{cba}\partial_b \mathring{v}_a(X) = \frac{1}{\Delta}x_c^i \mathring{\omega}^c(X).\end{aligned}$$

So, we arrived at the above-considered problem of calculation of kinetic energy in terms of vorticity.

The first term in the functional (9.136) is transformed by means of (3.23):

$$\rho\int\limits_{t_0}^{t_1}\int\limits_{\mathring{V}} X_i^a \mathring{v}_a(X)x_t^i \Delta \, d^3X \, dt = \rho\int\limits_{t_0}^{t_1}\int\limits_{\mathring{V}} \frac{1}{2}e_{ijk}x_a^i x_b^j e^{abc}\mathring{v}_c(X)x_t^k \, d^3X \, dt. \tag{9.139}$$

For further transformation we need the easily verifiable identities

$$\begin{aligned}e_{ijk}x_a^i x_b^j e^{abc}\mathring{v}_c(X)x_t^k &= \partial_b\left(e_{ijk}x_a^i x^j e^{abc}\mathring{v}_c(X)x_t^k\right)\\ &-e_{ijk}x_a^i x^j e^{abc}\partial_b\mathring{v}_c(X)x_t^k - e_{ijk}x_a^i x^j e^{abc}\mathring{v}_c(X)\frac{dx_b^k}{dt},\\ 2e_{ijk}x_a^i x^j e^{abc}\mathring{v}_c(X)\frac{dx_b^k}{dt} &= \frac{d}{dt}e_{ijk}x_a^i x^j e^{abc}\mathring{v}_c(X)x_b^k - e_{ijk}x_a^i x_t^j e^{abc}\mathring{v}_c(X)x_b^k,\end{aligned}$$

which yield

$$\begin{aligned}\frac{3}{2}e_{ijk}x_a^i x_b^j e^{abc}\mathring{v}_c(X)x_t^k &= \partial_b\left(e_{ijk}x_a^i x^j e^{abc}\mathring{v}_c(X)x_t^k\right) - \frac{1}{2}\frac{d}{dt}e_{ijk}x_a^i x^j e^{abc}\mathring{v}_c(X)x_b^k\\ &+e_{ijk}x^j x_t^i x_a^k e^{abc}\partial_b\mathring{v}_c(X).\end{aligned} \tag{9.140}$$

The integral of the first term,

$$\ell=\int\limits_{t_0}^{t_1}\int\limits_{\mathring{V}}\partial_b\left(e_{ijk}x_a^i x^j e^{abc}\mathring{v}_c(X)x_t^k\right)d\mathring{V}dt,$$

in general is not zero; however its variation is zero. Indeed,

$$\delta\ell = \delta \int_{t_0}^{t_1} \int_{\partial \mathring{V}} e_{ijk} x_a^i x^j e^{abc} \mathring{v}_c(X) x_t^k \mathring{n}_b d\mathring{A} dt,$$

where $d\mathring{A}$ is the area element, and $\mathring{n}_b$ the normal vector in the initial state. Let ξ^μ be parameters on $\partial\mathring{V}$, Greek indices run values 1,2, and $X^a = \mathring{r}^a(\xi^\mu)$ the parametric equations of $\partial\mathring{V}$, $\mathring{r}_\mu^a \equiv \partial\mathring{r}^a/\partial\xi^\mu \equiv \partial_\mu \mathring{r}^a$. The identity holds true:

$$\delta(e_{ijk} x_a^i x^j e^{abc} \mathring{v}_c(X) x_t^k \mathring{n}_b) = 3e_{ijk} \delta x^j \mathring{r}_\mu^c x_\nu^i e^{\mu\nu} \mathring{v}_c x_t^k - \mathring{\nabla}_\mu (e_{ijk} \delta x^i \mathring{r}_\nu^c x^j e^{\mu\nu} \mathring{v}_c x_t^k)$$
$$+\partial_t (e_{ijk} x_a^i x^j e^{abc} \mathring{v}_c \delta x^k \mathring{n}_b), \qquad (9.141)$$

where $\mathring{\nabla}_\mu$ denotes the covariant derivative over the boundary in the initial state. To check the validity of (9.6) we use the fact that $e^{abc}\mathring{n}_b = \frac{1}{\sqrt{\mathring{a}}} e^{\mu\nu} \mathring{r}_\nu^a \mathring{r}_\mu^c$, $\mathring{a}$ being the determinant of the surface metric tensor; then

$$\sqrt{\mathring{a}}\delta(e_{ijk} x_a^i x^j e^{abc} \mathring{v}_c(X) x_t^k \mathring{n}_b) = \delta(e_{ijk} x_\nu^i x^j \mathring{v}_c \mathring{r}_\mu^c x_t^k e^{\mu\nu}) = e_{ijk} x_\nu^i \mathring{v}_c \mathring{r}_\mu^c e^{\mu\nu} x_t^k \delta x^j$$
$$+e_{ijk} x^j \mathring{v}_c \mathring{r}_\mu^c x_t^k e^{\mu\nu} \partial_\nu \delta x^i + e_{ijk} x_\nu^i x^j \mathring{v}_c \mathring{r}_\mu^c e^{\mu\nu} (\delta x^k)_t =$$
$$3e_{ijk} x_\nu^i \mathring{v}_c \mathring{r}_\mu^c e^{\mu\nu} x_t^k \delta x^j + \partial_\nu (e_{ijk} x^j \mathring{v}_c \mathring{r}_\mu^c x_t^k e^{\mu\nu} \delta x^i) + (e_{ijk} x_\nu^i x^j \mathring{v}_c \mathring{r}_\mu^c e^{\mu\nu} \delta x^k)_t$$
$$-e_{ijk} x^j \mathring{v}_{c,a} \mathring{r}_\nu^c \mathring{r}_\mu^c x_t^k e^{\mu\nu} \delta x^i - e_{ijk} x^j \mathring{v}_c \mathring{r}_\mu^c x_{t,\nu}^k e^{\mu\nu} \delta x^i - e_{ijk} x_{\nu,t}^i x^j \mathring{v}_c \mathring{r}_\mu^c e^{\mu\nu} \delta x^k.$$

The last two terms cancel out, the term $e_{ijk} x^j \mathring{v}_{c,a} \mathring{r}_\nu^c \mathring{r}_\mu^c x_t^k e^{\mu\nu} \delta x^i$ is zero because, as we assumed, the normal component of vorticity, $\mathring{v}_{c,a} \mathring{r}_\mu^a e^{\mu\nu} \mathring{r}_\nu^c$, is zero, and, taking into account (14.38), we arrive at (9.6).

The integral of the second term in the right hand side of (9.6) is zero, integral of the third term goes to the time ends and does not affect equations and boundary conditions. Hence,

$$\delta\ell = \int_{t_0}^{t_1} \int_{\partial \mathring{V}} 3e_{ijk} \delta x^j \mathring{r}_\mu^c x_\nu^i e^{\mu\nu} \mathring{v}_c x_t^k d\mathring{A} dt. \qquad (9.142)$$

The integrand in (9.142) is proportional to the volume of the parallelogram formed by three vectors, δx^j, x_ν^i and x_t^k. At the boundary, all these vectors are tangent to the boundary, therefore the volume is zero, and, thus, $\delta\ell = 0$ as claimed. So, up to the time end terms and constant terms,

$$\rho \int_{t_0}^{t_1} \int_{\mathring{V}} X_i^a \mathring{v}_a(X) x_t^i \Delta \, d^3X \, dt = \rho \int_{t_0}^{t_1} \int_{\mathring{V}} \frac{1}{3} e_{ijk} x^j x_t^i x_a^k \frac{\mathring{\omega}^a}{\Delta} \, d^3X \, dt. \qquad (9.143)$$

We arrived at the action functional of the vortex line dynamics.

9.7 Quasi-Two-Dimensional and Two-Dimensional Vortex Flows

We call a flow quasi-two-dimensional if each vortex line crosses the plane $x^3 = const.$ at one point (Fig. 9.3). In such a flow rotational motion occurs mostly in the (x^1, x^2)-plane.

The variational principles formulated cannot be applied to quasi-two-dimensional flows because the vortex lines are not closed: if $\eta = 0$ and $\eta = l$ are the values of the parameter η for the points lying in cross-sections $x^3 = 0$ and $x^3 = L$, then $r^3(t, l, X) - r^3(t, 0, X) = L \neq 0$, and the divergence terms vanished for the closed vortex lines are no longer zero. Due to that the functional $\mathcal{A}$ must be modified[8]:

$$\mathcal{A} = \frac{1}{3}\rho \int_V e_{ijk} r^j \frac{\partial r^i}{\partial t} \frac{\partial r^k}{\partial \eta} \mathring{\omega} d^2 X d\eta + \frac{L}{6}\rho \int e_{\alpha\beta} r^\beta (t, 0, X) \frac{\partial r^\alpha (t, 0, X)}{\partial t} \mathring{\omega} d^2 X \tag{9.144}$$

The variation of the second integral in (9.144) cancels the divergence terms which appear in variation of the first integral.

Formula (9.144) can be simplified by a special choice of the parameter on the vortex lines. Identifying η with x,

$$x = r^3(t, \eta, X) \equiv \eta,$$

we describe the positions of vortex lines by two functions $x^\alpha = r^\alpha(t, x, X)$. We consider the flows in cylindrical domain $\Omega \times [0, L]$ which are periodic in the x- direction with the period L. Then functions $r^\alpha(t, x, X)$ are periodic:

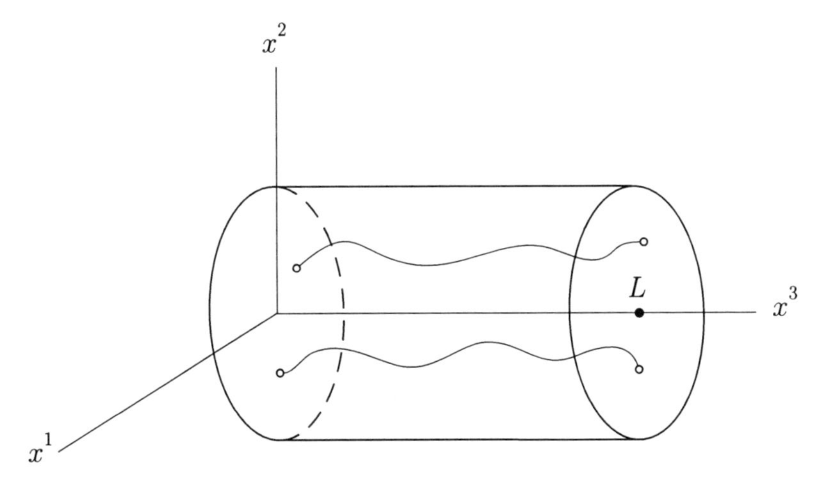

Fig. 9.3 Quasi-two-dimensional flow

[8] Greek indices run values 1, 2 and mark projections on axes x^1 and x^2.

$$r^{\alpha}(t,0,X)=r^{\alpha}(t,L,X). \tag{9.145}$$

The functional $\mathcal{A}$ takes the form

$$\mathcal{A}=\frac{1}{3}\rho\int\limits_V\left(-e_{\alpha\beta}\frac{\partial r^{\alpha}}{\partial t}\frac{\partial r^{\beta}}{\partial\eta}x+e_{\alpha\beta}r^{\beta}\frac{\partial r^{\alpha}}{\partial t}\right)\mathring{\omega}(X)d^2Xdx$$
$$+\frac{L}{6}\rho\int e_{\alpha\beta}r^{\beta}(t,0,X)\frac{\partial r^{\alpha}(t,0,X)}{\partial t}\mathring{\omega}d^2X. \tag{9.146}$$

Note the identity

$$e_{\alpha\beta}\frac{\partial r^{\alpha}}{\partial t}\frac{\partial r^{\beta}}{\partial\eta}x=-\frac{1}{2}e_{\alpha\beta}r^{\beta}\frac{\partial r^{\alpha}}{\partial t}+\frac{\partial}{\partial x}\left(\frac{1}{2}e_{\alpha\beta}r^{\beta}\frac{\partial r^{\alpha}}{\partial t}x\right)-\frac{\partial}{\partial t}\left(\frac{1}{2}e_{\alpha\beta}r^{\beta}\frac{\partial r^{\alpha}}{\partial x}x\right). \tag{9.147}$$

Plugging (9.147) to (9.146) and dropping the divergence term

$$\int\frac{\partial}{\partial t}\left(\frac{1}{2}e_{\alpha\beta}r^{\beta}\frac{\partial r^{\alpha}}{\partial x}x\right)\mathring{\omega}(X)d^2Xdx,$$

we obtain the final expression for $\mathcal{A}$:

$$\mathcal{A}=\frac{1}{2}\rho\int\limits_V e_{\alpha\beta}r^{\beta}(t,x,X)\frac{\partial r^{\alpha}(t,x,X)}{\partial t}\mathring{\omega}(X)d^2Xdx. \tag{9.148}$$

Note the different factors: $1/2$ in (9.148) and $1/3$ in (9.118).

The variational principle for kinetic energy must also be modified because the condition used for closed flows at the boundary, $\vec{v}\cdot\vec{n}=0$, does not hold on the planes $x=0$ and $x=L$.

Denote by $v, v^{\alpha}, \omega, \omega^{\alpha}$ and ψ, ψ^{α} the axial and the transversal components of velocity, vorticity, and stream function vector, respectively, and by $[\varphi]$ the difference of the values of function φ at $x=L$ and $x=0$:

$$[\varphi]\equiv\varphi(L,y^{\alpha})-\varphi(0,y^{\alpha}).$$

We assume that the stream function vector and the axial component of vorticity are periodic in the axial direction and the normal component of vorticity at $\partial\Omega\times[0,L]$ is zero:

$$[\psi]=\left[\psi^{\alpha}\right]=0, \tag{9.149}$$

$$\omega^{\alpha}n_{\alpha}=0 \quad \text{on } \partial\Omega\times[0,L], \quad [\omega]=0. \tag{9.150}$$

Since $v=e^{\alpha\beta}\partial_{\alpha}\psi_{\beta}$, the axial component of velocity is also periodic:

$$[v]=0. \tag{9.151}$$

The transversal components of velocity are not required to be periodic on the set of the admissible flow fields but, as we will see, they are periodic for the minimizing flow.

Due to the special geometry of the flow, it is convenient to change slightly the constraints (9.121) and (9.122) for the stream vector field. We set the constraints

$$\partial_\alpha \psi^\alpha = 0, \tag{9.152}$$

$$\psi^\alpha n_\alpha = 0 \quad \text{on } \partial\Omega \times [0, L], \tag{9.153}$$

$$\int_{\partial\Omega} \psi ds = 0 \quad 0 \leq x \leq L. \tag{9.154}$$

These constraints provide a one-to-one correspondence between velocity and stream vector field. Indeed, according to (9.152) and (9.153), the potential field, $\varphi_{,k}$, which can be added to ψ_k without changing the velocity field, should obey the boundary-value problem

$$\Delta\varphi = 0 \quad \text{in } \partial\Omega \times [0, L], \quad \frac{\partial\varphi}{\partial n} = 0 \quad \text{on } \partial\Omega \times [0, L].$$

Here Δ is the two-dimensional Laplace's operator. Hence, φ can be an arbitrary function of x. This function affects only the third component of the stream vector, ψ. The constraint (9.154) yields $\partial_x \varphi = 0$. Thus, the gradient invariance is eliminated.

The dual variational principle (9.156) takes the form

$$-\mathcal{K} = \min_{\substack{\psi, \psi_\alpha \in \\ (9.149),(9.152)-(9.154)}} \rho \int \left[\frac{1}{2} \left(curl\, \vec{\psi} \right)^2 - \vec{\psi} \cdot \vec{\omega} \right] dV.$$

It is easy to see that the minimizer satisfies the equations

$$\frac{\partial^2 \psi}{\partial x^\alpha \partial x} - \Delta_3 \psi_\alpha = \omega_\alpha, \quad -\Delta\psi = \omega, \tag{9.155}$$

Δ_3 being the three-dimensional Laplace's operator. Equations (9.155) can also be written as

$$e^{\alpha\beta} \left(\partial_\beta \left(curl\, \vec{\psi} \right)_3 - \frac{\partial}{\partial x} \left(curl\, \vec{\psi} \right)_\beta \right) = \omega^\alpha, \quad e^{\alpha\beta} \left(\partial_\beta \left(curl\, \vec{\psi} \right)_\alpha \right) = \omega. \tag{9.156}$$

Equations (9.156) indicate that $curl\, \vec{\psi}$ can be identified with velocity.

Note that the one-to-one correspondence between vorticity and velocity established for closed domains, does not hold for periodic flows: one can add arbitrary constant axial velocity without changing the vorticity field. The constraints imposed select a unique stream vector and, thus, a unique velocity field. One can show that this corresponds to a special choice of the inertial frame, namely the frame in which, for a given vorticity field, kinetic energy is minimum.

One remark is now in order. First, the constraints (9.152) and (9.153) mean that the functions ψ^α can be expressed in terms of one function, χ,

$$\psi^\alpha = e^{\alpha\beta}\partial_\beta\chi, \tag{9.157}$$

and

$$\frac{\partial\chi}{\partial n} = 0 \quad \text{on } \partial\Omega. \tag{9.158}$$

The velocity is determined by two functions, ψ and χ:

$$v^\alpha = e^{\alpha\beta}\left(\partial_\beta\psi - \partial_x\psi_\beta\right) = e^{\alpha\beta}\partial_\beta\psi + \partial^\alpha\partial_x\chi, \quad v = e^{\alpha\beta}\partial_\alpha\psi_\beta = -\Delta\chi.$$

The impermeability condition simplifies to $d\psi/ds = 0$, and without loss of generality, for simply connected region Ω, due to (9.154),

$$\psi = 0 \quad \text{on } \partial\Omega. \tag{9.159}$$

Remarkably, the interaction terms between ψ and χ in kinetic energy vanishes:

$$\int_\Omega\int_0^L \left(v^2 + v_\alpha v^\alpha\right) d^2y dx =$$

$$= \int_\Omega\int_0^L \left(\partial_\alpha\psi\partial^\alpha\psi + \partial_\alpha\partial_x\chi\partial^\alpha\partial_x\chi + (\Delta\chi)^2\right) d^2y dx. \tag{9.160}$$

Function χ is determined up to an arbitrary function of x. One can eliminate this arbitrariness by putting additionally

$$\int_\Omega \chi d^2y = 0, \quad 0 \le x \le L. \tag{9.161}$$

In the case of two-dimensional flows functions $r^\alpha(t, x, X)$ do not depend on x, and the functional $\mathcal{A}$ becomes

$$\mathcal{A} = \frac{\rho L}{2}\int_V e_{\alpha\beta} r^\beta(t, X)\frac{\partial r^\alpha(t, X)}{\partial t}\mathring{\omega}(X) d^2X.$$

Up to divergence terms, it can also be written as

$$\mathcal{A} = \rho L \int_V r^2(t,X) \frac{\partial r^1(t,X)}{\partial t} \mathring{\omega}(X) d^2X. \tag{9.162}$$

Further we drop the factor L in $\mathcal{A}$ and $\mathcal{K}$.

To compute the kinetic energy we note that only one component of stream vector, $\psi^3 \equiv \psi$, is not zero and

$$v^\alpha = e^{\alpha\beta}\psi_{,\beta}. \tag{9.163}$$

At the boundary, $\psi = 0$. The dual variational principle $I(\psi)$ takes the form

$$-\mathcal{K} = \min_{\psi\in(9.159)} \rho \left[\frac{1}{2} \int_\Omega \partial_\alpha\psi\partial^\alpha\psi d^2x - \int_\Omega \psi\omega d^2x \right], \tag{9.164}$$

and kinetic energy can be expressed in terms of the corresponding Green's function:

$$\mathcal{K} = \frac{\rho}{2} \int_V G(x,x')\,\omega(x)\,\omega(x')\,d^2x d^2x'.$$

We arrive at the following

Variational principle. *The true two-dimensional vortex motion of ideal incompressible fluid is a stationary point of the action functional*

$$I = \int_{t_0}^{t_1} [\mathcal{A} - \mathcal{K}]\,dt,$$

$$\mathcal{A} = \rho \int_\Omega r^2(t,X) \frac{\partial r^1(t,X)}{\partial t} \mathring{\omega}(X) d^2X, \tag{9.165}$$

$$\mathcal{K} = \frac{1}{2}\rho \int_\Omega \int_\Omega G(r(t,X), r(t,X'))\,\mathring{\omega}(X)\,\mathring{\omega}(X')\,d^2Xd^2X'.$$

Let vorticity be concentrated in small vicinities, $\Omega_1, \ldots, \Omega_N$, of points, $X_1, \ldots, X_N$. In Lagrangian coordinates, without loss of generality these vicinities can be viewed as circles of small radius ε. For $\varepsilon \to 0$, in the leading approximation the shortened action $\mathcal{A}$ transforms into the sum

$$\mathcal{A} = \sum_{s=1}^{N} \rho\gamma_{(s)} r^2_{(s)}(t) \frac{dr^1_{(s)}(t)}{dt},$$

where

$$r^{\alpha}_{(s)}(t) = r^{\alpha}(t, X_s)$$

and

$$\gamma_{(s)} = \int_{\Omega_s} \mathring{\omega}(X)\, d^2X.$$

This suggests that the dynamics of the system can be described by the motion in the region Ω of N points, $r_{(1)}, \ldots, r_{(N)}$. To justify such an expectation, we have to check that the kinetic energy also becomes a function only of $r_{(1)}, \ldots, r_{(N)}$ as $\varepsilon \to 0$. For kinetic energy we have the double sum:

$$\mathcal{K} = \frac{1}{2}\rho \sum_{s=1}^{N}\sum_{m=1}^{N} \int_{\Omega_s}\int_{\Omega_m} G\left(r(t, X), r\left(t, X'\right)\right) \mathring{\omega}(X)\, \mathring{\omega}\left(X'\right) d^2X d^2X'.$$

As $\varepsilon \to 0$, the terms of the sum with $s \neq m$ are approximated by a function of $r_{(s)}$ and $r_{(m)}$:

$$\frac{1}{2}\gamma_{(s)}\gamma_{(m)} G\left(r_{(s)}(t), r_{(m)}(t)\right).$$

Such approximation assumes that the points $r_{(s)}(t)$ and $r_{(m)}(t)$ remain on the distances much bigger than ε in the course of motion. To obtain an approximation of the terms with $s = m$ we note that, as $\left|r - r'\right| \to 0$,

$$G(r, r') = \frac{1}{2\pi} \ln \frac{1}{|r - r'|} + 2g(r),$$

where $g(r)$ is a smooth function uniquely determined by the region Ω. Therefore,

$$\frac{1}{2}\int_{\Omega_s}\int_{\Omega_s} G\left(r(t, X), r\left(t, X'\right)\right) \mathring{\omega}(X)\, \mathring{\omega}\left(X'\right) d^2X d^2X' = \gamma_{(s)}^2 g(r_{(s)})$$
$$+\frac{1}{2}\int_{\Omega_s}\int_{\Omega_s} \frac{1}{2\pi} \ln \frac{1}{|r(t, X) - r(t, X')|} \mathring{\omega}(X)\, \mathring{\omega}\left(X'\right) d^2X d^2X'. \tag{9.166}$$

The integral in the right hand side of (9.166) is on the order $\gamma_{(s)}^2 \ln \frac{1}{\varepsilon}$, and, in fact, much bigger than the other terms. However, it depends only on the details of the motion inside the vortex blob Ω_s and does not feel translations being thus independent on $r_{(s)}$. Therefore, the dynamics which involves only the positions of the vortex blobs, $r_{(1)}, \ldots, r_{(N)}$, is self-consistent. It is called point vortex dynamics. The self-energy of the mth point vortex (9.166) is infinite in the limit $\varepsilon \to 0$; however the "infinite term" does not depend on the translational motion of the vortex.

Collecting all the essential terms of the action functional, we obtain the following.

Variational principle. *The true motion of point vortices is a stationary point of the action functional*

$$I = \int_{t_0}^{t_1} [\mathcal{A} - \mathcal{K}]\, dt,$$

$$\mathcal{A} = \sum_{s=1}^{N} \rho \gamma_{(s)} r_{(s)}^2(t) \frac{dr_{(s)}^1(t)}{dt}, \tag{9.167}$$

$$\mathcal{K} = \sum_{s \neq m} \frac{1}{2} \rho \gamma_{(s)} \gamma_{(m)} G\left(r_{(s)}(t), r_{(m)}(t)\right) + \sum_{s=1}^{N} \rho \gamma_{(s)}^2 g(r_{(s)}(t)).$$

Note that function $g(r)$ tends to $-\infty$ as the point r approaches the boundary of region Ω, and therefore the vortices never reach the boundary. The functional $\mathcal{K}$ in (9.167) differs from the true kinetic energy by a large positive self-energy term. Not surprisingly, it can take the negative values. If one takes into account the shape change of the vortex blobs, then the additional terms characterizing the dynamics of the shape enter the action functional and the kinetic energy.

Sometimes it is convenient to deal with positive energy and keep the self-energy terms. This can be done, for example, by including in energy higher derivatives with a small parameter, ε:

$$-\mathcal{K} = \min_{\psi \in (9.159)} \rho \left[\int_{\Omega} \left(\frac{1}{2} \partial_\alpha \psi \partial^\alpha \psi + \frac{1}{2} \varepsilon^2 \partial_{\alpha\beta} \psi \partial^{\alpha\beta} \psi \right) d^2x - \sum_{s=1}^{N} \cdot \gamma_{(s)} \psi\left(r_{(s)}\right) \right].$$

Then one can show that kinetic energy is finite and converges to the true energy as $\varepsilon \to 0$. The parameter, ε, plays the role of the size of the vortex core. The results obtained for such a model are meaningful if they do not depend on ε.

9.8 Dynamics of Vortex Filaments in Unbounded Space

Vortex line dynamics deals only with the necessary degrees of freedom of fluid motion eliminating the “slave” degrees of freedom of potential flow. As an example, consider dynamics of a vortex filament, a thin fluid tube such that vorticity is negligible outside of this tube while inside the tube it is predominantly directed along its axis. If the size of the cross-section is much smaller than the characteristic radius of the tube, $\mathcal{R}$, then the motion of the vortex filament can be modeled by motion of a curve, Γ. Accordingly, the degrees of freedom of the system are the functions

$$x^i = r^i(t, \eta), \tag{9.168}$$

which determine the current position of Γ. It seems natural to identify Γ with the center line of the vortex tube and take the vortex intensity of the filament , $\mathring{\omega}(X)$, as a δ-function,

$$\mathring{\omega}(X) = \gamma\delta(X - X_0), \tag{9.169}$$

X_0 being the Lagrangian coordinates of Γ. That certainly makes sense in computation of the shortened action, $\mathcal{A}$, which is meaningful for δ-type vorticity:

$$\mathcal{A} = \frac{\rho\gamma}{3}\int\limits_V e_{ijk} r^j(t,\eta)\frac{\partial r^i(t,\eta)}{\partial t}\frac{\partial r^k(t,\eta)}{\partial \eta}d\eta. \tag{9.170}$$

However, there is an obstacle: kinetic energy is infinite for the δ-type vorticities. Moreover, as we will see, the "infinite term" is principally different from that for point vortices: the infinite self-energy of a point vortex does not affect its translational motion while for the vortex filament the self-energy of a filament segment provides the leading contribution to its translational velocity. Therefore, a more delicate analysis is needed. We start with a more adequate description of the filament kinematics.

Kinematics of vortex filament. Let $s(t,\eta)$ and $\tau^i(t,s)$ be the arc length along the filament,

$$\frac{\partial s(t,\eta)}{\partial \eta} = \sqrt{r^i_{,\eta} r_{i,\eta}}, \quad r^i_{,\eta} \equiv \frac{\partial r^i(t,\eta)}{\partial \eta}, \tag{9.171}$$

and the unit tangent vector,

$$\tau^i = \frac{\partial r^i(t,s)}{\partial s} = \frac{r^i_{,\eta}}{\sqrt{r^k_{,\eta} r_{k,\eta}}}. \tag{9.172}$$

We endow the line Γ with a couple of unit vectors, $\tau^i_1(t,\eta)$ and $\tau^i_2(t,\eta)$, which form together with the unit tangent vector to Γ an orthonormal triad:

$$\tau^i\tau_i = 1, \quad , \tau_i\tau^i_\mu = 0, \quad \tau^i_\mu\tau_{i\nu} = \delta_{\mu\nu}. \tag{9.173}$$

Greek indices run values 1, 2 and mark the vectors τ^i_1 and τ^i_2. The local basis $\{\tau^i, \tau^i_1, \tau^i_2\}$ is Cartesian, and therefore the tensor components with upper and lower indices i, j, k and μ, ν coincide.

We assume that the vortex filament has a circular cross-section, the radius of which, a, does not change over Γ but may depend on time: $a = a(t)$.

Consider motion of vortex lines of a special form:

$$x^i(t,X,\eta) = r^i(t,\eta) + a(t)\tau^i_\mu(t,\eta)X^\mu. \tag{9.174}$$

Here Lagrangian coordinates, X^μ, change within a unit circle,

$$X^\mu X_\mu \leq 1.$$

The vectors, $\tau^i_\mu(t,\eta)$, in (9.174) will be chosen in a special way. To describe it we first note the relations for the derivatives of the triad over Γ in terms of the curvatures of Γ, $\varkappa_\mu$, $\varkappa$:

$$\frac{\partial \tau^i}{\partial s} = \varkappa^\mu \tau^i_\mu, \quad \frac{\partial \tau^i_\mu}{\partial s} = -\varkappa_\mu \tau^i + \varkappa e_{\mu\cdot}^{\cdot\nu} \tau^i_\nu, \tag{9.175}$$
$$\varkappa_\mu \equiv \tau^i_\mu \frac{\partial \tau_i}{\partial s}, \quad \varkappa \equiv \frac{1}{2} e^{\mu\nu} \tau^i_\nu \frac{\partial \tau_{i\mu}}{\partial s}.$$

The derivation of these relations can be found further in Chap. 15[9] where they are used to characterize the deformations of elastic beams. Similarly to (9.175), differentiating the triad over time, one can write

$$\frac{\partial \tau^i(t,\eta)}{\partial t} = \Omega^\mu \tau^i_\mu, \quad \frac{\partial \tau^i_\mu(t,\eta)}{\partial t} = -\Omega_\mu \tau^i + \Omega e_{\mu\cdot}^{\cdot\nu} \tau^i_\nu, \tag{9.176}$$

where

$$\Omega_\mu \equiv \tau^i_\mu \frac{\partial \tau_i}{\partial t}, \quad \Omega \equiv \frac{1}{2} e^{\mu\nu} \tau^i_\nu \frac{\partial \tau_{i\mu}}{\partial t}.$$

If the rate of rotation, Ω, and the positions of Γ, $r^i(t,\eta)$, are known, then the second equation (9.176) may be considered as a system of partial differential equations to determine $\tau^i_\mu(t,\eta)$:

$$\frac{\partial \tau^i_\mu(t,\eta)}{\partial t} = -\left(\frac{\partial \tau^j(t,\eta)}{\partial t} \tau_{j\mu}\right) \tau^i(t,\eta) + \Omega(t,\eta) e_{\mu\cdot}^{\cdot\nu} \tau^i_\nu. \tag{9.177}$$

Here $\tau^i(t,\eta)$ are expressed in terms of $r^i(t,\eta)$ by (9.172). For given initial positions of τ^i_μ and a given motion of Γ, the further evolution of the vectors, $\tau^i_\mu(t,\eta)$, is defined by (9.177) uniquely. We make a special choice of τ^i_μ as follows. First, we choose some initial vectors, $\tau^i_\mu(0,\eta)$, and specify the evolution of τ^i_μ by setting $\Omega = 0$. We denote these uniquely defined vectors by $\mathring{\tau}^i_\mu(t,\eta)$. Then we introduce the vectors $\tau^i_\mu(t,\eta)$ which differ from $\mathring{\tau}^i_\mu$ by a rotation on the same angle at each point of Γ:

$$\tau^i_\mu(t,\eta) = o^\nu_\mu(t) \mathring{\tau}^i_\nu(t,\eta). \tag{9.178}$$

Here $o^\nu_\mu(t)$ is an orthogonal matrix.

[9] See (15.4); we need to employ here other notation for curvatures because the letter ω is been used for vorticity.

The rate of rotation, Ω, of vectors τ^i_μ is constant over Γ. Indeed,

$$\begin{aligned}\Omega &= \frac{1}{2} e^{\mu\nu} \tau^i_\nu \frac{\partial \tau_{i\mu}}{\partial t} = \frac{1}{2} e^{\mu\nu} o^\lambda_\nu \mathring{\tau}^i_\lambda \frac{\partial}{\partial t} o^\sigma_\mu \mathring{\tau}_{i\sigma} = \frac{1}{2} e^{\mu\nu} o^\lambda_\nu \mathring{\tau}^i_\lambda (\dot{o}^\sigma_\mu \mathring{\tau}_{i\sigma} + o^\sigma_\mu \frac{\partial \mathring{\tau}_{i\sigma}}{\partial t}) \\ &= \frac{1}{2} e^{\mu\nu} o_{\nu\sigma} \dot{o}^\sigma_\mu + \frac{1}{2} e^{\sigma\lambda} \mathring{\tau}^i_\lambda \frac{\partial \mathring{\tau}_{i\sigma}}{\partial t} = \frac{1}{2} e^{\mu\nu} o_{\nu\sigma} \dot{o}^\sigma_\mu .\end{aligned}$$

Here we used the fact that the rate of rotation of $\mathring{\tau}^i_\nu$ is zero.

A two-dimensional orthogonal matrix, $o^\nu_\mu(t)$, has one degree of freedom, the angle of rotation of the cross-section, $\varphi(t)$:

$$\left\| o^\nu_\mu \right\| = \left\| \begin{matrix} \cos\varphi & -\sin\varphi \\ \sin\varphi & \cos\varphi \end{matrix} \right\| .$$

It is easy to check that[10]

$$\Omega = \frac{d\varphi}{dt}. \tag{9.179}$$

So, in addition to $r^i(t,\eta)$, the vortex filament is endowed with two degrees of freedom, the cross-sectional radius, $a(t)$, and the angle of rotation of the filament cross-sections, $\varphi(t)$.

Note the following relations for variations (they are similar to (15.12)):

$$\delta\varphi = \frac{1}{2} e^{\mu\nu} \tau_{i\nu} \delta\tau^i_\mu = \frac{1}{2} e^{\mu\nu} o_{\nu\sigma} \delta o^\sigma_\mu, \qquad \delta\tau^i_\mu = -(\tau_{j\mu} \delta\tau^j)\tau^i + e^{\cdot\nu}_{\mu\cdot} \tau^i_\nu \delta\varphi. \tag{9.180}$$

In what follows we assume that the radius of the filament cross-section is much smaller than the total length of the filament, L, and the characteristic radius of curvature of the filament, $\mathcal{R}$, which is defined as

$$\mathcal{R} = \left(\max_s \sqrt{\varkappa_\mu \varkappa^\mu + \varkappa^2} \right)^{-1} .$$

So, the small parameters of the problem are

$$\frac{a}{L} \ll 1, \quad \frac{a}{\mathcal{R}} \ll 1, \quad a\varkappa \ll 1, \quad a\varkappa_1 \ll 1, \quad a\varkappa_2 \ll 1.$$

To transform the integrals from Lagrangian to Eulerian variables we need the formula for the determinant of the transformation,

$$\Delta = e_{ijk} \frac{\partial x^i}{\partial \eta} \frac{\partial x^j}{\partial X^1} \frac{\partial x^k}{\partial X^2} = a^2 (1 + a\varkappa_\mu X^\mu) \frac{\partial s}{\partial \eta} .$$

---

[10] Here the upper and low indices of o^ν_μ are interpreted as the row and colomn numbers, respectively (i.e. $o^1_2 = -\sin\varphi$, $o^2_1 = \sin\varphi$). Otherwise, the sign in (9.179) is negative.

In the leading approximation,

$$\Delta = a^2 \frac{\partial s}{\partial \eta}. \tag{9.181}$$

According to (9.106) and (9.181), vorticity in leading approximation is

$$\omega^i = \frac{\partial r^i(t, \eta, X)}{\partial \eta} \frac{\mathring{\omega}(X)}{\Delta} = \tau^i \frac{\mathring{\omega}(X)}{a^2}.$$

On the other hand, the circulation of velocity, γ, is linked to vorticity in the leading approximation as

$$\tau^i \omega_i \pi a^2 = \gamma.$$

Hence, inside the vortex tube,

$$\mathring{\omega}(X) = \gamma/\pi, \tag{9.182}$$

and $\mathring{\omega}(X) = 0$ beyond the tube.

Kinetic energy. We assume that the flow is unbounded and fluid is at rest at infinity. To determine the kinetic energy of the flow we solve (9.91). Since vorticity, ω^i, is not equal to zero only in some bounded region V, and Green' function of Laplace's operator is $1/4\pi \left|x - x'\right|$:[11]

$$\psi^i(x) = \int_V \frac{\omega^i(x')d^3x'}{4\pi\, |x - x'|}. \tag{9.183}$$

For a vortex filament, $\omega^i n_i = 0$ at ∂V, and therefore functions $\psi^i(x)$ (9.183) automatically satisfy the conditions $\partial_i \psi^i = 0$:

$$\begin{aligned}\partial_i \psi^i(x) &= \int_V \omega^i(x') \frac{\partial}{\partial x^i} \frac{1}{4\pi\, |x - x'|} d^3x' = -\int_V \omega^i(x') \frac{\partial}{\partial x'^i} \frac{1}{4\pi\, |x - x'|} d^3x' \\ &= \int_V \frac{1}{4\pi\, |x - x'|} \frac{\partial \omega^i(x')}{\partial x'^i} d^3x' = 0.\end{aligned}$$

We find the kinetic energy of the flow in terms of vorticity from Clapeyron's theorem (5.45) and (9.183):

$$\mathcal{K} = \frac{1}{2}\rho \int_V \psi_i(x)\omega^i(x)d^3x = \frac{1}{2}\rho \int_V \int_V \frac{\omega^i(x)\omega_i(x')}{4\pi\, |x - x'|} d^3x d^3x'. \tag{9.184}$$

[11] We suppress in this subsection the dependence on time.

Further we consider only the case of uniform vorticity, when $\tau^i \omega_i$ is constant over the filament cross-sections up to terms of the order $a/\mathcal{R}$; the same accuracy will be maintained in all further relations.

We are going to show that for any ε, which is much larger than a and much smaller than the characteristic radius of Γ, kinetic energy considered on motions (9.174) is the following functional of $r^i(t,\eta)$ and $a(t)$:

$$\mathcal{K} = \frac{\rho\gamma^2}{8\pi} \oint_\Gamma \oint_\Gamma \frac{\tau^i(s)\tau_i(s')}{|\Delta r| + \varepsilon} ds ds' + \frac{\rho\gamma^2}{8\pi}\left(2\ln\frac{2\varepsilon}{a} + \frac{1}{2}\right) L. \tag{9.185}$$

Here and in what follows the logarithmic terms are treated as the terms of order unity. The length of the filament in (9.185), L, is a functional of $r^i(t,\eta)$:

$$L = \oint_\Gamma \sqrt{r^i_{,\eta} r_{i,\eta}}\, d\eta, \qquad r^i_{,\eta} \equiv \frac{\partial r^i(\eta,t)}{\partial \eta}. \tag{9.186}$$

To prove (9.185) we split the integral (9.184) into the sum of two integrals:

$$\mathcal{K} = \frac{\rho}{2} \int_V \int_V \frac{\omega^i(x)(\omega_i(x') - \omega_i(x))}{4\pi\, |x - x'|} d^3x d^3x' + \frac{\rho\gamma^2}{2(\pi a^2)^2} \int_V \int_V \frac{d^3x d^3x'}{4\pi\, |x - x'|}. \tag{9.187}$$

The first integral in (9.187) is not singular. As $a \to 0$, it converges to

$$\frac{\rho\gamma^2}{2} \oint_\Gamma \oint_\Gamma \frac{\tau^i(s)(\tau_i(s') - \tau_i(s))}{4\pi\, |\Delta r|} ds ds',$$

where

$$|\Delta r| \equiv \sqrt{\Delta r^i\ \Delta r_i}, \qquad \Delta r^i \equiv r^i(s,t) - r^i(s',t).$$

The second integral we again split into a sum of two: for some ℓ such that $a \ll \ell \ll \mathcal{R}$, we present the second integral in the form:

$$\frac{\rho\gamma^2}{2(\pi a^2)^2} \int_V \int_{V,\, \rho(s',s)\geq \ell} \frac{d^3x d^3x'}{4\pi\, |x - x'|} + \frac{\rho\gamma^2}{2(\pi a^2)^2} \int_V \int_{V,\, \rho(s',s)\leq \ell} \frac{d^3x d^3x'}{4\pi\, |x - x'|}, \tag{9.188}$$

where $\rho(s',s)$ the shortest distance along Γ between the points s' and s. The first integral in (9.188) is not singular as $a \to 0$ and converges to

$$\frac{\rho\gamma^2}{2} \oint_\Gamma \oint_{\rho(s',s)\geq \ell} \frac{ds ds'}{4\pi\, |\Delta r|}.$$

The second integral in (9.188), after transformation to Lagrangian coordinates, is

$$\frac{\rho\gamma^2}{8\pi^3}\oint_\Gamma ds\int_{X^\mu X_\mu\leq 1}\int_{X'^\mu X'_\mu\leq 1}2\int_0^\ell\frac{d\xi d^2Xd^2X'}{\sqrt{a^2(X^\mu-X'^\mu)(X_\mu-X'_\mu)+\xi^2}}.$$

Integrating over ξ and using that $\ell\gg a$, we have for this integral

$$\frac{\rho\gamma^2}{8\pi^3}\oint_\Gamma ds\int_{X^\mu X_\mu\leq 1}\int_{X'^\mu X'_\mu\leq 1}2\ln\frac{2\ell}{a\,|X-X'|}d^2Xd^2X'$$

$$=\frac{\rho\gamma^2}{8\pi^3}\oint_\Gamma ds2(\pi^2\ln\frac{2\ell}{a}+J),\qquad J\equiv\int_{X^\mu X_\mu\leq 1}\int_{X'^\mu X'_\mu\leq 1}\ln\frac{1}{|X-X'|}d^2Xd^2X'.$$

The number, J, as easy to see[12], is equal to $\pi^2/4$. Collecting the results, we obtain,

$$\begin{aligned}\mathcal{K}=&\frac{\rho\gamma^2}{2}\oint_\Gamma\oint_\Gamma\frac{\tau^i(s)(\tau_i(s')-\tau_i(s))}{4\pi\,|\Delta\,r|}dsds'\\&+\frac{\rho\gamma^2}{2}\oint_\Gamma\oint_{\rho(s',s)\geq\ell}\frac{dsds'}{4\pi\,|\Delta\,r|}+\frac{\rho\gamma^2}{8\pi}\left(2\ln\frac{2\ell}{a}+\frac{1}{2}\right)L\end{aligned}$$

---

[12] There are several ways to compute J. Perhaps, the simplest one is to use that $G(X,X')=\frac{1}{2\pi}\ln\frac{1}{|X-X'|}$ is Green's function of Laplace's operator: $\Delta G(X,X')=-\delta(X-X')$. First, we write J in the form:

$$J=\int_{X^\mu X_\mu\leq 1}\int_{X'^\mu X'_\mu\leq 1}\ln\frac{1}{|X-X'|}\frac{1}{4}\frac{\partial^2}{\partial X'^\mu\partial X'_\mu}\left(X'^\mu X'_\mu-1\right)d^2Xd^2X'$$

Integrating by parts twice we have,

$$\begin{aligned}J=&\frac{1}{2}\int_{X^\mu X_\mu\leq 1}\int_{|X'|=1}\ln\frac{1}{|X-X'|}ds'\\&+\int_{X^\mu X_\mu\leq 1}\int_{X'^\mu X'_\mu\leq 1}\frac{1}{4}\left(X'^\mu X'_\mu-1\right)\frac{\partial^2}{\partial X'^\mu\partial X'_\mu}\ln\frac{1}{|X-X'|}d^2Xd^2X'.\end{aligned}$$

The integral of $\ln\frac{1}{|X-X'|}$ over the unit circle is zero: this can be checked by writing this integral in terms of complex variable,

$$\int_{|X'|=1}\ln\frac{1}{|X-X'|}ds'=\int_{|z'|=1}\ln\frac{1}{|z-z'|}\frac{dz'}{iz'}=-\operatorname{Im}\int_{|z'|=1}\ln\frac{1}{z-z'}\frac{dz'}{z'},$$

and transforming the contour of integration to a contour surrounding the singular point, $z'=0$, and a contour going around the singular point, $z'=z$. Thus,

$$J=\int_{X^\mu X_\mu\leq 1}\frac{1}{4}\left(1-X^\mu X_\mu\right)2\pi d^2X=\frac{\pi^2}{4}.$$

or, equivalently,

$$\mathcal{K} = \frac{\rho\gamma^2}{8\pi} \oint_\Gamma \oint_{\rho(s',s)\geq\ell} \frac{\tau^i(s)\tau_i(s')}{|\Delta r|} ds ds' + \frac{\rho\gamma^2}{8\pi} \oint_\Gamma \oint_{\rho(s',s)\leq\ell} \frac{\tau^i(s)(\tau_i(s') - \tau_i(s))}{|\Delta r|} ds ds'$$
$$+ \frac{\rho\gamma^2}{8\pi} \left(2 \ln \frac{2\ell}{a} + \frac{1}{2}\right) L$$

The second integral is zero within the accepted accuracy. Thus,

$$\mathcal{K} = \frac{\rho\gamma^2}{8\pi} \oint_\Gamma \oint_{\rho(s',s)\geq\ell} \frac{\tau^i(s)\tau_i(s')}{|\Delta r|} ds ds' + \frac{\rho\gamma^2}{8\pi} \left(2 \ln \frac{2\ell}{a} + \frac{1}{2}\right) L \tag{9.189}$$

On the other hand, for $\varepsilon \ll \ell$,

$$\frac{\rho\gamma^2}{8\pi} \oint_\Gamma \oint_\Gamma \frac{\tau^i(s)\tau_i(s')}{|\Delta r| + \varepsilon} ds ds' = \frac{\rho\gamma^2}{8\pi} \oint_\Gamma \int_{\rho(s',s)\geq\ell} \frac{\tau^i(s)\tau_i(s')}{|\Delta r| + \varepsilon} ds ds'$$
$$+ \frac{\rho\gamma^2}{8\pi} \oint_\Gamma \int_{\rho(s',s)\leq\ell} \frac{\tau^i(s)\tau_i(s')}{|\Delta r| + \varepsilon} ds ds'$$

In the first integral ε can be dropped. In the second integral, expanding $\tau_i(s')$ in Taylor series in vicinity of the point s, we see that only the first term of the expansion provides a noticeable contribution. Therefore,

$$\frac{\rho\gamma^2}{8\pi} \oint_\Gamma \oint_\Gamma \frac{\tau^i(s)\tau_i(s')}{|\Delta r| + \varepsilon} ds ds' = \frac{\rho\gamma^2}{8\pi} \oint_\Gamma \int_{\rho(s',s)\geq\ell} \frac{\tau^i(s)\tau_i(s')}{|\Delta r|} ds ds' + \frac{\rho\gamma^2}{8\pi} L 2 \ln \frac{\ell}{\varepsilon} \tag{9.190}$$

Comparing (9.189) and (9.190) we arrive at (9.185).

Functional A**.** Computation of the functional $\mathcal{A}$ on the motions (9.174) is a cumbersome task. A simpler way is to use the relation (9.126), which holds for any motion, $r^i(t, \eta, X)$:

$$\delta \int_{t_0}^{t_1} \mathcal{A} dt = \rho \int_{t_0}^{t_1} \int_V e_{ijk} \delta r^j(t, \eta, X) \frac{\partial r^i(t, \eta, X)}{\partial t} \frac{\partial r^k(t, \eta, X)}{\partial \eta} \mathring{\omega}(X) d^2 X d\eta dt.$$

In particular, for the motion (9.174),

$$\delta\int_{t_0}^{t_1} \mathcal{A}dt = \rho\mathring{\omega}\int_V e_{ijk}(\delta r^j + \delta(a\tau^j_\mu)X^\mu)(r^i_{,t} + (a\tau^i_\nu)_{,t}X^\nu)(r^k_{,t} + (a\tau^k_\lambda)_{,\eta}X^\lambda)d^2Xd\eta$$

Using here (9.180) and (9.182) along with the formula,

$$\int_{X^\mu X_\mu \le 1} X^\mu X^\nu d^2X = \frac{\pi}{4}\delta^{\mu\nu},$$

we have

$$\delta\int_{t_0}^{t_1} \mathcal{A}dt = \rho\gamma\int_{t_0}^{t_1}\left[\oint_\Gamma\left(e_{ijk}\frac{\partial r^i}{\partial t}\tau^k + \frac{\Omega a^2}{4}\frac{d\tau_j}{ds}\right)\delta r^j ds + \frac{1}{4}\frac{d(a^2L)}{dt}\delta\varphi - \frac{1}{2}\frac{d\varphi}{dt}aL\delta a\right]dt. \tag{9.191}$$

Functional $\mathcal{A}$ can be restored from (9.191) up to divergence terms:

$$\mathcal{A} = \frac{\rho\gamma}{3}\int_V e_{ijk}r^j(t,\eta)\frac{\partial r^i(t,\eta)}{\partial t}\frac{\partial r^k(t,\eta)}{\partial\eta}d\eta - \frac{\rho\gamma a^2}{4}L\frac{d\varphi}{dt}. \tag{9.192}$$

Dynamical equations of vortex filament. The action functional considered on the vortex motions (9.174) becomes a functional of positions of Γ, $r^i(t,\eta)$, and functions $a(t)$ and $\varphi(t)$: $I = I(r^i(t,\eta), a(t), \varphi(t))$. Assume that these functions are given at initial and final instants. Then the following variational principle holds:

Variational principle. *The true motion of the vortex filament is a stationary point of the action functional,*

$$\int_{t_0}^{t_1} [\mathcal{A} - \mathcal{K}]dt.$$

To derive the corresponding dynamical equations we find the variation of kinetic energy (9.185),

$$\delta\int_{t_0}^{t_1} \mathcal{K}dt = \int_{t_0}^{t_1}\left(\oint_\Gamma \frac{\delta\mathcal{K}}{\delta r^j}\delta r^j ds - \frac{\rho\gamma^2 L}{4\pi a}\delta a\right)dt. \tag{9.193}$$

Here we introduced the notations,

$$\frac{\delta \mathcal{K}}{\delta r^j} = \rho \gamma e_{ijk} v^i \tau^k - \frac{\rho \gamma^2}{8\pi} \frac{d\tau_j}{ds} \left(2 \ln \frac{2\varepsilon}{a} + \frac{1}{2} \right),$$

$$v^i \equiv \frac{\gamma}{4\pi} e^{ijk} \oint_{\Gamma} \frac{\tau_j(t, s') \triangle r_k}{(|\triangle r| + \varepsilon)^2 |\triangle r|} ds'. \tag{9.194}$$

The derivative, $d\tau_j/ds$, can be also written as

$$\frac{d\tau_j}{ds} = -\frac{e_{ijk} b^i \tau^k}{R} \tag{9.195}$$

where b^i is binormal and R curvature. Therefore,

$$\frac{\delta \mathcal{K}}{\delta r^j} = \rho \gamma e_{ijk} \left[v^i + \frac{\gamma}{8\pi} \frac{b^i}{R} \left(2 \ln \frac{2\varepsilon}{a} + \frac{1}{2} \right) \right] \tau^k.$$

Equating (9.191) and (9.193), we obtain the governing equations of the vortex filament dynamics:

$$e_{ijk} \left[\frac{\partial r^i(t, \eta)}{\partial t} - v^i - \frac{\gamma}{8\pi} \frac{b^i}{R} \left(2 \ln \frac{2\varepsilon}{a} + \frac{1}{2} \right) - \frac{a^2}{4} \frac{d\varphi}{dt} \frac{b^i}{R} \right] \tau^k = 0. \tag{9.196}$$

$$\frac{d\varphi}{dt} = \frac{\gamma}{2\pi a^2}, \qquad \frac{d}{dt} a^2 L = 0. \tag{9.197}$$

The first equation (9.197) determines the angular velocity of cross-sections, the second one means conservation of the filament volume,

$$\pi a^2 L = \mathring{V}, \tag{9.198}$$

$\mathring{V}$ being the initial volume of the filament. If a and $d\varphi/dt$ are eliminated from (9.196) by means of (9.198), then (9.196) along with the expression of L in terms of $r^i(t, \eta)$ (9.186) become a closed system of equations for the filament positions. One can check that this system is asymptotically equivalent to the one derived by the asymptotic analysis of Euler equations.

An interesting question arises: Is there a variational principle on the set of filament positions only? The answer is positive due to a very simple structure of the equations for a and φ. Indeed, the last term in the functional $\mathcal{A}$ (9.192) can be interpreted as the one generated by Lagrange multiplier, $\gamma\varphi/4\pi$, for the constraint

(9.198) after integration by parts. Therefore, the following variational principle holds:

Variational principle. *The true motion of the vortex filament is a stationary point of the action functional,*

$$\int_{t_0}^{t_1}\left(\frac{\rho\gamma}{3}\int_V e_{ijk}r^j(t,\eta)\frac{\partial r^i(t,\eta)}{\partial t}\frac{\partial r^k(t,\eta)}{\partial\eta}d\eta-\mathcal{K}(r^i(t,\eta))\right)dt,$$

where $\mathcal{K}(r^i(t,\eta))$ is the functional,

$$\mathcal{K}(r^i(t,\eta))=\frac{\rho\gamma^2}{8\pi}\oint_\Gamma\oint_\Gamma\frac{\tau^i(s)\tau_i(s')}{|\Delta r|+\varepsilon}dsds'+\frac{\rho\gamma^2}{8\pi}\left(\ln\frac{4\pi\varepsilon^2L}{\mathring{V}}+\frac{1}{2}\right)L, \tag{9.199}$$

and L is the functional (9.186).

The functional (9.199) is obtained from the functional (9.185) by eliminating a by means of the incompressibility condition (9.198).

Self-induction approximation. In kinetic energy, the logarithmic term may be dominant. Then, neglecting in (9.199) the terms of the order unity, one obtains for kinetic energy the expression,

$$\mathcal{K}=\frac{\rho\gamma^2}{8\pi}L\ln\frac{4\pi\varepsilon^2L}{\mathring{V}}. \tag{9.200}$$

Variation of such kinetic energy is

$$\int_{t_0}^{t_1}\mathcal{K}dt=-\int_{t_0}^{t_1}\frac{\rho\gamma^2}{8\pi}\left(\ln\frac{4\pi\varepsilon^2L}{\mathring{V}}+1\right)\frac{d\tau_j}{ds}\delta r^jds,$$

We keep here unity near logarithm for consistency of the model "forgetting" that the terms of the same order where neglected to obtain the simple formula (9.200). Using (9.195), we have

$$\int_{t_0}^{t_1}\mathcal{K}dt=\int_{t_0}^{t_1}\frac{\rho\gamma^2}{8\pi}\left(\ln\frac{4\pi\varepsilon^2L}{\mathring{V}}+1\right)\frac{e_{ijk}b^i\tau^k}{R}\delta r^jds.$$

Hence, the dynamical equations are:

$$e_{ijk}\left[\frac{\partial r^i(t,\eta)}{\partial t}-\frac{\gamma}{8\pi}\frac{b^i}{R}\left(\ln\frac{4\pi\varepsilon^2L}{\mathring{V}}+1\right)\right]\tau^k=0.$$

Another form of these equations is

$$\frac{\partial r^i(t,\eta)}{\partial t} = \frac{\gamma}{8\pi}\frac{b^i}{R}\left(\ln\frac{4\pi\varepsilon^2 L}{\overset{\circ}{V}} + 1\right) + \lambda\tau^i, \tag{9.201}$$

where λ is an arbitrary function.

The length of the filament is conserved in the self-induction approximation:

$$\frac{dL}{dt} = \oint \tau_i \frac{dr^i_{,t}}{ds} ds = -\oint r^i_{,t} \frac{d\tau_i}{ds} ds = 0. \tag{9.202}$$

The integrand in (9.202) is zero because $d\tau_i/ds$ is proportional to the normal of the curve, Γ, while, according to (9.201), $r^i_{,t}$ is orthogonal to the normal.

Due to the conservation of the filament length, L, the cross-sectional radius, a, does not change as well. Therefore, kinetic energy (9.200) may be written in an asymptotically equivalent form,

$$\mathcal{K} = \frac{\rho\gamma^2}{4\pi} L \ln\frac{L}{a},$$

where L is a functional (9.186), and a is a constant. The corresponding dynamic equations are:

$$\frac{\partial r^i(t,\eta)}{\partial t} = \frac{\gamma}{4\pi}\frac{b^i}{R}\left(\ln\frac{L}{a} + 1\right) + \lambda\tau^i.$$

They are called the equations of the self-induction approximation.

9.9 Vortex Sheets

Another case where the presence of a geometrical small parameter yields considerable simplifications is the dynamics of vortex sheets. A vortex sheet is a thin region of non-zero vorticity, a vicinity of some surface, Ω, with the thickness of the region, h, being much smaller than the characteristic radius of the surface, $\mathcal{R}$; besides, vorticity, up to small corrections of order $h/\mathcal{R}$, is tangent to Ω. Motion of the vortex sheet is modeled by the motion of the surface, Ω. The derivation of the governing variational principle in this case is simpler than for vortex filaments because the singularities are weaker. We begin with a more precise setting of the problem.

Denote the Lagrangian coordinates of the fluid particles in the vortex region by ξ^α, ξ, Greek indices run values 1, 2, $\xi^\alpha \in \overset{\circ}{\Omega}$, $-h/2 \le \xi \le h/2$. Motion of the surface, Ω, is described by the functions

$$x^i = r^i(t, \xi^\alpha).$$

We present the positions of the particles in the vortex region in the form

$$x^i(t, \xi^\alpha, \xi) = r^i(t, \xi^\alpha) + x'^i(t, \xi^\alpha, \xi), \tag{9.203}$$

without loss of generality, we can set the condition

$$\int_{-h/2}^{h/2} x'^i(t, \xi^\alpha, \xi) d\xi = 0. \tag{9.204}$$

Assume that vorticity has only two non-zero components, $\mathring{\omega}^\alpha$, and these components do not depend on ξ : $\mathring{\omega}^\alpha = \mathring{\omega}^\alpha(\xi^\beta)$. Since $\mathring{\omega}^3 = e^{3\alpha\beta}\partial_\alpha \mathring{v}_\beta$ is zero, $\mathring{v}_\beta$ is a potential vector: $\mathring{v}_\beta = \partial_\beta \chi$, which can be vanished by the redefining function φ in (9.14). Hence, only the third component of vector $\mathring{v}_a$ is not zero; denote it by $\Gamma(\xi^\alpha)/h$. The non-zero Lagrangian components of vorticity are

$$\mathring{\omega}^\alpha = e^{\alpha\beta}\partial_\beta \Gamma / h. \tag{9.205}$$

Accordingly, for the Eulerian components of this vector we have

$$\omega^i = r^i_\alpha e^{\alpha\beta}\partial_\beta \Gamma / h, \qquad r^i_\alpha \equiv \frac{\partial r^i(t, \xi^\alpha)}{\partial \xi^\alpha}. \tag{9.206}$$

We are going to show that, under some additional assumptions formulated further, dynamics of a vortex sheet in unbounded space is governed by

Migdal variational principle. *The true motion of a vortex sheet is a stationary point of the functional*

$$I(r(t, \xi)) = \int_{t_0}^{t_1} (\mathcal{A} - \mathcal{K}) dt, \tag{9.207}$$

$$\mathcal{A} = \rho \int_{\mathring{\Omega}} e_{ijk} r^i_1 r^j_2 r^k_t \Gamma d^2\xi, \tag{9.208}$$

$$\mathcal{K} = \frac{\rho}{8\pi} \int_{\mathring{\Omega}} \int_{\mathring{\Omega}} \frac{r^i_\alpha e^{\alpha\beta}\partial_\beta \Gamma r'^i_{\alpha'} e^{\alpha'\beta'}\partial_{\beta'}\Gamma' d^2\xi d^2\xi'}{|r - r'|}. \tag{9.209}$$

Here the prime marks the auxiliary variables of integration, ξ'^α, and $r' \equiv r(t, \xi'^\alpha)$, $\Gamma' \equiv \Gamma(\xi'^\alpha)$, $\partial_{\beta'} \equiv \partial / \partial \xi'^\beta$.

To derive this variational principle from the variational principle of dynamics of vortex lines we use the identity (9.140). According to this identity, dropping the full time derivative, which does not affect the equations, we can write for $\mathcal{A}$,

$$\mathcal{A} = \int_{\mathring{V}} \frac{1}{2} \rho e_{ijk} x_a^i x_b^j e^{abc} \mathring{v}_c x_t^k d\xi^1 d\xi^2 d\xi + \text{integrals over } \partial \mathring{V}.$$

For vortex sheets, the integrals over the vortex sheet faces vanish, because the vector, $\mathring{v}_c$, has only one non-zero component, $\mathring{v}_3$, and on the faces the factors, $e^{abc} \mathring{v}_c \mathring{n}_b$, is zero. If the vortex sheet is closed, then there are no other boundary contributions. If the sheet is not closed, then, in general, there is an additional integral over $\partial \mathring{\Omega}$. For simplicity, we make this integral equal to zero by an additional assumption that $\mathring{v}_3 = 0$ at the edge of the vortex sheet. Since only $\mathring{v}_3$ is not equal to zero,

$$\mathcal{A} = \rho \int_{\mathring{V}} e_{ijk} x_1^i x_2^j x_t^k \mathring{v}_3 d\xi^1 d\xi^2 d\xi$$

$$= \rho \int_{\mathring{V}} e_{ijk} (r_1^i + x_1'^i)(r_2^j + x_2'^j)(r_t^k + x_t'^k) \mathring{v}_3 d\xi^1 d\xi^2 d\xi. \tag{9.210}$$

Due to (9.204), the functional (9.210) differs from the functional (9.208) by quadratic and cubic terms with respect to derivatives of $x'^i(t, \xi^\alpha, \xi)$. An implicit condition of applicability of the Migdal variational principle is smallness of these terms in comparison with (9.208). Formula for kinetic energy (9.209) follows from (9.184) in the limit $h \to 0$.

9.10 Symmetry of the Action Functional and the Integrals of Motion

In this section the groups of symmetries of the action functional of ideal incompressible fluid are found. They give rise to the integrals of fluid motion. We begin our consideration by showing that the conservation of the velocity circulations stems from the invariance of kinetic energy with respect to the relabeling group of transformations.

Relabeling group. Consider the Hamilton variational principle: the true motion of an ideal incompressible fluid is a stationary point of the action functional,

$$I(x, t, X) = \int_{t_0}^{t_1} \int_V \frac{1}{2} \rho \frac{\partial x^i(t, X)}{\partial t} \frac{\partial x_i(t, X)}{\partial t} d^3 X, \tag{9.211}$$

on the set of all functions $x(t, X)$ such that their initial and final values are prescribed,

$$x(t_0, X) = x_0(X), \qquad x(t_1, X) = x_1(X), \tag{9.212}$$

the fluid does not detach from or penetrate through the wall,

$$x(t, X) \in V \quad \text{if } X \in V, \tag{9.213}$$

and the motion is incompressible,

$$\det \left\| \frac{\partial x}{\partial X} \right\| = 1. \tag{9.214}$$

Let us rename the particles: $X \to Y(X)$, and, for a given motion,

$$x = x(t, X)$$

consider another motion,

$$x = x'(t, X) \equiv x(t, Y(X)). \tag{9.215}$$

Condition that (9.213) is obviously satisfied. To satisfy (9.214) we set

$$\det \left\| \frac{\partial Y}{\partial X} \right\| = 1. \tag{9.216}$$

The new motion (9.215) does not obey (9.212), but this is not necessary for our purposes.

The action functional has the same values for both motions, $x = x(t, X)$ and $x = x'(t, X)$. Indeed,

$$\begin{aligned} I(x'(t, X)) &= \int_{t_0}^{t_1} \int_V \frac{1}{2} \rho \frac{\partial x'^i(t, X)}{\partial t} \frac{\partial x'_i(t, X)}{\partial t} d^3 X = \\ &= \int_{t_0}^{t_1} \int_V \frac{1}{2} \rho \frac{\partial x^i(t, Y(X))}{\partial t} \frac{\partial x_i(t, Y(X))}{\partial t} d^3 X = \\ &= \int_{t_0}^{t_1} \int_V \frac{1}{2} \rho \frac{\partial x^i(t, Y)}{\partial t} \frac{\partial x_i(t, Y)}{\partial t} \det \left\| \frac{\partial X}{\partial Y} \right\| d^3 Y. \end{aligned} \tag{9.217}$$

Taking into account (9.216) and changing the notation for the integration variables from Y to X we see that the integral (9.217) coincides with (9.211). Therefore,

$$\delta I = I(x'(t, X)) - I(x(t, X)) \equiv 0. \tag{9.218}$$

Let the relabeling be infinitesimal, i.e. $Y = X + \delta X$. Then

$$\delta x^i = x'^i(t, X) - x^i(t, X) = \frac{\partial x^i}{\partial X^a} \delta X^a. \tag{9.219}$$

The variation of functional (9.211) is

$$\delta I = \int_{t_0}^{t_1} \int_V \rho v_i \frac{d\delta x^i}{dt} d^3X =$$
$$= \left[\int_V \rho v_i \delta x^i d^3X \right]_{t_0}^{t_1} - \int_{t_0}^{t_1} \int_V \left[\frac{d}{dt} \rho v_i(t,X) \right] \delta x^i \, d^3X. \tag{9.220}$$

Assume that the motion $x(t, X)$ obeys the momentum equations. Then the last integral in (9.220) is zero because

$$\int_{t_0}^{t_1} \int_V \left[\frac{d}{dt} \rho v_i(t,X) \right] \delta x^i \, d^3X = \int_{t_0}^{t_1} \int_V \left(-\frac{\partial p}{\partial x^i} \delta x^i \right) dV =$$
$$= - \int_{t_0}^{t_1} \int_{\partial V} p \delta x^i n_i dA = 0. \tag{9.221}$$

Here we used (9.6) and, integrating by parts, took into account that

$$\frac{\partial \delta x^i}{\partial x^i} = 0 \text{ in } V, \qquad \text{and } \delta x^i n_i = 0 \quad \text{on } \partial V,$$

due to (9.214), (9.213), (9.215) and (9.216). Hence,

$$\delta I = \left[\int_V \rho v_i \delta x^i d^3X \right]_{t_0}^{t_1} = 0,$$

and this equation holds for any t_0, t_1. Thus, for any t_0, t_1,

$$\left. \int_V \rho v_i \delta x^i d^3X \right|_{t=t_0} = \left. \int_V \rho v_i \delta x^i d^3X \right|_{t=t_1},$$

or

$$\left. \int_V \rho v_i \frac{\partial x^i}{\partial X^a} \delta X^a d^3X \right|_{t=t_0} = \left. \int_V \rho v_i \frac{\partial x^i}{\partial X^a} \delta X^a d^3X \right|_{t=t_1}. \tag{9.222}$$

The functions δX^a are not arbitrary. Due to (9.216) they obey the equation

$$\frac{\partial \delta X^a}{\partial X^a} = 0.$$

This equation is satisfied if δX^a is a vector field concentrated on any line γ with a constant projection on this line (that is similar to conservation of Lagrangian components of vorticity considered in Sect. 9.6). Denoting a parameter along the line by η one obtains from (9.222)

$$\int\limits_{\gamma} \rho v_i \frac{\partial x^i}{\partial \eta} d\eta \Bigg|_{t=t_0} = \int\limits_{\gamma} \rho v_i \frac{\partial x^i}{\partial \eta} d\eta \Bigg|_{t=t_1} = \int\limits_{\gamma} \rho v_i dx^i,$$

i.e. the conservation of the velocity circulation along any closed fluid contour.

Isovorticity group in 2D. The variational principles for the functionals (9.165) and (9.117) differ from the least action principle by the elimination of many symmetries and, consequently, many integrals of motion (the velocity circulations). Nevertheless, certain symmetries (and integrals of motion) still remain. For example, Euler equations of the functional (9.165) yield incompressibility of motion (at each material point X, $\det \|\partial x/\partial X\| = const$). The question arises: what are the underlying symmetry groups for these integrals? Here we show that, for the two-dimensional case, this is the group relabeling the particles with the same vorticity. We call it the isovorticity group. More precisely, consider the action functional for two-dimensional flows:

$$I(r) = \int\limits_{t_0}^{t_1} dt \left[\rho \int\limits_{V} y(t, X) \frac{\partial x(t, X)}{\partial t} \mathring{\omega}(X) d^2X - \mathcal{K} \right]$$

$$\mathcal{K} = \frac{1}{2}\rho \int\limits_{V}\int\limits_{V} G\left(r(t, X), r\left(t, X'\right)\right) \mathring{\omega}(X) \mathring{\omega}\left(X'\right) d^2X d^2X' \qquad (9.223)$$

Let us show that the action functional has the same value for two motions $x = r(t, X)$ and $x = r'(t, X)$ if

$$r'(t, X) = r(t, Y(X)),$$

and the relabeling $X \to Y(X)$ conserves the vorticity:[13]

[13] Equation (9.224) can also be written in the form

$$\mathring{\omega}(X) d^2X = \mathring{\omega}(Y) d^2Y$$

emphasizing the vorticity conservation as a measure.

$$\mathring{\omega}(X) = \mathring{\omega}(Y(X)) \left| \frac{\partial Y}{\partial X} \right|, \quad \left| \frac{\partial Y}{\partial X} \right| \equiv \det \left\| \frac{\partial Y}{\partial X} \right\|. \tag{9.224}$$

Incompressibility of motion follows from this symmetry group. Indeed, the kinetic energy is an invariant under this transformation:

$$\mathcal{K}(r'(t,X)) = \frac{1}{2}\rho \int_V \int_V G(r'(t,\bar{X}), r'(t,X))\mathring{\omega}(X)\mathring{\omega}(\bar{X})d^2Xd^2\bar{X} =$$

$$= \frac{1}{2}\rho \int_V \int_V G(r(t,Y(X)), r(t,Y(\bar{X}))\mathring{\omega}(X)\mathring{\omega}(\bar{X})d^2Xd^2\bar{X} =$$

$$= \frac{1}{2}\rho \int_V \int_V G(r(t,Y(\bar{X})), r(t,Y(\bar{X})))\mathring{\omega}(Y(X))\mathring{\omega}(Y(\bar{X})) \left| \frac{\partial Y}{\partial X} \right| \left| \frac{\partial Y}{\partial X} \right| d^2Xd^2\bar{X} =$$

$$= \frac{1}{2}\rho \int_V \int_V G(r(t,Y), r(t,\bar{Y}))\mathring{\omega}(Y)\mathring{\omega}(\bar{Y})d^2Yd^2\bar{Y} = \mathcal{K}(r(t,X))$$

The first integral in (9.223) is also invariant:

$$\int_V y'(t,X)\frac{\partial x'(t,X)}{\partial t}\mathring{\omega}(X)d^2X = \int_V y(t,Y(X))\frac{\partial x(t,Y(X))}{\partial t}\mathring{\omega}(X)d^2X =$$

$$\int_V y(t,Y(X))\frac{\partial x(t,Y(X))}{\partial t}\mathring{\omega}(Y(X)) \left| \frac{\partial Y}{\partial X} \right| d^2X = \int_V y(t,Y)\frac{\partial x(t,Y)}{\partial t}\mathring{\omega}(Y)d^2Y.$$

In the same way as for the least action principle we obtain that

$$\int_V y(t,X)\delta x(t,X)\mathring{\omega}(X)d^2X = \int_V y(t,X)\frac{\partial x(t,X)}{\partial X^\mu}\delta X^\mu \mathring{\omega}(X)d^2X = const. \tag{9.225}$$

The functions δX^μ in (9.225) obey the constraint which follows from (9.224):

$$\mathring{\omega}(X) = \mathring{\omega}(X+\delta X) \left| \frac{\partial(X+\delta X)}{\partial X} \right| = \left(\mathring{\omega}(X) + \frac{\partial \mathring{\omega}}{\partial X^\mu}\delta X^\mu \right) \left(1 + \frac{\partial \delta X^\mu}{\partial X^\mu} \right).$$

Keeping only the leading terms we obtain

$$\frac{\partial \mathring{\omega}\delta X^\mu}{\partial X^\mu} = 0.$$

Thus,

$$\mathring{\omega}\delta X^{\mu}=e^{\mu\nu}\frac{\partial\chi(X)}{\partial X^{\nu}},\tag{9.226}$$

where χ is an arbitrary function. Therefore, the function χ is constant at the boundary. The vector, δX^{μ}, is tangent to the boundary. Assuming that V is a simply connected region, we set $\chi=0$ at the boundary. Plugging (9.226) into (9.225) and integrating by parts we obtain

$$\int\limits_{V}e^{\mu\nu}\frac{\partial y(t,X)}{\partial X^{\mu}}\frac{\partial x(t,X)}{\partial X^{\nu}}\chi(X)d^{2}X=const.$$

The factor at χ is the Jacobian of transformation from Lagrangian to Eulerian coordinates. Since $\chi(X)$ is an arbitrary function, the Jacobian is constant at each particle, as was claimed.

Isovorticity group in 3D. The symmetry group of the functional (9.117) is a relabeling group (i.e. the group of transformation (9.215)) which conserves vorticity in the following sense:

$$\mathring{\omega}^{a}(X)\sqrt{\mathring{g}(X)}\frac{\partial Y^{b}(X)}{\partial X^{a}}=\mathring{\omega}^{b}(Y(X))\sqrt{\mathring{g}(Y(X))}\left|\frac{\partial Y}{\partial X}\right|.\tag{9.227}$$

This can be checked by inspection. For an infinitesimal transformation, the admissible variations $\delta X^{a}=Y^{a}-X^{a}$ obey the equation[14]

$$\frac{\partial}{\partial X^{a}}\left[\sqrt{\mathring{g}}\left(\mathring{\omega}^{a}\delta X^{b}-\mathring{\omega}^{b}\delta X^{a}\right)\right]=0.\tag{9.228}$$

The general solution of this equation is

$$\mathring{\omega}^{a}\delta X^{b}-\mathring{\omega}^{b}\delta X^{a}=\frac{1}{\sqrt{\mathring{g}}}e^{abc}\frac{\partial\chi(X)}{\partial X^{c}}.\tag{9.229}$$

---

[14] Indeed, from (9.227)

$$\mathring{\omega}^{a}(X)\sqrt{\mathring{g}(X)}\frac{\partial\left(X^{b}+\delta X^{b}\right)}{\partial X^{a}}=\mathring{\omega}^{b}\left(X^{a}+\delta X^{a}\right)\sqrt{\mathring{g}\left(X^{a}+\delta X^{a}\right)}\left(1+\frac{\partial\delta X^{a}}{\partial X^{a}}\right).$$

Keeping the terms of the first order, we obtain in the left hand side,

$$\mathring{\omega}^{a}(X)\sqrt{\mathring{g}(X)}\frac{\partial\delta X^{b}}{\partial X^{a}}=\frac{\partial}{\partial X^{a}}\left(\mathring{\omega}^{a}(X)\sqrt{\mathring{g}(X)}\delta X^{b}\right),$$

and in the right hand side

$$\frac{\partial\sqrt{\mathring{g}}\mathring{\omega}^{b}}{\partial X^{a}}\delta X^{a}+\mathring{\omega}^{b}\sqrt{\mathring{g}}\frac{\partial\delta X^{a}}{\partial X^{a}}=\frac{\partial}{\partial X^{a}}\left(\mathring{\omega}^{b}(X)\sqrt{\mathring{g}(X)}\delta X^{a}\right).$$

That yields (9.228).

Equation (9.229) can be solved with respect to $\partial\chi/\partial X^c$:

$$\frac{\partial\chi}{\partial X^c} = \sqrt{\mathring{g}}\, e_{abc}\mathring{\omega}^a \delta X^b. \tag{9.230}$$

If δX^a is proportional to $\mathring{\omega}^a$, i.e. one relabels the particles on the same vortex lines, then $\partial\chi/\partial X^c \equiv 0$. We consider the symmetries with respect to the relabeling of the neighboring vortex lines, i.e. $\delta X^a \neq \lambda\mathring{\omega}^a$ at all points.

Both vectors $\mathring{\omega}^a$ and δX^a are tangent to the boundary. Projecting (9.230) on the tangent directions to the boundary we find that χ is constant at the boundary, and, without loss of generality, can be set equal to zero. It also follows from (9.230) that the vectors $\partial\chi/\partial X^c$ and $\mathring{\omega}^c$ are orthogonal:

$$\mathring{\omega}^c \frac{\partial\chi}{\partial X^c} = 0. \tag{9.231}$$

Therefore, χ is constant along the vortex lines. In the same way as in the previous two cases, from the invariance of the action functional we obtain

$$\int e_{ijk} x^i (t, X)\, \delta x^j (t, X) \frac{\partial x^k (t, X)}{\partial X^a} \mathring{\omega}^a (X) \sqrt{\mathring{g}} d^3 X = const.$$

Here

$$\delta x^j = \frac{\partial x^j}{\partial X^b}\delta X^b.$$

Hence,

$$\begin{aligned}\int e_{ijk} x^i \frac{\partial x^j}{\partial X^a}\frac{\partial x^k}{\partial X^b}\delta X^b \mathring{\omega}^a \sqrt{\mathring{g}} d^3 X &= \int e_{ijk} x^i \frac{\partial x^j}{\partial X^a}\frac{\partial x^k}{\partial X^b}\frac{1}{2}\left(\mathring{\omega}^a \delta X^b - \mathring{\omega}^b \delta X^a\right)\sqrt{\mathring{g}} d^3 X \\ &= \int e_{ijk} x^i \frac{\partial x^j}{\partial X^a}\frac{\partial x^k}{\partial X^b} e^{abc}\frac{\partial\chi}{\partial X^c} d^3 X.\end{aligned}$$

Integrating by parts, we obtain the following expression for this integral:

$$-\int e_{ijk}\frac{\partial x^i}{\partial X^c}\frac{\partial x^j}{\partial X^a}\frac{\partial x^k}{\partial X^b} e^{abc}\chi (X)\, d^3 X = -3! \int \left|\frac{\partial x}{\partial X}\right| \chi (X)\, d^3 X.$$

In the vortex line coordinate system χ is constant along the vortex lines. The function χ is arbitrary as a function of the vortex line. Thus, for each vortex line, the integral

$$\int \left|\frac{\partial x}{\partial X}\right| d\eta \tag{9.232}$$

remains unchanged in the course of the motion.

The existence of this invariant of motion can be derived directly from (9.130). Indeed, let us take this equation in the form (9.114). Computing the divergence of (9.114) and using the fact that velocity and vorticity fields are divergence-free we have

$$\frac{\partial \dot{r}^i}{\partial x^i} = \omega^i \frac{\partial \lambda}{\partial x^i}. \tag{9.233}$$

The left hand side of (9.233) is time derivative of $\ln \Delta$ (see (3.71)). Thus, in Lagrangian coordinates,

$$\frac{d\Delta}{dt} = \Delta \mathring{\omega}^3 \frac{\partial \lambda}{\partial \eta}. \tag{9.234}$$

The product $\Delta\omega^3$ is a function of only X. Integrating (9.234) over a closed vortex line we obtain the integrals of motion: for each vortex line,

$$\frac{d}{dt} \int \left| \frac{\partial x}{\partial X} \right| \partial \eta = 0. \tag{9.235}$$

The integrals (9.235) mean that the volume of any vortex tube comprised of vortex lines does not change in the course of motion, as it must be for incompressible fluid. Note that the integrals (9.235) do not constrain the motion of any finite number of vortex lines.

The invariance of the action functional with respect to relabeling of the particles on the same vortex line yields a "degeneracy" of (9.130): contraction of (9.130) with the vorticity vector is an identity.

9.11 Variational Principles for Open Flows

To obtain an extension of the least action principle to open flows, consider motion in four-dimensional space-time. A symbolic picture of motion in a closed container is shown in Fig. 9.4a. Each trajectory connects the given initial and final positions of a fluid particle. A typical open flow is shown in Fig. 9.4b. Each trajectory also

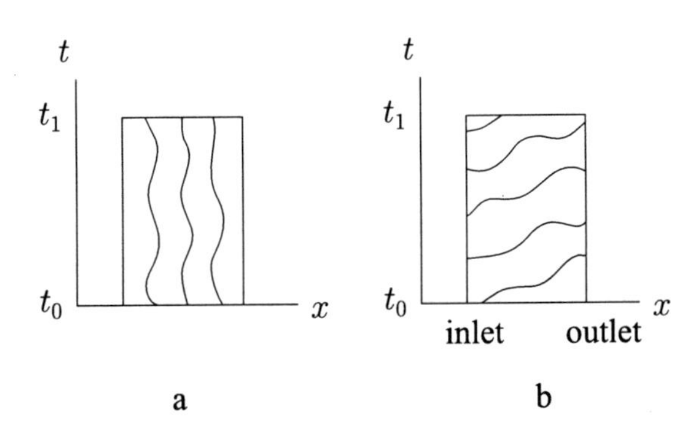

Fig. 9.4 Geometry of the particle trajectories in space-time for closed (**a**) and open (**b**) flows

connects the initial and final positions, but now some of the initial positions are at the inlet of the flow, while part of the final positions is at the outlet of the flow. In closed flows, the most natural choice of Lagrangian coordinates is an identification of Lagrangian coordinates with Eulerian ones at the initial instant, $X^1\left(x^i, t_0\right) = x^1, X^2\left(x^i, t_0\right) = x^2, X^3\left(x^i, t_0\right) = x^3$. In open flows, it is natural to identify one of the Lagrangian coordinates with the moment of the appearance of the particle in the container, while the two others can be the coordinates of the point at the inlet, where the particle appears for the first time.

Consider now the initial and the final positions of the fluid particles at the inlet and outlet as given. The stationary points of the action functional are sought on the set of functions $X^a\left(x^i, t\right)$, which satisfy constraints (9.33), (9.34) and (9.35) along with the following conditions at the inlet and the outlet:

$$X^a\left(x^i, t\right) = X^a_{in}\left(x^i, t\right) \quad \text{at the inlet}$$
$$X^a\left(x^i, t\right) = X^a_{out}\left(x^i, t\right) \quad \text{at the outlet}$$

(X^a_{in} and X^a_{out} are some prescribed functions). Then, the second term in (9.38) vanishes even if there is a flow through ∂V. Thus, the action functional has a stationary point at the real motion of an ideal fluid.

Chapter 10
Ideal Compressible Fluid

Ideal compressible fluid can be considered as an elastic body the internal energy of which depends only on the mass density of the body. Therefore, the variational principles formulated for elastic bodies are valid for compressible fluids as well. However, they deserve special consideration because, due to a simplified energy structure, they are enriched by new interesting features.

10.1 Variational Principles in Lagrangian Coordinates

Consider a vessel filled with ideal compressible fluid. The region occupied by the vessel is denoted by $\tilde{V}$. If the wall of the vessel deforms with time, then we write $\tilde{V}(t)$. The motion of the vessel is given. Moving fluid may detach from the wall of the vessel, i.e. the region, $V(t)$, occupied by the fluid, in general, does not coincide with the region $\tilde{V}(t)$. We consider such a case later, and begin with the discussion of non-detaching flows, when $V(t) = \tilde{V}(t)$. The initial and final particle positions are given.

The functions $x(X, t)$ are assumed to be twice continuously differentiable (thus, the shock waves are excluded from consideration). The entropy of the particles is a known function of the Lagrangian coordinates, X. The internal energy, U, is a given function of mass density, entropy and Lagrangian coordinates: $U = U(\rho, S, X)$. The explicit dependence of U on Lagrangian coordinates can be caused by heterogeneity of the fluid. For homogeneous fluid, $U = U(\rho, S)$. If the flow is isentropic, i.e. entropy does not change over particles, internal energy may be considered as a function of density only, $U = U(\rho)$. For the general case of inhomogeneous fluid we write $U = U(\rho, X)$, implying that the dependence of entropy on Lagrangian coordinate is incorporated in this function. The external mass forces are assumed to possess a potential, $\Phi(x)$.

Least action principle. *The true motion of ideal compressible fluid is a stationary point of the functional*

V.L. Berdichevsky, *Variational Principles of Continuum Mechanics*,
Interaction of Mechanics and Mathematics, DOI 10.1007/978-3-540-88467-5_10,

$$I(x(X,t)) = \int_{t_0}^{t_1}\int_V \rho\left(\frac{1}{2}\frac{\partial x^i(t,X)}{\partial t}\frac{\partial x_i(t,X)}{\partial t} - U(\rho,X) - \Phi(x(X,t))\right)dVdt, \tag{10.1}$$

on the set of functions, $x(X,t)$, which satisfy the conditions

$$x^i(X^a,t) = \mathring{x}^i(X^a), \quad x^i(X^a,t_1) = \overset{1}{x}{}^i(X^a),$$
$$x^i(X^a,t) \in \partial\tilde{V}(t) \text{ for } X^a \in \partial\mathring{V}. \tag{10.2}$$

This statement is equivalent to

Mopertuis-Lagrange variational principle. *The true motion of ideal compressible fluid is a stationary point of the kinetic energy functional,*

$$\int_{t_0}^{t_1}\int_V \frac{1}{2}\rho\frac{\partial x^i(t,X)}{\partial t}\frac{\partial x_i(t,X)}{\partial t}dVdt,$$

on the set of all trajectories satisfying the conditions (10.2) and the law of conservation of energy,

$$\int_V \rho\left(\frac{1}{2}\frac{\partial x^i(t,X)}{\partial t}\frac{\partial x_i(t,X)}{\partial t} + U(\rho,X) + \Phi(x(X,t))\right)dV = \mathcal{E} = const. \tag{10.3}$$

In the Mopertuis-Lagrange principle, the instant t_1 at which the fluid arrives at its finial state is not fixed and must be varied.

The equivalence of the two variational principles can be proven in the same way as in the mechanics of the systems with a finite number of degrees of freedom.

From the energy equation (10.3),

$$\int_V \frac{1}{2}\rho\frac{\partial x^i(t,X)}{\partial t}\frac{\partial x_i(t,X)}{\partial t}dV = \mathcal{E} - \int_V \rho\left[U(\rho,X) + \Phi(x(X,t))\right]dV,$$

one can find the "time differential",

$$dt = \frac{\sqrt{\frac{1}{2}\int_V \rho dx_i dx^i dV}}{\sqrt{\mathcal{E} - \int_V \rho\left[U(\rho,X) + \Phi(x(X,t))\right]dV}}.$$

Here $dx(X)$ is the differential along a path in the space of functions $x(X)$. Replacing integration over time in the action functional by the integration along a path in the space of functions $x(X)$, we put the Mopertuis-Lagrange principle in the form of the Jacobi variational principle:

Jacobi variational principle. *The true motion of ideal compressible fluid starting from the state,* $\overset{\circ}{x}(X)$, *and the finishing at the state,* $\overset{1}{x}(X)$, *is a stationary point of the functional,*

$$\int_{\overset{\circ}{x}(X)}^{\overset{1}{x}(X)} \sqrt{\mathcal{E} - \int_V \rho \left[U(\rho, X) + \Phi(x(X,t)) \right] dV} \sqrt{\frac{1}{2} \int_V \rho dx_i dx^i dV}, \tag{10.4}$$

on the set of all trajectories satisfying the law of energy conservation (10.3).

10.2 General Features of Dynamics of Compressible Fluid

Modifications of the least action principle considered further employ the remarkable structure of solutions of the equations of ideal compressible fluid, to a discussion of which we proceed. This structure is, in essence, similar to that of incompressible fluid reviewed in Sect. 9.2.

The closed system of equations of ideal compressible fluid consists of momentum equations,

$$\frac{dv_i}{dt} = -\frac{1}{\rho} \frac{\partial p}{\partial x^i} - \frac{\partial \Phi(x)}{\partial x^i}, \tag{10.5}$$

the constitutive equation for pressure,

$$p = \rho^2 \frac{\partial U(\rho, X)}{\partial \rho}, \tag{10.6}$$

and the continuity equation in Lagrangian form,

$$\rho \sqrt{\hat{g}} = \rho_0 \sqrt{g_0}, \qquad \sqrt{\hat{g}} = \Delta = \det \left\| x_a^i \right\|,$$

or in the differential form,

$$\frac{\partial \rho}{\partial t} + \frac{\partial \rho v^i}{\partial x^i} = 0. \tag{10.7}$$

We are going to show that the following statements hold:

1. If fluid is homogeneous and the flow is isentropic, then, at each fluid particle, the contravariant Lagrangian components of vorticity, ω^a, divided by mass density, do not change in time:

$$\frac{\omega^a}{\rho} = \frac{\overset{\circ}{\omega}{}^a}{\rho_0}. \tag{10.8}$$

In particular, this means that

a. Each vortex line consists of the same fluid particles
b. The vorticity cannot be generated or vanished in the course of motion
c. A fluid region, where the flow was potential initially, moves with a potential velocity field

2. If fluid is inhomogeneous, then vorticity may be generated:

$$\frac{d}{dt}\frac{\omega^a}{\rho}=\frac{1}{\rho_0\sqrt{\mathring{g}}}e^{abc}\partial_b\left(\left.\frac{\partial U(\rho,X)}{\partial X^c}\right|_{\rho=const}\right). \tag{10.9}$$

In particular, if the fluid is homogeneous, i.e. the dependence of internal energy on Lagrangian coordinates is caused only by the dependence of entropy on X: $U = U(\rho, S)$, $S = S(X)$, and entropy is not constant, then, as follows from (10.9), generation of vorticity is governed by the equation:

$$\frac{d}{dt}\frac{\omega^a}{\rho}=\frac{1}{\rho_0\sqrt{\mathring{g}}}e^{abc}\partial_b\left(T\frac{\partial S}{\partial X^c}\right)=\frac{1}{\rho_0\sqrt{\mathring{g}}}e^{abc}\frac{\partial T}{\partial X^b}\frac{\partial S}{\partial X^c}. \tag{10.10}$$

In an observer's frame this equation takes the form:

$$\frac{d}{dt}\frac{\omega^i}{\rho}-\frac{\omega^k}{\rho}v^i_{,k}=\frac{1}{\rho}e^{ijk}T_{,j}S_{,k}.$$

So, vorticity may develop only if the gradients of temperature and entropy are not collinear.

3. In isentropic flows of homogeneous fluids the covariant Lagrangian components of velocity can be written as

$$v_a=\frac{\partial\varphi(t,X)}{\partial X^a}+\mathring{v}_a(X).$$

where φ is some function of coordinates and time while $\mathring{v}_a$ depend only on the Lagrangian coordinates. Accordingly, in an observer's frame,

$$v_i=\frac{\partial\varphi}{\partial x^i}+X^a_i\mathring{v}_a(X) \tag{10.11}$$

4. In case of homogeneous fluids, any solution of (10.5), (10.6) and (10.7) can be expressed in terms of four scalar functions, φ, α, β and μ,

$$v_i=\frac{\partial\varphi}{\partial x^i}+\alpha\frac{\partial\beta}{\partial x^i}+\mu\frac{\partial S}{\partial x^i}, \tag{10.12}$$

which satisfy the equations

$$\frac{d\alpha}{dt} = 0, \quad \frac{d\beta}{dt} = 0, \tag{10.13}$$

$$\frac{d\mu}{dt} = T \equiv \frac{\partial U}{\partial S}. \tag{10.14}$$

Pressure can be found from a generalized Cauchy-Lagrange integral :

$$\frac{d\varphi}{dt} = -\left(\frac{p}{\rho} + U + \Phi - \frac{v^2}{2}\right) \tag{10.15}$$

The derivation of these statements proceeds as follows. Let us contract the momentum equations (10.5) with x_a^i, i.e. project these equations onto the axes of the Lagrangian coordinate system. On the right hand side, taking into account the constitutive equation (10.6), we get

$$-\frac{1}{\rho}\left(\frac{\partial p}{\partial x^i} + \rho\frac{\partial \Phi}{\partial x^i}\right)x_a^i = -\left(\frac{1}{\rho}\frac{\partial p}{\partial X^a} + \frac{\partial \Phi}{\partial X^a}\right) =$$

$$= -\left(\frac{\partial}{\partial X^a}\left(\frac{p}{\rho} + \Phi\right) + \frac{p}{\rho^2}\frac{\partial \rho}{\partial X^a}\right) = -\frac{\partial}{\partial X^a}\left(\frac{p}{\rho} + U + \Phi\right) + \partial_a U.$$

By $\partial_a U$, we denote the derivative,

$$\partial_a U = \left.\frac{\partial U(\rho, X^a)}{\partial X^a}\right|_{\rho = const}.$$

The left hand side of momentum equation is transformed using (9.12). We obtain

$$\frac{dv_a}{dt} = -\frac{\partial}{\partial X^a}\left(\frac{p}{\rho} + U + \varphi - \frac{v^2}{2}\right) + \partial_a U. \tag{10.16}$$

Let us introduce the function φ by (10.15). Denoting the difference $v_a - \partial\varphi/\partial X^a$ by χ_a,

$$v_a = \frac{\partial \varphi}{\partial X^a} + \chi_a, \tag{10.17}$$

we can write the momentum equations as

$$\frac{d\chi_a}{dt} = \partial_a U. \tag{10.18}$$

Any solution of the momentum equations can be written in the form (10.15), (10.16), (10.17) and (10.18), and, conversely, for any functions φ and χ_a satisfying the (10.15) and (10.18), the velocity v_a (10.17) satisfies the momentum equations.

According to (10.17), the curls of v_a and χ_a coincide:

$$e^{abc}\partial_b v_c = e^{abc}\partial_b \chi_c. \tag{10.19}$$

On the other hand,

$$e^{abc}\partial_b v_c = \sqrt{\hat{g}}\varepsilon^{abc}\partial_b v_c = \sqrt{\hat{g}}\omega^a = \rho_0\sqrt{\mathring{g}}\frac{\omega^a}{\rho}. \tag{10.20}$$

Equation (10.9) follows from (10.18), (10.19) and (10.20).

Equation (10.9) is obtained from (10.10) by contraction of (10.10) with x^i_a and the use of the first formula (10.7) and the relation,

$$e^{ijk} = \frac{1}{\sqrt{\hat{g}}} x^i_a x^j_b x^k_c e^{abc}.$$

For the homogeneous fluid,

$$\partial_a U = T\frac{\partial S}{\partial X^a},$$

and (10.18) takes the form

$$\frac{d\chi_a}{dt} = T\frac{\partial S}{\partial X^a}. \tag{10.21}$$

We define a function μ by (10.14). Since S does not depend on time, the solution of (10.21) is

$$\chi_a = \mathring{v}_a(X) + \mu\frac{\partial S}{\partial X^a},$$

where $\mathring{v}_a$ are some function of the Lagrangian coordinates. For isentropic flows this equation along with (10.17) yield (10.11).

According to the theorem on the canonical presentation of linear differential forms (see Sect. 9.2), for a linear differential form, $\mathring{v}_a dX^a$, in three-dimensional space, there exist functions $\tilde{\varphi}$, α and β, such that $\mathring{v}_a dX^a = d\tilde{\varphi} + \alpha d\beta$, or

$$\mathring{v}_a = \frac{\partial\tilde{\varphi}}{\partial X^a} + \alpha\frac{\partial\beta}{\partial X^a}.$$

Since $\mathring{v}_a$ are function of the Lagrangian coordinates, the functions $\tilde{\varphi}$, α, and β also depend only on the Lagrangian coordinates. Denote the sum $\varphi + \tilde{\varphi}$ by φ. Then, the equation for φ (10.15) does not change, while the Lagrangian components of velocity (10.17) become

$$v_a = \frac{\partial\varphi}{\partial X^a} + \alpha\frac{\partial\beta}{\partial X^a} + \mu\frac{\partial S}{\partial X^a}. \tag{10.22}$$

Projecting (10.22) onto the observer's frame (i.e. contracting it with X^a_i), we obtain (10.12). The dependence of α and β only on the Lagrangian coordinates is expressed by (10.13).

Note that, according to the above-mentioned theorem on the canonical presentation of the linear differential form, the velocity v_i can also be written as

$v_i = \bar{\varphi}_{,i} + \bar{\alpha}\bar{\beta}_{,i}$. In such presentation, however, $\bar{\varphi}$, $\bar{\alpha}$ and $\bar{\beta}$ are, generally speaking, functions of both time and Lagrangian coordinates.

For isentropic flows of compressible fluids, as for incompressible fluid, the velocity is presented in terms of Clebsch's potentials:

$$v_i = \frac{\partial\varphi}{\partial x^i} + \alpha\frac{\partial\beta}{\partial x^i}. \tag{10.23}$$

In terms of the Clebsch's potentials, the generalized Cauchy-Lagrange integral (10.15) can be written in several equivalent forms:

$$\begin{aligned} \frac{d\varphi}{dt} + \frac{p}{\rho} + U + \Phi - \frac{v^2}{2} &= \frac{d\varphi}{dt} + \alpha\frac{d\beta}{dt} + \frac{p}{\rho} + U + \Phi - \frac{v^2}{2} = \\ &= \frac{\partial\varphi}{\partial t} + \alpha\frac{\partial\beta}{\partial t} + v^k\left(\frac{\partial\varphi}{\partial x^k} + \alpha\frac{\partial\beta}{\partial x^k}\right) + \frac{p}{\rho} + U + \Phi - \frac{v^2}{2} = \\ &= \frac{\partial\varphi}{\partial t} + \alpha\frac{\partial\beta}{\partial t} + \frac{v^2}{2} + \frac{p}{\rho} + U + \Phi = 0. \end{aligned} \tag{10.24}$$

For potential flows (α or β is zero), the relations (10.24) become the Cauchy-Lagrange integral

$$\frac{\partial\varphi}{\partial t} + \frac{v^2}{2} + \frac{p}{\rho} + U + \Phi = 0.$$

If $U = 0$ for $\rho = 0$, then the sum $p/\rho + U$ can also be written as

$$\frac{p}{\rho} + U = \int_0^\rho \frac{dp}{\rho}.$$

Isentropic flows belong to a larger class of the so-called barotropic flows – the flows for which pressure is a function of density only. It is easy to see that for barotropic flows (10.23), (10.13) and (10.24) remain valid if the function U is interpreted as the function $\tilde{U}$ defined by the equation $p(\rho) = \rho^2 d\tilde{U}/d\rho$. In particular, $\tilde{U} = U$ for isentropic flows and $\tilde{U} = F$ for isothermal flows (i.e. the flows with $T \equiv const$). In further consideration of the barotropic flows we keep the formulas obtained for the isentropic flows without changes, supposing that U in this formulas means not the internal energy but the function $\tilde{U}$.

10.3 Variational Principles in Eulerian Coordinates

Variation of Lagrangian coordinates. There is a straightforward way to obtain a variational form of the governing equations of ideal compressible fluid: the action functional,

$$I = \int_{t_0}^{t_1} \int_{\tilde{V}(t)} L dV dt \tag{10.25}$$

$$L = \rho \left(\frac{v^2}{2} - U(\rho, X(x,t)) - \Phi(x) \right), \tag{10.26}$$

must be considered on the set of functions, $X(x,t)$. Then in (10.26) mass density is given by the formula

$$\rho = \rho_0 \left(X^a \left(x^i, t \right) \right) \frac{\det \left\| \frac{\partial X^a}{\partial x^i} \right\|}{\det \left\| \frac{\partial X^a}{\partial \mathring{x}^i} \right\|}, \tag{10.27}$$

while velocity is defined in terms of $X(x,t)$ by (9.30) (or (9.31)). The initial and final values of $X(t,x)$, $\mathring{X}(x)$ and $\overset{1}{X}(x)$ are given.

The functions $\mathring{X}(x)$ and $\overset{1}{X}(x)$ are defined in the regions $\mathring{V} = V(t_0)$ and $\overset{1}{V} = V(t_1)$, respectively. It is assumed that there exists a mapping $\mathring{V} \to \overset{1}{V}$, for which the functions $\mathring{X}^a\left(x^i\right)$ and $\overset{1}{X}^a\left(x^i\right)$ coincide.

The equation of the region $\mathring{V}$ at the initial moment can be written as

$$f(X) = 0.$$

The impermeability of the wall means that the equation

$$f(X(x,t)) = 0 \tag{10.28}$$

is the equation of the boundary of the given region, $\tilde{V}(t)$.

Least action principle. *The true motion of ideal compressible fluid is a stationary point of the functional (10.25) on the set of admissible functions $X(x,t)$.*

This variational principle is justified in the same way as in Sect. 9.3; one needs only to modify (9.41):

$$\begin{aligned}\partial_i L = -\rho \frac{\partial \Phi}{\partial x^i}, \quad X_i^a \frac{\partial L}{\partial X_k^a} - L\delta_i^k = -\rho v_i v^k + \rho \frac{\partial L}{\partial \rho} \delta_i^k - L\delta_i^k = \\ = -\rho v_i v^k - \rho^2 \frac{\partial U}{\partial \rho} \delta_i^k, \quad \frac{\partial L}{\partial X_t^a} X_i^a = -\rho v_i.\end{aligned} \tag{10.29}$$

Substituting (10.29) into (9.40) yields the momentum equations of ideal compressible fluid (10.5).

Variations of density and velocity. In variational principles in Lagrangian coordinates, the density is a dependent field. It is defined in terms of the particle trajectories from the continuity equation. In Eulerian coordinates, density is linked to velocity by the continuity equation

$$\frac{\partial \rho}{\partial t} + \frac{\partial \rho v^i}{\partial x^i} = 0. \tag{10.30}$$

Admissible velocity fields satisfy the condition that there exist functions $X^a\left(x^i, t\right)$ (Lagrangian coordinates), which are a solution of the system of equations

$$\frac{\partial X^a}{\partial t} + v^k \frac{\partial X^a}{\partial x^k} = 0 \tag{10.31}$$

with the initial conditions

$$X^a\left(x^k, t_0\right) = \overset{\circ}{X}{}^a\left(x^k\right), \tag{10.32}$$

such that they take on the assigned values at the time $t = t_1$,

$$X^a\left(x^k, t\right) = \overset{1}{X}{}^a\left(x^k\right), \tag{10.33}$$

and for all t and x^k,

$$\det\left\|\frac{\partial X^a}{\partial x^k}\right\| \neq 0. \tag{10.34}$$

The definition of density (10.27) and the constraints (10.32) and (10.33) show that the density fields at the initial and final instants should be considered as known:[1]

$$\rho\left(x^i, t_0\right) = \rho_0\left(x^i\right), \quad \rho\left(x^i, t_1\right) = \rho_1\left(x^i\right). \tag{10.35}$$

By our assumption, the flow does not detach from the wall. Thus,

$$v^i n_i = c_x \ \text{ on } \partial\tilde{V}(t), \tag{10.36}$$

where c_x is the normal velocity of the wall in the observer's frame.
Lin variational principle. *The stationary points of the functional*

$$\int_{t_0}^{t_1}\int_{\tilde{V}(t)} \rho\left(\frac{v^2}{2} - U(\rho, X) - \Phi(x)\right) dV\, dt \tag{10.37}$$

on the set of functions $\rho\left(x^i, t\right)$, $v^i\left(x^i, t\right)$, and $X^a\left(x^i, t\right)$, satisfying the conditions (10.30), (10.31), (10.32), (10.33), (10.34), (10.35) and (10.36), are the solutions of the momentum equations of ideal compressible fluid.

---

[1] In fact, one can consider density ρ_1 as determined by the mapping $\overset{\circ}{V} \to \overset{1}{V}$.

To justify this variational principle, we get rid of the constraints by means of the Lagrange multipliers. Denoting the Lagrange multipliers for the constraints (10.30) and (10.31) by φ and $\rho\chi_a$, respectively, we obtain the functional

$$I\left(\rho, v^i, \varphi, \chi_a\right) = \int_{t_0}^{t_1}\int_{\tilde{V}(t)} \left[\rho\left(\frac{v^2}{2} - U(\rho, X) - \Phi(x)\right) + \right.$$

$$\left. +\varphi\left(\frac{\partial\rho}{\partial t} + \frac{\partial\rho v^i}{\rho x^i}\right) + \rho\chi_a\left(\frac{\partial X^a}{\partial t} + v^k\frac{\partial X^a}{\partial x^k}\right)\right] dVdt. \tag{10.38}$$

This functional should be varied with the constraints (10.32), (10.33), (10.35) and (10.36). The constraint (10.34) is not essential in the sense that it provides sufficient freedom for variations of the functions involved, if it is satisfied at the stationary point.

Varying the functional (10.38) with respect to φ and ψ_a gives the continuity equation and (10.31). Taking the variation with respect to ρ we obtain the generalized Cauchy-Lagrange integral (10.15):

$$\frac{v^2}{2} - U - \rho\frac{\partial U}{\partial\rho} - \Phi - \frac{d\varphi}{dt} = 0.$$

Taking the variation with respect to v_i yields the expression for velocity through the "potentials" (10.17):

$$v_i = \frac{\partial\varphi}{\partial x^i} + \chi_a\frac{\partial X^a}{\partial x^i},$$

while the variation with respect to X^a gives (10.18):

$$-\partial_a U + \frac{d\chi_a}{dt} = 0.$$

The surface integrals that appear in integration by parts are equal to zero due to non-detachment condition and the initial/final instant constraints.

Weakening of time end conditions. At the initial and final instants let some system of hypersurfaces be fixed in regions $\mathring{V}$ and $\overset{1}{V}$. The hypersurfaces are the level surfaces of some function $\mathring{\beta}(x)$ in $\mathring{V}$ and function $\overset{1}{\beta}(x)$ in $\overset{1}{V}$. The hypersurfaces $\overset{1}{\beta}(x) = c$ are considered as the positions of the hypersurfaces $\mathring{\beta}(x) = c$ at the final instant; besides, in the initial and final instants density takes same given values, $\rho_0(x)$ and $\rho_1(x)$.

According to (4.33), the variations δx^i satisfy the relations

$$\frac{\partial\left(\rho_0 \delta x^i\right)}{\partial x^i}=0,\quad \delta x^i \frac{\partial \overset{\circ}{\beta}}{\partial x^i}=0 \quad \text{for } t=t_0, \qquad \frac{\partial\left(\rho_1 \delta x^i\right)}{\partial x^i}=0,\quad \delta x^i \frac{\partial \overset{1}{\beta}}{\partial x^i}=0 \quad \text{for } t=t_1 .$$

The stationarity of the action functional yields the equations

$$\begin{aligned} v_i &= \frac{\partial \overset{\circ}{\varphi}}{\partial x^i}+\overset{\circ}{\alpha} \frac{\partial \overset{\circ}{\beta}}{\partial x^i} \quad \text{for } t=t_0, \\ v_i &= \frac{\partial \overset{1}{\varphi}}{\partial x^i}+\overset{1}{\alpha} \frac{\partial \overset{1}{\beta}}{\partial x^i} \quad \text{for } t=t_1, \end{aligned}$$

where φ and $\rho\alpha$ with indices are the corresponding Lagrange multipliers.

The condition that the hypersurfaces $\overset{\circ}{\beta}(x)=c$ transform to the hypersurfaces $\overset{1}{\beta}(x)=c$ in the course of motion can be formulated in the following way: there exists a function, $\beta(x, t)$, such that

$$\frac{\partial \beta}{\partial t}+v^k \frac{\partial \beta}{\partial x^k}=0, \tag{10.39}$$

$$\left.\beta\right|_{t=t_0}=\overset{\circ}{\beta}, \quad \left.\beta\right|_{t=t_1}=\overset{1}{\beta}. \tag{10.40}$$

Equations (10.39) and (10.40) are the constraints on the admissible velocity fields. Remarkably, for barotropic flows, the following variational principle, which we shall formulate in the Eulerian coordinates, is valid.

Variational principle. *The stationary points of the functional,*

$$\int_{t_0}^{t_1} \int_{\tilde{V}(t)} \rho\left(\frac{v^2}{2}-U(\rho, X)-\Phi(x)\right) d V d t, \tag{10.41}$$

on the set of functions $\rho(x, t)$, $v^i(x, t)$ selected by the constraints (10.30), (10.35), (10.36), (10.39) and (10.40), satisfy the momentum equations of ideal compressible fluid.

Indeed, introducing the Lagrange multipliers φ and $\rho\alpha$ for the constraints (*10.30*) and (10.39), respectively, we obtain the functional

$$\int_{t_0}^{t_1} \int_{\tilde{V}(t)}\left[\rho\left(\frac{v^2}{2}-U(\rho)-\Phi(x)\right)+\varphi\left(\frac{\partial \rho}{\partial t}+\frac{\partial \rho v^i}{\rho x^i}\right)-\rho \alpha\left(\frac{\partial \beta}{\partial t}+v^k \frac{\partial \beta}{\partial x^k}\right)\right] d V d t. \tag{10.42}$$

Varying with respect to φ and α results in (10.30) and (10.39); varying with respect to ρ yields the Cauchy-Lagrange integral

$$\frac{v^2}{2} - U(\rho) - \rho \frac{\partial U}{\partial \rho} - \Phi(x) - \frac{d\varphi}{dt} = 0;$$

varying with respect to v^i results in the expression of velocity in terms of the Clebsch potentials,

$$v^i = \frac{\partial \varphi}{\partial x^i} + \alpha \frac{\partial \beta}{\partial x^i};$$

varying with respect to β gives the conservation of α along the particle trajectories

$$\frac{d\alpha}{dt} = \frac{\partial \alpha}{\partial t} + v^k \frac{\partial \alpha}{\partial x^k} = 0. \tag{10.43}$$

These equations describe a barotropic flow of compressible fluid.

The surface integrals are equal to zero due to the boundary condition (10.36) and the relations

$$\delta \rho = \delta \beta = 0 \quad \text{for } t = t_0, t_1, \tag{10.44}$$

which follow from (10.35) and (10.40).

Until the end of the section, we will consider the barotropic flows only.

Bateman principles. Let us integrate by parts the second term in (10.42). We get

$$\int_{t_0}^{t_1} \int_{\tilde{V}(t)} \left[\rho \left(\frac{v^2}{2} - U(\rho) - \Phi(x) \right) - \rho \left(\frac{\partial \varphi}{\partial t} + v^k \frac{\partial \varphi}{\partial x^k} \right) - \rho \alpha \left(\frac{\partial \beta}{\partial t} + v^k \frac{\partial \beta}{\partial x^k} \right) \right] dV dt +$$
$$+ \int_{\overset{1}{V}} \rho_1 \varphi dV - \int_{\overset{\circ}{V}} \rho_0 \varphi dV. \tag{10.45}$$

It is easy to check that variation of the functional (10.45) with respect to φ, yields not only the continuity equation but also the non-detachment condition (10.36). Therefore, we get

First Bateman variational principle. *The stationary points of the functional (10.45) on the set of functions $\rho(x,t)$, $v^i(x,t)$, $\varphi(x,t)$, $\alpha(x,t)$, $\beta(x,t)$, satisfying the conditions (10.35) and (10.40) are the solutions of the continuity equation and the momentum equations which obey the non-detachment conditions (10.36).*

According to the Cauchy-Lagrange integral (10.15), the integrand of the four-dimensional integral in (10.45), $\mathcal{P}$, coincides with pressure at the stationary points; it differs from the integrand of the original action functional by a divergent term,

$$\frac{\partial}{\partial t}(\rho \varphi) + \frac{\partial}{\partial x^i}(\rho v^i \varphi).$$

The integrand, $\mathcal{P}$, is a convex function of v^i and does not contain derivatives of v^i. Therefore, finding the extremum over v^i is equivalent to minimization of $\mathcal{P}$ with respect to v^i. The calculations result in

Second Bateman variational principle. *The stationary points of the functional*

$$\int_{t_0}^{t_1}\int_{\tilde{V}(t)}\left[-\rho\left(\frac{v^2}{2}+U(\rho)+\Phi(x)+\frac{\partial\varphi}{\partial t}+\alpha\frac{\partial\beta}{\partial t}\right)\right]dVdt+\int_{\overset{1}{V}}\rho_1\varphi dV-\int_{\overset{\circ}{V}}\rho_0\varphi dV, \tag{10.46}$$

where

$$v_i=\frac{\partial\varphi}{\partial x^i}+\alpha\frac{\partial\beta}{\partial x^i}, \tag{10.47}$$

on the set of functions $\rho(x,t)$, $\varphi(x,t)$, $\alpha(x,t)$, $\beta(x,t)$, satisfying the conditions (10.35) and (10.40), are the solutions of the continuity and momentum equations with the boundary condition (10.36).

The integrand of the four-dimensional integral in (10.46) will be denoted by $\mathcal{P}$ as well:

$$\mathcal{P}=-\rho\left(\frac{v^2}{2}+U(\rho)+\Phi(x)+\frac{\partial\varphi}{\partial t}+\alpha\frac{\partial\beta}{\partial t}\right).$$

As in the first Bateman's principle, it coincides with pressure at the stationary points.

The integrand $\mathcal{P}$ depends algebraically on ρ. Therefore, finding the extremum of the functional (10.46) with respect to ρ is equivalent to calculating the extremum over ρ of the function

$$\mathcal{P}\left(\rho^*,\rho\right)=\rho^*\rho-\rho U(\rho).$$

Here, we introduced the notation

$$-\rho^*=\frac{v^2}{2}+\Phi(x)+\frac{\partial\varphi}{\partial t}+\alpha\frac{\partial\beta}{\partial t}. \tag{10.48}$$

The function $\rho U(\rho)$ is a strictly convex function of ρ (see Example 8 in Sect. 5.7). For any fixed ρ^*, the function $\mathcal{P}(\rho^*,\rho)$ is strictly concave with respect to ρ and has a unique maximum. The extremum of $\mathcal{P}(\rho^*,\rho)$ can only be the maximum. Denote the maximum of $\mathcal{P}(\rho^*,\rho)$ with respect to ρ by $\hat{\mathcal{P}}(\rho^*)$:

$$\hat{\mathcal{P}}\left(\rho^*\right)=\max_{\rho\geq 0}\left(\rho^*\rho-\rho U(\rho)\right). \tag{10.49}$$

So, $\hat{\mathcal{P}}(\rho^*)$ is the Young-Fenchel transformation of the function $\rho U(\rho)$.

After calculating $\hat{\mathcal{P}}(\rho^*)$, we get

Dual Bateman variational principle.[2] *The stationary points of the functional*

$$\int_{t_0}^{t_1}\int_{\tilde{V}(t)} \hat{\mathcal{P}}(\rho^*)dV\,dt + \int_{\overset{1}{V}} \rho_1\varphi d^3x - \int_{\overset{\circ}{V}} \rho_0\varphi dV, \tag{10.50}$$

$$-\rho^* \equiv \frac{1}{2}\left(\partial_i\varphi + \alpha\partial_i\beta\right)\left(\partial^i\varphi + \alpha\partial^i\beta\right) + \frac{\partial\varphi}{\partial t} + \alpha\frac{\partial\beta}{\partial t} + \Phi(x).$$

on the set of functions $\varphi(x,t)$, $\alpha(x,t)$, $\beta(x,t)$ selected by the constraints (10.40) satisfy the continuity equations, the momentum equations, and the non-detachment condition.

At the extremals, $\hat{\mathcal{P}}(\rho^*)$ coincides with the pressure of the fluid.

Function $\hat{\mathcal{P}}(\rho^*)$ can be explicitly found for isentropic flows of the ideal gas. In this case it is a power function:

$$\hat{\mathcal{P}}\left(\rho^*\right) = k\left(\rho^*\right)^{\frac{\gamma}{\gamma-1}}, \tag{10.51}$$

k being a constant. Indeed, for an ideal gas,

$$U(\rho) = \frac{A\rho^{\gamma-1}}{\gamma-1},$$

where the constant A is related to the initial state of the gas, while γ is a characteristic of the gas. We have

$$p = \rho^2\frac{\partial U}{\partial\rho} = A\rho^\gamma, \quad \rho^* = \frac{d}{dt}\rho U(\rho) = \frac{A\gamma\rho^{\gamma-1}}{\gamma-1}, \quad \hat{\mathcal{P}}\left(\rho^*(\rho)\right) = A\rho^\gamma = p.$$

Hence, $\hat{\mathcal{P}}(\rho^*)$ is given by (10.51), where

$$k = A\left(\frac{\gamma-1}{\gamma A}\right)^{\frac{\gamma}{\gamma-1}}.$$

10.4 Potential Flows

Let us expand further the set of admissible functions by discarding one of the constraints on $\beta(x,t)$ (10.40). Then the constraint (10.39) becomes unessential. We arrive at

[2] The term "dual variational principle" is due to the fact that in this principle, only the Lagrange multipliers are varied.

Kelvin variational principle. *The stationary points of the functional*

$$\int_{t_0}^{t_1}\int_{\tilde{V}(t)} \rho\left(\frac{v^2}{2} - U(\rho) - \Phi(x)\right) dV dt,$$

on the set of functions $\rho(x,t)$ and $v^i(x,t)$ satisfying the continuity equation (10.30), the boundary condition (10.36) and the initial-final conditions (10.35) are the potential flows of the fluid.

Kelvin principle is equivalent to the variational principle for the functional (10.41), in which one of the constraints is dropped. In this case $\partial\beta$ becomes arbitrary either at $t = t_0$ or at $t = t_1$. That yields $\alpha = 0$ for $t = t_0$ or for $t = t_1$, which, together with (10.43), shows that α is equal to zero identically. The corresponding extremals are potential flows.

Dual variational principle. *On the set of all functions $\varphi(x,t)$, the stationary points of the functional (10.50), for which*

$$-\rho^* = \frac{1}{2}\frac{\partial\varphi}{\partial x^i}\frac{\partial\varphi}{\partial x_i} + \frac{\partial\varphi}{\partial t} + \Phi(x), \tag{10.52}$$

satisfy the continuity equation, the momentum equations, and the non-detachment condition, if density, velocity and pressure are defined in terms of the potential, φ, by the formulas

$$\rho = \frac{\partial\mathcal{P}(\rho^*)}{\partial\rho^*}, \quad v_i = \frac{\partial\varphi}{\partial x^i}, \quad p = \hat{\mathcal{P}}(\rho^*). \tag{10.53}$$

The first formula is a consequence of the definition of the function $\hat{\mathcal{P}}(\rho^*)$ (10.49).

The variational principle follows from dual Bateman variational principle because, after dropping the constraints on $\beta(x,t)$ at one of the time ends, function α becomes zero at the stationary value of the action functional. This variational principle can also be obtained directly from Kelvin principle using the general scheme of constructing the dual variational principles considered in Sect. 5.6.

Motion of the fluids with free surfaces. If the region $V(t)$, occupied by the fluid, does not coincide with the region $\tilde{V}(t)$, then $V(t)$ should be varied subject to the constraint $V(t) \subset \tilde{V}(t)$. Let us consider as an example, a generalization of Luke variational principle for compressible fluids moving over a plane. Denote the Cartesian coordinates in the plane by x^α, and the distance from the free surface to the plane at the point x^α by $h(x^\alpha, t)$. Let the fluid be in a vessel $\tilde{V} = \{x^\alpha, x : x^\alpha \in \Omega, 0 \le x < +\infty\}$, where Ω is some bounded region in the plane $\{x^1, x^2\}$.

Variational principle. *The true potential flow of ideal compressible barotropic fluid over a plane is a stationary point of the functional*

$$\int_{t_0}^{t_1}\int_{\Omega}\int_0^{h(x^\alpha,t)} \left(\hat{\mathcal{P}}(\rho^*) - \bar{p}(t)\right) dx dx^1 dx^2 dt + \int_{\Omega}\left(\int_0^{h(x^\alpha,t)} \rho_1 \varphi dx - \int_0^{h_0(x^\alpha)} \rho_0 \varphi dx\right) dx^1 dx^2, \tag{10.54}$$

where

$$-\rho^* = \frac{1}{2}\frac{\partial \varphi}{\partial x^i}\frac{\partial \varphi}{\partial x_i} + \frac{\partial \varphi}{\partial t} + \Phi(x),$$

on the set of functions $\varphi(x^\alpha, x, t)$ *and* $h(x^\alpha, t)$ *selected by the constraints*

$$h(x^\alpha, t_0) = h_0(x^\alpha), \quad h(x^\alpha, t_1) = h_1(x^\alpha), \quad h \geq 0. \tag{10.55}$$

Here $\mathcal{P}(\rho^*)$ is the function (10.49), $\bar{p}(t)$ is a given pressure on the free surface, ρ_0 and ρ_1 are the given density distributions at the instants t_0 and t_1, respectively. If the values of the potential at times t_0 and t_1 are given, then the last term in (10.54) is an additive constant.

10.5 Incompressible Fluid as a Limit Case of Compressible Fluid

The ideal incompressible fluid can be considered as the limit in a sequence of the models of compressible fluids, for which $U(\rho, S) \to +\infty$ if at every ρ except the point ρ_0, where the internal energy is finite. In this section, we derive the least action variational principle for incompressible fluid by the asymptotic analysis of the action functional of compressible fluid. We will use the Lagrangian coordinates as independent variables.

The action functional of the ideal compressible fluid is

$$I = \int_{t_0}^{t_1}\int_{\mathring{V}} \rho\left(\frac{1}{2}v_i v^i - U(\rho, S) - \Phi(x)\right) d\mathring{V} dt. \tag{10.56}$$

The admissible functions $x^i(X^a, t)$ take on the assigned values at the initial and final instants,

$$x^i\left(X^a, t_0\right) = \mathring{x}^i\left(X^a\right), \quad x^i\left(X^a, t_1\right) = \overset{1}{x}{}^i\left(X^a\right), \tag{10.57}$$

and satisfy the non-detachment condition

$$x^i\left(X^a, t\right) \in \partial\tilde{V}(t) \quad \text{for } X^a \in \partial\mathring{V}, \tag{10.58}$$

where for each t, $\tilde{V}(t)$ is a given space region. Mass density, ρ, is expressed in terms of the functions, $x^i(X^a, t)$, as

$$\rho = \rho_0 \left(X^a\right) \frac{\det \left\| \frac{\partial \overset{\circ}{x}^i}{\partial X^a} \right\|}{\det \left\| \frac{\partial x^i}{\partial X^a} \right\|}, \tag{10.59}$$

where $U(\rho, S)$, $\Phi(x)$, $\rho_0(X)$, and $S(X)$ are given functions of their arguments, $\overset{\circ}{V}$ is a fixed region in the space of Lagrangian variables, X.

According to the least action principle, the true trajectories are the stationary points of the functional (10.56) on the set of functions selected by the constraints (10.57) and (10.58).

For definiteness, let us take the function U in the form

$$U = \frac{1}{2} a^2 \frac{(\rho - \rho_0)^2}{\rho_0^2}, \quad a^2 = const.$$

The constant, a, has the meaning of the speed of sound. We consider motions for which velocity is much smaller than a. Formally, one can treat it as a limit,[3] $a \to +\infty$.

The leading (in the asymptotic sense) term of the action functional is

$$\int_{t_0}^{t_1} \int_{\overset{\circ}{V}} \frac{1}{2} a^2 \frac{(\rho - \rho_0)^2}{\rho_0} dV dt. \tag{10.60}$$

The minimum of the functional (10.60) is equal to zero. The minimum is reached at the function, $\rho = \rho_0(X)$, or, according to (10.59), for

$$\det \left\| \frac{\partial x^i}{\partial X^a} \right\| = \det \left\| \frac{\partial \overset{\circ}{x}^i}{\partial X^a} \right\|. \tag{10.61}$$

The given values of the functions $x^i(X^a, t)$ at the initial and final instants should be consistent with the condition (10.61):

$$\det \left\| \frac{\partial \overset{1}{x}^i}{\partial X^a} \right\| = \det \left\| \frac{\partial \overset{\circ}{x}^i}{\partial X^a} \right\|. \tag{10.62}$$

[3] The general scheme of the variational-asymptotic method deals with the functionals depending on small parameters. For a verbatim application of the general scheme, a small parameter, $\varepsilon = 1/a$, should be used, and, instead of the functional, I, the functional

$$\frac{1}{a^2} I = \int_{t_0}^{t_1} \int_{\overset{\circ}{V}} \rho \left(\frac{1}{2a^2} v_i v^i - \frac{1}{2} \frac{(\rho - \rho_0)^2}{\rho_0^2} - \frac{1}{a^2} \Phi(x) \right) d\overset{\circ}{V} dt,$$

should be considered. In this section, we employ a natural reformulation of the variational-asymptotic method for the case of large parameters.

We assume that (10.62) holds true.

The stationary points of the leading term of the functional are functions $x^i\left(X^a, t\right)$ satisfying the constraints (10.57), (10.58) and (10.61). These functions comprise the set $\mathcal{M}_0$ in the general scheme of the variational-asymptotic method.

At the next step of the variational-asymptotic method, we present the particle trajectories as a sum:

$$x^i\left(X^a, t\right) = \bar{x}^i\left(X^a, t\right) + x'^i\left(X^a, t\right)$$

where $\bar{x}^i\left(X^a, t\right) \in \mathcal{M}_0$ and x'^i are small. Accordingly,

$$\rho = \rho_0 + \rho', \qquad \rho' = -\rho_0 \mathring{\partial}_i x'^i.$$

Keeping in the functional the leading terms containing the functions, x'^i, we get

$$\int_{t_0}^{t_1} \int_{\mathring{V}} \rho_0 \left[\left(\bar{v}_i + v'_i\right) \frac{dx'^i}{dt} - \frac{1}{2} a^2 \left(\mathring{\partial}_i x'^i\right)^2 - \Phi_{,i} x'^i - \left(\frac{1}{2} \bar{v}^k \bar{v}_k - \Phi\right) \mathring{\partial}_i x'^i \right] d\mathring{V} dt,$$

$\bar{v}^i \equiv d\bar{x}^i\left(X^a, t\right)/dt$, $v'^i \equiv dx'^i\left(X^a, t\right)/dt$. The leading terms must have the same order, therefore

$$(\text{the magnitude of } x'^i) \sim \frac{v}{a} \ell,$$

v and ℓ being the characteristic length and characteristic velocity of the problem, respectively.[4] Consequently, if the functional I is considered on the set $\mathcal{M}_0$, the corrections will be small (on the order of a^{-2}) and, in the first approximation, the variational problem is reduced to minimizing the functional

$$\int_{t_0}^{t_1} \int_{\mathring{V}} \rho_0 \left(\frac{1}{2} \bar{v}_i \bar{v}^i - \Phi\left(\bar{x}\right) \right) d\mathring{V} dt, \tag{10.63}$$

on the set of functions $\bar{x}^i\left(X^a, t\right)$ satisfying the constraints (10.57), (10.58) and (10.61). This is indeed the least action principle for the ideal incompressible fluid in Lagrangian coordinates considered in Sect. 9.1. Writing down the next approximation, one obtains the variational principle of the theory of sound waves. We leave this as an exercise for the reader.

[4] We assume that $\nabla \Phi \sim \nabla v^i v_i$.

Chapter 11
Steady Motion of Ideal Fluid and Elastic Body

All the preceding variational formulations, including those in Eulerian coordinates, dealt with the case when the flow region, $V(t)$, contains the same particles. If this region does not move in the observer's frame and its boundaries are penetrable to the media (i.e. some particles leave the region while other particles enter it), then the variational principles considered are not true. However, one may expect that the same functionals, perhaps, modified by additional terms describing the energy flux through penetrable parts of the boundary, have the stationary value on the true motion. Such expectation is supported by the variational principles for open flows of Sect. 9.11. In this chapter we show that this is indeed the case for steady open flows of the ideal fluids and elastic bodies. We also consider simplifications caused by steadiness of the flow in variational principles for closed flows.

11.1 The Kinematics of Steady Flow

The flows all the characteristics of which, including velocity and density, do not change in time at every space point are called steady flows. For steady flows, the particle trajectories, $x^i(X^a, t)$, are the solutions of the system of ordinary differential equations,

$$\frac{dx^i(X^a, t)}{dt} = v^i(x^k). \tag{11.1}$$

The system of equations is autonomous, i.e. its right hand side does not depend on time. Therefore, motion occurs over a two-parametric family of stream lines, the lines that are tangent to the velocity field. In three-dimensional space, the stream lines of a divergence-free vector field can be either closed or cover densely some regions. We consider further only the case when the stream lines can be described by parametric equations,

$$x^i = r^i(X^\mu, \zeta),$$

V.L. Berdichevsky, *Variational Principles of Continuum Mechanics*,
Interaction of Mechanics and Mathematics, DOI 10.1007/978-3-540-88467-5_11,

where $r^i(X^\mu,\zeta)$ are smooth functions, X^μ, $\mu = 1, 2$, mark the stream line, and ζ is the parameter along the stream line. This includes the case of closed stream lines for closed flows and the case of smooth stream lines for open flows. It is convenient to choose the parameter along the stream lines in such a way as to make $r^i_{,\zeta}$ equal to velocity:

$$r^i_{,\zeta} = v^i. \tag{11.2}$$

Any particle trajectory starting at a stream line belongs to this stream line. Therefore X^μ are two of the three Lagrangian coordinates. Let us show that, for any steady flow, the dependence of particle trajectories on the third Lagrangian coordinate, $X^3 \equiv X$, can be written simply as

$$x^i = r^i(X^\mu, X + t). \tag{11.3}$$

Indeed, the motion of particles over a stream line, X^μ, is described by the function, $\zeta(X^\mu, X, t)$. According to (11.2), $\partial\zeta(X^\mu, X, t)/\partial t = 1$. We identify the Lagrangian coordinate, X, with the parameter, ζ, at the initial instant. Thus, $\zeta = X + t$.

The inversions of the formulas (11.3) are

$$X^\mu = X^\mu\left(x^i\right), \quad X = \zeta\left(x^i\right) - t, \tag{11.4}$$

where $\zeta\left(x^i\right)$ is some function of the Eulerian coordinates. Therefore, the steady motion is defined either by providing three functions of the Lagrangian coordinates $r^i(X^\mu, \zeta)$, or three function of the Eulerian coordinates $X^\mu\left(x^i\right)$, and $\zeta\left(x^i\right)$. The functions $X^\mu\left(x^i\right)$ are called the stream functions of the flow. The variational principles for the steady motion should naturally be formulated in the Eulerian coordinates; therefore, we will further use the functions $X^\mu\left(x^i\right)$ and $\zeta\left(x^i\right)$ as the field variables.

Let us express velocity and density in terms of functions $X^\mu\left(x^i\right)$ and $\zeta\left(x^i\right)$. To this end we write down the equations $x^i_a X^b_i = \delta^b_a$ for $a = 3$, $b = 1, 2, 3$. Since $x^i_3 = v^i$, we get

$$v^i X^\alpha_{,i} = 0, \quad v^i \zeta_{,i} = 1. \tag{11.5}$$

The first two equations (11.5) mean that the vector v^i is perpendicular to the vectors $X^1_{,i}$ and $X^2_{,i}$. Consequently, it is proportional to their vector product

$$v^i = a e^{ijk} X^1_{,j} X^2_{,k}. \tag{11.6}$$

The coefficient a is found from the last equation (11.5):

$$\frac{1}{a} = e^{ijk} X^1_{,j} X^2_{,k} \zeta_{,i}. \tag{11.7}$$

For steady flows density ρ does not change at any space point. In particular, on some surface $\mathcal{P}$ intersecting the stream lines, ρ is a function only of X^μ: $\rho = \tilde{\rho}(X^\mu)$. Let us show that at any space point, ρ is given by the formula

$$\rho = \sigma(X^\mu)\, e^{ijk} X^1_{,j} X^2_{,k} \zeta_{,i} \tag{11.8}$$

where the function $\sigma(X^\mu)$ is defined by the density values on the surface $\mathcal{P}$:

$$\sigma(X^\mu) = \tilde{\rho}(X^\mu) e^{ijk} X^1_{,j} X^2_{,k} \zeta_{,i} \ \text{on}\ \mathcal{P}. \tag{11.9}$$

Indeed, due to (11.6), (11.7) and (11.8), the mass flux ρv^i can be written as

$$\rho v^i = \sigma(X^\mu)\, e^{ijk} X^1_{,j} X^2_{,k}.$$

Therefore, the function ρ (11.8) satisfies the continuity equation,

$$\left(\rho v^i\right)_{,i} = \left(\sigma(X^\mu)\, e^{ijk} X^1_{,j} X^2_{,k}\right)_{,i} = 0.$$

The solution of the continuity equation which takes on given values on $\mathcal{P}$ is unique, as is easy to see.

If a coordinate transformation $X^\mu \to X'^\mu$ is made, the coefficient σ acquires a factor $\det \left\| \partial X^\mu / \partial X'^\nu \right\|$. Therefore, it can be made equal to unity by an appropriate choice of the coordinates X^μ. Further, we assume that $\sigma = 1$, and that

$$\rho v^i = e^{ijk} X^1_{,j} X^2_{,k}, \quad \rho = e^{ijk} X^1_{,j} X^2_{,k} \zeta_{,i}. \tag{11.10}$$

The steady flows have an important feature: the components of the inverse distortion, X^a_i, and, consequently, the components of the distortion, x^i_a, and its determinant, Δ, do not depend on time at each point. This follows from the formulas (11.4).

In the course of motion, the particles cross the surface $\mathcal{P}$. As the initial values of particle's characteristics, one can take their values at $\mathcal{P}$. In particular, let us define $\mathring{x}^i_a$ by the equality $\mathring{x}^i_a = x^i_a$ on $\mathcal{P}$. We will assume $\mathring{x}^i_a$ to be known functions of X^μ. Note that such a definition of $\mathring{x}^i_a$ differs from the definition used in Chaps. 3 and 4: $\mathring{x}^i_a = x^i_a$ for $t = t_0$. We accept it to preserve the property of $\mathring{x}^i_a$ to have zero variations: $\delta \mathring{x}^i_a = 0$. The variations of the object $x^i_a \big|_{t=0}$ do not equal to zero when the stream lines are varied.

11.2 Steady Motion with Impenetrable Boundaries

Consider a steady motion of continua in some region V. The particles do not enter or leave the region, and $v^i n_i = 0$ everywhere on ∂V.

The particles which are on the boundary move over the boundary. In terms of the functions $X^\mu\left(x^i\right)$, this condition can be written in the following way: there exists a function $f(X^\mu)$, such that for $x^i \in \partial V$

$$f\left(X^{\mu}\left(x^{i}\right)\right)=0. \tag{11.11}$$

The equation $f\left(X^{\mu}\right)=0$ defines the boundary of the region ∂V in the Lagrangian coordinates.

According to the least action principle in Eulerian coordinates, the following equality takes place:

$$\int\left(\frac{\partial L}{\partial X_{t}^{a}}\frac{\partial}{\partial t}\partial X^{a}+\frac{\partial L}{\partial X_{i}^{a}}\frac{\partial}{\partial x^{i}}\partial X^{a}+\frac{\partial L}{\partial X^{a}}\partial X^{a}\right)dVdt=0, \tag{11.12}$$

where

$$L=\rho\left(\frac{v^{2}}{2}-U\left(X_{i}^{a},X^{a}\right)-\Phi\left(x^{i}\right)\right).$$

Recall that the symbol ∂ means the variation at a fixed space point.

Functions U and L do not depend explicitly on X, otherwise, they would change with time at a fixed space point, but U and L may depend on X^{μ}.

According to (10.29),

$$\frac{\partial L}{\partial X_{t}^{a}}=-\rho v_{i}x_{a}^{i}$$

and, since for the steady flow ρ, v^{i} and x_{a}^{i} do not depend on time for every space point, and $\partial X^{a}=0$ at $t=t_{0},t_{1}$, the first term in (11.12) is equal to zero. The other two terms contain only ∂X^{a} and their derivatives with respect to the spatial coordinates. Therefore, the integral (11.12) is not affected by the constraints on ∂X^{a} for $t=t_{0},t_{1}$, and, at every instant, the equation holds:

$$\int\limits_{V}\left(\frac{\partial L}{\partial X^{a}}\partial X^{a}+\frac{\partial L}{\partial X_{i}^{a}}\partial X_{i}^{a}\right)dV=0. \tag{11.13}$$

Now, note that in varying the functions (11.4), ∂X^{a} do not depend on time, while the dependence of ∂X^{a} on space coordinates can be arbitrary. If we limit the admissible variations by the variations ∂X^{μ} and $\partial\zeta$, for which $\partial X^{\mu}=0$, $\partial\mathring{x}_{a}^{i}=0$ on $\mathcal{P}$, then $\delta\mathring{g}_{ab}=0$, $\delta\rho_{0}=0$, and the left hand side of (11.13) represents the variation of the integral of L. Besides, as follows from (10.29), the equations

$$\frac{\partial L}{\partial X^{a}}-\frac{\partial}{\partial x^{i}}\frac{\partial L}{\partial X_{i}^{a}}=0 \tag{11.14}$$

are the equations of the steady flow of the compressible fluid. Therefore, the following variational principle is valid.

Variational principle. *Consider the functional*

$$\int_V \rho\left(\frac{1}{2}v_i v^i - U(\rho, X^\mu) - \Phi(x)\right)dV, \tag{11.15}$$

where ρ and v^i are expressed through $X^\mu(x)$ and $\zeta(x)$ according to the formulas (11.10). The stationary points of this functional on the set of all functions $X^\mu(x^i)$ and $\zeta(x^i)$ obey the condition

$$\partial \det \left\| X_i^a \right\| = 0 \,\text{on}\, \mathcal{P} \tag{11.16}$$

and the constraint (11.11), satisfy the momentum equations of ideal compressible fluid.

Let us give a direct derivation of the momentum equations. It is convenient to write the integrand as

$$L = L\left(p^i, \rho, X^\mu, x\right) = \frac{p^i p_i}{2\rho} - \rho U(\rho, X^\mu) - \rho\Phi(x), \quad p^i \equiv \rho v^i,$$

because the momentum, $p^i = \rho v^i$, is expressed in terms of the field variables by simple relations (11.10). We have[1]

$$\delta I = \int_V \left(\frac{\partial L}{\partial p^i}\left(e^{ijk}\left(\partial X^1\right)_{,j} X^2_{,k} + e^{ijk} X^1_{,j}\left(\partial X^2\right)_{,k}\right) + \frac{\partial L}{\partial \rho}\left(e^{ijk}\left(\partial X^1\right)_{,j} X^2_{,k}\zeta_{,i} + \right.\right.$$
$$\left.\left. + e^{ijk} X^1_{,j}\left(\partial X^2\right)_{,k}\zeta_{,i} + e^{ijk} X^1_{,j} X^2_{,k}\left(\partial\zeta\right)_{,i}\right) - \rho T\frac{\partial S}{\partial X^\mu}\partial X^\mu\right) dV. \tag{11.17}$$

We have for the derivatives of Lagrangian:

$$\frac{\partial L}{\partial p^i} = v^i, \quad \frac{\partial L}{\partial \rho} = -\left(\frac{v^2}{2} + U + \rho\frac{\partial U}{\partial \rho} + \Phi\right).$$

Denote $\partial L/\partial \rho$ by $-R$. Varying ζ, we obtain the equation

$$e^{ijk}(R)_{,i} X^1_{,j} X^2_{,k} = 0.$$

It shows that the determinant of the matrix of the derivatives of the functions R, X^1, and X^2 is equal to zero. That means that R is a function of X^1 and X^2 only:

$$R = R\left(X^1, X^2\right). \tag{11.18}$$

[1] The fluid is assumed to be homogeneous; thus,

$$\frac{\partial U}{\partial X^\mu} = T\frac{\partial S}{\partial X^\mu}.$$

The coefficient at $\partial\zeta$ in the surface integral, which is obtained after integration by parts, is equal to zero because the vector $e^{ijk}X^1_{,j}X^2_{,k}$ is equal to the velocity and is, consequently, orthogonal to the normal vector.

Using (11.18) and the equality $v_i e^{ijk}X^1_{,j}X^2_{,k} = \rho v^2$, which follows from (11.10), the equations resulting from variations of X^1 and X^2 can be written as

$$\left(-v_{i,j}e^{ijk}\zeta_{,i}\left(R_{,j} - TS_{,j}\right)\right)X^2_{,k} = 0,$$
$$\left(-v_{i,k}e^{ijk}\zeta_{,i}\left(R_{,k} - TS_{,k}\right)\right)X^1_{,j} = 0.$$

These equations show that the vector in the parenthesis is proportional to velocity. Denoting the corresponding factor by μ, we can write

$$-v_{i,j}e^{ijk} + e^{ijk}\left(R_{,j} - TS_{,j}\right)\zeta_{,i} = \mu v^k. \tag{11.19}$$

The three equations (11.19), in fact, contain the two equations for the required functions while the third one serves to determine μ. The coefficient μ can be excluded by projecting (11.19) on a plane perpendicular to the velocity, i.e. by contracting (11.19) with $e_{krs}v^r$. Using (3.19) and the equalities $v^i\zeta_{,i} = 1$, $v^i R_{,i} = 0$, $v^i S_{,i} = 0$, we get the equations

$$v_{i,j}v^j - \partial_i\left(\frac{v^2}{2}\right) + R_{,i} - TS_{,i} = 0,$$

which, as can be seen from the definition of R,

$$R = \frac{v^2}{2} + U + \rho\frac{\partial U}{\partial \rho} + \Phi, \tag{11.20}$$

are the momentum equation of the compressible fluid.

At the variation, ∂X^1, on the boundary of region V, the coefficient is

$$v_i e^{ijk}n_j X^2_{,k} - Re^{ijk}n_j X^2_{,k}\zeta_{,i}.$$

Since the normal vector n_i is proportional to $\frac{\partial f}{\partial X^1}X^1_{,i} + \frac{\partial f}{\partial X^2}X^2_{,i}$, this coefficient is of the form $a\frac{\partial f}{\partial X^1}$. The coefficient at the variation, ∂X^2, is of the form $a\frac{\partial f}{\partial X^2}$. Since $\frac{\partial f}{\partial X^1}\partial X^1 + \frac{\partial f}{\partial X^2}\partial X^2 = 0$ on ∂V, the variation X^α on ∂V does not give additional relations.

Let us formulate the analogous variational principle for the ideal incompressible homogeneous fluid (it can be derived, from the considered one, by the variational-asymptotic method).

Let the density be a constant: $\rho = \rho_0 = const$. Then the functions X^μ and ζ satisfy the constraint

$$e^{ijk}X^1_{,j}X^2_{,k}\zeta_{,i} \equiv const. \tag{11.21}$$

Based on (11.10) and (3.19), the kinetic energy is

$$\frac{\rho v^2}{2} = \frac{1}{2\rho}\left[X^1_{,i}X^{1,i}X^2_{,j}X^{2,j} - \left(X^1_{,i}X^{2,i}\right)^2\right].$$

The integral of Φ over the given region V plays a role of an additive constant and, thus, can be dropped. Therefore, the following variational principle is true.
Variational principle. *The stationary points of the functional*

$$\frac{1}{2\rho_0}\int_V \left[X^1_{,i}X^{1,i}X^2_{,j}X^{2,j} - \left(X^1_{,i}X^{2,i}\right)^2\right]dV$$

on the set of functions $X^1\left(x^i\right)$, $X^2\left(x^i\right)$, *and* $\zeta\left(x^i\right)$ *subject to the constraints (11.11) and (11.21), satisfy the momentum equations of ideal incompressible fluid.*

A similar variational principle is valid for elastic bodies.
Variational principle. *The stationary points of the functional*

$$\int_V \rho\left(\frac{v^2}{2} - U - \Phi(x)\right)dV \tag{11.22}$$

(where ρ *and* v^i *are expressed in terms of functions,* $X^\mu(x)$ *and* $\zeta(x)$, *according to the formulas (11.10), and* U *is a known function of* $X^\mu_{,i}$, $\zeta_{,i}$, *and* X^μ*) on the set of functions* $X^\mu(x)$ *and* $\zeta(x)$, *subject to the constraints*

$$\text{on } \mathcal{P}:\ X^\mu = \mathring{X}^\mu(x),\quad \zeta = \mathring{\zeta}(x),\quad n^i\zeta_{,i} = \mathring{a}(x),\quad n^iX^\mu_{,i} = \mathring{a}^\mu(x)$$

and the constraint (11.11), satisfy the momentum equations of the steady motion of the elastic body inside the volume V *and zero boundary conditions for tangent components of the surface force.*

The constraints on the surface $\mathcal{P}$ are set forth in order to have zero variations of $\mathring{x}^i_\alpha$ and to make the left hand side of (11.13) a variation of the integral of L. In deriving the equations, one should take into account that the constitutive equations for the Cauchy stress tensor in terms of functions $X^a\left(x^i\right)$ have the form

$$\sigma^i_j = -\rho\frac{\partial U}{\partial X^a_i}X^a_j.$$

11.3 Open Steady Flows of Ideal Fluid

Now let a part of the boundary, S, of the region V be penetrable to the particles (Fig. 11.1). In this case, the variation of the functional (11.15) is not equal to zero. It is natural to postulate the following variational equation:

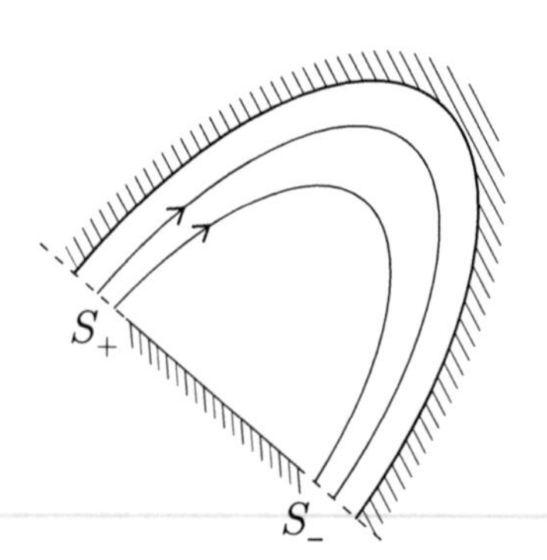

Fig. 11.1 An open flow

$$\delta \int_V \rho \left(\frac{v^2}{2} - U(\rho, X^\mu) - \Phi(x) \right) dV = \int_S \left(\lambda \partial \zeta + \lambda_\mu \partial X^\mu \right) dA. \tag{11.23}$$

The functional on the right hand side of (11.23) should be given.

In what follows, we consider the case when the functions $X^\mu(x)$ on S are known:

$$X^\mu(x) = a^\mu(x) \quad \text{on } S. \tag{11.24}$$

If the parts of the boundary through which the particles enter and leave V are denoted by S_+ and S_-, respectively, the condition (11.24) means that a mapping $S_+ \to S_-$ is given: the starting point of the stream line on S_+ maps to the end of this line on S_-. According to (11.24), $\partial X^\mu = 0$ on S.

The functional on the right-hand side of (11.23) is specified by prescribing either $\zeta(x)$ on S :

$$\zeta(x) = a(x) \quad \text{on } S, \tag{11.25}$$

or the coefficient λ. Since the functional (11.15) is invariant with respect to the shifts $\zeta \to \zeta + \varphi(X^\mu)$, where φ is an arbitrary function of its argument, without loss of generality one can set $\zeta(x) = 0$ on S_-. Then, assigning $\zeta(x)$ on S_+ fixes the velocity with which the particles go over the stream lines.

To figure out the physical meaning of λ, consider the coefficient at $\partial\zeta$ on the boundary. As follows from (11.17), it has the value

$$-Re^{ijk} n_i X^1_{,j} X^2_{,k}. \tag{11.26}$$

It includes the derivatives of X^μ only over S; therefore, assigning λ with known X^μ on S is equivalent to assigning R on S as a given function of X^μ:

$$R = \tilde{R}(X^\mu) \quad \text{on } S. \tag{11.27}$$

If X^μ and ζ are given on S, then the right-hand side of (11.23) is equal to zero, and we get the following

Variational principle. *The steady motion of ideal compressible fluid is a stationary point of the functional*

$$\delta \int_V \rho \left(\frac{v^2}{2} - U(\rho, X^\mu) - \Phi(x) \right) dV$$

on the set of functions $X^\mu(x^i)$ and $\zeta(x)$, satisfying the conditions (11.24) and (11.25) on S, and the condition (11.11) on $\partial V - S$.

If X^μ and λ are given on S, then the functional on the right-hand side of (11.23) is holonomic and we have the following

Variational principle. *The steady motion of ideal incompressible fluid is a stationary point of the functional*

$$\int_V \rho \left(\frac{v^2}{2} - U(\rho, X^\mu) - \Phi(x) \right) dV + \int_S \tilde{R} e^{ijk} n_i X^1_{,j} X^2_{,k} \zeta dA \qquad (11.28)$$

on the set of functions $X^\mu(x)$ and $\zeta(x)$, satisfying the condition (11.24) on S and the condition (11.11) on $\partial V - S$.

The surface integral in (11.28) can be written as

$$\int_{S_+} (\tilde{R}_+(X^\mu)\zeta_+ - \tilde{R}_-(X^\mu)\zeta_-) dX^1 dX^2,$$

where the indices $+$ and $-$ mark the quantities on S_+ and S_-, respectively. The volume integral in (11.28) is invariant with respect to shifts of ζ by an arbitrary function of X^μ. Therefore, in order for the variational problem to be well-posed, the surface integral has to be equal to zero for all functions $\zeta_+ = \zeta_- = \varphi(X^\mu)$. Thus, there is a necessary condition which has to be satisfied by the given functions:

$$\tilde{R}_+(X^\mu) = \tilde{R}_-(X^\mu).$$

Using the first equation (11.10), the surface integral in (11.28) can be written as

$$\int_S \rho v^i n_i \tilde{R} \zeta dA.$$

The function $\tilde{R}$ can be extended on the entire region V by setting $\tilde{R} = \tilde{R}(X^\mu(x))$. Then, taking into account that $v^i n_i = 0$ on $\partial V - S$, $(\rho v^i)_{,i} = 0$, $v^i \tilde{R}_{,i} = 0$, $v^i \zeta_{,i} = 1$, the surface integral in (11.28) can be rewritten as

$$\int_S \rho v^i n_i \tilde{R} \zeta dA = \int_{\partial V} \rho v^i n_i \tilde{R} \zeta dA = \int_V \rho \tilde{R} dV.$$

Note that now varying ζ is equivalent to varying the density, ρ: although ρ depends on the derivative of ζ along the stream line, the values of ζ are not given at the boundary, and this derivative (or the density ρ) can be chosen as a new independent function (see the discussion of a similar issue on the independent variation of velocity in Sect. 9.3). Since $p_i = \rho v_i$ do not depend on ζ, we get

Giese-Kraiko variational principle. *The steady flow of ideal compressible fluid is a stationary point of the functional*

$$\int_V \left(\frac{p_i p^i}{2\rho} - \rho U(\rho, X^\mu) - \rho\Phi(x) + \rho\tilde{R}(X^\mu) \right) dV \tag{11.29}$$

on the set of all functions $\rho(x)$ and $X^\mu(x)$ satisfying the constraints (11.24) on S and the constraints (11.11) on $\partial V - S$.

In (11.29), $p_i = \rho v^i$ are expressed through X^μ according to the formulas (11.10).

Varying with respect to ρ yields an algebraic problem of calculating the stationary value of the function

$$\frac{p_i p^i}{2\rho} - \rho U(\rho, S) - \rho\left(\Phi - \tilde{R}\right).$$

The stationary value of ρ is found from the equation

$$\frac{p_i p^i}{2\rho^2} + U + \rho\frac{\partial U}{\partial \rho} + \Phi = \tilde{R}, \tag{11.30}$$

which has the meaning of the Bernoulli integral.

After excluding $p_i = \rho v_i$ and ρ by means of (11.30) and (11.10), the integrand in (11.30) becomes a function of X^μ and the derivatives of X^μ only, and, as easy to see, it is equal to $\rho v^2 + p$. Therefore, we obtain

Giese variational principle. *The steady flow of ideal compressible fluid is a stationary points of the functional,*

$$\int_V \left(\rho v^2 + p\right) dV, \tag{11.31}$$

on the set of all functions $X^\mu(x)$ satisfying the constraints (11.24) on S and the constraints (11.11) on $\partial V - S$.

As an example, let us calculate the functional (11.31) for a barotropic motion of the ideal gas, when

$$U = A\frac{\rho^{\gamma-1}}{\gamma - 1}.$$

Equation (11.30) becomes

$$\frac{p^i p_i}{2\rho^2} + \frac{\gamma A}{\gamma - 1}\rho^{\gamma-1} = \tilde{R} - \Phi.$$

From this equation we find ρ as a function of the velocity

$$\rho = \left(\frac{\gamma - 1}{\gamma A}\right)^{\frac{1}{\gamma-1}} \left(\tilde{R} - \Phi - \frac{v^2}{2}\right)^{\frac{1}{\gamma-1}} \tag{11.32}$$

Therefore,

$$\rho^2 v^2 = \left(\frac{\gamma - 1}{\gamma a}\right)^{\frac{2}{\gamma-1}} \left(\tilde{R} - \Phi - \frac{v^2}{2}\right)^{\frac{2}{\gamma-1}} v^2.$$

On the other hand, from the first equation (11.10),

$$\rho^2 v^2 = \rho v^i \rho v_i = X^1_{,i} X^{1,i} X^2_{,j} X^{2,j} - \left(X^1_{,i} X^{2,i}\right)^2.$$

Equating both exptressions for $\rho^2 v^2$, we get a nonlinear equation which determines v^2 as a function of $X^\mu_{,i}$:

$$\left(\frac{\gamma - 1}{\gamma a}\right)^{\frac{2}{\gamma-1}} \left(\tilde{R} - \Phi - \frac{v^2}{2}\right)^{\frac{2}{\gamma-1}} v^2 = X^1_{,i} X^{1,i} X^2_{,j} X^{2,j} - \left(X^1_{,i} X^{2,i}\right)^2. \tag{11.33}$$

From (11.32) and the constitutive equation $p = a\rho^\gamma$, up to a constant factor, the integrand in (11.31) has the form

$$L = \left(R^* - \Phi - \frac{v^2}{2}\right)^{\frac{1}{\gamma-1}} + a\left(R^* - \Phi - \frac{v^2}{2}\right)^{\frac{\gamma}{\gamma-1}}$$

where v^2 is the solution of the equation (11.33) .

11.4 Two-Dimensional Flows

Consider in the Cartesian coordinate system x, y, z two-dimensional flows , i.e. the flows for which $X^1 = \psi(x, y)$, $X^2 = z$, $\zeta = \zeta(x, y)$.

Equation (11.10) becomes

$$\rho v_x = \psi_{,y}, \quad \rho v_y = -\psi_{,x}, \quad v_z = 0.$$

All variational principles formulated above can be applied for two-dimensional flows. For example, in the case of the isoenergetic motion,[2] the Giese-Kraiko principle transforms to

Lin-Rubinov variational principle. *Steady two-dimensional flow of ideal gas is a stationary point of the functional*

$$\int_V \left[\frac{1}{2\rho} \left(\psi_{,x}^2 + \psi_{,y}^2 \right) - \frac{a\rho^\gamma}{\gamma - 1} + \tilde{R}\rho \right] dxdy$$

on the set of functions $\psi(x, y)$, which are constant on $\partial V - S$ and take on the assigned values on S.[3]

Analogously, the variational principles for axis-symmetric flows can be obtained: one has to take into account that in the cylindrical coordinate system r, φ, z: $X^1 = \psi(r, z)$, $X^2 = \varphi$, $\zeta = \zeta(r, z)$.

11.5 Variational Principles on the Set of Equivortical Flows

All the variational principles for steady flows considered above are "true variational principles" in the sense that the set of admissible functions is fixed and does not depend on a stationary point. The variational principles of this section stand apart: they are concerned with the variational features of stationary points on the sets that are determined by the stationary points. Such variational principles are closer to the non-holonomic variational problems: for each element u, its own set of admissible variations, δu, is introduced.

We begin from the consideration of incompressible fluids. First, note that the Lagrangian components of vorticity do not change over the stream lines: from (9.15) and (11.2), $\mathring{v}_c = r_c^i r_{i,\zeta} - \partial_c \varphi$, therefore,

$$\mathring{\omega}^a(X^\mu, X) = e^{abc} \partial_b \mathring{v}_c = e^{abc} r_c^i r_{ib,\zeta}. \tag{11.34}$$

Differentiating this relation over time we get

$$0 = e^{abc} r_c^i r_{ib,\zeta\zeta}.$$

On the other hand, the derivative of $\mathring{\omega}^a$ over X^3, is

$$\mathring{\omega}_{,3}^a = e^{abc} r_c^i r_{ib,\zeta\zeta}.$$

Hence, $\mathring{\omega}_{,3}^a = 0$.

---

[2] Isoenergetic is the motion for which $\tilde{R} \equiv const$.

[3] Here, V is a region in the plane x, y, and S is a part of its boundary.

For a steady motion and a given vorticity, $\mathring{\omega}^a$, kinetic energy is a functional of $r(X, \zeta)$,

$$\mathcal{K}(r(X,\zeta)) = \rho_0 \int_V \frac{1}{2} V^i(x|r(X,\zeta)) V_i(x|r(X,\zeta)) d^3x. \tag{11.35}$$

Velocity does not depend on time, and so does kinetic energy.

Consider the functional $\mathcal{A}$ on steady motion,

$$\mathcal{A} = \frac{1}{3}\rho_0 \int\limits_{\mathring{V}} e_{ijk} r^j (X^\mu, X+t) r^i_\zeta (X^\mu, X+t) r^k_a (X^\mu, X+t) \mathring{\omega}^a (X^\mu) d^3X.$$

Transforming the integral to Eulerian variables,

$$\mathcal{A} = \frac{1}{3}\rho_0 \int\limits_V e_{ijk} r^j r^i_\zeta r^k_a \mathring{\omega}^a (X^\mu(x)) \frac{1}{\Delta} d^3x,$$

we see that $\mathcal{A}$ does not depend on time. Consider the functional

$$\bar{J} = \mathcal{A} - \mathcal{K}, \tag{11.36}$$

$$\mathcal{A} = \frac{1}{3}\rho_0 \int\limits_{\mathring{V}} e_{ijk} r^j (X^\mu, \zeta) r^i_\zeta (X^\mu, \zeta) r^k_a (X^\mu, \zeta) \mathring{\omega}^a (X^\mu) d^2X d\zeta. \tag{11.37}$$

For a given vorticity, $\mathring{\omega}^a$, $\bar{J}$ is a functional of diffeomorphisms that map the set of Lagrangian coordinates, $\mathring{V}$, to a given region, V.

Variational principle. *The true stream lines of the steady vortex flow are the stationary points of the functional (11.36) on the set of all compressible diffeomorphisms $\mathring{V} \to V$.*

We check the validity of this statement by direct derivation. We have

$$\begin{aligned}
&\delta \left[e_{ijk} r^j r^i_\zeta r^k_a \mathring{\omega}^a (X^\mu) \right] \\
&= e_{ijk} \delta r^j r^i_\zeta r^k_a \mathring{\omega}^a (X^\mu) + e_{ijk} r^j (\delta r^i)_\zeta r^k_a \mathring{\omega}^a (X^\mu) + e_{ijk} r^j r^i_\zeta (\delta r^k)_a \mathring{\omega}^a (X^\mu) \\
&= e_{ijk} \delta r^j r^i_\zeta r^k_a \mathring{\omega}^a (X^\mu) + (e_{ijk} r^j \delta r^i r^k_a \mathring{\omega}^a (X^\mu))_\zeta - e_{ijk} r^j_\zeta \delta r^i r^k_a \mathring{\omega}^a (X^\mu) \\
&\quad - e_{ijk} r^j \delta r^i r^k_{a,\zeta} \mathring{\omega}^a (X^\mu) (e_{ijk} r^j r^i_\zeta \delta r^k \mathring{\omega}^a (X^\mu))_a - e_{ijk} r^j_a r^i_\zeta \delta r^k \mathring{\omega}^a (X^\mu) \\
&\quad - e_{ijk} r^j r^i_{\zeta,a} \delta r^k \mathring{\omega}^a (X^\mu) - e_{ijk} r^j r^i_\zeta \delta r^k \mathring{\omega}^a_{,a} \\
&= 3 e_{ijk} \delta r^j r^i_\zeta r^k_a \mathring{\omega}^a (X^\mu) + (e_{ijk} r^j \delta r^i r^k_a \mathring{\omega}^a (X^\mu))_\zeta + (e_{ijk} r^j r^i_\zeta \delta r^k \mathring{\omega}^a)_a.
\end{aligned}$$

Here we used that $\mathring{\omega}^a_{,a} = 0$. Integral of the divergence terms vanish (recall that $\mathring{\omega}^a \mathring{n}_a = 0$ at the boundary, and the stream lines are closed). Finally,

$$\delta \mathcal{A} = \rho_0 \int_{\mathring{V}} e_{ijk} \delta r^j r^i_\zeta r^k_a \mathring{\omega}^a (X^\mu) d^2 X d\zeta. \tag{11.38}$$

Variation of kinetic energy is (see (9.128))

$$\delta \mathcal{K} = -\rho_0 \int_{\mathring{V}} e_{ijk} \delta r^j V^i r^k_a \mathring{\omega}^a (X^\mu) d^2 X d\zeta. \tag{11.39}$$

We obtain the equation

$$e_{ijk}(r^i_\zeta - V^i)\omega^k = 0. \tag{11.40}$$

This is equivalent to the equation of steady flows,

$$e^{ijk}\partial_j(v^m \partial_m v_k) = 0,$$

where by v_k we mean the field $V_k(x|\, r(X, \zeta))$. Indeed, (11.40) can be written as

$$r^i_\zeta = V^i + \lambda \omega^i, \tag{11.41}$$

where λ is an arbitrary function. The curl of acceleration is

$$e^{ijk}\partial_j(v^m \partial_m v_k) = v^m \partial_m \omega^i + e^{ijk}\partial_j v^m \partial_m v_k.$$

From (9.110),

$$e^{ijk}\partial_j(v^m \partial_m v_k) = v^m \partial_m \omega^i - \omega^k \partial^i v_k + \omega^i \partial^k v_k. \tag{11.42}$$

Divergence of velocity, $\partial^k v_k$, is zero because velocity field is found from the kinematic problem of Sect. 9.5 (note that divergence of r^i_ζ is, in general, non-zero). For the first term in the right hand side of (11.42) we have

$$v^m \partial_m \omega^i = \left(r^m_\zeta - \lambda \omega^m\right) \partial_m \left(r^i_a(X^\mu, \zeta) \frac{\mathring{\omega}^a(X^\mu)}{\Delta} \right) = \frac{\partial}{\partial \zeta} r^i_a(X^\mu, \zeta) \frac{\mathring{\omega}^a(X^\mu)}{\Delta}$$

$$-\lambda \omega^m \partial_m \omega^i = r^i_{\zeta,a} \frac{\mathring{\omega}^a(X^\mu)}{\Delta} - \frac{\omega^i}{\Delta} \frac{\partial \Delta}{\partial \zeta} - \lambda \omega^m \partial_m \omega^i = (V^i + \lambda \omega^i)_{,m} \omega^m - \frac{\omega^i}{\Delta} \frac{\partial \Delta}{\partial \zeta}$$

$$-\lambda \omega^m \partial_m \omega^i = = v^i_{,m} \omega^m + \lambda_{,m} \omega^i \omega^m - \frac{\omega^i}{\Delta} \frac{\partial \Delta}{\partial \zeta}.$$

The derivative of the determinant can be expressed in terms of derivatives of r^i_ζ :

$$\frac{\partial \Delta}{\partial \zeta} = \frac{\partial \Delta}{\partial r^i_a} \frac{\partial r^i_a}{\partial \zeta} = \Delta \frac{\partial X^a}{\partial x^i} \frac{\partial r^i_\zeta}{\partial X^a} = \Delta \frac{\partial r^i_\zeta}{\partial x^i}.$$

According to (11.41),

$$\frac{\partial r^i_\zeta}{\partial x^i} = \omega^i \frac{\partial \lambda}{\partial x^i}.$$

Thus

$$v^m \partial_m \omega^i = v^i_{,m} \omega^m,$$

and the right hand side of (11.42) vanishes, i.e. momentum equations hold true at the stationary points as claimed.

The idea to consider equivortical flows, i.e. the flows with the same vorticity, $\mathring{\omega}^a$, as a set of admissible functions, was suggested by V.I. Arnold.

Arnold variational principle. *The true stream lines of the steady vortex flow are the stationary points of kinetic energy (11.35) on the set of all incompressible diffeomorphisms* $\mathring{V} \to V$.

This result follows from (11.39): for incompressible flows,

$$\frac{\partial \delta r^i}{\partial x^i} = 0,$$

and therefore there are functions, χ_k, such that

$$\delta r^i = e^{ijk} \partial_j \chi_k \qquad \text{or} \qquad e_{ijk} \delta r^i = \partial_j \chi_k - \partial_k \chi_j. \tag{11.43}$$

Hence,

$$\delta \mathcal{K} = \rho_0 \int_{\mathring{V}} \left(v^j \omega^k - v^k \omega^j \right) \partial_j \chi_k d^2 X d\zeta,$$

and, due to arbitrariness of χ_k,

$$\left(v^j \omega^k - v^k \omega^j \right)_{,j} = 0. \tag{11.44}$$

The left hand side of (11.44) is the curl of the momentum equations of ideal incompressible fluid. The boundary terms are zero due to conditions $v^j n_j = \omega^j n_j = 0$.

Comparison of Arnold variational principle and the variational principle for the functional $\bar{J}$ suggests that the following statement holds:

Variational principle. *Let* $x^i = \check{r}^i(X^\mu, \zeta)$ *be the parametric equations of stream lines of an incompressible steady ideal fluid flow with vorticity,* $\mathring{\omega}^a(X^\mu)$, *and the parameter* ζ *be chosen in such a way that* $\check{r}^i_\zeta$ *is fluid velocity. Then* $\check{r}^i(X^\mu, \zeta)$ *is a stationary point of the functional* $\mathcal{A}$ *(11.37) on the set of all incompressible diffeomorphisms* $\mathring{V} \to V$.

Indeed, from (11.38) and (11.43),[4]

$$\delta \mathcal{A} = \rho_0 \int\limits_V e_{ijk} \delta r^j \check{r}^i_\zeta \check{\omega}^k d^3x = -\rho_0 \int\limits_V (\check{r}^i_\zeta \check{\omega}^k - \check{r}^k_\zeta \check{\omega}^i) \partial_i \chi_k d^3x,$$

and $\delta \mathcal{A} = 0$ yields the equations,

$$(\check{r}^i_\zeta \check{\omega}^k - \check{r}^k_\zeta \check{\omega}^i)_{,i} = 0.$$

which are equivalent to the equations of ideal incompressible fluid (note that $\partial_i \check{r}^i_\zeta = 0$).

Arnold's variational principle can be generalized to compressible fluids.

Arnold-Grinfeld variational principle. *The stationary points of the total energy functional of a barotropic fluid*

$$\int\limits_V \rho \left(\frac{v^2}{2} + U(\rho) + \Phi(x) \right) dV$$

on the set of all equivortical flows conserving the mass are the steady flows of ideal compressible fluid.

In this variational principle, velocity and density are varied. Conservation of mass is understood in the natural sense: admissible density and velocity satisfy the equations $\left(\rho v^i\right)_{,i} = 0$, and the mapping conserving vorticity also conserves mass.

At first glance, this variational principle contradicts the variational principle for the functional (11.15), which claims that the difference, not the sum, of kinetic and internal energy has the stationary value. The resolution of this "paradox" is that there are two different functionals of kinetic energy, the functional

$$\mathcal{K}(r(X,\zeta)) = \int_V \frac{\rho}{2} V^i(x|r(X,\zeta)) V_i(x|r(X,\zeta)) d^3x \tag{11.45}$$

and the functional

$$\mathcal{K}_1(r(X,\zeta)) = \int_{\mathring{V}} \frac{\rho}{2} r^i_{,\zeta} r_{i,\zeta} \Delta d^2X d\zeta. \tag{11.46}$$

They coincide at the stationary point but differ at admissible motions. In Arnold and Arnold-Grinfeld variational principles, the functional $\mathcal{K}(r(X,\zeta))$, i.e. the functional on equivortical motions, is employed while in the functional (11.15) kinetic energy is understood as $\mathcal{K}_1(r(X,\zeta))$. The variations of $\mathcal{K}(r(X,\zeta))$ and $\mathcal{K}_1(r(X,\zeta))$ differ by sign:

[4] $\check{\omega}^i$ is vorticity at the stationary point.

$$\delta\mathcal{K}(r(X,\zeta)) = -\delta\mathcal{K}_1(r(X,\zeta)). \tag{11.47}$$

This is why the functional (11.15) is the difference of kinetic and potential energies, while the variational principle on the set of equivortical flows claims stationarity of the total energy.

To prove (11.47), we note that

$$\delta\mathcal{K}_1(r(X,\zeta)) = \delta\int_{\mathring{V}}\frac{\rho_0}{2}r^i_{,\zeta}r_{i,\zeta}\sqrt{\mathring{g}}d^2Xd\zeta = \int_{\mathring{V}}\rho_0 r_{i,\zeta}(\delta r^i)_{,\zeta}\sqrt{\mathring{g}}d^2Xd\zeta,$$

and, due to closedness of stream lines and independence of the product, $\rho_0\sqrt{\mathring{g}} = \sigma$, on ζ (see (11.8)):

$$\delta\mathcal{K}_1 = -\int_{\mathring{V}}(\rho_0\sqrt{\mathring{g}}v_i)_{,\zeta}\delta r^i d^2Xd\zeta = -\int_V \rho v_{i,j}r^j_{,\zeta}\delta r^i dV = -\int_V \rho v_{i,j}v^j\delta r^i dV. \tag{11.48}$$

For variation of $\mathcal{K}(r(X,\zeta))$ we have

$$\begin{aligned}\delta\mathcal{K}(r(X,\zeta)) &= \delta\int_V\frac{\rho}{2}V^i(x|r(X,\zeta))V_i(x|r(X,\zeta))d^3x\\ &= \delta\int_{\mathring{V}}\frac{\rho_0}{2}V^i(x|r(X,\zeta))V_i(x|r(X,\zeta))\sqrt{\mathring{g}}d^2Xd\zeta\\ &= \int_{\mathring{V}}\rho_0 V_i\delta V^i\sqrt{\mathring{g}}d^2Xd\zeta = \int_V \rho V_i\delta V^i dV.\end{aligned}$$

Variation δV^i is caused by variation of vorticity:

$$\delta\omega^i = \delta\left(r^i_a\frac{\mathring{\omega}^a}{\Delta}\right) = (\delta r^i)_{,a}\frac{\mathring{\omega}^a}{\Delta} - r^i_a\frac{\mathring{\omega}^a}{\Delta}\frac{\partial\delta r^k}{\partial x^k} = (\delta r^i)_{,k}\omega^k - \omega^i(\delta r^k)_{,k}.$$

Hence, the variation of Eulerian components of vorticity at a space point, $\eth\omega^i$, are

$$\eth\omega^i = \delta\omega^i - \delta x^k\omega^i_{,k} = (\delta r^i)_{,k}\omega^k - \omega^i(\delta r^k)_{,k} - \delta x^k\omega^i_{,k} = (\delta r^i\omega^k - \omega^i\delta r^k)_{,k}.$$

Variation of velocity at a space point, $\eth V^i = \delta V^i - \delta x^k V^i_{,k}$, are linked to $\eth\omega^i$ as

$$e^{ikj}\partial_k\eth V_j = \eth\omega^i = (\delta r^i\omega^k - \omega^i\delta r^k)_{,k}.$$

The general solution to this equation, $\eth V_j$, is a sum of the partial solution, $e_{ikj}\delta r^i\omega^k$, and an arbitrary potential field, $\varphi_{,j}$:

$$\eth V_j = e_{ikj}\delta r^i\omega^k + \varphi_{,j}.$$

Hence,

$$\delta\mathcal{K}(r(X,\zeta)) = \int_V \rho V^j(e_{ikj}\delta r^i\omega^k + \varphi_{.j} + \delta x^k V_{j,k})dV =$$
$$\int_V \rho V^j\left[-\delta r^i(V_{j,i} - V_{i,j}) + \partial_j\varphi + \delta x^k V_{j,k}\right]dV = \int_V \rho V^j\left[\delta r^i V_{i,j} + \varphi_{.j}\right]dV. \tag{11.49}$$

The last term in (11.49) vanishes because at the true motion $(\rho V^j)_{,j} = 0$, and $\rho V^j n_j = 0$. Finally,

$$\delta\mathcal{K}(r(X,\zeta)) = \int_V \rho V^j V_{i,j}\delta r^i dV. \tag{11.50}$$

At the stationary point,[5] $v^i = V^i$, and (11.47) follows from (11.48) and (11.50).

This reasoning yields the following variational principle for elastic bodies.

Variational principle. *Let $x^i = \check{r}^i(X^\mu,\zeta)$ be the parametric equations of stream lines of a steady motion of homogeneous elastic body. Then $\check{r}^i(X^\mu,\zeta)$ is a stationary point of the functional of total energy,*

$$\mathcal{K}(r(X,\zeta)) + \int_{\mathring{V}} \rho_0\left(U\left(r^i_a\right) + \Phi\left(r(X,\zeta)\right)\right)d\mathring{V}, \tag{11.51}$$

on the set of all mappings, $\mathring{V} \to V$, that have the same vorticity as $\check{r}^i(X^\mu,\zeta)$.

If there are external forces at the boundary, then the corresponding linear functional must be added to (11.51).

11.6 Potential Flows

This section is concerned with the extremal features of the steady potential flows of the compressible fluid. For such flows, the closed system of equations comprises the continuity equation

$$\frac{\partial\rho v^i}{\partial x^i} = 0 \quad \text{in } V, \tag{11.52}$$

the potentiality condition

$$v^i = \frac{\partial\varphi}{\partial x^i}, \tag{11.53}$$

[5] For admissible motions, $v^i \neq V^i$.

the Cauchy-Lagrange integral

$$\frac{1}{2}v^2 + \frac{d(\rho U(\rho))}{d\rho} + \Phi\left(x^i\right) = k = const, \tag{11.54}$$

and the boundary condition

$$\rho v^i n_i = h \quad \text{on } \partial V. \tag{11.55}$$

It is necessary for the consistency of the system of equations that the total boundary mass flux, h, vanishes

$$\int_{\partial V} h dA = 0. \tag{11.56}$$

Denote by ρ^* the function of $\nabla\varphi$ and x :

$$\rho^* \equiv k - \frac{1}{2}v^2 - \Phi(x), \quad v_i = \frac{\partial \varphi}{\partial x^i}.$$

Here k is the constant from the Cauchy-Lagrange integral. The Cauchy-Lagrange integral can be viewed as the relation following from maximization of $\rho^*\rho - \rho U(\rho)$ with respect to ρ. The corresponding maximum value has the meaning of pressure. As before, we denote it by $\mathcal{P}(\rho^*)$.

Bateman-Dirichlet variational principle. *On the set of all functions $\varphi(x)$, the stationary points of the functional*

$$J = \int_V \mathcal{P}\left(\rho^*\right) dV + \int_{\partial V} \varphi h dA \tag{11.57}$$

satisfy the equations of steady potential flows of compressible gas (11.52) and (11.55).

The Cauchy-Lagrange integral is automatically valid due to the definition of $\mathcal{P}(\rho^*)$.

The function, $\mathcal{P}(\rho^*)$, is a convex function of ρ^* as the Young-Fenchel transformation of the convex function, $\rho U(\rho)$. The same cannot be said of the dependence of $\mathcal{P}$ on the potential gradient, $\varphi_i \equiv \partial\varphi/\partial x^i$. Indeed, let us calculate the second derivatives of $\mathcal{P}$ with respect to φ_i:

$$\frac{\partial \mathcal{P}}{\partial \varphi_i} = \frac{\partial \mathcal{P}}{\partial \rho^*}\frac{\partial \rho^*}{\partial \varphi_i} = -\rho\varphi_i,$$

$$\frac{\partial^2 \mathcal{P}}{\partial \varphi_i \partial \varphi_j} = -\rho\delta_{ij} - \frac{d\rho}{dp}\frac{\partial \mathcal{P}}{\partial \varphi_j} = -\rho\left(\delta_{ij} - \frac{1}{c^2}\varphi_i\varphi_j\right).$$

Here, c denotes the speed of sound: $c^2 = dp/d\rho$. The matrix $\left\| \frac{\partial^2 \mathcal{P}}{\partial \varphi_i \partial \varphi_j} \right\|$ is negative for the subsonic flows ($v^2 < c^2$) and positive for supersonic flows ($v^2 > c^2$). The subsonic flow can be selected by replacing in (11.57) $\mathcal{P}(\rho^*)$ by $\tilde{\mathcal{P}}(\rho^*)$:

$$\tilde{\mathcal{P}}(\rho^*) = \begin{cases} \mathcal{P}(\rho^*), & v^2 \leq c^2 \\ -\infty & v^2 > c^2 \end{cases}$$

The function $\tilde{\mathcal{P}}(\rho^*)$ is apparently a strictly concave function of φ_i. Therefore, after substituting $\mathcal{P}(\rho^*)$ by $\tilde{\mathcal{P}}(\rho^*)$ in (11.57), the corresponding functional has a unique stationary point, and at this stationary point the functional reaches its maximum.

The solvability condition of the boundary value problem (11.56) apparently coincides with the necessary condition for the functional to be bounded from above. It is obtained by considering the shifts of φ for a constant.

Let us construct the dual variational principle. Denote the Young-Fenchel transformation of the function-$\tilde{\mathcal{P}}$, which is convex with respect to φ_i, by $\mathcal{P}^*(p^i)$:

$$\mathcal{P}^*(p^i) = \max_{\varphi_i} \left(p^i \varphi_i + \tilde{\mathcal{P}}(\varphi_i) \right).$$

The dual variables have the meaning of momentum. If the flow corresponding to a given momentum p^i is subsonic, then $\mathcal{P}^*$ is equal to $p + \rho \frac{v^2}{2}$ expressed in terms of momentum p^i.

For the dual principle, we have

$$\max_{\varphi} \left[\int_V \tilde{\mathcal{P}}(\rho^*) \, dV + \int_{\partial V} \varphi h \, dA \right] = -\min_{\varphi} \left[-\int_V \tilde{\mathcal{P}}(\rho^*) \, dV - \int_{\partial V} \varphi h \, dA \right] =$$

$$-\min_{\varphi} \max_{p^i} \left[\int_V (p^i \varphi_i - \mathcal{P}^*(p^i)) dV - \int_{\partial V} \varphi h \, dA \right] =$$

$$-\max_{p^i} \min_{\varphi} \left[\int_V (p^i \varphi_i - \mathcal{P}^*(p^i)) dV - \int_{\partial V} \varphi h \, dA \right] =$$

$$-\max_{p^i} \left[-\int_V \mathcal{P}^*(p^i) \, dV \right] = \min_{p^i} \left[-\int_V \mathcal{P}^*(p^i) dV \right]. \tag{11.58}$$

The minimum in (11.58) is calculated over all p^i satisfying the constraints

$$\frac{\partial p^i}{\partial x^i} = 0 \quad \text{in } V, \qquad p^i n_i = h \quad \text{on } \partial V. \tag{11.59}$$

Thus, we get

Bateman-Kelvin variational principle. *The minimizing element of the functional,*

$$\int_V \mathcal{P}^*(p^i)dV,$$

considered on the set of vector fields, p^i, satisfying the constraints (11.59), *corresponds to a subsonic flow of the compressible gas.*

11.7 Regularization of Functionals in Unbounded Domains

As a rule, the flows in unbounded regions have diverging energy functionals. In order to make the variational principles sensible, it is necessary to modify (to regularize) the energy functional without changing Euler equations. We describe the regularization method for an example of a two-dimensional potential flow of compressible gas following Shiffman [277].

Consider subsonic steady potential flow of a compressible gas in two-dimensional plane, R_2. The gas flows around some body, B. The velocity potential, φ, is a function of two variables, x and y, defined in the exterior, $R_2 - B$, of the body, B. The Lagrangian of the problem, L, is a function of $\varphi_x = \partial\varphi/\partial x$ and $\varphi_y = \partial\varphi/\partial y$. Let $L\left(\varphi_x, \varphi_y\right)$ be a strictly convex function of φ_x and φ_y satisfying the condition

$$ka_i a^i \le \frac{\partial^2 L}{\partial\varphi_i \partial\varphi_j} a_i a_j \le K a_i a^i \tag{11.60}$$

for any a_i and all φ_x and φ_y. Here, k and K are constants, and, for brevity, the notation $\varphi_1 = \varphi_x$ and $\varphi_2 = \varphi_y$ is used. One can check that the function $L = -\mathcal{P}(\rho^*)$ satisfies this condition in the case of the subsonic flows.

At infinity, velocity is supposed to tend to some limit values, φ_i^∞. We formulate this condition by subjecting the admissible functions, φ, to the constraint

$$\Phi = \int_{R_2 - B} \left(\varphi_i - \varphi_i^\infty\right)\left(\varphi^i - \varphi^{i\infty}\right)dxdy < +\infty. \tag{11.61}$$

The integral (11.61) converges only if $\varphi_i \to \varphi_i^\infty$ at infinity. Note that condition (11.61) excludes non-zero circulation of velocity around the body B: for nonzero circulation $\varphi_i - \varphi_i^\infty \sim 1/r$, and the integral (11.61) diverges.

The functional

$$\int_{R_2 - B} L\left(\varphi_x, \varphi_y\right)dxdy$$

is obviously diverging for any admissible field. We need to replace it by another functional, which yields the same Euler equations and remains meaningful for $\varphi_i \to \varphi_i^\infty$ at infinity. Consider the variational problem

$$I = \int\limits_{R_2 - B} \left[L\left(\varphi_x, \varphi_y\right) - L\left(\varphi_x^{\infty}, \varphi_y^{\infty}\right) - L_{\varphi_x}\left(\varphi_x^{\infty}, \varphi_y^{\infty}\right)\left(\varphi_x - \varphi_x^{\infty}\right) - \right.$$

$$\left. - L_{\varphi_y}\left(\varphi_x^{\infty}, \varphi_y^{\infty}\right)\left(\varphi_y - \varphi_y^{\infty}\right)\right] dxdy \rightarrow \min .$$

This problem yields the same Euler equations. At the same time, the functional I is bounded from above and below, because

$$I = \int\limits_0^1 L_{\varphi_i \varphi_j}\Big|_{\varphi_i^{\infty} + \tau\left(\varphi_i - \varphi_i^{\infty}\right)} \left(\varphi_i - \varphi_i^{\infty}\right)\left(\varphi_j - \varphi_j^{\infty}\right)(1 - \tau) d\tau dxdy. \tag{11.62}$$

From (11.62) and (11.60), it follows that

$$\frac{1}{2} k \Phi \leq I \leq \frac{1}{2} K \Phi .$$

Therefore, the minimization problem for the functional I is well-posed.

Regularization in other variational problems is based on the same idea.

Chapter 12
Principle of Least Dissipation

The variational principles of ideal fluid and elastic body are all based on ignoring the dissipation. If the dissipation is not negligible, the governing equations do not have a variational structure, they possess a quasi-variational structure (see Sect. 2.6). In the another extreme case, when the dissipation plays the key role while the inertia and the internal energy effects are negligible, the variational structure of the governing equations appears again, but this is the variational structure of the non-equilibrium processes. This chapter is concerned with the corresponding variational principles. As has been mentioned in Sect. 2.6, these variational principles, in contrast to the least action principle, reflect the special features of the models used rather than the laws of Nature.

12.1 Heat Conduction

Consider an adiabatically isolated body, V. The heat propagation in the body is governed by the following system of equations: the first law of thermodynamics, linking the internal energy rate with the heat flux, $\vec{q}$,

$$\rho \frac{\partial U}{\partial t} = -div\,\vec{q}, \tag{12.1}$$

the constitutive equation

$$U = U(S), \tag{12.2}$$

and Fourier law

$$q^i = D^{ij} \frac{\partial}{\partial x^j} \frac{1}{T}, \quad T = \frac{dU(S)}{dS} \tag{12.3}$$

Due to Onsager's relations,

$$D^{ij} = D^{ji}. \tag{12.4}$$

V.L. Berdichevsky, *Variational Principles of Continuum Mechanics*,
Interaction of Mechanics and Mathematics, DOI 10.1007/978-3-540-88467-5_12,

For adiabatically isolated body, there is no heat flux through the boundary:

$$q^i n_i = 0 \quad \text{on } \partial V. \tag{12.5}$$

In linear non-equilibrium thermodynamics one assumes that temperature deviates slightly from a constant value, T_0 :

$$T = T_0 + T', \quad T' \ll T_0.$$

Therefore, Fourier law in linear approximation can be written as

$$q^i = -\frac{D_0^{ij}}{T_0^2}\partial_j T', \quad D_0^{ij} = D^{ij}\big|_{T=T_0} \tag{12.6}$$

while for time derivative of internal energy we have

$$\rho\frac{\partial U}{\partial t} = \rho\left.\frac{dU}{dT}\right|_{T=T_0}\frac{\partial T'}{\partial t} = c_v\frac{\partial T'}{\partial t}, \quad c_v \equiv \rho\left.\frac{dU}{dT}\right|_{T=T_0}.$$

Finally, the first law of thermodynamics transforms into a linear equation for T' :

$$c_v\frac{\partial T'}{\partial t} = \partial_i\left(\frac{D_0^{ij}}{T_0^2}\partial_j T'\right).$$

This equation contains D_0^{ij}, the values of D^{ij} at $T = T_0$. What follows pertains to the case of linear non-equilibrium thermodynamics and to the special models of non-equilibrium thermodynamics when temperature may change considerably, but, as in the linear case, heat conductivities, D^{ij}, do not depend on temperature.

Let us find the total dissipation[1] in the continuum at an instant t:

$$\begin{aligned}\mathcal{D} &= \frac{d}{dt}\int\limits_V \rho S dV = \int\limits_V \rho\frac{\partial S}{\partial t}dV = \int\limits_V \frac{1}{T}\rho\frac{\partial U}{\partial t}dV \\ &= \int\limits_V \frac{1}{T}div\,\vec{q}\,dV = \int\limits_V \vec{q}\nabla\frac{1}{T}\,dV = \int\limits_V D^{ij}\frac{\partial}{\partial x^i}\frac{1}{T}\frac{\partial}{\partial x^j}\frac{1}{T}\,dV.\end{aligned} \tag{12.7}$$

Here we used (12.1), (12.2), (12.3), (12.4) and (12.5). In terms of coldness, $\beta \equiv 1/T$, the dissipation is

$$\mathcal{D} = \int\limits_V D^{ij}\partial_i\beta\partial_j\beta\,dV. \tag{12.8}$$

[1] Another term used for this quantity is the dissipation rate.

In (12.8) the coefficients may depend on space points but do not depend on β, and the quadratic form, $D^{ij}\xi_i\xi_j$, is non-negative. In thermodynamic equilibrium, $\beta \equiv$ const and dissipation is equal to zero. Since dissipation is non-negative, one can say that dissipation is minimum in thermodynamic equilibrium. It is remarkable that dissipation is also minimum in a non-equilibrium steady process when we maintain temperature in some part of the body.

Least dissipation principle. *If temperature is maintained in a part, V_2, of region V, then dissipation considered as a functional of temperature (or coldness) attains its minimum value for steady heat conduction in region $V_1 = V - V_2$.*

Indeed,

$$\mathcal{D} = \int_{V_1} D^{ij}\partial_i\beta\partial_j\beta\, dV + \int_{V_2} D^{ij}\partial_i\beta\partial_j\beta\, dV. \tag{12.9}$$

The second integral is known. Varying the first integral with respect to β, we obtain the equation of steady heat conduction,

$$\partial_i q^i = 0, \tag{12.10}$$

where q^i are given by (12.3). The admissible temperature fields are assumed to be smooth. The values of β, given in the region, V_2, also determine the values of β on the surface, Σ, separating V_1 and V_2. Since the first integral in (12.9) feels the values of β at the boundary of V_1, the values of β on Σ must be considered as known. On the surface $\partial V_1 - \Sigma$ the admissible functions, $\beta(x)$, are arbitrary. This yields the boundary condition of adiabatic isolation, $q^i n_i = 0$ at $\partial V_1 - \Sigma$.

The second integral (12.9) plays the role of an additive constant and may be omitted. Dropping also index 1 at the region V we arrive at the minimization problem for the functional

$$\mathcal{D}(\beta) = \int_V D^{ij}\partial_i\beta\partial_j\beta\, dV$$

on the set of all functions β taking the prescribed values on Σ :

$$\beta = \beta_{(b)} \quad \text{on } \Sigma.$$

Applying the general scheme of construction of the dual variational principles, we obtain the following

Dual variational principle. *The true heat flux gives the maximum value to the functional*

$$2\int_\Sigma q^i n_i \beta_{(b)} dA - \int_V D_{ij}^{-1} q^i q^j\, dV \tag{12.11}$$

on the set of all vector fields, q^i, selected by (12.10).

Here D_{ij}^{-1} is the inverse tensor to D^{ij}. The maximum value of the functional (12.11) is equal to dissipation.

Note that the minimum dissipation principle does not hold if heat conductivities depend on temperature. This emphasizes its restricted physical meaning.

If the heat flux is given on the surface Σ, then the true coldness field provides minimum to the functional

$$\int_V D^{ij}\partial_i\beta\partial_j\beta\, dV - 2\int_\Sigma q_n\beta dA. \tag{12.12}$$

This functional can be obtained from the dissipation functional (12.9) in the following way. Let us fix β on Σ. Then the true β-field gives minimum to the second term in (12.9). Presenting this term by means of the dual variational problem we have

$$\mathcal{D} = \int_{V_1} D^{ij}\partial_i\beta\partial_j\beta\, dV + 2\int_\Sigma q^i n_i\beta dA - \int_{V_2} D_{ij}^{-1}q^i q^j\, dV. \tag{12.13}$$

It remains to note that the unit normal vector in (12.13) looks outside V_2, and, therefore, q_n in (12.12) is $q_n = -q^i n_i$. Therefore, the variational principle for the functional (12.12) can be also interpreted as the minimum dissipation principle when the heat flux vector in V_2 is fixed (it must be admissible, i.e. satisfy (12.10) in V_2).

12.2 Creeping Motion of Viscous Fluid

Consider a slow flow of viscous fluid in a container, V. The velocity is assumed to be so small that inertia forces are negligible compared to the friction forces. The velocity of the fluid is prescribed on the boundary,

$$v^i = v^i_{(b)} \quad \text{on } \partial V. \tag{12.14}$$

The flow is incompressible,

$$\frac{\partial v^i}{\partial x^i} = 0, \tag{12.15}$$

while the fluid is homogeneous and isotropic. Then the velocity field is the solution of the Stokes' equations,

$$-\frac{\partial p}{\partial x^i} + \mu\Delta v_i = 0 \quad \text{in } V, \tag{12.16}$$

with the boundary conditions (12.14). The dependence of the velocity field on time appears if the boundary values, or the boundary itself, slowly depend on time. Then the following variational principle holds.

Least dissipation principle. *The true motion of viscous fluid delivers the minimum value to the total dissipation*

$$\mathcal{D} = 2\int_V \mu v_{(i,j)} v^{(i,j)} dV \tag{12.17}$$

on the set of all velocity fields selected by the constraints (12.14) and (12.15).

Indeed, introducing the Lagrange multiplier, p, for the constraint (12.15) we get the functional

$$\int_V \left(2\mu v_{(i,j)} v^{(i,j)} - p v^i_{,i}\right) dV.$$

Its Euler equations with respect to velocity is (12.16).

The functional $\mathcal{D}$ differs from the true dissipation by the factor in the integrand: the dissipation is

$$2\int \frac{\mu}{T} v_{(i,j)} v^{(i,j)} dV.$$

We will assume in this and the next section that temperature is constant and, writing the dissipation, drop the factor $1/T$.

The minimum property of dissipation is a characteristic of the model rather than a manifestation of a deep physical feature of nonequilibrium processes. This becomes especially clear if we consider a class of models of viscous flow with non-linear potential stress-strain rate relations. For such models, a dissipation potential, $D\left(e'_{ij}\right)$, exists such that[2]

$$\sigma'^{ij} = \frac{\partial D}{\partial e'_{ij}}, \quad e_{ij} \equiv v_{(i,j)} = \frac{1}{2}\left(\frac{\partial v_i}{\partial x^j} + \frac{\partial v_j}{\partial x^i}\right). \tag{12.18}$$

The momentum equations are

$$-\frac{\partial p}{\partial x_i} + \partial_j \sigma'^{ij} = 0. \tag{12.19}$$

These equations can be obtained by minimization of the functional

$$\int_V D\left(v_{(i,j)}\right) dV \tag{12.20}$$

---

[2] Primes denote tensor deviators: $\sigma'^{ij} = \sigma^{ij} - \frac{1}{3}\sigma^k_k \delta^{ij}$, $e'_{ij} = e_{ij} - \frac{1}{3}e^k_k \delta^{ij}$.

on the set of incompressible velocity fields obeying the boundary conditions (12.14). If D is a quadratic form,

$$D = \mu e'_{ij} e'^{ij}, \tag{12.21}$$

then the functional (12.20) differs from the total dissipation

$$\int_V \sigma'^{ij} e'_{ij} dV$$

by a factor, i.e. the minimum dissipation principle holds. The same is true for any homogeneous function like, for example, a function

$$D = \mu \left(e'_{ij} e'^{ij} \right)^{\frac{1}{2}\left(1+\frac{1}{m}\right)}$$

considered in plasticity theory. However, if the dissipative potential is not a homogeneous function, like, for example, the dissipative potential of visco-plastic medium,

$$D = k\sqrt{e'_{ij} e'^{ij}} + \mu e'_{ij} e'^{ij}$$

then the functional to be minimized, (12.20), differs from the total dissipation.

Further, we consider the case of the media with general constitutive equations (12.18). We assume that the dissipative potential D is a strictly convex function of the strain rate tensor, e_{ij}.

Let the boundary of region V comprises two surfaces, S and Σ; surface forces, f_i, are given on S, while velocities are given on Σ:

$$v^i = v^i_{(b)} \quad \text{on } \Sigma. \tag{12.22}$$

Besides, some body forces, F_i, acts on the fluid.

Variational principle. *The true motion provides the minimum value to the functional*

$$I(v) = \int_V D\left(e_{ij}\right) dV - l(v), \tag{12.23}$$

$$l(v) = \int_V F_i v^i dV + \int_S f_i v^i dA \tag{12.24}$$

on the set of all velocity fields satisfying the constraints (12.15) and (12.22),

If there are no kinematic constraints (12.22) ($\Sigma = \varnothing$ and $S = \partial V$), then the functional (12.23) has a kernel – the set of all velocity fields corresponding to the rigid motions. For such fields, $D = 0$. The necessary condition for the functional $I(v)$ to be bounded from below is the vanishing of the resultant and the total moment of the forces acting on the fluid:

$$\int\limits_V F_i dV + \int\limits_{\partial V} f_i dA = 0, \quad \int\limits_V e^{ijk} F_j x_k dV + \int\limits_{\partial V} e^{ijk} f_j x_k dA = 0. \qquad (12.25)$$

The sufficiency of the condition (12.25) or (12.22) for boundedness below can be proven in the same way as for elastic bodies. The uniqueness theorem follows from the strict convexity of the dissipative potential and the convexity of the set of the admissible functions.

Now we construct the dual variational principle. Denoting the dual variables, the components of the viscous stress tensor, by τ^{ij} and the Young-Fenchel transformation of the function $D\left(e'_{ij}\right)$ by $D^*\left(\tau^{ij}\right)$ (in the space of the deviators of the strain rate tensor),

$$D^*\left(\tau^{ij}\right) = \max_{e'_{ij}} \left(\tau^{ij} e'_{ij} - D\left(e'_{ij}\right)\right),$$

we can write

$$D\left(e'_{ij}\right) = \max\left(\tau^{ij} e'_{ij} - D^*\left(\tau^{ij}\right)\right).$$

The maximum is calculated over all τ^{ij} of the deviator space,

$$\tau^{ij} = \tau^{ji}, \quad \tau_i^i = 0. \qquad (12.26)$$

Rewriting the variational principle as the minimax principle

$$\check{I} = \min_{\substack{v \in (12.22), \\ (12.15)}} \max_{\tau^{ij} \in (12.26)} \left(\int\limits_V \left(\tau^{ij} e'_{ij} - D^*\left(\tau^{ij}\right)\right) dV - l(v) \right).$$

and getting rid of the constraint (12.15) by introducing a Lagrange multiplier, the pressure $p(x)$, we have

$$\check{I} = \min_{v \in (12.22)} \max_{p, \tau \in (12.26)} \left(\Phi(v, p, \tau) - \int\limits_V D^*\left(\tau^{ij}\right) dV \right).$$

Here,

$$\Phi(v, p, \tau) = \int\limits_V \left(-p g^{ij} + \tau^{ij}\right) \frac{\partial v_i}{\partial x^j} dV - l(v).$$

Following to the general scheme of Sect. 5.8, one can check that the order of calculation of maximum and minimum can be changed, and, if

$$-\frac{\partial p}{\partial x_i}+\frac{\partial \tau^{ij}}{\partial x^j}+F^i=0 \quad \text{in } V, \quad \left(-pg^{ij}+\tau^{ij}\right)n_j=f^i \quad \text{on } S, \tag{12.27}$$

then

$$\min_{v_i\in(12.22)}\Phi(v,p,\tau)=\int_\Sigma\left(-p\delta_i^j+\tau_i^j\right)n_j v_{(b)}^i d\sigma\equiv l^*(p,\tau).$$

If p and τ^{ij} does not satisfy the constraints (12.27), then

$$\min_{v_i\in(12.22)}\Phi(v,p,\tau)=-\infty.$$

After minimization over v^i, we need to find maximum with respect to p, and τ^{ij}. Therefore, all fields p,τ^{ij} which do not satisfy the constraints (12.27) should be excluded. Thus, we get

Dual variational principle. *The true stress field provides the maximum for the functional*

$$J(p,\tau)=l^*(p,\tau)-\int_V D^*\left(\tau^{ij}\right)dV$$

on all fields p and τ^{ij} satisfying the constraints (12.27). Moreover,

$$\max J(p,\tau)=\min I(v).$$

12.3 Ideal Plasticity

In metals, elastic deformation is usually negligible compared to plastic deformation (on the order of 10^{-4} compared to, say, 10^{-2}). Therefore, one may identify the plastic strain rate with the total strain rate

$$\dot\varepsilon_{ij}^{(p)}=e_{ij}=\frac{1}{2}\left(\frac{\partial v_i}{\partial x^j}+\frac{\partial v_j}{\partial x^i}\right). \tag{12.28}$$

Since the plastic deformation usually preserves the volume, motion is incompressible:

$$\frac{\partial v^i}{\partial x^i}=0. \tag{12.29}$$

For slow motion, one may ignore the inertia effects, and momentum equations become

$$\frac{\partial \sigma^{ij}}{\partial x^j} = 0. \tag{12.30}$$

Besides, for slow motion heat conduction makes temperature constant over the specimen and equal to the ambient temperature. Thus,

$$\rho T \frac{dS}{dt} = \sigma^{ij} e_{ij} \equiv D.$$

In ideal plasticity, D is assumed to be a homogeneous function of first order with respect to e'_{ij} : $D = D\left(e'_{ij}\right)$; $D\left(\lambda e'_{ij}\right) = |\lambda| D\left(e'_{ij}\right)$. The total dissipation becomes a functional of the velocity field:

$$\mathcal{D} = \int\limits_V \sigma^{ij} e_{ij} dV = \int\limits_V D\left(e_{ij}\right) dV. \tag{12.31}$$

If velocity is prescribed on the boundary,

$$v^i = v^i_{(b)} \quad \text{on } \partial V, \tag{12.32}$$

then the variational principle holds:
Least dissipation principle. *The true velocity field provides the minimum value to the dissipation functional on the set of all incompressible velocity fields with the prescribed boundary values.*

At the minimizer, the stresses

$$\sigma^{ij} = -p\delta^{ij} + \sigma'^{ij}, \quad \sigma'^{ij} = \frac{\partial D}{\partial e'_{ij}}$$

satisfy the equilibrium equations, p being the Lagrange multiplier for the incompressibility condition (12.29).

Ideal plastic body is a special case of non-linear viscous media. For an ideal plastic body the dissipation is a homogeneous function of the first order. This feature of dissipation makes an ideal plastic body drastically different from viscous fluid. To see that, let us construct the dual variational principle. By the definition of the Young-Fenchel transformation of the dissipation density, D:

$$D^*\left(\sigma'^{ij}\right) = \max_{e'_{ij}} \left(\sigma'^{ij} e'_{ij} - D\left(e'_{ij}\right)\right). \tag{12.33}$$

First, we compute D^* for von Mises model, when

$$D = k\sqrt{e'_{ij} e'^{ij}}.$$

We have to find

$$D^*\left(\sigma'^{ij}\right) = \max_{e'_{ij}}\left(\sigma'^{ij}e'_{ij} - k\sqrt{e'_{ij}e'^{ij}}\right).$$

Any element in the strain rate space can be presented as a product of an element, $\mathring{e}'_{ij}$, on the unit sphere

$$\mathring{e}'_{ij}\mathring{e}'^{ij} = 1$$

and a positive number, λ :

$$e'_{ij} = \lambda\mathring{e}'_{ij}.$$

Therefore, the maximization problem can be split into the succession of the two problems

$$D^*\left(\sigma'^{ij}\right) = \max_{\lambda\geq 0}\max_{\mathring{e}'_{ij}}\lambda\left(\sigma'^{ij}\mathring{e}'_{ij} - k\right). \tag{12.34}$$

Maximum of a linear function, $\sigma'^{ij}\mathring{e}'_{ij}$, on the sphere is achieved when $\mathring{e}'_{ij}$ is proportional to σ'^{ij} : the linear function is the scalar product of a unit vector, $\mathring{e}'$, and a given vector, σ', in a five-dimensional space of deviator tensors; it reaches its maximum when the angle between the two vectors is equal to zero:

$$\mathring{e}'_{ij} = \varkappa\sigma'_{ij}, \quad \varkappa^2\sigma'_{ij}\sigma'^{ij} = 1$$

Hence,

$$D^*\left(\sigma'^{ij}\right) = \max_{\lambda\geq 0}\lambda\left(\sqrt{\sigma'_{ij}\sigma'^{ij}} - k\right).$$

Obviously, D^* may have only two values, 0 and $+\infty$, depending on whether $\sqrt{\sigma'_{ij}\sigma'^{ij}} - k$ smaller or greater than zero:

$$D^*\left(\sigma'^{ij}\right) = \begin{cases} 0 & \text{if } \sigma'_{ij}\sigma'^{ij} \leq k^2 \\ +\infty & \text{if } \sigma'_{ij}\sigma'^{ij} > k^2 \end{cases}. \tag{12.35}$$

According to the general scheme of Sect. 5.6 the dual functional is

$$\int_{\partial V}\left(-p\delta^{ij} + \sigma'^{ij}\right)n_j v_{i(b)}dA - \int_V D^*\left(\sigma'_{ij}\right)dV.$$

We have to maximize this functional, and thus all stress fields for which $D^* = +\infty$ must be excluded. Therefore, we admit only the stress field such that

$$\sigma'_{ij}\sigma'^{ij} \leq k^2.$$

The surface $\sigma'_{ij}\sigma'^{ij} = k^2$ is called yield surface. We arrive at
Dual variational principle. *The true stress field provides the maximum value to the linear functional*

$$\int_{\partial V} \left(-p\delta^{ij} + \sigma'^{ij}\right) n_j v_{i(b)} dA$$

on the set of all stress fields which obey the equilibrium equations and lie inside the yield surface.

For a general homogeneous convex function of the first order, $D^*\left(e'_{ij}\right)$, which is smooth everywhere (except the origin where such a function is always singular), Young-Fenchel transformation is similar to (12.35):

$$D^*\left(\sigma'^{ij}\right) = \begin{cases} 0 & \text{if } f\left(\sigma'^{ij}\right) \leq 0 \\ +\infty & \text{if } f\left(\sigma'^{ij}\right) > 0 \end{cases}$$

where $f\left(\sigma'^{ij}\right)$, the yield function, is some smooth convex function. Accordingly, in the dual variational principle the admissible stresses obey the condition $f\left(\sigma'^{ij}\right) \leq 0$. In ideal plasticity theory some non-smooth convex functions are also used; such cases can be treated as limits in a sequence of smooth dissipation functions.

12.4 Fluctuations and Variations in Steady Non-Equilibrium Processes

We have seen in Chap. 2 a deep relationship between fluctuations and variations: Einstein's formula for probability density of macroscopic variables (2.38) shows that in thermodynamic equilibrium the most probable state corresponds to the maximum value of entropy, i.e. it yields the first Gibbs principle. The question arises whether a similar fact is true for non-equilibrium processes. In this section we give some arguments in favor of the positive answer.

Consider a steady non-equilibrium process in some region, V. The major example will be heat conduction. Let the process be characterized by a finite set of variables, $y = (y_1, \ldots, y_m)$. In case of heat conduction we partition the region in a large number of small subregions and characterize the process in each subregion by a finite number of variables. The variables, y, fluctuate and have some probability density function, $f(y)$. We define entropy of the non-equilibrium process by the formula

$$f(y) = const\, e^{S(y)}. \tag{12.36}$$

Since the system is macroscopic, fluctuations of y are small. Besides, it is observed a certain macroscopic state, $\hat{y}$. The most probable values of y provide maximum to $S(y)$. Therefore, there must be a universal

Variational principle. *The true values of the characteristics of a steady non-equilibrium process correspond to the maximum value of the non-equilibrium entropy.*

Unfortunately, we do not have a reason to state that there is a universal function of y, $S(y)$, which does not depend on boundary conditions. In principle, for different true states, $\hat{y}$, the non-equilibrium entropy could be different: $S = S(\hat{y}, y)$. Function $S(\hat{y}, y)$ must have a maximum over y at $y = \hat{y}$. If $S(\hat{y}, y)$ does not depend on $\hat{y}$, we would have a "true" variational principle.

Let us try to construct the non-equilibrium entropy for heat conduction. First of all, we accept that the non-equilibrium entropy coincides with the equilibrium one for an equilibrium process. The equilibrium entropy for two bodies being in thermal contact was considered in Sect. 2.5. The thermodynamic state of the body is characterized by its energy. Denote the equilibrium entropies of the two bodies by $S_1(E_1)$ and $S_2(E_2)$, E_1 and E_2 being the energies of the bodies. The total energy, $E = E_1 + E_2$, is conserved because the system, body 1+body 2, is isolated. The total equilibrium entropy, $S(E_1, E_2)$, is the sum of equilibrium entropies of the bodies:

$$S(E_1, E_2) = S_1(E_1) + S_2(E_2).$$

The energies, E_1 and E_2, fluctuate. Their probability density function is

$$f(E_1, E_2) = const\, \delta(E - E_1 - E_2)\, e^{S_1(E_1)+S_2(E_2)}. \tag{12.37}$$

The most probable values of energies correspond to the maximum of entropy, $S_1(E_1) + S_2(E_2)$, under the constraint $E_1 + E_2 = E$. As we have seen in Sect. 2.5, this yields the equality of the temperatures of the bodies at equilibrium.

Expanding this reasoning to a continuum, we choose as the characteristic of the thermodynamic state of the system the field of the density of internal energy per unit mass, $U(x)$. The equilibrium entropy density per unit mass is denoted, as before, by S. The thermodynamic properties of the system are described by the function, $S = S(U)$. Internal energy fluctuates. The fluctuations occur in the functional space of the functions, $U(x)$. The corresponding probability density is a functional, $f(U)$. Generalizing (12.37) we can write

$$f(U) = const\, \delta\left(E - \int_V \rho U dV\right) e^{\mathcal{S}(U)}, \tag{12.38}$$

$$\mathcal{S}(U) = \int_V \rho S(U) dV. \tag{12.39}$$

Again, maximum of entropy under constraint,

$$\int_V \rho U dV = E,$$

corresponds to constant temperature over the body, as it must be in thermodynamic equilibrium.

Let the process now be non-equilibrium. We characterize it by the two fields, energy density, $U(x)$, and the heat flux, $q^i(x)$. For definiteness, let the heat flux be prescribed at the boundary,

$$q^i n_i\big|_{\partial V} = q_{(b)}. \tag{12.40}$$

We wish to construct a functional, $S(U, q)$, which has the maximum value at the solution of the boundary value problem,

$$\partial_i \hat{q}^i = 0, \qquad \hat{q}^i = \varkappa \partial^i \frac{1}{T}, \qquad \frac{1}{T} \equiv \frac{\partial S(\hat{U})}{\partial \hat{U}}, \qquad \hat{q}^i n_i\big|_{\partial V} = q_{(b)}. \tag{12.41}$$

Besides, for isolated bodies, i.e. for $q_{(b)} = 0$, the non-equilibrium entropy, $S(U, q)$, should coincide with the equilibrium one (12.39). We will consider the general case, when the heat conductivity, $\varkappa$, depends on temperature, or, for our choice of primary thermodynamic characteristic, on energy, $\varkappa = \varkappa(U)$. Then, as was mentioned in Sect. 12.1, the least dissipation principle is not valid, while we expect to obtain a meaningful variational principle for such a case.

Fluctuations of energy and the heat flux are not arbitrary; they are linked by the energy equation:

$$\rho \frac{\partial U}{\partial t} = -\frac{\partial q^i}{\partial x^i}. \tag{12.42}$$

Let the initial value of U be $\hat{U}$, and Δt be so small that q^i can be viewed as practically constant during this time interval. Denote by δU and δq^i the differences,

$$\delta U = U - \hat{U}, \qquad \delta q^i = q^i - \hat{q}^i.$$

Then from (12.42),

$$\rho \delta U = -\frac{\partial \delta q^i}{\partial x^i} \Delta t. \tag{12.43}$$

We see that the fluctuations of energy are completely determined by the fluctuations of the heat flux.

Consider the functional

$$\mathcal{S}(U,q) = \int_V \left(\rho S(U) - \frac{\rho}{2\varkappa(\hat{U})} q^i q_i \Delta t \right) dV. \tag{12.44}$$

For $q^i = 0$ it transforms to the functional (12.39). We consider the functional (12.44) in a small vicinity of the field, $\hat{q}^i$, which is divergence-free. Variations of heat flux, δq^i, are not divergence-free. The non-zero divergence of δq^i yields the variation of energy. At the boundary $\delta q^i n_i = 0$. The dependence of $\mathcal{S}(U,q)$ on the true fields, $\hat{U}$ and $\hat{q}^i$, enters through the small deviations of U and q^i from $\hat{U}$ and $\hat{q}^i$.

Variation of the functional, $\mathcal{S}(U,q)$, is

$$\delta\mathcal{S}(U,q) = \int_V \rho \left(\frac{1}{T} \delta U - \frac{1}{\varkappa(\hat{U})} q^i \delta q_i \Delta t \right) dV. \tag{12.45}$$

Plugging in (12.45) the expression of δU in terms of δq^i (12.43), and setting $\delta\mathcal{S} = 0$ for arbitrary δq^i, we arrive at the equations of steady heat conduction. Similarly, the variational principles for other non-equilibrium processes can be formulated.

The variational principle described is not a "true" variational principle: the functional depends on the stationary point. This makes it similar to the variational principles for vortex flows of Sects. 9.6 and 11.5.

Chapter 13
Motion of Rigid Bodies in Fluids

13.1 Motion of a Rigid Body in Creeping Flow of Viscous Fluid

Consider in some vessel, V, a linearly viscous isotropic incompressible fluid. The fluid occupies the entire vessel and contains a rigid body which can move in the fluid (Fig. 13.1). At some instant, the body occupies a region B and has a translational velocity, u^i:

$$v^i = u^i \quad \text{on } \partial B. \tag{13.1}$$

At the walls of the vessel, the fluid does not slip:

$$v^i = 0 \quad \text{on } \partial V. \tag{13.2}$$

If the region V is unbounded, then the no-slip condition at the boundary is replaced by the condition that the fluid is at rest at infinity.

Least dissipation principle. *The true fluid motion minimizes the dissipation*

$$\mathcal{D}(v) = \int_{V-B} 2\mu e_{ij} e^{ij} dV, \quad e_{ij} = \frac{1}{2}\left(\frac{\partial v_i}{\partial x^j} + \frac{\partial v_j}{\partial x^i}\right), \tag{13.3}$$

on the set of all incompressible velocity fields,

$$\frac{\partial v^i}{\partial x^i} = 0, \tag{13.4}$$

satisfying the boundary conditions (13.1), (13.2).

The minimizing element, $\check{v}^i$, obeys the momentum equation,

$$\frac{\partial \sigma^{ij}}{\partial x^j} = 0, \quad \sigma^{ij} = -\check{p} g^{ij} + \mu\left(\partial^i \check{v}^j + \partial^j \check{v}^i\right), \tag{13.5}$$

where $\check{p}$ is the Lagrange multiplier for the incompressibility constraint (13.4).

V.L. Berdichevsky, *Variational Principles of Continuum Mechanics*,
Interaction of Mechanics and Mathematics, DOI 10.1007/978-3-540-88467-5_13,

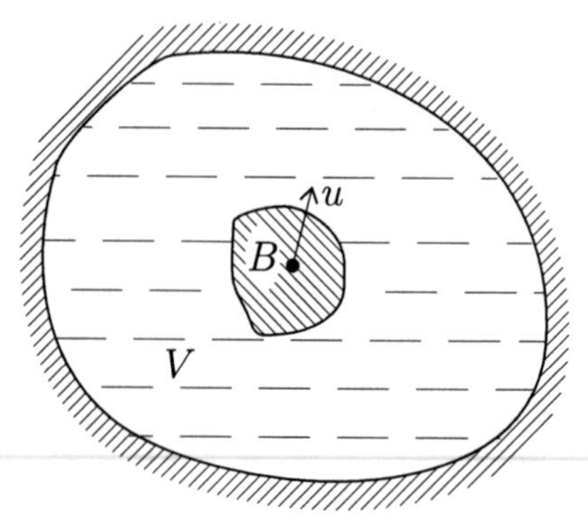

Fig. 13.1 Notation for motion of rigid body in a vessel

The system of equations for the minimizer, the so-called Stokes approximation, is linear. Therefore $\check{v}^i$ depends linearly on u^i. Hence, the minimum value, $\check{\mathcal{D}}$, of dissipation $\mathcal{D}(v)$ is a quadratic function of u_i:

$$\check{\mathcal{D}} = D^{ij} u_i u_j.$$

To determine the meaning of the coefficients of this quadratic form we note that, due to (13.1), (13.2), (13.4) and (13.5),

$$\check{\mathcal{D}} = \int\limits_{V-B} 2\mu \check{v}_{(i,j)} \check{v}^{(i,j)} dV = \int\limits_{V-B} \sigma^{ij} \check{v}_{i,j} dV = \int\limits_{\partial B} \sigma^{ij} n_j \check{v}_i dA = F^i u_i.$$

Here we introduced the notation,

$$F^i \equiv \int\limits_{\partial B} \sigma^{ij} n_j dA. \tag{13.6}$$

The normal vector in (13.6) looks outside the region occupied by the fluid, and therefore F^i are the components of the force with which the body acts on the fluid. The negative force, $-F^i$, is the force acting on the body. Since F^i is linear with respect to u_i,

$$F^i = D^{ij} u_j.$$

The force is potential,

$$F^i = \frac{\partial \mathbb{D}}{\partial u_i},$$

with the dissipative potential, $\mathbb{D}$, equal to

$$\mathbb{D} = \frac{1}{2} \mathcal{D}.$$

The estimates of the minimum value of dissipation yield the estimates of the force. Consider a few elementary consequences of the least dissipation principle.

1. For a steady flow of the viscous fluid, the force found within the Stokes approximation gives a low bound for the force calculated by means of the Navier-Stokes equations.

Indeed, the value of dissipation on any velocity field, including the velocity field found from the Navier-Stokes equations, is greater than that for the velocity field minimizing the functional $\mathcal{D}(v)$ and corresponding to the Stokes theory. It remains to show that for steady flow, $F^i u_i$ coincides with dissipation in the framework of the Navier-Stokes theory. If v_i is a solution of the Navier-Stokes equations and F^i_{NS} is the force computed from the Navier-Stokes theory, then

$$\int\limits_{V-B} 2\mu v_{(i,j)} v^{(i,j)} dV = \int\limits_{V-B} (2\mu v^{(i,j)} - p\delta^{ij}) v_{i,j} dV$$

$$= F^i_{NS} u_i - \int\limits_{V-B} (2\mu v^{(i,j)} - p\delta^{ij})_{,j} v_i dV = F^i_{NS} u_i - \int\limits_{V-B} \rho v^j v^i_{,j} v_i dV$$

$$= F^i_{NS} u_i - \int\limits_{V-B} \frac{1}{2} (\rho v^j v^2)_{,j} dV = F^i_{NS} u_i - \int_{\partial B} \frac{1}{2} \rho u^j u^2 n_j dA.$$

Here we used the boundary conditions of the Navier-Stokes theory: $v^i = u^i$ on ∂B. The last integral is, obviously, zero. Thus, the dissipation is equal to $F^i_{NS} u_i$.

2. The dissipation, and, consequently, the force increases as the size of the body increases.

Indeed, let a body B' be added to the body B. Denote by $\mathcal{D}_B(v)$ and $\check{\mathcal{D}}_{B+B'}(v)$ the dissipations corresponding to the bodies B and B', respectively, and by $v^i_{B+B'}$ the velocity field of the fluid motion around the body $B + B'$.Consider the velocity field

$$\tilde{v}^i = \begin{cases} v^i_{B+B'}, & x \in V - (B + B') \\ u^i & x \in B' \end{cases}$$

The velocity field $\tilde{v}^i$ is defined outside the body B and satisfies the conditions on ∂B and ∂V. Consequently, it is admissible in the minimization problem for the functional $\mathcal{D}_B(v)$. We have

$$\check{\mathcal{D}}_B \leq \mathcal{D}_B(\tilde{v}) = \check{\mathcal{D}}_{B+B'},$$

as claimed.

An immediate consequence of that feature is the following statement.

3. The force acting on the body in the Stokes flow increases if any additional body is placed in the flow and kept at rest.

If the rigid body, B, moves arbitrarily, then

$$v_i = u_i + e_{ijk}\omega^j \left(x^k - r^k\right) \quad \text{on } \partial B, \tag{13.7}$$

where r^k are the coordinates of the point of the body with the velocity u_i, and ω^j the angular velocity. It can be checked that $\check{\mathcal{D}}$ is the quadratic form with respect to u_i and ω_i, and the derivatives

$$F^i = \frac{1}{2}\frac{\partial \check{\mathcal{D}}}{\partial u_i}, \qquad M^i = \frac{1}{2}\frac{\partial \check{\mathcal{D}}}{\partial \omega_i}, \tag{13.8}$$

have the meaning of the resultant and the moment of the forces with which the body acts on the fluid. The moments have the same variational features as those mentioned for the resultants.

The inversion of formulas (13.8) is

$$u_i = \frac{\partial \mathcal{D}^*(F, M)}{\partial F^i}, \qquad \omega_i = \frac{\partial \mathcal{D}^*(F, M)}{\partial M^i}, \tag{13.9}$$

where $\mathcal{D}^*(F, M)$ is the Legendre transformation of $\check{\mathcal{D}}/2$ with respect to u_i and ω_i. Clearly, $\mathcal{D}^*$ also depends on the position vector of the rigid body, r^k, and on the orthogonal matrix, α^i_a, specifying the orientation of the body. Therefore, we write further $\mathcal{D}^* = \mathcal{D}^*(r, \alpha, F, M)$. Let us derive the variational formula for $\mathcal{D}^*(r, \alpha, F, M)$.

Consider the variational principle dual to the least dissipation principle. Following the general scheme of Sect. 5.6 we have

$$\begin{aligned}
\check{\mathcal{D}} &= \min_{v^i \in (13.7),(13.2),(13.4)} \max_{\sigma'^{ij}=\sigma'^{ji}} \int_{V-B} \left(\sigma'^{ij}\partial_i v_j - \frac{1}{4\mu}\sigma'^{ij}\sigma'_{ij}\right) dV \\
&= \min_{v^i \in (13.7),(13.2)} \max_{\sigma'^{ij}=\sigma'^{ji},\, p} \int_{V-B} \left((\sigma'^{ij} - p\delta^{ij})\partial_i v_j - \frac{1}{4\mu}\sigma'^{ij}\sigma'_{ij}\right) dV,
\end{aligned}$$

where p is the Lagrange multiplier for the constraint (13.4). Denoting the sum, $\sigma'^{ij} - p\delta^{ij}$, by σ^{ij}, and switching the order of computation of minimum and maximum, we obtain

$$\check{\mathcal{D}} = \max_{\sigma^{ij}} \min_{v^i \in (13.7),(13.2)} \int_{V-B} \left(\sigma^{ij}\partial_i v_j - \frac{1}{4\mu}\sigma'^{ij}\sigma'_{ij}\right) dV.$$

Hence, the admissible fields, σ^{ij}, obey the constraints,

$$\frac{\partial \sigma^{ij}}{\partial x^j} = 0, \tag{13.10}$$

and

$$\check{\mathcal{D}} = \max_{\sigma^{ij} \in (13.10)} \left(\int_{\partial B} \sigma^{im}(u_i + e_{ijk}\omega^j\left(x^k - r^k\right))n_m dA - \int_{V-B} \frac{1}{4\mu}\sigma'^{ij}\sigma'_{ij} dV \right). \tag{13.11}$$

Using the notation

$$F^i = \int_{\partial B} \sigma^{im} n_m dA, \qquad M_j = \int_{\partial B} \sigma^{im} e_{ijk} \left(x^k - r^k\right) n_m dA, \tag{13.12}$$

we can write the variational problem (13.11) as

$$\check{\mathcal{D}} = \max_{\sigma^{ij} \in (13.10)} \left(F^i u_i + M^i \omega_i - \int_{V-B} \frac{1}{4\mu} \sigma'^{ij} \sigma'_{ij} dV \right). \tag{13.13}$$

Consider the following function of the force and the moment:

$$\mathcal{D}^*(r, \alpha, F, M) = \min_{\sigma^{ij} \in (13.10),(13.12)} \int_V \frac{1}{4\mu} \sigma'^{ij} \sigma'_{ij} dV. \tag{13.14}$$

It follows from (13.13) and (13.14) that

$$\check{\mathcal{D}} = \max_{F^i, M^i} \left(F^i u_i + M^i \omega_i - \mathcal{D}^*(r, \alpha, F, M) \right).$$

Therefore (13.9) holds true, while to find $\mathcal{D}^*(r, \alpha, F, M)$ one has to solve the variational problem (13.14).

Since $u_i = dr_i/dt$, and $\omega_i = \frac{1}{2} e_{ijk} \alpha^{ja} d\alpha^k_a/dt$ (see (3.54)), (13.9) becomes a system of ordinary differential equations governing motion of a rigid body in viscous fluid:

$$\frac{dr_i}{dt} = \frac{\partial \mathcal{D}^*(r, \alpha, F, M)}{\partial F^i}, \qquad \frac{1}{2} e_{ijk} \alpha^{ja} \frac{d\alpha^k_a}{dt} = \frac{\partial \mathcal{D}^*(r, \alpha, F, M)}{\partial M^i}. \tag{13.15}$$

If there is a system of rigid bodies, with kinematic parameters, $r^i_{(1)}, \ldots, r^i_{(m)}, \alpha^i_{(1)a}, \ldots, \alpha^i_{(m)a}$, and kth body is subject to the external forces with the resultant, $F^i_{(k)}$, and moment, $M^i_{(k)}$, then $\mathcal{D}^*$ is a function of the sets of arguments, $r = (r^i_{(1)}, \ldots, r^i_{(m)})$, $\alpha = (\alpha^i_{(1)a}, \ldots, \alpha^i_{(m)a})$, $F = (F^i_{(1)}, \ldots, F^i_{(m)})$, $M = (M^i_{(1)}, \ldots, M^i_{(m)})$, and the system of differential equations takes the form

$$\frac{dr_{(k)i}}{dt} = \frac{\partial \mathcal{D}^*(r, \alpha, F, M)}{\partial F^i_{(k)}}, \quad \frac{1}{2} e_{ijk} \alpha^{ja}_{(k)} \frac{d\alpha^k_{(k)a}}{dt} = \frac{\partial \mathcal{D}^*(r, \alpha, F, M)}{\partial M^i_{(k)}}, \quad k = 1, \ldots, m. \tag{13.16}$$

The "potential" structure of this system indicates that there are quite peculiar interactions between the bodies moving in viscous fluid. The system of equations (13.16) is the basis for theoretical studies of properties of suspensions.

13.2 Motion of a Body in Ideal Incompressible Fluid

The Thomson-Tait equations. Constructing the equations governing the motion of a deformable body in a potential flow of ideal incompressible fluid is a non-elementary issue if the starting point is momentum equations. The derivation is considerably simplified if the variational approach is used.

We assume that the body occupies region $B(t)$ at time t, B will denote the region run by the Lagrangian coordinates of the body, X^a. The motion and the deformation of the body are described by a finite number of parameters $q^k(t)$ $(k = 1, \ldots, n)$:

$$x^i(X^a, t) = \chi^i(X^a, q^k) \quad \text{for} \quad X^a \in \partial B; \quad q^k = q^k(t).$$

The velocity of the boundary points depends linearly on $\dot{q}^k$:

$$\dot{x}^i(X^a, t) = \frac{\partial \chi^i}{\partial q^k}\dot{q}^k \quad \text{for } X^a \in \partial B. \tag{13.17}$$

The characteristics of the rigid motion of the body, $r^i(t)$ and $\alpha^i_a(t)$, are included in the set of parameters q^k.

Suppose that the fluid occupies a moving and possibly deforming vessel $V(t)$ and in the process of motion does not detach from the walls of the vessel and the boundary of the body. The deformation and the motion of the vessel is also given by a finite number of parameters, $b^k(t)$; the velocity of the vessel wall depends linearly on $\dot{b}^k$. The sets of the parameters, q^k and b^k, will be denoted by q and b, respectively.

Let $\Phi(x)$ be the potential of the external body forces, $\mathcal{K}_B$ and $\mathcal{U}_B$ the kinetic and the internal energies of the body, and $\mathcal{K}_B = \mathcal{K}_B(q, \dot{q})$, $\mathcal{U}_B = \mathcal{U}_B(q)$.

Least action principle. *The true motion of the body and the fluid is a stationary point of the functional*

$$\int_{t_0}^{t_1}\left(\int_{V(t)-B(t)}\left(\tfrac{1}{2}\rho v^i v_i - \rho\Phi(x)\right)dV + \mathcal{K}_B - \mathcal{U}_B\right)dt \tag{13.18}$$

on the set of functions $x(X, t)$ and $q(t)$ with the prescribed initial and final values.

Assigning the initial positions of the fluid particles fixes the choice of the Lagrangian coordinates. Assume that the final positions of the fluid particles are chosen in such a way as to cause a potential flow in the vessel. The possibility of such a choice is guaranteed by the solvability of the corresponding problems.

For potential flow, the velocity field of the fluid at each instant is defined by the geometry of the regions $V(t)$ and $B(t)$, i.e. by the values of the parameters q and b and the velocities of the fluid on $\partial V(t)$ and $\partial B(t)$, i.e. by $\dot{q}$ and $\dot{b}$. The kinetic energy of the fluid, $\mathcal{K}_F$, is found by solving the variational problem

$$\mathcal{K}_F = \min_v \int\limits_{V(t)-B(t)} \frac{1}{2}\rho v^i v_i dV, \tag{13.19}$$

where the minimum is sought over all the velocity fields satisfying the conditions

$$\partial_i v^i = 0, \quad v^i n_i\big|_{\partial B(t)} = \frac{\partial \chi^i}{\partial q^k}\dot{q}^k n_i, \quad v^i n_i\big|_{\partial V(t)} = c_k \dot{b}^k, \tag{13.20}$$

where $c_k \dot{b}^k$ is the normal component of the velocity of the fluid on $\partial V(t)$. The kinetic energy of fluid motion is a function of $q, \dot{q}, b, \dot{b}$: $\mathcal{K}_F = \mathcal{K}_F(q, \dot{q}, b, \dot{b})$.

Suppose that the variational problem (13.19) and (13.20) is solved and the function $\mathcal{K}_F\left(q, \dot{q}, b, \dot{b}\right)$ is found. Then the functional (13.18) becomes

$$\int\limits_{t_0}^{t_1} L\left(q, \dot{q}, b, \dot{b}\right) dt, \tag{13.21}$$

where

$$L = \mathcal{K}_F + \mathcal{K}_B - \mathcal{U}_B - \bar{\Phi}(q, b),$$
$$\bar{\Phi}(q, b) = \int\limits_{V(t)-B(t)} \rho \Phi(x) dV.$$

The least action principle yields the equations of motion

$$\frac{\partial L}{\partial q^k} - \frac{d}{dt}\frac{\partial L}{\partial \dot{q}^k} = 0. \tag{13.22}$$

For the case of motion of rigid bodies, these equations were established by Thomson and Tait [293]. Consider some examples.

Example 1. Consider the motion of a rigid body in an unbounded fluid being at rest at infinity. The position of the body is defined by the position vector $r^i(t)$ and an orthogonal matrix $\alpha^i_a(t)$. The constraints (13.20) become

$$v^i_{,i} = 0 \ \text{ in } R_3 - B(t), \quad v^i n_i = \left(r^i + e^{ijk}\omega_j (x_k - r_k)\right) n_i \ \text{ on } \partial B(t), \tag{13.23}$$

where $\omega_k = \frac{1}{2} e_{klm} \alpha^l_a \dot{\alpha}^{ma}$. The kinetic energy of the fluid is finite only if $v^i \to 0$ as $x \to \infty$; therefore, the condition that the fluid is at rest at infinity is automatically included in the variational formulation and does not need to be mentioned explicitly.

Let X^a be Cartesian coordinates rigidly linked to the body. The constraints (13.23) become

$$v^a_{,a} = 0 \ \text{ in } R_3 - B, \quad v^a n_a = \left(u^a + e^{abc}\omega_b X_c\right) n_a, \tag{13.24}$$

where $u^a = \alpha_i^a \dot{r}^i$, $\omega_b = \frac{1}{2} e_{bac} \alpha_i^c \dot{\alpha}^a$, and B is the region in which the Lagrangian coordinates of the body particles change. The kinetic energy of the fluid is quadratic with respect to u^a and ω^a is

$$\mathcal{K}_F = \frac{1}{2} \left(M^{ab} u_a u_b + 2R^{ab} u_a \omega_b + J^{ab} \omega_a \omega_b \right). \tag{13.25}$$

The coefficients M^{ab}, R^{ab}, and J^{ab} depend only on the geometry of the region B. The kinetic energy of the rigid body also has the form (13.25). The coefficients of the quadratic form $\mathcal{K} = \mathcal{K}_F + \mathcal{K}_B$ will be denoted by $\bar{M}^{ab}$, $\bar{R}^{ab}$, and $\bar{J}^{ab}$. The explicit dependence of $\mathcal{K}$ on the characteristics of motion is:

$$\mathcal{K} = \frac{1}{2} \left(\bar{M}^{ab} \alpha_a^i \alpha_b^j \dot{r}_i \dot{r}_j + \bar{R}^{ab} \alpha_a^i \alpha_b^j \dot{r}_i e_{jkl} \alpha_c^k \dot{\alpha}^{lc} + \frac{1}{4} \bar{J}^{ab} \alpha_a^i \alpha_b^j e_{imn} \alpha_c^m \dot{\alpha}^{nc} e_{jm'n'} \alpha_{c'}^{m'} \dot{\alpha}^{n'c'} \right).$$

Euler equations for the action functional are the equations of motion of a free rigid body with the effective characteristics $\bar{M}^{ab}$, $\bar{R}^{ab}$, $\bar{J}^{ab}$.

If L^{ab} and $\bar{M}^{ab}$ are diagonal,

$$M^{ab} = m_{(a)} \delta^{ab}, \quad \bar{M}^{ab} = \bar{M} \delta^{ab}$$

and m is the mass of the body, then

$$\bar{M} = m + m_{(a)}.$$

This looks like an effective increase of the mass of the body; the parameter $m_{(a)}$ is called the attached mass. In general case, the tensor M^{ab} is called the tensor of attached mass, and its trace, $\frac{1}{3} M_a^a$, the averaged attached mass.

Example 2. In the previous example, the kinetic energy did not depend on the position of the body. This is caused by the invariance of the kinetic energy with respect to translations, which, in turn, appears due to the absence of the external boundaries of the flow. Let us consider a problem where the flow has an external boundary.

Let the fluid occupy the half-space $x \geq 0$ and a ball of radius a move in the fluid along the x axis. The system has one degree of freedom – the distance q from the center of the ball to the plane $x = 0$. The flow potential , φ, is defined by the boundary value problem

$$\Delta \varphi = 0, \qquad \frac{\partial \varphi}{\partial n} = 0 \quad \text{for } x = 0, \qquad \frac{\partial \varphi}{\partial n} = \dot{q} n_x \quad \text{on the ball surface.}$$

At infinity $\nabla \varphi = 0$ because kinetic energy of the fluid is finite.

In the absence of the wall, φ is the dipole potential:

$$\varphi = \frac{1}{2} a^3 \dot{q} \frac{\partial}{\partial x} \frac{1}{|r - r_0|}.$$

Here, r and r_0 are the position vectors of a space point and the center of the ball, respectively, and $|r|$ the magnitude of vector, r. For large q/a, the harmonic function

$$\varphi = \frac{1}{2} a^3 \dot{q} \frac{\partial}{\partial x} \left(\frac{1}{|r - r_0|} - \frac{1}{|r + r_0|} \right) \tag{13.26}$$

satisfies the boundary condition at the wall and approximately satisfies the boundary condition on the ball. One can show that (13.26) is the leading term of the expansion with respect to a small parameter a/q. We will use only this term. Then kinetic energy is

$$\mathcal{K}_F = \frac{\pi \rho a^3}{3} \left(1 + \frac{3}{8} \frac{a^3}{q^3} \right) \dot{q}^2. \tag{13.27}$$

In particular, it follows from (13.27) that for a ball in an unbounded fluid flow, the tensor of attached mass is spherical and has the value, $M^{ab} = \frac{2}{3} \rho \pi a^3 \delta^{ab}$.

According to (13.22), the motion equation for a ball of mass m is governed by the equation

$$\frac{d}{dt} \left[m + \frac{2\pi a^3}{3} \rho \left(1 + \frac{3}{8} \frac{a^3}{q^3} \right) \right] \dot{q} + \frac{3\pi \rho a^6}{8 q^4} \dot{q}^2 = 0.$$

It takes much more effort to derive this equation considering the balance of forces.

Example 3. The body B may consist of several simply connected components. Then (13.22) describes the interactions of the bodies in fluids. For example, let us consider two balls moving in an unbounded fluid which is at rest at infinity. The system has six degrees of freedom $r_1^i(t)$ and $r_2^i(t)$ – the coordinates of the centers of the balls. In order to calculate the kinetic energy of the fluid, $\mathcal{K}_F$, we have to solve the boundary value problem

$$\Delta \varphi = 0 \quad \text{in } R_3 - B_1 - B_2, \qquad \frac{\partial \varphi}{\partial n} = \dot{r}_1^i n_i \quad \text{on } \partial B_1, \qquad \frac{\partial \varphi}{\partial n} = \dot{r}_2^i n_i \quad \text{on } \partial B_2 \tag{13.28}$$

with the condition that kinetic energy of the fluid is finite. Kinetic energy is a quadratic form:

$$\mathcal{K}_F = \frac{1}{2} \left(\underset{1}{M} \dot{r}_1^i \dot{r}_1^j + 2 N_{ij} \dot{r}_1^i \dot{r}_2 + \underset{2}{M} \dot{r}_2^i \dot{r}_2^j \right) \tag{13.29}$$

The equations (13.28) are invariant with respect to translations, therefore, in the quadratic form (13.29) the coefficients, $\underset{1}{M}_{ij}$, N_{ij} and $\underset{2}{M}_{ij}$, depend only on the difference, $r_1^i - r_2^i$.

The exact solution of the problem (13.28) can be obtained in the form of series; however, the corresponding expressions for the coefficients in (13.29) are so complicated that they can hardly help in the analysis of the motion of the balls. We give here the first terms of an asymptotic expansion when the ratios of the radii of the balls, a_1, a_2, to the distance between their centers, $R = |r_1 - r_2|$, are small. In the first approximation, the balls do not interact, and

$$\underset{1}{M}_{ij} = \frac{2\pi\rho}{3} a_1^3 \delta_{ij}, \quad N_{ij} = 0, \quad \underset{2}{M}_{ij} = \frac{2\pi\rho}{3} a_2^3 \delta_{ij}.$$

The approximation which takes into account corrections on the order of $(a_1/R)^3$, $(a_2/R)^3$, is as follows: the expressions for $\underset{1}{M}_{ij}$ and $\underset{2}{M}_{ij}$ do not change, while N_{ij} becomes

$$N_{ij} = \frac{\pi\rho a_1^3 a_2^3}{R^3} \left(\delta_{ij} - \frac{3}{R^2} R_i R_j \right), \quad R^i = r_1^i - r_2^i.$$

Polia-Shiffer's theorem. Even in the case of the bodies moving in unbounded fluid, $\mathcal{K}_F$ was found analytically only for ellipsoids. Therefore, for bodies with a more complex shape and any bodies in bounded regions, the approximation methods are used.

As an example of application of the variational formula for the kinetic energy, consider an elegant statement hypothesized by Polia and proved by Shiffer:

The averaged attached mass of the body moving in an unbounded fluid being at rest at infinity is not smaller than the attached mass of the sphere of the same volume.

Actually, a stronger statement holds: each eigenvalue of the tensor of attached mass is not smaller than the attached mass of the sphere; it yields the assertion on the averaged attached mass.

Denote the region occupied by the body by B and the exterior of B by V. According to the Dirichlet principle,

$$\mathcal{K}_F = \max_{\varphi} \mathcal{K}^* (\varphi),$$

$$\frac{1}{\rho} \mathcal{K}^* (\varphi) = \int\limits_{\partial B} \varphi u_i n^i d\sigma - \frac{1}{2} \int\limits_V \frac{\partial \varphi}{\partial x^i} \frac{\partial \varphi}{\partial x_i} dV.$$

The unit normal vector is directed outward of the fluid.

Choosing various functions φ, we get low estimates of $\mathcal{K}_F$:

$$\mathcal{K}^* (\varphi) \leq \mathcal{K}_F. \tag{13.30}$$

Let us take φ as

$$\varphi = a^i \frac{\partial \chi}{\partial x^i}, \tag{13.31}$$

where χ is the gravity potential of the body B,

$$\chi = \int_B \frac{d^3x'}{|x - x'|},$$

and a^i are some constants. We will choose a^i later to optimize the estimate. We can expect that the trial functions (13.31) provide a good estimate, because for an appropriate choice of constants a^i, formula (13.31) gives the exact solution for fluid motion past an ellipsoid.

Substituting the trial functions (13.31) in the functional, $\mathcal{K}^*(\varphi)$, and integrating by parts we obtain

$$\frac{1}{\rho}\mathcal{K}^*(\varphi) = \int_{\partial B_{(e)}} \varphi \left(u_i - \frac{1}{2}\frac{\partial \varphi}{\partial x^i} \right) n^i dA. \quad (13.32)$$

The index (e) emphasizes that in computation of the integral (13.32) the limit of the integrand is taken when ∂B is approached from the exterior side of the body, the fluid side. In deriving (13.32), it is taken into account that the integral over V of $\varphi \Delta \varphi$ is equal to zero since φ is a harmonic function.

Let us replace the limit value of the integrand from the exterior of B by the limit value from the interior of B in order to reduce the surface integral to an integral over the region B by means of the divergence theorem. We have to take into account that the function $\varphi = a^i \chi_{,i}$ is continuous on ∂B, while the derivatives, $\partial\varphi/\partial x^i$, have a discontinuity, and

$$\left.\frac{\partial \varphi}{\partial x^k}\right|_{(e)} - \left.\frac{\partial \varphi}{\partial x^k}\right|_{(i)} = 4\pi n_k n^j a_j. \quad (13.33)$$

The index (i) denotes the limit values at ∂B from the interior side of the body.

Replacing in (13.32) the limit values of the functions from the fluid side by the limit values of the functions from the body side, we get

$$\frac{1}{\rho}\mathcal{K}^*(\varphi) = \int_{\partial B_{(i)}} \varphi \left(u_i - 2\pi n_i n^j a_j - \frac{1}{2}\frac{\partial \varphi}{\partial x^i} \right) n^i dA. \quad (13.34)$$

Applying to (13.34) the divergence theorem, we find

$$\frac{1}{\rho}\mathcal{K}^*(\varphi) = \int_B \left[-\frac{\partial \varphi}{\partial x^i} u^i + 2\pi a^i \frac{\partial \varphi}{\partial x^i} + \frac{1}{2}\frac{\partial \varphi}{\partial x^i}\frac{\partial \varphi}{\partial x_i} \right] dV.$$

The term $\varphi \Delta \varphi = \varphi a^k \frac{\partial}{\partial x^k} \Delta \chi$ is equal to zero in B, since the gravity potential χ satisfies the equation $\Delta \chi = const.$ in the region B.

Let us decrease the last term using the Cauchy inequality (5.20):[1]

$$\frac{1}{|B|}\int_B \frac{\partial\varphi}{\partial x^i}dV\int_B \frac{\partial\varphi}{\partial x_i}dV \leq \int_B \frac{\partial\varphi}{\partial x^i}\frac{\partial\varphi}{\partial x_i}dV. \tag{13.35}$$

Introducing the notation

$$A_{ij} = -\int_B \chi_{,ij}dV,$$

we get the estimate

$$\frac{1}{\rho}\mathcal{K}^*(\varphi) \geq A_{ij}u^i a^j - \frac{1}{2}\left(4\pi A_{ij} - \frac{1}{|B|}A_{im}A_j^m\right)a^i a^j. \tag{13.36}$$

It is important that for an ellipsoid, $\partial\varphi/\partial x^i$ are constant in B; therefore in (13.35), and, consequently, in (13.36) (for a^i which maximize the right hand side) the equality is attained.

Let us choose a system of coordinates x^i oriented along the principle axes of the tensor A_{ij}. Then

$$\frac{1}{\rho}\mathcal{K}^*(\varphi) \geq A_1u_1a_1 + A_2u_2a_2 + A_3u_3a_3 - \frac{1}{2}\left(4\pi A_1 - \frac{1}{|B|}A_1^2\right)a_1^2 - \\ -\frac{1}{2}\left(4\pi A_2 - \frac{1}{|B|}A_2^2\right)a_2^2 - \frac{1}{2}\left(4\pi A_3 - \frac{1}{|B|}A_3^2\right)a_3^2, \tag{13.37}$$

$A_1 \equiv A_{11}$, $A_2 \equiv A_{22}$, $A_3 \equiv A_{33}$. In this system of coordinates, the maximum of the right hand side is attained for

$$a_1 = \frac{u_1}{4\pi - \frac{A_1}{|B|}}, \quad a_2 = \frac{u_1}{4\pi - \frac{A_2}{|B|}}, \quad a_3 = \frac{u_1}{4\pi - \frac{A_3}{|B|}}.$$

Therefore,

$$\frac{1}{\rho}\mathcal{K}^*(\varphi) \geq \frac{1}{2}\left(\frac{A_1}{4\pi - \frac{A_1}{|B|}}u_1^2 + \frac{A_2}{4\pi - \frac{A_2}{|B|}}u_2^2 + \frac{A_3}{4\pi - \frac{A_3}{|B|}}u_3^2\right).$$

[1] We set in (5.20) $g = 1$ to obtain

$$\left(\int_B f dV\right)^2 \leq |B|\int_B f^2 dV,$$

then apply this inequality to each component of the vector $\nabla\varphi$, and take the sum of the three inequalities such obtained.

The following inequality holds:

$$\frac{A_1}{4\pi - \frac{A_1}{|B|}} + \frac{A_2}{4\pi - \frac{A_2}{|B|}} + \frac{A_3}{4\pi - \frac{A_3}{|B|}} \geq \frac{A_1 + A_2 + A_3}{4\pi - \frac{A_1+A_2+A_3}{3|B|}}. \tag{13.38}$$

Indeed, the function

$$f(x) = \frac{x}{4\pi - \frac{x}{|B|}}$$

is strictly convex because, for $0 \leq x \leq |B|$, $\partial^2 f/\partial x^2 > 0$. Let us consider the minimum of the strictly convex function $f(A_1)+f(A_2)+f(A_3)$ with the constraint $A_1 + A_2 + A_3 = const = c$. Introducing the Lagrange multiplier, we see that the minimum is reached for $A_1 = A_2 = A_3$. Consequently, at the minimum point

$$A_1 = A_2 = A_3 = \frac{c}{3}, \quad f(A_1) + f(A_2) + f(A_3) = 3f\left(\frac{c}{3}\right) = \frac{c}{4\pi - \frac{c}{3|B|}}.$$

For arbitrary A_1, A_2, A_3 this yields, (13.38).

From the definition of A_{ij} it follows that

$$A_1 + A_2 + A_3 = \int_B \Delta\chi dV = 4\pi |B|.$$

Therefore,

$$\frac{1}{\rho}\mathcal{K}^*(\varphi) \geq \frac{3}{4}|B|\left(u_1^2 + u_2^2 + u_3^2\right),$$

and each eigenvalue of the tensor of the attached mass is not smaller than $\frac{3}{2}\rho|B|$, i.e. the attached mass of the sphere. In particular,

$$\frac{1}{3}M_i^i \geq \frac{1}{2}|B| = \frac{1}{3}M_{i(\text{ball})}^i.$$

13.3 Motion of a Body in a Viscous Fluid

The equations for forces and moments acting on the body moving in a potential flow of the ideal incompressible fluid and in Stokes flow of a viscous fluid possess a variational structure, as has been shown in the two preceding sections. The force acting on the body in ideal fluid, is[2]

[2] Since in this section the kinetic energy of the body does not appear and only the kinetic energy of the fluid is encountered, the corresponding index at kinetic energy is omitted.

$$F_i = \frac{\partial \mathcal{K}}{\partial r^i} - \frac{d}{dt}\frac{\partial \mathcal{K}}{\partial \dot{r}^i}, \tag{13.39}$$

while in viscous fluid

$$F_i = -\frac{\partial \mathbb{D}}{\partial \dot{r}^i}. \tag{13.40}$$

In these equations, $\mathcal{K}$ and $\mathbb{D}$ are functions of the instantaneous characteristics of the motion only.

The following question arises: is there a universal relation between the generalized forces acting on the body with the kinetic energy of the fluid $\mathcal{K}$ and the dissipative potential $\mathbb{D}$, when they depend not only on the instantaneous characteristics of the motion, but also on the "history" of the motion?

It turns out that, with some assumptions (which shall be introduced further), a generalization of (13.39) and (13.40) is the variational equation

$$Q_k \delta q^k = \delta \mathcal{K} - \frac{d}{dt}\underset{\cdot}{\delta} \mathcal{K} - \underset{\cdot}{\delta} \mathbb{D}. \tag{13.41}$$

Here, the generalized coordinates q^k characterize the motion and the deformation of the body, Q_k are the corresponding generalized forces, $\mathcal{K}$ and $\mathbb{D}$ are the functionals of the motion history. The dependence of $\mathcal{K}$ and $\mathbb{D}$ on the motion history can be described by the two groups of variables (separated by a semi-colon)[3]:

$$\mathcal{K} = \mathcal{K}_{\tau=0}^{\tau=t}\left(q\left(\tau\right); \dot{q}\left(\tau\right)\right), \quad \mathbb{D} = \mathbb{D}_{\tau=0}^{\tau=t}\left(q(\tau); \dot{q}\left(\tau\right)\right).$$

The actual method of distinguishing the two groups of variables is described below. In (13.41), δ is the variation operator with respect to both variables, and $\underset{\cdot}{\delta}$ is the operator which performs the variation with respect to the second group of arguments and subsequently substitutes $\delta\dot{q}$ by δq.

If $\mathcal{K}$ is a function of $r^i(t)$ and $\dot{r}^i(t)$ only: $\mathcal{K} = \mathcal{K}\left(r^i; \dot{r}^i\right)$ and $\mathbb{D} = 0$, then (13.39) follows from (13.41):

$$F_i \delta r^i = \frac{\partial \mathcal{K}}{\partial r^i}\delta r^i + \frac{\partial \mathcal{K}}{\partial \dot{r}^i}\delta \dot{r}^i - \frac{d}{dt}\frac{\partial \mathcal{K}}{\partial \dot{r}^i}\delta r^i = \left(\frac{\partial \mathcal{K}}{\partial r^i} - \frac{d}{dt}\frac{\partial \mathcal{K}}{\partial \dot{r}^i}\right)\delta r^i.$$

If $\mathcal{K} = 0$ and $\mathbb{D} = \mathbb{D}\left(r^i; \dot{r}^i\right)$, then (13.40) follows from (13.41) as well.

The variational equation (13.41) defines n functionals $Q_1, \ldots, Q_n$ through the two functionals $\mathcal{K}$ and $\mathbb{D}$.

Let us prove (13.41). Let V and B be regions in the space of variables X^a which are run by the Lagrangian coordinates of the fluid and the body, respectively. The

[3] The symbol $\mathcal{K}_{\tau=0}^{\tau=t}(q)$ denotes the functionals of functions $q(\tau)$ defined on the segment $0 \le \tau \le t$.

fluid is assumed to be viscous and incompressible, and the boundary of the vessel V does not move. For given particle positions of the body and the fluid at the initial and final instants, and given motion of the boundary of the region V, we have the variational equation

$$\int_0^{t_1}\left(\delta\int_{V-B}\frac{\rho}{2}v_iv^i dV+\int_{\partial B}P_i\delta x^i dA-\int_V\frac{\partial D}{\partial e_{ij}}\partial_j\delta x_i dV\right)dt=0,$$

where P_i are the components of the surface force acting on the fluid from the body, D is the density of the fluid dissipative potential (the fluid does not have to be Newtonian).

The variational equation can be written as an equality valid at any time t

$$\int_0^{t}\left(\delta\int_{V-B}\rho\frac{v^2}{2}dV+\int_{\partial B}P_i\delta x^i dA-\int_V\frac{\partial D}{\partial e_{ij}}\partial_j\delta x_i dV\right)dt-\left.\int_{V-B}\rho v_i\delta x^i dV\right|_t=0. \tag{13.42}$$

Let us prescribe some motion of the body, $q=q(t)$. Then, the motion of the fluid can by found from the system of equations

$$\rho\frac{dv_i}{dt}=-\partial_i p+\partial_j\frac{\partial D}{\partial e_{ij}}, \tag{13.43}$$

$$\det\left\|\frac{\partial x^i}{\partial X^a}\right\|=\det\left\|\frac{\partial \mathring{x}^i}{\partial X^a}\right\|, \tag{13.44}$$

which are supplemented by the boundary conditions

$$x^i\left(X^a,t\right)=\overset{*}{x}^i\left(X^a\right)\ \text{on}\ \partial V,\quad x^i\left(X^a,t\right)=\chi^i\left(X^a,q^k\right)\ \text{on}\ \partial B. \tag{13.45}$$

Assume that the body and the fluid are at rest at the initial instant:

$$x^i\left(X^a,0\right)=\mathring{x}^i\left(X^a\right),\quad v^i\left(X^a,0\right)=0. \tag{13.46}$$

For the consistency of the initial and the boundary conditions on ∂B at $t=0$, we require that

$$\dot{q}^k(0)=0.$$

We assume that the system of equations (13.43), (13.44), (13.45) and (13.46) determines uniquely the fluid motion on the considered time interval. Then, in principle, one can find the particle trajectories of the fluid, and, for each X^a and t, they are some functionals of the motion history of the body B, i.e. of $q(\tau)$ for $0\le\tau\le t$.

The variational equation (13.42) is an identity due to (13.43), (13.44) and (13.45) and the equality,

$$\left(-p\delta_i^j + \frac{\partial D}{\partial e_j^i}\right) n_j = P_i \quad \text{on } \partial B,$$

for arbitrary δx^i vanishing at $t = 0$ and on $\partial V \times [0, t]$. In particular, the particle trajectories for the fluid, satisfying (13.43), (13.44), (13.45) and (13.46), can be substituted into (13.42), and δx^i can be interpreted as its variation caused by an infinitesimally small variation of the body motion. Then the variations δx^i are some functionals of $q(\tau)$ and $\delta q(\tau)$ at every point X, t for $0 \le \tau \le t$, which are linear with respect to $\delta q(\tau)$. Let us write this as follows:

$$\delta x^i = \mathop{l^i}_{\tau=0}^{\tau=t} (X, t|q(\tau); \delta q(\tau)).$$

At the boundary of the body,

$$\delta x^i = \frac{\partial \chi^i}{\partial q^k} \delta q^k.$$

The variational equation (13.42) becomes

$$\int_0^t \left[\delta \mathcal{K} - Q_k \delta q^k - \int_V \frac{\partial D}{\partial e_{ij}} \partial_j \delta x_i \, dV \right] - \left. \int_{V-B} \rho v_i \delta x^i dV \right|_t = 0, \tag{13.47}$$

where $\mathcal{K}$, the kinetic energy of the fluid calculated for the solutions of the problem (13.43), (13.44), (13.45) and (13.46), is some functional of $q(\tau)$, δ is the variation of the body motion, and Q_k are the generalized forces:

$$Q_k = -\int_{\partial B} P_i \frac{\partial \chi^i}{\partial q^k} dA.$$

It remains to check the relation

$$v^i(X, t) = \mathop{l^i}_{\tau=0}^{\tau=t} (X, t|q(\tau); \dot{q}(\tau)), \tag{13.48}$$

to finish the proof. Indeed, if the equality (13.48) is true, then the functionals

$$\mathcal{K} = \int_{V-B} \frac{\rho}{2} v^i v_i dV, \quad \mathbb{D} = \int_V D\left(\partial_{(i} v_{j)}\right) dV,$$

which are calculated for the velocity (13.48), become the functionals of the two groups of variables.

Let us introduce the operator $\underset{\circ}{\delta}$ which takes the variation with respect to the second group of variables and subsequently substitutes $\delta\dot{q}(\tau)$ by $\delta q(\tau)$. If follows from (13.48) that

$$\underset{\circ}{\delta} v^i = \delta x^i.$$

The operator $\underset{\circ}{\delta}$, as the operator of taking the variation with respect to q, commutes with the differential operators:

$$\underset{\circ}{\delta}\partial_i v_j = \partial_i \underset{\circ}{\delta} v_j = \partial_i \delta x_j.$$

Therefore, the variational equation (13.47) can be written as

$$\int_0^t \left[\delta\mathcal{K} - Q_k \delta q^k - \underset{\circ}{\delta}\mathbb{D}\right] dt - \underset{\circ}{\delta}\mathcal{K} = 0. \tag{13.49}$$

Taking the derivative of (13.49) with respect to time results in (13.41).

So, we have to prove the relation (13.48). Let us take the variation of (13.43) and (13.44). We get some system of equations for δx^i in the region $V - B$. Let us write it as $L\delta\mathbf{x} = \mathbf{0}$. The operator L is a linear differential operator with variable coefficients. Taking the variation of the initial conditions results in

$$\delta x^i = 0, \quad \frac{d\delta x^i}{dt} = 0 \quad \text{in } V - B \quad \text{for } t = 0. \tag{13.50}$$

At the boundary of the fluid,

$$\delta x^i = 0 \quad \text{on } \partial V, \quad \delta x^i = \frac{\partial \chi^i}{\partial q^k} \quad \text{on } \partial B. \tag{13.51}$$

The equations, $L\delta\mathbf{x} = \mathbf{0}$, together with the initial and the boundary conditions (13.50) and (13.51) define the functionals $\underset{\tau=0}{\overset{\tau=t}{l^i}}(X, t|q(\tau); \delta q(\tau))$.

Let us differentiate (13.43), (13.44) and (13.45) with respect to time. We get the equations $L\mathbf{v} = 0$ with the boundary conditions

$$v^i = 0 \quad \text{on } \partial V, \quad v^i = \frac{\partial \chi^i}{\partial q^k}\dot{q}^k \quad \text{on } \partial B, \tag{13.52}$$

and the initial conditions

$$v^i = 0 \quad \text{at } t = 0.$$

Let $\ddot{q}^k = 0$ for $t = 0$. Then,

$$\frac{dv^i}{dt} = 0 \quad \text{for } t = 0. \tag{13.53}$$

Indeed, it follows from (13.43) that curl $d\mathbf{v}/dt = 0$ for $t = 0$. Differentiating the continuity equation $v^i_{,i} = 0$ with respect to time and setting $t = 0$, we get div $d\mathbf{v}/dt = 0$. Consequently, $d\mathbf{v}/dt = \text{grad}\varphi$ and $\Delta\varphi = 0$. On the boundary of the fluid, taking the derivative of (13.52), we have $d\mathbf{v}/dt = \text{grad}\varphi = 0$ at ∂V and ∂B for $t = 0$. Therefore, $\varphi = const$ and (13.53) holds.

The assumption that the motion starts from rest is essential, because otherwise (13.48) is not valid.

It is easy to check that the above reasoning can be applied verbatim for viscoelastic fluid, for which the internal energy is a function of the distortion. Then the internal energy $\mathcal{U}$, which is a functional of the previous motion of the body, must be included in the variational equation as well:

$$Q_k \delta q^k = \delta (\mathcal{K} - \mathcal{U}) - \frac{d}{dt} \dot{\delta} \mathcal{K} - \dot{\delta} \mathbb{D}. \tag{13.54}$$

The problem of calculating forces and moments acting on the rigid body in viscous fluid is extremely difficult. Therefore, it is sensible to use the variational equation (13.41) to determine forces and moments, by postulating the functionals $\mathcal{K}$ and $\mathbb{D}$ from phenomenological reasonings. Note that there is a universal dependence between $\mathcal{K}$ and $\mathbb{D}$ due to which they cannot be prescribed arbitrarily. Indeed, let us set $\delta q^k = 0$ at time t in (13.41). Then, for any functions $\delta q^k(\tau)$ which, along with their first and second derivatives, are equal to zero at the initial time and time t, the following relation holds:

$$-\delta \mathcal{K} + \frac{d}{dt} \dot{\delta} \mathcal{K} + \dot{\delta} \mathbb{D} = 0. \tag{13.55}$$

If $\mathcal{K}$ and $\mathbb{D}$ are some functions of $q, \dot{q}$, then (13.55) is automatically satisfied. However, if memory is taken into consideration, the relation (13.55) puts the constraint on the possible functionals, $\mathcal{K}$ and $\mathbb{D}$.

Consider some examples.

Translational unsteady motion of a sphere. Consider a sphere of radius a in an unbounded flow of viscous incompressible fluid. For $t = 0$, the fluid and the ball are at rest. Then the sphere begins to move along a line with the velocity $u(t)$. Let us find the force acting on the sphere, supposing that the nonlinear terms in the momentum equations for the fluid can be ignored. Let R, θ, φ be the spherical coordinates rigidly connected to the ball, v_R, v_θ, v_φ the velocity components relative to an inertial system projected on R, θ and φ–axes, and the velocity of the sphere is directed along the ray $\theta = 0$. The solution of the problem is of the form [172]

$$v_R = -\frac{1}{R^2 \sin\theta}\frac{\partial \psi}{\partial \theta}, \quad v_\theta = \frac{1}{R\sin\theta}\frac{\partial\psi}{\partial R}, \quad v_\varphi = 0,$$

$$\psi = \frac{1}{2\pi i}\int\limits_{\sigma-i\infty}^{\sigma+i\infty} \frac{au(s)}{2}\left\{3e^{-(R-a)\sqrt{\frac{s}{\nu}}}\left(\sqrt{\frac{s}{\nu}}+\frac{\nu}{sa}\right)-\frac{1}{R}\left(a^2+3a\sqrt{\frac{\nu}{s}}+\frac{3\nu}{s}\right)\right.$$
$$\left. e^{st}\sin^2\theta\right\}ds,$$

$$u(s) = \int\limits_0^{+\infty} u(t)e^{st}dt, \quad \nu = \frac{\mu}{\rho},$$

$$p = \frac{\rho a^3 u'(t)}{2R^2} + \frac{3a^2\rho\sqrt{\frac{\nu}{\pi}}}{2R^2}\int\limits_0^t \frac{u'(\tau)d\tau}{\sqrt{t-\tau}} + \frac{3\mu a}{2R^2}u(t).$$

where $u'(t) = du(t)/dt$. Calculations yield the following expressions for the kinetic energy and the dissipative potential:

$$\mathcal{K} = \frac{\rho\pi a^3}{3}u^2(t) + \frac{3}{2}\rho\sqrt{\pi\nu}a^2\int\limits_0^t\int\limits_0^t \frac{u(t-\tau)u(t-\xi)}{(\tau+\xi)^{\frac{3}{2}}}d\xi d\tau,$$

$$\mathbb{D} = 3\pi\mu a u^2(t) + 3\rho\sqrt{\pi\nu}a^2\int\limits_0^t\int\limits_0^t \frac{u'(t-\tau)u'(t-\xi)}{(\tau+\xi)^{\frac{1}{2}}}d\xi d\tau. \quad (13.56)$$

Since in the system of coordinates attached to the sphere, the geometry of the region does not change, $\mathcal{K}$ and $\mathbb{D}$ do not depend on the first group of variables. It is easy to check that the functionals (13.56) satisfy the equality (13.55). From (13.41) and (13.56) follows the known equation for the force acting on the sphere (Basset's formula):

$$F = -6\pi\mu a u(t) - \frac{2}{3}\pi\rho a^3 u'(t) - 6\rho\sqrt{\pi\nu}a^2\int\limits_0^t \frac{u'(t-\tau)}{\sqrt{\tau}}d\tau.$$

Bubble vibrations. Consider a spherical cavity of radius a in an unbounded viscous incompressible non-heat-conducting fluid. There is a surface tension on the surface of the cavity with the coefficient of the surface tension, σ. The radius of the cavity may change with time. The cavity is filled with a gas, and its energy density per unit mass is denoted by $U_g(\rho_g, S)$. The motions of the gas in the cavity is assumed to be adiabatic, while the change of the gas density, ρ_g, over the cavity negligibly small. There is no exchange of mass between the gas and the fluid, hence the total mass of the gas does not change with time

$$\frac{4}{3}\pi a^3 \rho_g = m = const. \tag{13.57}$$

Let us derive the equation describing the change of the bubble radius a with time from the variational equation for the fluid-bubble system:

$$\delta \int_{t_0}^{t_1} \left[\int_{V-B} \rho \left(\frac{v_i v^i}{2} - U \right) dV - 4\pi a^2 \sigma - \int_{\Sigma} \bar{p} \delta x^i n_i dA + \delta \int_{V} DdV \right] dt = 0, \tag{13.58}$$

where V is the volume bounded by a sphere, Σ, of a large radius (further, we will tend it to ∞), $\bar{p}$ is the pressure given at Σ (in the limit, $\bar{p}$ becomes p_∞); the variations are assumed to be equal to zero for $t = t_0, t_1$.

Let us calculate the functionals in the variational equation. Due to the incompressibility and symmetry, the motion of the fluid is determined by the continuity equation

$$v_i = -\dot{a}a^2 \frac{\partial}{\partial x^i} \frac{1}{r}, \qquad r = \sqrt{x^i x_i}.$$

Therefore, after integration of $\mathcal{K}$ and $\mathbb{D}$ over the exterior of the bubble, we get

$$\mathcal{K} = 2\pi a^3 \rho \dot{a}^2, \quad \mathbb{D} = 8\pi \mu a \dot{a}^2.$$

The kinetic energy of the gas, as well as the change in the entropy of the fluid, related to the viscous dissipation of the gas, will be ignored. The internal energy of the gas in the bubble is

$$U_B = \frac{4}{3}\pi a^3 \rho_g U_g (\rho_g).$$

For the motion of the fluid corresponding to the bubble expansion,

$$\delta x^i = -\left(a^2 \frac{\partial}{\partial x^i} \frac{1}{r} \right) \delta a.$$

Therefore,

$$\int_{\Sigma} \bar{p} \delta x^i n_i dA = \bar{p} 4\pi a^2 \delta a$$

and the integral over Σ in (13.58) does not depend on the radius of the sphere Σ. Tending the radius of the sphere to infinity, we see that the work of the external forces at infinity is

$$p_\infty 4\pi a^2 \delta a = \delta\left(\frac{4}{3}\pi a^3 p_\infty\right).$$

Finally, the variational equation becomes

$$\delta \int_{t_0}^{t_1} \left(2\pi a^2 \rho \dot{a}^2 - \tfrac{4}{3}\pi a^3 \left(\rho_g U_g\left(\rho_g\right) + p_\infty\right) - 4\pi a^2 \sigma\right) dt - \delta \int_{t_0}^{t_1} \left(8\pi \mu a^2 \dot{a}^2\right) dt = 0. \tag{13.59}$$

Varying with respect to a, and taking into account (13.57), we obtain the equation

$$a\ddot{a} + \frac{3}{2}\dot{a}^2 = \frac{1}{\rho}\left(p_g - p_\infty - \frac{2\sigma}{a} - \frac{4\mu\dot{a}}{a}\right), \tag{13.60}$$

where $p_g = \rho_g^2 \partial U_g / \partial \rho_g$ is the pressure inside the bubble. This equation governs the bubble vibrations. It was first derived from other reasoning by Rayleigh. Note that for a moving bubble, the kinetic energy of the fluid caused by translational motion of the bubble with velocity u, $\frac{1}{3}\pi a^3 \rho u^2$, must be included in the total kinetic energy of the system. Accordingly, the term $-\frac{1}{4}u^2$ appears in the right hand side of (13.60). It describes the interaction between the translational and vibrational motions of the bubble.

Appendices

Appendices A, B, C contain some interesting variational principles that are beyond the main scope of the book. Appendices D and E provide some details to the issues that have been considered.

A. Holonomic Variational Equations

Consider a functional, $\delta\Omega$, of two variables, u and δu, which is linear with respect to δu. The equation

$$\delta\Omega = 0, \tag{A.1}$$

is called variational equation. The variational equation is holonomic if $\delta\Omega$ is the variation of some functional $I(u)$. The question arises, what are the conditions under which the variational equation is holonomic? In this appendix such necessary and sufficient conditions are formulated and discussed.

Let us start with the finite-dimensional case. A "finite-dimensional model" of $\delta\Omega$ is a linear differential form of the type $F_\varkappa(u)\,\delta u^\varkappa$, where $u = \{u^\varkappa\} \in R_n$:

$$\delta\Omega = F_\varkappa(u)\,\delta u^\varkappa.$$

The usual mathematical notation for such forms is $\Omega(\delta)$; we prefer the "thermodynamic notation" $\delta\Omega$. The question is: what are the conditions for $\delta\Omega$ to be a variation of some function $\varphi(u)$, $\delta\Omega = \delta(\varphi)$, or, equivalently, what are the conditions for the vector, $F_\varkappa(u)$, to be potential, $F_\varkappa = \partial\varphi/\partial u^\varkappa$?

Let functions $F_\varkappa(u)$ be continuous and differentiable in some region A in R_n. Define in A two infinitesimally small fields $\delta u^\varkappa$ and $\delta' u^\varkappa$. Denote by $\delta'\delta\Omega$ the variation of $\delta\Omega$ along the field $\delta' u^\varkappa$. By definition, $\delta'\delta\Omega$ is

$$\delta'\delta\Omega = F_\varkappa\left(u + \delta' u\right)\delta u^\varkappa - F_\varkappa(u)\,\delta u^\varkappa, \tag{A.2}$$

where one keeps only the terms of leading order. Thus,

$$\delta'\delta\Omega = \frac{\partial F_\varkappa(u)}{\partial u^{\varkappa'}}\delta' u^{\varkappa'}\delta u^\varkappa. \tag{A.3}$$

The following theorem holds: in order for $\delta\Omega$ to be a variation of some function in the region A, it is necessary and sufficient that the equation

$$\delta'\delta\Omega = \delta\delta'\Omega \tag{A.4}$$

is satisfied at every point of the region. It follows from (A.3) and (A.4), the well-known conditions for the potentiality of the vector field, $F_{\varkappa}(u)$,

$$\frac{\partial F_{\varkappa}(u)}{\partial u^{\varkappa'}} = \frac{\partial F_{\varkappa'}(u)}{\partial u^{\varkappa}}. \tag{A.5}$$

The conditions (A.5) are local, i.e. compliance with those conditions in a neighborhood of a certain point guarantees the existence in this neighborhood of such a function, $\varphi(u)$, that $\delta\Omega = \delta(\varphi)$. The extension of $\varphi(u)$ onto the whole region is not necessarily unique. The uniqueness is guaranteed only if A is simply-connected.

The idea of the proof of (A.4) is the following. In order for $\delta\Omega$ to be a variation of some function, it is necessary and sufficient for the integral of $\delta\Omega$ over any closed curve Γ in A to be zero:

$$\oint_{\Gamma} \delta\Omega = 0. \tag{A.6}$$

Suppose that in A, every closed curve is a boundary of some two-dimensional surface, Σ. At every point of Σ, we take two linearly independent vectors continuously changing along Σ. Denote the infinitesimally small increments along these vectors by $\delta u^{\varkappa}$ and $\delta' u^{\varkappa}$. According to the Stokes theorem,

$$\oint_{\Gamma} \delta\Omega = \int_{\Sigma} \left(\delta'\delta\Omega - \delta\delta'\Omega\right). \tag{A.7}$$

Due to the arbitrariness of the curve Γ and the surface Σ, it follows from (A.6) and (A.7) that $\delta\Omega$ is holonomic if and only if (A.4) holds at any point, $u^{\varkappa}$, for any infinitesimally small increments $\delta u^{\varkappa}$ and $\delta' u^{\varkappa}$.

It turns out that this statement can be generalized to infinite-dimensional spaces [298, 299], and (A.4) is the criterion that the functional $\delta\Omega$ is holonomic.

Example 1. Consider the functional $\delta\Omega$ of the form

$$\delta\Omega = \int_{V} F\left(x, u, u_i, u_{ij}\right) \delta u d^n x, \tag{A.8}$$

where $x = \left\{x^i\right\} \in R_n$, $u_i \equiv u_{,i}$, $u_{ij} \equiv u_{,ij}$, and F is a twice continuously differentiable function with respect to its arguments. The function $F\left(x, u, u_i, u_{ij}\right)$ can be considered as an operator acting on $u(x)$. Operator F is called potential if $\delta\Omega$ is a variation of some functional. Let us obtain the conditions for $\delta\Omega$ to be holonomic, or, equivalently, for the operator, F, to be potential. For simplicity we will assume that the admissible functions are zero on ∂V:

$$\delta u = 0 \quad \text{on } \partial V. \tag{A.9}$$

The variation of the functional $\delta\Omega$ is

$$\delta'\delta\Omega = \int_V \left(\frac{\partial F}{\partial u}\delta' u + \frac{\partial F}{\partial u_i}\left(\delta' u\right)_{,i} + \frac{\partial F}{\partial u_{ij}}\left(\delta' u\right)_{,ij} \right) \delta u d^n x.$$

According to (A.4), the following equality holds:

$$\int_V \left(\frac{\partial F}{\partial u}\delta' u + \frac{\partial F}{\partial u_i}\left(\delta' u\right)_{,i} + \frac{\partial F}{\partial u_{ij}}\left(\delta' u\right)_{,ij} \right) \delta u d^n x$$
$$= \int_V \left(\frac{\partial F}{\partial u}\delta u + \frac{\partial F}{\partial u_i}(\delta u)_{,i} + \frac{\partial F}{\partial u_{ij}}(\delta u)_{,ij} \right) \delta' u d^n x. \tag{A.10}$$

Let us put the left-hand side of (A.10) into the same form as the right-hand side. To this end, we need to move derivatives from $\delta' u$ to δu using integration by parts. Taking into account that the integrals over the boundary of V are zero due to (A.9), we can write

$$\int_V \frac{\partial F}{\partial u_i}\left(\delta' u\right)_{,i} \delta u d^n x = -\int_V \left(\frac{\partial F}{\partial u_i}\delta u \right)_{,i} \delta' u d^n x,$$
$$\int_V \frac{\partial F}{\partial u_{ij}}\left(\delta' u\right)_{,ij} \delta u d^n x = \int_V \left(\frac{\partial F}{\partial u_{ij}}\delta u \right)_{,ij} \delta' u d^n x. \tag{A.11}$$

Due to the arbitrariness of $\delta' u$ in V, it follows from (A.10) and (A.11) that the equality is true:

$$\frac{\partial F}{\partial u}\delta u - \left(\frac{\partial F}{\partial u_i}\delta u \right)_{,i} + \left(\frac{\partial F}{\partial u_{ij}}\delta u \right)_{,ij} =$$
$$= \frac{\partial F}{\partial u}\delta u + \frac{\partial F}{\partial u_i}(\delta u)_{,i} + \frac{\partial F}{\partial u_{ij}}(\delta u)_{,ij}\,. \tag{A.12}$$

This equality holds at any point of the region V for an arbitrary function δu. At every point, δu and all derivatives of δu can be considered independent. Therefore, the coefficients at δu, $(\delta u)_{,i}$, and $(\delta u)_{,ij}$ should vanish. The coefficient at $(\delta u)_{,ij}$ is equal to zero identically. Setting the coefficient at $(\delta u)_{,i}$ equal to zero, we get the condition

$$\frac{\partial F}{\partial u_i} - \frac{\partial}{\partial x^j}\frac{\partial F}{\partial u_{ij}} = 0. \tag{A.13}$$

The coefficient at δu is equal to zero due to (A.13). So, the necessary and sufficient condition for the functional (A.8) to be holonomic is (A.13). It is easy to check that when $F = \delta L/\delta u$, where $L = L(x, u, u_i)$, the condition (A.13) is satisfied identically.

It follows from (A.13) that when $F = F(x, u)$, functional (A.8) is always holonomic, and when $F = F(x, u, u_i)$ $(\partial F/\partial u_i \not\equiv 0)$ it is always non-holonomic.

Let us show that in the holonomic case F can depend on the second derivatives only linearly. Indeed, a more detailed version of (A.13) is

$$\frac{\partial F}{\partial u^i} - \frac{\partial^2 F}{\partial x^j \partial u_{ij}} - \frac{\partial^2 F}{\partial u \partial u_{ij}} u_j - \frac{\partial^2 F}{\partial u_k \partial u_{ij}} u_{kj} - \frac{\partial^2 F}{\partial u_{ij} \partial u_{kl}} u_{klj} = 0.$$

The third derivatives of u are present only in the last term, and, since u and the derivatives of u can take on arbitrary values at any given point,

$$\frac{\partial^2 F}{\partial u_{ij} \partial u_{kl}} = 0,$$

which proves the statement made.

Holonomic integral functionals. Similarly, one can investigate when the functionals of the form

$$\delta\Omega = \int_V F_\varkappa\left(x^i, u^\varkappa, u_i^\varkappa, \ldots, u_{i_1 \ldots i_N}^\varkappa\right) \delta u^\varkappa d^n x \tag{A.14}$$

are holonomic. In (A.14),

$$u_{i_1 \ldots i_k}^\varkappa \equiv \frac{\partial^k u^\varkappa}{\partial x^{i_1} \ldots \partial x^{i_k}},$$

and $F_\varkappa$ are functions of $u^\varkappa$ and their derivatives up to the order N, differentiable the necessary number of times.

We assume that $u^\varkappa$ on ∂V satisfy the conditions

$$u|_{\partial V} = \tilde{u}_0^\varkappa, \quad \left.\frac{\partial u^\varkappa}{\partial n}\right|_{\partial V} = \tilde{u}_1^\varkappa, \quad \left.\frac{\partial^2 u^\varkappa}{\partial n^2}\right|_{\partial V} = \tilde{u}_2^\varkappa, \ldots \tag{A.15}$$

where $\partial/\partial n$ is the normal derivative on ∂V. The number of given derivatives depends on N. When N is even, this number is $N/2$ (including the derivative of the zeroth order), and when N is odd, the number is $(N+1)/2$. The boundary conditions (A.15) are equivalent to prescribing on ∂V functions, $u^\varkappa$, and the corresponding number of their derivatives. Therefore,

$$\delta u^{\varkappa} = 0, \quad \delta u_i^{\varkappa} = 0, \quad \delta u_{i_1 i_2}^{\varkappa} = 0, \ldots \quad \text{on } \partial V. \tag{A.16}$$

The condition (A.4) for the functional (A.14) yields the equation

$$\frac{\partial F_{\varkappa}}{\partial u_{i_1 \ldots i_k}^{\varkappa'}} = \sum_{s=0}^{N-k} (-1)^{k+s} C_{k+s}^{s} \frac{\partial^s}{\partial x^{i_{k+1}} \ldots \partial x^{i_{k+s}}} \frac{\partial F_{\varkappa'}}{\partial u_{i_1 \ldots i_k i_{k+1} \ldots i_{k+s}}^{\varkappa}}, \tag{A.17}$$

which has to be satisfied identically for any functions $u^{\varkappa}\left(x^i\right)$. Here, $C_k^m = \frac{k!}{m!(m-k)!}$, and to write the equations in a compact form, the following convention is accepted: if a quantity, $T^{i_1 \ldots i_k}$, is encountered in the sum for $k = 0$, that means that this quantity does not have any indices; in particular, $u_{i_1 \ldots i_k}^{\varkappa}$ for $k = 0$, coincides with the zeroth derivative of $u^{\varkappa}$, i.e. $u^{\varkappa}$.

Example 1 (continued). In the case of the functional (A.8), the conditions (A.17) contains three equations corresponding to $k = 0, 1, 2$. Equations for $k = 2$, are identities; for $k = 0, 1$ the equations are

$$\begin{aligned} \frac{\partial F}{\partial u} &= \frac{\partial F}{\partial u} - \frac{\partial}{\partial x^i} \frac{\partial F}{\partial u_i} + \frac{\partial^2}{\partial x^i \partial x^j} \frac{\partial F}{\partial u_{ij}}, \\ \frac{\partial F}{\partial u_i} &= -\frac{\partial F}{\partial u_i} + 2 \frac{\partial}{\partial x^j} \frac{\partial F}{\partial u_{ij}}. \end{aligned} \tag{A.18}$$

The second equation (A.18) can be rewritten as (A.13), while the first is an identity due to (A.13).

Example 2. Let $F_{\varkappa}$ depend on the derivatives of $u^{\varkappa}$ not higher than the second order. Then (A.17) is reduced to the three equations corresponding to $k = 0, 1, 2$:

$$\begin{aligned} \frac{\partial F_{\varkappa}}{\partial u^{\varkappa'}} &= \frac{\delta F_{\varkappa'}}{\delta u^{\varkappa}}, \quad \left(\frac{\delta F_{\varkappa'}}{\delta u^{\varkappa}} \equiv \frac{\partial F_{\varkappa'}}{\partial u^{\varkappa}} - \frac{\partial}{\partial x^i} \frac{\partial F_{\varkappa'}}{\partial u_i^{\varkappa}} + \frac{\partial^2}{\partial x^i \partial x^j} \frac{\partial F_{\varkappa'}}{\partial u_{ij}} \right), \\ \frac{\partial F_{\varkappa}}{\partial u_i^{\varkappa'}} &= -\frac{\partial F_{\varkappa'}}{\partial u_i^{\varkappa}} + 2 \frac{\partial}{\partial x^j} \frac{\partial F_{\varkappa'}}{\partial u_{ij}^{\varkappa}}, \\ \frac{\partial F_{\varkappa}}{\partial u_{ij}^{\varkappa'}} &= \frac{\partial F_{\varkappa'}}{\partial u_{ij}^{\varkappa}}. \end{aligned} \tag{A.19}$$

Using the third relation (A.19), the second one can be written as

$$2 \frac{\partial}{\partial x^i} \frac{\partial F_{\varkappa'}}{\partial u_{ij}^{\varkappa}} = \frac{\partial}{\partial x^i} \left(\frac{\partial F_{\varkappa'}}{\partial u_{ij}^{\varkappa}} + \frac{\partial F_{\varkappa'}}{\partial u_{ij}^{\varkappa}} \right).$$

It can be put in a symmetric form:

$$\frac{\delta F_{\varkappa}}{\delta u_i^{\varkappa'}} + \frac{\delta F_{\varkappa'}}{\delta u_i^{\varkappa}} = 0.$$

Here we used the notation for the variational derivative,

$$\frac{\delta F}{\delta u_i} = \frac{\partial F}{\partial u_i} - \frac{\partial}{\partial x^i} \frac{\partial F}{\partial u_{ij}}.$$

Similarly transforming the first equation (A.19), we get the following holonomicity conditions:

$$\begin{aligned} &\frac{\partial F_\varkappa}{\partial u_i^{\varkappa'}} - \frac{\partial F_{\varkappa'}}{\partial u_i^{\varkappa}} = \frac{1}{2} \frac{\partial}{\partial x^i} \left(\frac{\delta F_\varkappa}{\delta u_i^{\varkappa'}} - \frac{\delta F_{\varkappa'}}{\delta u_i^{\varkappa}} \right), \\ &\frac{\delta F_\varkappa}{\delta u_i^{\varkappa'}} + \frac{\delta F_{\varkappa'}}{\delta u_i^{\varkappa}} = 0, \\ &\frac{\partial F_\varkappa}{\partial u_{ij}^{\varkappa'}} - \frac{\partial F_{\varkappa'}}{\partial u_{ij}^{\varkappa}} = 0. \end{aligned} \tag{A.20}$$

Note that the general holonomicity conditions (A.17) can be written more compactly by means of the following construction. Consider $u^\varkappa$ as functions of $n+1$ variables, x^i $(i = 1, \ldots, n)$ and an auxiliary variable, τ $\left(u^\varkappa\left(x^i\right) \equiv u^\varkappa\left(x^i, \tau\right)\right.$ for $\tau = 0)$. Let us introduce the "Lagrangian":

$$L = F_\varkappa \left(x^i, u^\varkappa, u_i^\varkappa, \ldots\right) \frac{\partial u^\varkappa}{\partial \tau}.$$

Then (A.17) are equivalent to

$$\frac{\delta L}{\delta u^\varkappa} = 0, \tag{A.21}$$

where the variational derivative is taken with respect to the functions $u^\varkappa\left(x^i, \tau\right)$, and then τ is set equal to zero.

The potential. If the conditions for the functional $\delta\Omega$ to be holonomic are satisfied, then the "potential," $I(u)$, can be found from by the equation

$$I(u) = \int_{u_0}^{u} \delta\Omega,$$

where the integral is taken over some path in the functional space connecting a fixed element, u_0, with the current element, u. Usually, the potential of holonomic functionals is easy to guess. The holonomicity condition is useful in cases when the existence of the potential is not obvious.

Expanding the functional space. Some functionals can be made holonomic by expanding the set of functions on which they are considered. A number of examples of this kind will be given in the next appendix. Here we consider one common indication for such a possibility.

Let two continua occupy regions V_1 and V_2, respectively. These two regions have a common piece of boundary, Σ.

The variational equation for the first continuum has the form

$$\delta(I_1) + \delta A_1 = 0.$$

We need to decide whether this variational equation is holonomic. Let the variational equation for the second continuum and for the system "continuum 1 + continuum 2" be

$$\delta(I_2) + \delta A_2 = 0, \quad \delta(I_1 + I_2) + \delta A_{12} = 0.$$

The functional, δA_{12}, describes the action of the surrounding on the system "continuum 1 + continuum 2." From these variational equations,

$$\delta A_{12} = \delta A_1 + \delta A_2.$$

If the functional, δA_{12}, is holonomic, then the variational equation for the first continuum is also holonomic, since

$$\delta A_1 = \delta A_{12} + \delta(I_2).$$

The holonomicity is achieved by expanding the set of field variables by including in this set the field variables of the second continuum.

The variational equation for the first continuum is holonomic, in particular, for $\delta A_{12} = 0$. This is true, for example, in the case of an isolated system "continuum 1 + continuum 2," or in the case when the boundary of the second continuum is clamped.

Example 3. Consider an elastic body which has a common piece of boundary, Σ, with a rigid body, T. We assume that the elastic body is "glued" to T along Σ. The rest of the boundary of the elastic body could be subjected to external surface force.

The virtual work of forces, acting from the rigid body on the elastic body,

$$\delta A = \int_{t_0}^{t} \int_{\Sigma} f_i \delta x^i d\sigma dt, \tag{A.22}$$

can be expressed in terms of work on the infinitesimally small displacements and rotations of the rigid body, δr^i and $\delta\varphi^i$ (see (3.47)):

$$\delta A = \int_{t_0}^{t} \left(F_i \delta r^i + M_i \delta \varphi^i\right) dt \tag{A.23}$$

where F_i and M_i are the total resultant and total moments of forces acting on the elastic body:

$$F_i = \int_{\Sigma} f_i d\sigma, \quad M_i = \varepsilon_{ijk} \int_{\Sigma} f^j \left(x^k - r^k(t)\right) d\sigma.$$

If F_i and M_i are some prescribed functions of parameters, $r^i(t)$ and $\alpha_a^i(t)$, defining the motion of the rigid body, δA is, in general, non-holonomic. However, if the rigid body does not experience other external actions than those from the elastic body, then δA is holonomic on the expanded set of variables, including the functions, $r^i(t)$ and $\alpha_a^i(t)$, because the following variational equation holds:

$$\delta \int_{t_0}^{t_1} \mathcal{K}_T dt - \int_{t_0}^{t} \left(F_i \delta r^i + M_i \delta \varphi^i\right) dt = 0,$$

where $\mathcal{K}_T$ is the kinetic energy of the rigid body, a known function of dr^i/dt, α_a^i and $d\alpha_a^i/dt$. The potential of δA is the action of the rigid body.

This is an example of the expansion of the functional space suggested by a physical reasoning. Using "non-physical" expansions, one can put any system of equations in the form of Euler equations of some functional. An example of that was given in Sect. 1.8.

B. On Variational Formulation of Arbitrary Systems of Equations

Consider a system of equations and boundary conditions,

$$F_\varkappa\left(x^i, \frac{\partial u^\varkappa}{\partial x^i}, \frac{\partial^2 u^\varkappa}{\partial x^i \partial x^j}, \ldots\right) = 0 \quad \text{in } V,$$
$$f_\varkappa\left(x^i, \frac{\partial u^\varkappa}{\partial x^i}, \ldots\right) = 0 \quad \text{on } \partial V. \tag{B.1}$$

We are going to discuss the following question: in which cases this system of equations is Euler system of equations for some functional?

There is a certain difficulty in getting the answer caused by the possibility of transforming the system of equations by, e.g., multiplication on some functions of x^i, $u^\varkappa$, $\partial u^\varkappa/\partial x^i$, or by substitution of the required functions. As a result, the system may lose (or acquire) the property of being Euler system of equations.

Example. Consider on the segment $[0, a]$ the boundary value problem

$$F \equiv p(x)\frac{d^2u}{dx^2} + q(x)\frac{du}{dx} + r(x)u + f(x) = 0, \tag{B.2}$$

$$u(0) = u(a) = 0. \tag{B.3}$$

We form the functional

$$\delta\Omega = \int_0^a F\delta u dx. \tag{B.4}$$

If this functional is holonomic on the set of functions selected by the conditions (B.3), then (B.2) are the Euler equations.

From the holonomicity conditions (A.13) for the functional (B.4), we have

$$q(x) - \frac{d}{dx}p(x) = 0. \tag{B.5}$$

If the functions, $p(x)$ and $q(x)$, do not satisfy (B.5), then the functional (B.4) is not holonomic and (B.2) cannot be the Euler equation for a functional of the form

$$\int_0^a L\left(x, u, \frac{du}{dx}\right)dx.$$

Let us show that (B.2) can be transformed in such a way, that it becomes a Euler equation. Let us make the substitution $u \rightarrow v$:

$$u(x) = b(x)v(x).$$

The equation takes the form

$$p\frac{d^2}{dx^2}bv + q\frac{d}{dx}bv + rbv + f = 0. \tag{B.6}$$

Now we choose $b(x)$ in such a way that (B.6) satisfy the criterion (A.13). Then $b(x)$ must satisfy the equation

$$\frac{d}{dx}[b(x)p(x)] = 2\frac{db(x)}{dx}p(x) + q(x)b(x).$$

This equation has the following solution:

$$b(x) = cp(x)e^{-\int_0^x \frac{q(x)dx}{p(x)}}, \quad c = const. \tag{B.7}$$

For definiteness, we set $c = 1$.

The equations and boundary conditions for the function $v(x)$ become

$$G \equiv \frac{d}{dx}(b(x)\,p(x)) + \bar{r}v(x) + f(x) = 0, \quad \bar{r} \equiv rb + p\frac{d^2b}{dx^2} + q\frac{db}{dx}, \tag{B.8}$$

$$v(0) = v(a) = 0. \tag{B.9}$$

The functional

$$\int_0^a G\delta v dx$$

is holonomic on the set of functions satisfying (B.9). It is easy to guess the functional for which (B.8) is the Euler equation:

$$\int_0^a \left(b(x)\,p(x) \left(\frac{dv}{dx}\right)^2 - \bar{r}v^2 - 2f(x)\,v \right) dx. \tag{B.10}$$

The holonomicity was achieved by a change of the required function.

It could also be obtained by multiplying (B.2) by $1/b(x)$ where $b(x)$ is the function (B.7). Then, (B.2) becomes

$$\frac{d}{dx}\left(\frac{p}{b}\frac{du}{dx}\right) + \frac{r}{b}u + f = 0.$$

It is the Euler equation for the functional

$$\int_0^a \left[\frac{p}{b}\left(\frac{du}{dx}\right)^2 - \frac{r}{b}u^2 - 2fu \right] dx. \tag{B.11}$$

The functional (B.11) becomes the functional (B.10) after the substitution $u \to v$: $u = bv$.

Another complication is related to a hardly formalized requirement that the variational principle sought has some physical meaning. Without such a requirement, it is easy to construct various "non-physical" variational principles, which are of little or no interest. For example: the solution of (B.1) minimizes the functional

$$\int_V F_\varkappa F^\varkappa dv + \int_{\partial V} f_\varkappa f^\varkappa dA.$$

There are no sensible answers to the posed question. However, the search for these answers lead to the formulation of a number of nontrivial "non-physical" variational principles.

Morse-Feshbach principle. Consider the Cauchy problem for the heat conductivity equations

$$u_t = k\Delta u + f, \tag{B.12}$$

$$u|_{\partial V} = 0, \quad u|_{t=0} = 0. \tag{B.13}$$

in a four-dimensional region $V \times [0, T]$.

It is easy to obtain from (A.17) that there are no functionals of the form

$$\int_0^T \int_V L\left(x^i, t, \frac{\partial u}{\partial t}, \frac{\partial u}{\partial x^i}\right) dV dt$$

for which (B.12) would be the Euler equation.

Let us introduce the adjoint boundary value problem

$$-v_t = k\Delta v + g, \tag{B.14}$$

$$v|_{\partial V} = 0, \quad v|_{t=T} = 0. \tag{B.15}$$

and consider (B.12), (B.13), (B.14) and (B.15) as a system of equations for two functions, u and v.

Then, (B.12) and (B.13) are the Euler equations for the functional

$$\int_0^T \int_V \left(u_t v + k u_{,i} v^{,i} - f v - g u\right) dV dt \tag{B.16}$$

with the constraints (B.13) and (B.15).

Generalization. Let u, v be elements of some Hilbert space, H, (u, v) the scalar product in H, L – the linear operator acting from H to H, L^* – the conjugate operator, i.e. the operator which is defined by the equality

$$(Lu, v) = \left(u, L^* v\right)$$

which holds for any u and v.

Then the equations

$$Lu = f, \quad L^* v = g \tag{B.17}$$

are the stationary points of the functional

$$\Phi(u, v) = (Lu, v) - (f, v) - (g, u).$$

Indeed, (B.17) follow from the equations

$$\delta\Phi = (L\delta u, v) + (Lu, \delta v) - (f, \delta v) - (g, \delta u) = \left(L^* v - g, \delta u\right) + (Lu - f, \delta v) = 0.$$

Gurtin-Tonti principle. Consider the Cauchy problem

$$\frac{du(t)}{dt} = f(t), \tag{B.18}$$

$$u(0) = 0 \tag{B.19}$$

on the segment $[0, a]$. The functional

$$\int_0^a \left(\frac{du(t)}{dt} - f(t)\right) \delta u(t) dt$$

is obviously non-holonomic. Let us consider a different functional,

$$\delta\Omega = \int_0^a \left(\frac{du}{dt} - f\right)\Big|_{a-t} \delta u(t) dt \tag{B.20}$$

on the set of continuous differentiable functions selected by the condition (B.19). Here, $(du/dt - f)|_{a-t}$ is the value of the function $du/dt - f$ at the point $a - t$.

Let us check whether the holonomicity condition (A.4) is satisfied:

$$\delta'\delta\Omega = \int_0^a \frac{d\delta' u}{dt}\Big|_{a-t} \delta u(t) dt = -\int_0^a \frac{d\delta' u(\tau)}{d\tau} \delta u(a - \tau) d\tau =$$
$$= \int_0^a \delta' u(\tau) \frac{d\delta u}{d\tau}\Big|_{a-\tau} d\tau = \delta\delta'\Omega.$$

Consequently, the functional (B.20) is holonomic. It is easy to guess the corresponding potential,

$$I(u) = \frac{1}{2}\int_0^a \frac{du}{dt}\Big|_{a-t} u(t) dt - \int_0^a f(a - t) u(t) dt. \tag{B.21}$$

Note that the Lagrangian in the functional (B.21)

$$L = \frac{1}{2} \left. \frac{du}{dt} \right|_{a-t} u(t) - f(a-t) u(t)$$

is nonlocal.

The analogous principle holds for the Cauchy problem for the equations of second order:

$$\frac{d^2 u}{dt^2} - f(t) = 0, \quad 0 \leq t \leq T, \tag{B.22}$$

$$u(0) = 0, \quad \frac{du(0)}{dt} = 0. \tag{B.23}$$

On the set of functions $u(0) = 0$, the extremals of the functional

$$I(u) = \frac{1}{2} \int_0^T \left. \frac{du}{dt} \right|_{T-t} \left. \frac{du}{dt} \right|_t dt - \int_0^T f(T-t) u(t) dt \tag{B.24}$$

satisfy (B.22) and the second relation (B.23). Note that the condition $du(0)/dt = 0$ turns out to be the natural boundary condition for the functional (B.24), while the "natural" functional for (B.22) with a local Lagrangian

$$\frac{1}{2} \int_0^T \left(\frac{du}{dt} \right)^2 dt + \int_0^T f(t) u(t) dt$$

does not feel this boundary condition.

The variational principles for (B.18), (B.19), (B.22) and (B.23) can easily be extended to the Cauchy problems for linear parabolic and hyperbolic systems of equations.

C. A Variational Principle for Probability Density

The close relationship between first-order partial differential equations and ordinary differential equations is well known. In analytical mechanics, its counterpart is the relationship between the Hamilton-Jacobi equation and Hamiltonian equations. Since Hamiltonian equations can be obtained from the variational principle, it can be expected that a certain variational principle also exists for the Hamilton-Jacobi equation. That variational principle cannot be of the same type as the ordinary integral variational principles because Euler equations for them are partial differential equations of at least second order, while the Hamilton-Jacobi equation is a first-order equation. Here we formulate and prove the variational principle for the Hamilton-Jacobi equation. We show that a natural construction in terms of which the variational principle can be formulated is the Gibbs ensemble of statistical mechanics;

the varied functional is the mathematical expectation of the action of analytical mechanics, while the Hamilton-Jacobi equation is the Euler equation corresponding to varying the probability density function.

Consider a mechanical system with generalized coordinates $q^1, q^2, \ldots, q^n$, momenta $p_1, \ldots, p_n$, and the Hamiltonian $H(p, q, t)$, $p = \{p_1, \ldots, p_n\}$, $q = \{q^1, \ldots, q^n\}$. Let us take an ensemble of such systems with the probability density function $f(p, q, t)$. In this ensemble the velocities, $\dot{p}$, $\dot{q}$, of the points representing the system in the phase space, R, can be viewed as functions of p, q, t. The function $f(p, q, t)$ then satisfies the continuity equation

$$\frac{\partial f}{\partial t} + \frac{\partial \left(\dot{q}^i f\right)}{\partial q^i} + \frac{\partial \left(\dot{p}_i f\right)}{\partial p_i} = 0. \tag{C.1}$$

All indices run through the values $1, \ldots, n$.

We introduce some "Lagrangian coordinates" of the system's trajectories – the functions $\pi_a(p, q, t)$ $(a = 1, \ldots, n)$ which are conserved along the trajectories:

$$\frac{\partial \pi_a}{\partial t} + \dot{q}^i \frac{\partial \pi_a}{\partial q^i} + \dot{p}_i \frac{\partial \pi_a}{\partial p_i} = 0. \tag{C.2}$$

Consider the functional

$$I = \int_{t_0}^{t_1} \int_R \left[p_i \dot{q}^i(p, q, t) - H(p, q, t)\right] f(p, q, t)\, dp\, dq\, dt - \int_R \varphi(q, \pi)\, f(p, q, t_1)\, dp\, dq. \tag{C.3}$$

Here $\varphi(q, \pi)$ is considered to be a given function of the arguments q^i and π_a and $f(p, q, t)$ for each t is assumed to be decaying in R at infinity with a rate sufficient for the convergence of the integrals.

The first term in (C.3) has the meaning of the mathematical expectation of the action functional of analytical mechanics; the second term, as will be seen, is related to the momentum flux at $t = t_1$.

Consider the stationary points of the functional (C.3) on a set of the functions $f(p, q, t)$, $\dot{p}_i(p, q, t)$, $\dot{q}^i(p, q, t)$, and $\pi_a(p, q, t)$, subject to constraints (C.1), (C.2), and the initial conditions

$$f = \psi(q, \pi) \left|\frac{\partial \pi}{\partial p}\right| \quad \text{for } t = t_0, \tag{C.4}$$

where $\psi(q, \pi)$ is a given function, and $|\partial \pi / \partial p|$ is the determinant of the matrix $\|\partial \pi_a / \partial p_i\|$.

Variational principle. *At the stationary points of functional (C.3), the functions $\dot{p}_i$, $\dot{q}_i$ satisfy Hamiltonian equations; the Lagrange multiplier for the constraint (C.1) satisfies the Hamilton-Jacobi equation.*

Indeed, let us write down Euler equations of functional (C.3). Denote by α and $\theta_a f$ the Lagrange multipliers for the constraints (C.1) and (C.2). Euler equations are then found by varying the functional

$$\int_{t_0}^{t_1}\int_R \left[\left(p_i\dot{q}^i - H\right) f + \alpha\left(\frac{\partial f}{\partial t} + \frac{\partial\left(\dot{q}^i f\right)}{\partial q^i} + \frac{\partial\left(\dot{p}_i f\right)}{\partial p_i}\right)\right.$$
$$\left. + \theta^a f\left(\frac{\partial \pi_a}{\partial t} + \dot{q}^i\frac{\partial \pi_a}{\partial q^i} + \dot{p}_i\frac{\partial \pi_a}{\partial p_i}\right)\right] dpdqdt - \int_R \varphi\left(q,\pi\right) f\left(p,q,t_1\right) dpdq.$$

In R we obtain the following equations: by variation of $\dot{q}^i\left(p,q,t\right)$,

$$\frac{\partial \alpha}{\partial q^i} + \theta^a\frac{\partial \pi_a}{\partial q^i} = p_i, \tag{C.5}$$

by variation of $\dot{p}_i\left(p,q,t\right)$,

$$\frac{\partial \alpha}{\partial p_i} - \theta^a\frac{\partial \pi_a}{\partial p_i} = 0, \tag{C.6}$$

by variation of $f\left(p,q,t\right)$,

$$p_i\dot{q}^i - H - \frac{\partial \alpha}{\partial t} - \dot{q}^i\frac{\partial \alpha}{\partial q^i} - \dot{p}_i\frac{\partial \alpha}{\partial p_i} + \theta^a\left(\frac{\partial \pi_a}{\partial t} + \dot{q}^i\frac{\partial \pi_a}{\partial q^i} + \dot{p}_i\frac{\partial \pi_a}{\partial p_i}\right) = 0, \tag{C.7}$$

by variation of $\pi_a\left(p,q,t\right)$,

$$\frac{\partial\left(\theta^a f\right)}{\partial t} + \frac{\partial\left(\theta^a\dot{q}^i f\right)}{\partial q^i} + \frac{\partial\left(\theta^a\dot{p}_i f\right)}{\partial p_i} = 0. \tag{C.8}$$

The variation of α and θ_a yields (C.1) and (C.2). The equations hold in a region of space R where $f > 0$.

By virtue of the arbitrariness of the variations δf and $\delta\pi_a$ at $t = t_1$, we obtain the additional relations

$$\alpha = \varphi\left(q,\pi\right) \quad \text{at } t = t_1, \tag{C.9}$$

$$\theta^a - \tfrac{\partial \varphi}{\partial \pi_a} = 0 \quad \text{at } t = t_1. \tag{C.10}$$

Variations at $t = t_0$ yield the equality

$$\int_R \left(\alpha\delta f + \theta^a f\delta\pi_a\right)\Big|_{t=t_0} dpdq = 0. \tag{C.11}$$

It will be shown further that equality (C.11) is fulfilled automatically by virtue of (C.4) and (C.6).

We now proceed to the analysis of (C.5)–(C.10). Equation (C.7) can be re-written by using (C.5) and (C.6) in a simpler form:

$$\frac{\partial \alpha}{\partial t} - \theta^{a} \frac{\partial \pi_{a}}{\partial t} + H(p, q, t) = 0. \tag{C.12}$$

Suppose that functions $\pi(p, q, t)$ at each q, t determine a one-to-one mapping $p \leftrightarrow \pi$ so that p can be considered as functions of t, q, π: $p = p(\pi, q, t)$ and

$$|\partial \pi / \partial p| \neq 0. \tag{C.13}$$

Then all the functions of p, q, and t can also be considered as functions of π, q and t. In particular, $\alpha = \alpha(\pi, q, t)$. Equations (C.5) and (C.6) in terms of $\alpha(\pi, q, t)$ take a simpler form:

$$\partial_{i} \alpha = p_{i}, \tag{C.14}$$

$$\partial^{a} \alpha = \theta^{a}. \tag{C.15}$$

By $\partial_i \alpha$ and $\partial^a \alpha$ we denoted the partial derivatives of α with respect to q_i and π_a, at constant π, t and q, t, respectively. Equations (C.14) and (C.15) define a canonical transformation of $p, q \rightarrow \pi, \theta$ (see [4]).

By virtue of (C.14) and (C.15), (C.12) can be rewritten as a Hamilton-Jacobi equation for the function $\alpha(\pi, q, t)$:

$$\partial_{t} \alpha + H\left(\partial_{i} \alpha, q^{i}, t\right) = 0. \tag{C.16}$$

Here $\partial_t \alpha$ is the time derivative of α at constant π, q.

Equation (C.16) together with the initial condition (C.9) form the Cauchy problem for the function α.

Note that the assumption (C.13) is satisfied if the determinant of the matrix $\left\|\partial^2 \varphi / \partial q_i \partial \pi_a\right\|$ is nonzero and $t_1 - t_0$ is sufficiently small. Indeed, for small $t_1 - t_0$ and smooth φ and H, one can guarantee the unique existence of a smooth solution to the Cauchy problem (C.16) and (C.9) (see §47 in [4]). By virtue of (C.14), the determinant of the matrix $\|\partial p_i / \partial \pi_a\|$ coincides with the determinant of the matrix $\left\|\partial^2 \alpha / \partial q_i \partial \pi_a\right\|$; due to (C.9), at $t = t_1$ it coincides with the determinant of the matrix $\left\|\partial^2 \varphi / \partial q_i \partial \pi_a\right\|$. By continuity, it will be nonzero over a sufficiently small time interval.

Let us show now that at a stationary point the Hamiltonian equations hold:

$$\dot{p}_{i} = -\frac{\partial H}{\partial q^{i}}, \quad \dot{q}^{i} = -\frac{\partial H}{\partial p_{i}}. \tag{C.17}$$

We will need the auxiliary equations

$$\frac{\partial\theta^a}{\partial q^i}\frac{\partial\pi_a}{\partial q^j}-\frac{\partial\theta^a}{\partial q^j}\frac{\partial\pi_a}{\partial q^i}=0,\qquad \frac{\partial\theta^a}{\partial p_i}\frac{\partial\pi_a}{\partial p_j}-\frac{\partial\theta^a}{\partial p_j}\frac{\partial\pi_a}{\partial p_i}=0,$$
$$\frac{\partial\theta^a}{\partial p_j}\frac{\partial\pi_a}{\partial q^i}-\frac{\partial\theta^a}{\partial q^i}\frac{\partial\pi_a}{\partial p_j}=\delta_i^j. \tag{C.18}$$

These equations can be obtained by differentiating (C.5) and (C.6) with respect to p_i, q_i :

$$\frac{\partial^2\alpha}{\partial q^i\partial q^j}+\frac{\partial\theta^a}{\partial q^j}\frac{\partial\pi_a}{\partial q^i}+\theta^a\frac{\partial^2\pi_a}{\partial q^i\partial q^j}=0,$$
$$\frac{\partial^2\alpha}{\partial q^i\partial p_j}+\frac{\partial\theta^a}{\partial p_j}\frac{\partial\pi_a}{\partial q^i}+\theta^a\frac{\partial^2\pi_a}{\partial p_j\partial q^i}=\delta_i^j,$$
$$\frac{\partial^2\alpha}{\partial q^i\partial p_j}-\frac{\partial\theta^a}{\partial q^i}\frac{\partial\pi_a}{\partial p_j}-\theta^a\frac{\partial^2\pi_a}{\partial q^i\partial p_j}=0,$$
$$\frac{\partial^2\alpha}{\partial p_i\partial p_j}-\frac{\partial\theta^a}{\partial p_i}\frac{\partial\pi_a}{\partial p_j}-\theta^a\frac{\partial^2\pi_a}{\partial p_i\partial p_j}=0.$$

In the first equation we eliminate α and π_a by alternating over i, j. That gives the first equation (C.18). Similarly, from the last equation we obtain the second equation (C.18). Subtraction of the second and the third equations yields the third equation (C.18).

Let us differentiate (C.12) with respect to p and q. Making use of (C.6) and (C.7) we have

$$\frac{\partial\pi_a}{\partial q^i}\frac{\partial\theta^a}{\partial t}-\frac{\partial\theta^a}{\partial q^i}\frac{\partial\pi_a}{\partial t}=-\frac{\partial H}{\partial q^i},$$
$$\frac{\partial\pi_a}{\partial p_i}\frac{\partial\theta^a}{\partial t}-\frac{\partial\theta^a}{\partial p_i}\frac{\partial\pi_a}{\partial t}=-\frac{\partial H}{\partial p_i}.$$

Substituting here $\partial\pi_a/\partial t$ in terms of $\dot p, \dot q$ from (C.2) and $\partial\theta^a/\partial t$ from the equation

$$\frac{\partial\theta^a}{\partial t}+\dot q^i\frac{\partial\theta^a}{\partial q^i}+\dot p_i\frac{\partial\theta^a}{\partial p_i}=0$$

which follows from (C.8) and (C.1), and making use of (C.18), we obtain the Hamiltonian equations (C.17).

Equation (C.1), by virtue of Hamiltonian equations (C.17), is transformed into the Liouville equation

$$\frac{\partial f}{\partial t}+\frac{\partial H}{\partial p_i}\frac{\partial f}{\partial q^i}-\frac{\partial H}{\partial q^i}\frac{\partial f}{\partial p_i}=0.$$

The system of equations obtained can be solved as follows. First we solve the Cauchy problem (C.16) and (C.9). From its solution $\alpha(\pi, q, t)$, we find from (C.14) the functions $\pi_a(\pi, q, t)$. The values of these functions are substituted into the initial data (C.4), and the initial value $f(p, q, t_0)$ of the function f is found. The Cauchy problem is then solved for Liouville equation (C.1) and (C.16). The functions θ^a are reconstructed from $\alpha(\pi, q, t)$ using (C.15).

If $\varphi = q_i\pi_i$, then "the Lagrangian coordinates" π_a, by virtue of (C.14) have the meaning of momenta at the final time point $t = t_1$, and $\psi(q, \pi)$ have the meaning of the density of the simultaneous distribution of the initial coordinates and the final momenta.

Equation (C.10) is satisfied as an identity due to (C.15) and (C.9). We will show that (C.11) is also satisfied as an identity.

Let us find the variation of the function f at $t = t_0$. To this end we need the equality

$$\delta\left|\frac{\partial\pi}{\partial p}\right| = \left|\frac{\partial\pi}{\partial p}\right|\frac{\partial\delta\pi_a}{\partial\pi_a}, \tag{C.19}$$

which is obtained in the same way as (4.19).

According to the initial condition (C.4) and formula (C.19), we have for the variation of the function f,

$$\delta f = \frac{\partial\psi}{\partial\pi_a}\delta\pi_a\left|\frac{\partial\pi}{\partial p}\right| + \psi\left|\frac{\partial\pi}{\partial p}\right|\frac{\partial\delta\pi_a}{\partial\pi_a} = \left|\frac{\partial\pi}{\partial p}\right|\frac{\partial(\psi\delta\pi_a)}{\partial\pi_a}.$$

Therefore

$$\int_R \alpha\delta f dpdq = \int_R \alpha\left|\frac{\partial\pi}{\partial p}\right|\frac{\partial(\psi\delta\pi_a)}{\partial\pi_a}dpdq =$$

$$= \int_R \alpha\frac{\partial(\psi\delta\pi_a)}{\partial\pi_a}d\pi dq = -\int_R \psi\partial_a\alpha\delta\pi_a d\pi dq = -\int_R f\partial_a\alpha\delta\pi_a dpdq. \tag{C.20}$$

From equality (C.20) it follows that relation (C.11) is satisfied identically by virtue of (C.15). This completes the proof of the variational principle stated.

Note that the constraint (C.2) could be replaced by the constraint

$$\frac{\partial(\pi_a f)}{\partial t} + \frac{\partial(\pi_a\dot{q}_i f)}{\partial q^i} + \frac{\partial(\pi_a\dot{p}_i f)}{\partial p_i} = 0,$$

in which case the Lagrange multiplier for constraint (C.1) would have the meaning of Legendre transformation of the function $\alpha(\pi, q, t)$ with respect to π_a, and would be a function of θ, q, t.

If the function $\psi(q, \pi)$ is chosen as the δ-function, then the variational principle becomes the Hamilton principle for a mechanical system with given coordinates q

at $t = t_0$. The values of momenta at $t = t_1$ are obtained as the natural boundary condition. One can modify the variational principle for the case when the positions of the particles at $t = t_0$ and $t = t_1$ are given. Then, instead of n "Lagrangian coordinates" π_a, $2n$ Lagrangian coordinates should be introduced, half of which take the given values at $t = t_0$ and the other half at $t = t_1$.

D. Lagrange Variational Principle

To derive (1.40) we have first to find the variation of the action functional. To this end, it is convenient to get rid of the constraint (1.36) by means of a Lagrange multiplier. We consider an auxiliary variational problem which contains an extra unknown function, $\lambda(t)$, and has a modified action functional,

$$\tilde{I} = \int_{t_0}^{t_1} (K+\lambda(t)(K+U-E))dt. \tag{D.1}$$

We denote the integrand by L :

$$L(q,\dot{q},\lambda) \equiv K+\lambda(t)(K+U-E).$$

Note that this function is not necessarily equal to $K-U$.

At the stationary trajectory the variation of functional $\tilde{I}$ must vanish for all admissible variations δq, δt_1 and $\delta\lambda$.

We will work with a more general functional,

$$I(q(t),t_1) = \int_{t_0}^{t_1} L(q,\dot{q},t)dt, \tag{D.2}$$

not assuming that $L = K+\lambda(K+U-E)$. The difference from the Hamilton variational principle is that the upper limit of the integral is also varied.

Let us give infinitesimally small variation to the trajectory and the arrival time,

$$q'(t) = q(t) + \delta q(t), \; t_1' = t_1 + \delta t_1,$$

and consider the difference,

$$I\left(q'(t),t_1'\right) - I(q(t),t_1) = \int_{t_0}^{t_1'} L\left(q+\delta q, \frac{d(q+\delta q)}{dt}, t\right)dt - \int_{t_0}^{t_1} L\left(q, \frac{dq}{dt}, t\right)dt. \tag{D.3}$$

To obtain δI we have to keep only the terms of the first order in (D.3). Note that the domain of integration of the first integral in (D.3) differs from that of the second integral. To perform the calculation it is convenient to write down the first integral in (D.3) as an integral over the segment $[t_0, t_1]$. Therefore, we consider a mapping of $[t_0, t_1]$ onto $\left[t_0, t_1'\right]$:

$$t' = t'(t) = t + \delta t.$$

Here δt is an infinitesimally small smooth function of time which is zero at $t = t_0$ and equal to δt_1 at $t = t_1$. Thus,

$$\begin{aligned} I\left(q'(t), t_1'\right) &= \int_{t_0}^{t_1'} L\left(q\left(t'\right) + \delta q\left(t'\right), \frac{d(q\left(t'\right) + \delta q\left(t'\right))}{dt'}, t'\right) dt' \\ &= \int_{t_0}^{t_1} L\left(q\left(t'(t)\right) + \delta q\left(t'(t)\right), \frac{d(q\left(t'(t)\right) + \delta q\left(t'(t)\right))}{dt'}, t'(t)\right) \frac{dt'}{dt} dt. \end{aligned} \tag{D.4}$$

Let us introduce the total variation of $q(t)$, $\tilde{\delta} q$, as the difference of the values of the function, $q' = q + \delta q$, at the shifted point, $t'(t)$, and the function, q, at the point t :

$$\tilde{\delta} q = q'\left(t'(t)\right) - q(t).$$

Obviously, up to the terms of higher order,

$$\tilde{\delta} q = q'(t + \delta t) - q(t) = \delta q + \frac{dq}{dt} \delta t.$$

The operator δ commutes with the operator $\dfrac{d}{dt}$,

$$\delta \frac{dq}{dt} = \frac{dq'}{dt} - \frac{dq}{dt} = \frac{d(q + \delta q)}{dt} - \frac{dq}{dt} = \frac{d}{dt} \delta q. \tag{D.5}$$

The operator $\tilde{\delta}$ does not possess such a property: $\tilde{\delta} \dfrac{d}{dt}$ and $\dfrac{d}{dt} \tilde{\delta}$ are different and linked by the relation

$$\tilde{\delta} \frac{dq}{dt} = \frac{d}{dt} \tilde{\delta} q - \frac{dq}{dt} \frac{d\delta t}{dt}. \tag{D.6}$$

This relation follows from a chain of equalities:

$$\tilde{\delta}\frac{dq}{dt} = \left.\frac{dq'(t')}{dt'}\right|_{t'=t'(t)} - \frac{dq(t)}{dt}$$

$$= \frac{d(q+\tilde{\delta}q)}{dt}\frac{dt}{dt'} - \frac{dq}{dt} = \left(\frac{dq}{dt}+\frac{d\tilde{\delta}q}{dt}\right)\left(1+\frac{d\delta t}{dt}\right)^{-1} - \frac{dq}{dt}$$

$$= \left(\frac{dq}{dt}+\frac{d\tilde{\delta}q}{dt}\right)\left(1-\frac{d\delta t}{dt}\right) - \frac{dq}{dt} = \frac{d\tilde{\delta}q}{dt} - \frac{dq}{dt}\frac{d\delta t}{dt}.$$

Let us show that the variation of the functional I is

$$\delta I = \int_{t_0}^{t_1}\left[\frac{\delta L}{\delta q^i}\tilde{\delta}q^i - \frac{\delta L}{\delta q^i}\dot{q}^i\delta t + \frac{d}{dt}\left(\frac{\partial L}{\partial \dot{q}^i}\tilde{\delta}q^i - H\delta t\right)\right]dt \tag{D.7}$$

where the function H is defined as

$$H = \frac{\partial L}{\partial \dot{q}^i}\dot{q}^i - L. \tag{D.8}$$

Indeed, expanding the integrand in (D.4), we have

$$I\left(q'(t), t_1'\right) = \int_{t_0}^{t_1} L\left(q(t)+\tilde{\delta}q, \frac{dq}{dt}+\tilde{\delta}\frac{dq}{dt}, t+\delta t\right)\left(1+\frac{d\delta t}{dt}\right)dt$$

$$= \int_{t_0}^{t_1}\left[L\left(q, \frac{dq}{dt}, t\right) + \frac{\partial L}{\partial q^i}\tilde{\delta}q^i + \frac{\partial L}{\partial \dot{q}^i}\tilde{\delta}\frac{dq^i}{dt} + \frac{\partial L}{\partial t}\delta t\right]\left(1+\frac{d\delta t}{dt}\right)dt$$

$$= I(q(t), t_1) + \int_{t_0}^{t_1}\left[\frac{\partial L}{\partial q^i}\tilde{\delta}q^i + \frac{\partial L}{\partial \dot{q}^i}\left(\frac{d\tilde{\delta}q^i}{dt} - \dot{q}^i\frac{d\delta t}{dt}\right) + \frac{\partial L}{\partial t}\delta t + L\frac{d\delta t}{dt}\right]dt.$$

Integrating the second term in the integrand by parts, we find

$$\delta I = \int_{t_0}^{t_1}\left[\frac{\delta L}{\delta q^i}\tilde{\delta}q^i + \frac{d}{dt}\left(\frac{\partial L}{\partial \dot{q}^i}\tilde{\delta}q^i\right) + \frac{\partial L}{\partial t}\delta t - H\frac{d\delta t}{dt}\right]dt. \tag{D.9}$$

It may be checked by inspection that there is an identity,

$$\frac{dH}{dt} \equiv -\frac{\delta L}{\delta q^i}\dot{q}^i - \frac{\partial L}{\partial t}, \tag{D.10}$$

where $\frac{\partial L}{\partial t}$ is the partial derivative of $L(q, \dot{q}, t)$ with respect to t (with q and $\dot{q}$ held fixed) and $\frac{d}{dt}$ is the derivative which takes into account the change of all arguments with time. Substituting $\frac{\partial L}{\partial t}$ found from (D.10) into (D.9), we obtain (D.7).

After moving the terms with time derivative to the boundary, (D.7) takes the form

$$\delta I = \int_{t_0}^{t_1} \left[\frac{\delta L}{\delta q^i} \tilde{\delta} q^i - \frac{\delta L}{\delta q^i} \dot{q}^i \delta t \right] dt + \left[\frac{\partial L}{\partial \dot{q}^i} \tilde{\delta} q^i - H \delta t \right]_{t_0}^{t_1}. \quad \text{(D.11)}$$

The admissible variations $\tilde{\delta} q$ must be zero at the ends of the integration domain,

$$\tilde{\delta} q = 0 \text{ at } t = t_0, t_1, \quad \text{(D.12)}$$

because $q(t)$ are prescribed at the ends: both $q(t)$ and $q(t) + \tilde{\delta} q(t)$ must be equal to q_0 and q_1 at $t = t_0$ and $t = t_1$, respectively. Actually, the possibility of writing the constraints for variations in such a simple form is the major motivation for using the variation $\tilde{\delta} q$ instead of the variation δq. For zero $\tilde{\delta} q$ at the ends, the third term in (D.11) vanishes.

Variation δt must be zero at $t = t_0$. At all other points of the segment $[t_0, t_1]$ the variation δt is arbitrary.

Let us first take $\delta t \equiv 0$. Then, from the equality $\delta I = 0$, the expression for δI (D.7) and the main lemma of calculus of variations, we obtain the system of equations

$$\frac{\delta L}{\delta q^i} = 0. \quad \text{(D.13)}$$

Now allow δt to be non-zero. The first two terms of (D.11) are zero due to (D.13). Therefore, the equation, $\delta I = 0$, yields the condition

$$H = 0 \quad \text{for } t = t_1. \quad \text{(D.14)}$$

The equations obtained are valid for an arbitrary (smooth) function L. Let us return now to the Lagrange function of the Lagrange variational principle,

$$L = K + \lambda(t)(K + U - E). \quad \text{(D.15)}$$

We have a closed system of equations for $n + 1$ functions $q_i(t), \lambda(t)$ comprising the energy equation (1.36), (D.13) and the boundary conditions (1.37) and (D.14). Let us show that the function $\lambda(t)$ can be found explicitly. From (D.15), and the definition of H (D.8),

$$H = \frac{\partial K}{\partial \dot{q}^i}(1+\lambda)\dot{q}^i - K - \lambda(K + U - E).$$

Using the homogeneity of kinetic energy, (1.33), and the energy equation (1.36), we see that

$$H = (1 + 2\lambda) K.$$

According to (D.13) and the identity (D.10),

$$\frac{dH}{dt} = -\frac{\partial L}{\partial t}.$$

The explicit dependence of L on time is due only to the the dependence of L on λ, and therefore

$$\frac{\partial L}{\partial t} = \frac{\partial L}{\partial \lambda}\frac{d\lambda}{dt} = (K+U-E)\frac{d\lambda}{dt}.$$

Therefore, in addition, $dH/dt = 0$, and $H = (1 + 2\lambda) K = const$. The boundary condition (D.14) shows that this constant is equal to zero. Since $K \neq 0$, we have $1 + 2\lambda = const$. So, function $\lambda(t)$ is constant and is equal to $-\frac{1}{2}$. For $\lambda = -\frac{1}{2}$, function L, up to a constant, is half of the difference of the kinetic and the potential energy,

$$L = \frac{1}{2}(K-U) + \frac{1}{2}E. \quad \text{(D.16)}$$

For the function L (D.16), (D.13) are equivalent to Lagrange equations (1.40).

Apparently, the factor $\frac{1}{2}$ and the additive constant $\frac{1}{2}E$ are not essential in writing down the dynamical equations, and one can replace function L (D.16) by the function (1.41).

E. Microdynamics Yielding Classical Thermodynamics

As we have seen, the Hamiltonian structure of the equations of micro-mechanics yields the laws of equilibrium thermodynamics. The question arises: How important it is that the equations of micro-mechanics are Hamiltonian? In other words: Could non-Hamiltonian equations of microdynamics yield the equations of classical equilibrium thermodynamics? It turns out that the class of such equations is indeed slightly wider than the class of Hamiltonian equations.

To formulate the answer, it is convenient to introduce coordinates $x^\mu (\mu = 1, \ldots, 2n)$ in the phase space, $x^1 = p_1, \ldots, x^n = p_n, x^{n+1} = q^1, \ldots, x^{2n} = q^n$ and write down the Hamiltonian equations (1.63) in the form

$$\dot{x}^{\mu} = \omega^{\mu\nu} \frac{\partial H(x)}{\partial x^{\nu}}. \tag{E.1}$$

where $\omega^{\mu\nu}$ is a constant antisymmetric tensor:

$$\omega^{\mu\nu} = \begin{cases} 1 & \mu \geq n+1, \, \nu = \mu \\ -1 & \mu \leq n, \, \nu = n+\mu \\ 0 & \text{otherwise} \end{cases}$$

Tensor $\omega^{\mu\nu}$ has a non-zero determinant. Therefore, one can introduce an inverse tensor, i.e. the tensor obeying the equations

$$\Omega_{\nu\mu}\omega^{\mu\lambda} = \delta_{\nu}^{\lambda}, \tag{E.2}$$

with δ_{ν}^{λ} being Kronecker's tensor:

$$\delta_{\nu}^{\lambda} = \begin{cases} 1 \text{ if } \lambda = \nu \\ 0 \text{ if } \lambda \neq \nu \end{cases}$$

Tensor $\Omega_{\nu\mu}$ is antisymmetric. Contracting (E.1) with $\Omega_{\nu\mu}$ and using (E.2) we obtain another form of the Hamiltonian equations,

$$\Omega_{\nu\mu}\dot{x}^{\mu} = \frac{\partial H(x)}{\partial x^{\nu}}. \tag{E.3}$$

If one makes a general coordinate transformation then $\Omega_{\nu\mu}$ and $\omega^{\mu\lambda}$ transform in accordance with the tensor laws. If the Hamiltonian equations are written in arbitrary coordinates, the tensors $\omega^{\mu\nu}$ and $\Omega_{\mu\nu}$ are certain functions of those coordinates. These functions are not arbitrary because there are special coordinates $\{p, q\}$ for which (E.3) takes (locally) the form of the usual Hamiltonian equations (1.63). It turns out [5] that the sufficient and necessary condition for the system (E.3) to be Hamiltonian is

$$\frac{\partial \Omega_{\nu\mu}}{\partial x^{\lambda}} + \frac{\partial \Omega_{\lambda\nu}}{\partial x^{\mu}} + \frac{\partial \Omega_{\mu\lambda}}{\partial x^{\nu}} = 0. \tag{E.4}$$

Equations (E.4) are equivalent to the existence of functions $P_{\mu}(x)$ such that

$$\Omega_{\mu\nu} = \frac{\partial P_{\mu}(x)}{\partial x^{\nu}} - \frac{\partial P_{\nu}(x)}{\partial x^{\mu}}.$$

The coordinate system in which equations (E.3) takes the form (1.63) is the system where the differential expression $P_{\mu}(x)dx^{\mu}$ takes the canonical form $p_1 dq^1 + \ldots + p_n dq^n$. Some natural reasonings on the features of micromotion show that equations of micromotion should have the from (E.1) with $\omega^{\mu\nu}$ being functions of phase coordinates. These equations yield the laws of equilibrium thermodynamics if the tensor $\omega^{\mu\nu}$ obeys the condition [50]

$$\frac{\partial}{\partial x^{\mu}}\left(\frac{1}{\sqrt{\omega}}\omega^{\mu\nu}\right)=0 \tag{E.5}$$

where $\omega \equiv \det \|\omega^{\mu\nu}\|$. This condition is weaker than (E.4): one can show that (E.5) follows from (E.4) but not vice versa.

Bibliographic Comments

The diversity of topics considered makes it impossible to give a comprehensive literature review. The bibliographic comments given here are based on the papers which turned out to be in the view field of the author, and make no pretence at completeness.

Chapter 1

The history of variational principles can be learned from the collection of the original papers [248]. It is considered also in the books by E. Mach [195], A. Mayer [207], F. Klein [149], C. Lanczos [170], and H. Goldstine [116]. There are many excellent books to master in Hamiltonian mechanics; we mention just a few: Arnold [4], Arnold, Kozlov and Neistadt [6], Landau and Lifshits [171], and Lichtenberg and Liberman [182].

Chapter 2

There is a tremendous wealth of literature on thermodynamics. The key contributions to the "derivation of thermodynamics from mechanics" were made by Boltzmann, Gibbs, P. Hertz, Birkhoff, Khinchine and Kubo. The treatment of thermodynamics in this chapter closely follows the author's monograph [46], where a historic review and the references to the original papers can be found. A generalization of the Einstein formula for finite-dimensional Hamiltonian systems was obtained in [43], the quasi-Hamiltonian structure of equations of macrophysics was suggested in [44] and later justified in [50] where the notion of the secondary and the higher order thermodynamics was also introduced. A proposition that entropy in secondary thermodynamics is decaying in isolated stable systems was made in [54].

Chapters 3 and 4

The contents of these chapters is standard. As an additional reading one can use the text books by L.I. Sedov [268, 269], I.M. Gelfand and S.V. Fomin [110], A.J. McConnell [208], S.R. de Groot and P. Mazur [87] and a paper by L.I. Sedov [270]. A key point used in the development of models with high derivatives, a proper transformation of surface integrals, was made by Kirchhoff. It was applied in the 1960s to the construction of refined elastic models by Mindlin [215].

Chapter 5

Additional readings: S.G. Mikhlin [209, 210, 211, 212], A.D. Ioffe and V.M. Tikhomirov [138], I. Ekeland and R. Temam [95], R. Rockafellar [256], L.C. Young [320], J.L. Synge [290], L.S. Pontrjagin [249], R.E. Bellman and R.E. Kalaba [20], and K.A. Lurie [190]. See regarding inequalities [18, 124, 138, 238, 240], dual variational principles [7, 26, 81, 95, 131, 138, 232, 250, 274, 290] error estimates of approximate solutions [151, 209, 220, 221, 250, 280, 290], method of constraint unlocking [222], and Rayleigh-Ritz's method [81, 117, 209, 210, 211, 253]. The treatment of the dual variational problems in the book follows [26]. The idea to eliminate differential constraints in the dual variational principle was suggested by Hashin and Strikman in [126]. The treatment of this topic in the book is slightly different from [126] and the studies that followed. The variational-asymptotic method was formulated in the author's papers [28, 31]. It was further applied to various problems in [35, 37, 62, 68, 134, 173]. Nowadays, it is being actively developed the so-called Γ-convergence method (see [74, 75]), which is, in fact, a version of the variational-asymptotic method. Some exact results in variational problems with a small parameter were obtained by E.S. Levitin [178, 179, 180, 181]. A link between the minimization and the integration in functional spaces was mentioned and used in [48]. Its further applications can be found in [49, 52, 53, 56, 61, 65, 174]. In the book we do not consider various modifications of the notion of convexity. Besides, the minimization of non-convex functionals has not been discussed.[1] This can be learned from [12, 13, 14, 15, 16, 73, 74].

Chapter 6

There are many modifications of variational principles of linear elasticity theory. Our treatment is focused on three major statements: Gibbs principle, its dual version – Castigliano principle, and the corresponding minimax principle – Reissner principle. These three assertions are enough for most applications. The Hashin-Strikman variational principle is used to obtain the bounds for effective characteristics of micro-inhomogeneous bodies (see [126, 127, 318] and a review [214]). Regarding Korn's inequality [153] see [105, 137, 162, 210, 217, 241, 279].

Chapter 7

Dual variational principle in geometrically nonlinear elasticity was discussed in [30, 90, 176, 189, 203, 225, 226, 227, 228, 235, 237, 287, 288, 289, 312, 313, 314, 333, 334, 335, 336].

[1] An exception is the dual variational principle in nonlinear elasticity considered in Chap. 7.

The stationary principle for complementary energy of semi-linear material was obtained by L.M. Zubov [333]. W. Koiter [152] noted that the functional of complementary energy can be multi-valued, and it remained unclear how to choose the proper branch of the functional. Later Zubov described the character of ambiguity in the dependence of distortion on Piola-Kirchhoff's stress tensor [336]; a complete treatment of the problem was given in [38]. The extended stationary principle of complementary energy was suggested by Fraijs de Vebeke [101] and Christoffersen [80].

A review of the relations suggested for free energy is given by A.I. Lurie [189].

An issue of a priori constraints which should be imposed on free energy has been actively investigated in recent years; one of the first papers was that by Ball [12].

Young-Fenchel transformation of energy of semi-linear material and the corresponding dual variational principle were obtained in [66].

The conditions of phase equilibrium of elastic bodies were derived by Eshelby [98] and Grinfeld [118, 119]; an extension to dynamics is proposed by Truskinovsky [297].

Chapter 8

Action functional in dynamical problems is studied in practically all treatises on classical calculus of variation. Theory of eigenvibrations of linear systems was created by Rayleigh [253]. A comprehensive treatment of this subject was given by Courant [81]. Note also the monograph by Gould [117]. A generalization of Rayleigh's variational principle to nonlinear vibrations and a variational principle in Eulerian variables were given in [38].

Chapter 9

The first formulation of variational principle for ideal incompressible fluid was given by Lagrange [168]. He derived from this principle the dynamical equations of ideal fluid known now as Euler-Lagrange equations. Variational principle for fluid with a free surface was suggested by Riabuchinsky [255]. Surprisingly, the variational principle for a flow over a plane (Luke principle) was found only in 1967 [188]. The hypothesis on the extremal property of the attached mass was suggested by Pâolya [238] and proved by Shiffer [275, 276] (see also Payne [239]). A variational principle in terms of Clebsch's potentials was suggested by Seliger and Whitham [271]. The variational principle in dynamics of vortex lines was found by Berdichevsky [46, 47] and Kuznetsov and Ruban [166]. Further modifications were done in [49, 53]; an extension to dynamics of vortex lines in compressible fluid is suggested in [260]. A derivation of the variational principle of vortex line dynamics from the least action principle is given in [57]. For a vortex filament in unbounded domain the functional $\mathcal{A}$ first appeared in the paper by Rasetti and Regge [251]. It remained unclear though what is a proper choice of kinetic energy. Further analysis [58] showed that the Rasetti-Regge variational principle yields the correct equations in the leading approximation, when one keeps logarithmically large terms, and needs a correction, reproduced here, if the terms on the order of unity are also retained. Note that the Hamiltonian structure of the dynamical equations of a vortex filament in unbounded region, when one keeps only logarithmically large terms, was first established by Marsden and Weinstein [202] (see also a review in [5], Sect. 6.3B). The variational principle for open flows was formulated in [46]. Some exact results on minimization of the action functional of ideal incompressible fluid were obtained by Shnirelman [278].

Chapter 10

Variational principles in fluid mechanics were discussed in [76, 79, 88, 89, 91, 92, 94, 97, 135, 139, 141, 165, 175, 184, 186, 194, 198, 226, 242, 272, 273, 283, 285, 291, 292, 310, 311, 326, 327]. The Hamilton principle for ideal compressible fluid in Lagrangian variables was studied by G. Zemplen [329]; he also obtained the discontinuity conditions. A detailed treatment is given in the monograph by L. Lichtenstein [183]. Variational principles in Eulerian coordinates when the field functions are Lagrangian variables were constructed by Davydov [84], Mauersberger [204, 205, 206] and Rogula [258]. Herivel [128] noted that by varying velocity, density and entropy as functions of Eulerian coordinates subjected to continuity equation and the conservation of entropy in fluid particle, one obtains from the Hamilton principle the quasi-potential flows ($v_i = \partial_i \varphi + \mu \partial_i S$). Lin variational principle was suggested in [185]. Earlier, Davydov [84] noticed that instead of three constraints (10.31) one can use one constraint (10.39), thus obtaining the correct expression of velocity in terms of Clebsch's potentials. However, since the set of admissible functions was not clearly introduced the necessity to set the constraint (10.31) or (10.39), as was mentioned in [271], seemed puzzling. The extension of the set of admissible functions described in the text makes transparent the link between these two variational principles. As was mentioned by Moffat [218] (see also [76]), presentation of velocity in terms of Clebsch's potentials, though possible locally, yields some integral constraints on velocity.

Pressure as Lagrangian appears for the first time, perhaps, in Hargreave's paper [125]; a detailed study was given by Bateman [17].

The variational principle for the functional (10.54) [38] generalizes for compressible case the Luke principle (an attempt of such generalization in [304] was not successful).

Derivation of discontinuity conditions from variational principles was considered in [161, 165, 183, 186, 192, 193, 330, 331].

Chapter 11

Arnold's variational principle was suggested in [3], its generalization to compressible fluids in [120], other variational principles of Sect. 11.5 in [57], Giese variational principle in [112], Giese-Kraiko variational principle in [112, 161], Lin-Rubinov variational principle in [186], variational principles for potential compressible flows in [17], and other variational principles of this chapter in [38].

Chapter 12

Regarding the principle of least dissipation see, e.g., Rayleigh [252]. A recent review of variational principles of ideal plasticity is given by Kamenjarzh [145]. The idea that variations in variational principles are closely related to fluctuations in physical systems was suggested by Glansdorff and Prigogine [114], but they did not pursue it to obtain quantitative statements.

Chapter 13

Theory of motion of bodies in fluids was developed by Thompson and Tait [293]. It is discussed in detail by Lamb [169] and Milne-Thomson [213], where many examples of calculation of kinetic energy are given, and by Birkhoff [71]. Kirchhoff [148] found a number of exact solutions of equations of motion of rigid body in potential flows. Generalizations to vortex flows are given in [243, 244, 303]. Section 13.3 follows the paper [36]. An example with Basset force was prepared by S. Utkina.

Appendices

The proof of (A.4) in functional spaces can be found in monographs by Vainberg [298, 299]. Holonomicity condition (A.17) was obtained independently by Tonti [294] and the author [23]. Holonomicity condition (A.21) was suggested by B. Kupershmidt (see [197]). Holonomicity condition (A.20) was found in the case of functions of one variable by Helmholtz. "Non-physical" variational principles are considered in [121, 122, 133, 196, 259, 295]. Appendix C follows paper [40].

Bibliography

1. H.G. Allen. *Analysis and design of structural sandwich panels*. Pergamon Press, New York, 1969.
2. S.A. Ambartsumjan. *General theory of anisotropic shells*. Nauka, Moscow, 1974.
3. V.I. Arnold. A variational principle for three-dimensional stationary flows of the ideal fluid. *Journal of Applied Mathematics and Mechanics (PMM)*, 29(5):846–851, 1965.
4. V.I. Arnold. *Mathematical methods of classical mechanics*. Springer-Verlag, Berlin, 1989.
5. V.I. Arnold and B.A. Khesin. *Topological methods in hydrodynamics*. Springer, New York, 1998.
6. V.I. Arnold, V.V. Kozlov, and A.I. Neistadt. *Mathematical aspects of classical and celestial mechanics*, volume 3 of *Encycl. of Math. Sciences*. Springer, 1988.
7. A.M. Arthurs. *Complementary variational principles*. Clarendon Press, Oxford, 1980.
8. I.B. Babuska. Solution of interface problems by homogenization. *SIAM Journal of Mathematical Analysis*, 7:603–645, 1976.
9. N.S. Bakhvalov. Average characteristics of bodies with periodic structure. *Soviet Mathematics – Doklady*, 218(5):1046–1048, 1974.
10. N.S. Bakhvalov. Homogenization of partial differential operators with rapidly oscillating coefficients. *Soviet Mathematics – Doklady*, 16:351–355, 1975.
11. N.S. Bakhvalov and G. Panasenko. *Homogenisation: averaging processes in periodic media*. Kluwer-Academic Publishers Group, Norwell, MA/Dordrecht, 1989.
12. J.M. Ball. Constitutive inequalities and existence theorems in nonlinear elastostatics. In R.J. Knops, editor, *Nonlinear analysis and mechanics, Heriot Watt Sympoisum*, pp. 187–241. Pitman, London, 1977.
13. J.M. Ball. Convexity conditions and existence theorems in nonlinear elsticity. *Archive for Rational Mechanics and Analysis*, 63:337–403, 1977.
14. J.M. Ball. Strict convexity, strong ellipticity, and regularity in the calculus of variations. *Proceedings of Cambridge Philosophical Society*, 87:501–513, 1980.
15. J.M. Ball. Discontinuous equilibrium solutions and cavitation in nonlinear elasticity. *Philosophical Transactions of Royal Society London*, 306(ser. A):557–611, 1982.
16. J.M. Ball. The calculus of variations and material science. *Quarterly of Applied Mathematics*, 56:719–740, 1998.
17. H. Bateman. *Partial differential equations of mathematical physics*. Cambridge Univ. Press, Cambridge, 1959.
18. E.F. Beckenbach and R.E. Bellman. *Inequalities*. Springer, Berlin, 1961.
19. A.Yu. Beliaev. *Homogenization in the problems of filtration theory*. Nauka, Moscow, 2004.
20. R.E. Bellman and R.E. Kalaba. *Dynamic programming and modern control theory*. Academic Press, New York, 1965.
21. A. Bensoussan, J.L. Lions, and G. Papanicolaou. Surquelques phenomenes asymptotiques stationares. *Comptes Rendus de l'Academie des Sciences Paris*, 281:89–94, 1975.

22. A. Bensoussan, J.L. Lions, and G. Papanicolaou. *Asymptotlc analysis for periodic structures*. North Holland Publ., Amsterdam, 1978.
23. V.L. Berdichevsky. Variational equation of continuum mechanics. In *Problems of Solid Mechanics V. Novozhilov 60th birthday anniversary volume*. Sudostroenie, Leningrad, 1970.
24. V.L. Berdichevsky. Equations describing transverse vibration of thin elastic plates. *Mechanics of Solids*, (6):138–141, 1972.
25. V.L. Berdichevsky. Dynamic theory of thin elastic plates. *Mechanics of Solids*, (6):86–96, 1973.
26. V.L. Berdichevsky. A variational principle. *Soviet Physics – Doklady*, 19(4):188–190, 1974.
27. V.L. Berdichevsky. Spatial averaging of periodic structures. *Soviet Physics – Doklady*, 20(5):334–335, 1975.
28. V.L. Berdichevsky. Equations of the theory of anisotropic inhomogeneous rods. *Soviet Physics – Doklady*, 21(5):286–288, 1976.
29. V.L. Berdichevsky. High-frequency long-wave plate vibrations of plates. *Physics Dokladi*, 22(4):604–606, 1977.
30. V.L. Berdichevsky. Variational-asymptotic method of constructing the nonlinear shell theory. In G.K. Mikhailov W.T. Koiter, editor, *Proceedings of IUTAM Symposium on Shell Theory*, Tbilisi, pp. 137–161. North Holland Company, Amsterdam, 1979.
31. V.L. Berdichevsky. Variational-asymptotic method of constructing the shell theory. *Journal of Applied Mathematics and Mechanics (PMM)*, 42(4):711–736, 1979.
32. V.L. Berdichevsky. Mechanical equations for fluids with particles. In *Problems of averaging and development of continuum models*, pp. 10–35. Moscow University Publishing, Moscow, 1980.
33. V.L. Berdichevsky, V.A. Misiura. On interaction effect between extension and bending of cylindrical shells. In *Proceedings of XIII Conference on Plate and Shell Theory*, pp. 165–171. Erevan, 1980.
34. V.L. Berdichevsky and K.C. Le. High-frequency long-wave shell vibrations. *Journal of Applied Mathematics and Mechanics*. 44:737–744, 1980.
35. V.L. Berdichevsky. On the energy of an elastic beam. *Journal of Applied Mathematics and Mechanics (PMM)*, 45(4):518–529, 1981.
36. V.L. Berdichevsky. On the force acting on a body in viscous fluids. *Journal of Applied Mathematics and Mechanics (PMM)*, 45(4):628–631, 1981.
37. V.L. Berdichevsky. Variational principles in the problem of averaging random structures. *Soviet Physics – Doklady*, 26(11):1053–1055, 1981.
38. V.L. Berdichevsky. *Variational principles of continuum mechanics*. Nauka, Moscow, 1983.
39. V.L. Berdichevsky. Heat conduction of checkerboard structures. *Moscow University* Mechanics Bulletin, 40:15–25, 1985.
40. V.L. Berdichevsky. On a variational principle of statistical mechanics. *Moscow University* Mechanics Bulletin, 42:8–14, 1987.
41. V.L. Berdichevsky. On turbulence theory. In *L.I. Sedov 80th birthday anniversary volume*, pp. 106–110. Moscow University Publishing, Moscow, 1987.
42. V.L. Berdichevsky. The problem of averaging random structures in terms of distribution functions. *Journal of Applied Mathematics and Mechanics (PMM)*, 51(6):704–710, 1987.
43. V.L. Berdichevsky. A connection between thermodynamic entropy and probability. *Journal of Applied Mathematics and Mechanics (PMM)*, 52(6):947–957, 1988.
44. V.L. Berdichevsky. Law of evolution to equilibrium in nonlinear thermodynamics. *International Journal of Engineering Science*, 28(7):697–704, 1990.
45. V.L. Berdichevsky. Statistical mechanics of point vortices. *Physical Review E*, 51(5):4432–4452, 1995.
46. V.L. Berdichevsky. *Thermodynamics of chaos and order*. Addison Wesley Longman, 1997.
47. V.L. Berdichevsky. Statistical mechanics of vortex lines. *Physical Review E*, 57(3):2885–2905, 1998.
48. V.L. Berdichevsky. Distribution of minimum values of weakly stochastic functionals. In V.L. Berdichevsky, V.V. Jikov, and G. Papanicolaou, editors, *Homogenization*, pp. 141–186. World Scientific, Singapore, 1999.

49. V.L. Berdichevsky. On statistical mechanics of vortex lines. *International Journal of Engineering Science*, 40(123–129), 2002.
50. V.L. Berdichevsky. Structure of equations of macrophysics. *Physical Review E*, 68(066126):26, 2003.
51. V.L. Berdichevsky. Distribution of functionals of solutions in some stochastic variational problems. *Izvestia Vuzov, Yudovich anniversary volume*, pp. 56–59, 2004.
52. V.L. Berdichevsky. Homogenization in micro-plasticity. *Journal of Mechanics and Physics of Solids*, 53:2459–2469, 2005.
53. V.L. Berdichevsky. Averaged equations of ideal fluid turbulence. *Continuum Mechanics and Thermodynamics*, 19:133–175, 2007.
54. V.L. Berdichevsky. On entropy of microstructure and secondary thermodynamics. In D. Jeulin and S. Forest, editors, *Continuum Models and Discrete Systems, Proceeding of CMDS*, pp 25–29. Mines Paris, Paris Tech, 2008.
55. V.L. Berdichevsky. Distribution of minimum values of stochstic functionals. *Networks and Heterogeneous Media*, 3(3):437–460, 2008.
56. V.L. Berdichevsky. On statistical mechanics of vortex lines. In *Proc. IUTAM Symposium on Hamiltonian Dynamics, Vortex Structures and Turbulence*, pp. 205–210. Springer-Verlag, Berlin, 2008.
57. V.L. Berdichevsky. On variational features of vortex flows. *Continuum Mechanics and Thermodynamics*, 20:219–229, 2008.
58. V.L. Berdichevsky. A variational principle in dynamics of vortex filaments. *Physical Review E*, 78(3):036304, 2008.
59. V.L. Berdichevsky. An asymptotic theory of sandwich plates. *International Journal of Engineering Science*, 2009.
60. V.L. Berdichevsky. Nonlinear theory of hard-skin plates and shells. *International Journal of Engineering Science*, 2009.
61. V.L. Berdichevsky. Entropy of microstructure. *Journal of Mechanics and Physics of Solids*, doi:10.1016/j.jmps.2007.07.004, 2007.
62. V.L. Berdichevsky, E. Armanios, and A. Badir. Theory of anisotropic thin-walled closed-cross-section beams. *Composites Engineering*, 2(5–7):411–432, 1992.
63. V.L. Berdichevsky, V.V. Jikov, and G. Papanicolaou, editors. *Homogenization*. World Scientific, Singapore, 1999.
64. V.L. Berdichevsky and S. Kvashnina. On equations describing the transverse vibrations of elastic bars. *Journal of Applied Mathematics and Mechanics (PMM)*, 40(1):104–118, 1976.
65. V.L. Berdichevsky and K.C. Le. A theory of charge nucleation in two dimensions. *Physical Review E*, 66:026129, 2002.
66. V.L. Berdichevsky and V. Misyura. On a dual variational principle in geometrically nonlinear elasticity theory. *Journal of Applied Mathematics and Mechanics (PMM)*, 43(2):343–351, 1979.
67. V.L. Berdichevsky and V. Misyura. Effect of accuracy loss in classical shell theory. *Journal of Applied Mechanics*, 59(2 Pt. 2):5217–5223, 1992.
68. V.L. Berdichevsky and V.G. Soutyrine. Problem of an equivalent rod in nonlinear theory of springs. *Journal of Applied Mathematics and Mechanics (PMM)*, 47(2):197–205, 1983.
69. L.V. Berlyand. Asymptotic stress-strain distribution function for crystal with random point defects. In V.L. Berdichevsky, V.V. Jikov, and G. Papanicolaou, editors, *Homogenization*, pp. 179–192. World Scientific, Singapore, 1999.
70. A.S. Besicovitch. *Almost periodic functions*. Cambridge University Press, Cambridge, 1932.
71. G. Birkhoff. *Hydrodynamics: a study in logic, fact, and similitude*. Princeton University Press, Princeton, N.J., second edition, 1960.
72. H. Bohr. *Almost periodic functions*. Chelsea Pub. Co., New York, 1947.
73. A. Braides. Loss of polyconvexity by homogenization. *Archive for Rational Mechanics and Analysis*, 127:183–190, 1994.
74. A. Braides and A. Defranceschi. *Homogenization of multiple integrals*. Oxford University Press, Oxford, 1998.
75. A. Braides and L.M. Truskinovsky. Asymptotic expansions by Gamma-convergence. *Continuum Mechanics and Thermodynamics*, 20:21–62, 2008.

76. F.P. Bretherton. A note on Hamilton's principle for perfect fluids. *Journal of Fluid Mechanics*, 44:19–31, 1970.
77. B. Budiansky and J.L. Sanders. On the "best" first order linear shell theory. In *Progress in Applied Mechanics, The Prager anniversary volume*, pp. 129–140. Macmillan Comp., Norway, 1963.
78. L.J. Campbell and K. O'Neil. Statistics of two-dimensional point vortices and high-energy vortex states. *Journal of Statistical Physics*, 65:495–529, 1991.
79. P. Casal. Principles variationnels en fluide compressible et en magnetodynamique. *Rev. Mec. Gren. Anyl.*, XI:40, 1965.
80. J. Christoffersen. On Zubov principle of stationary complementary energy in the non-linear theory of elasticity. *DCAMM Report*, 44:31, 1973.
81. R. Courant and D. Hilbert. *Methods of mathematical physics*, volume I. Interscience Publisher Inc., New York, 1953.
82. H. Cramer and M.R. Leadbetter. *Stationary and related stochastic processes*, volume I. John Wiley & Sons Inc., New York, London, Sydney, 1967.
83. V.M. Darevsky. On basic relations in the theory of thin shells. *Journal of Applied Mathematics and Mechanics (PMM)*, 25(3):768–790, 1961.
84. B. Davydov. Variational principle and canonic equations for ideal fluid. *Soviet Physics – Doklady*, 69(1), 1949.
85. E. De Giorgi. Sulla convergenza di alcvune succesioni do integrali del tipo dell' area. *Rendiconti Naturali*, 6:277–294, 1975.
86. E. De Giorgi and S. Spagnolo. Sulla convergenze delli integrali dell energia per operatori elliptici del secondo ordine. *Bollettino della Unione Mathematica Italiana*, 8:391–411, 1973.
87. S.R. De Groot and P. Mazur. *Non-equilibrium thermodynamics*. North-Holland Publ. Co., Amsterdam, 1962.
88. I. Deval. Sur la dynamique des fluids perfaits et le principles d'Hamilton. *Bulletin de l'Académié de Belgique (Classe des Sciences)*, t.37:386–390, 1951.
89. Th. de Donder and F.H. Van der Dungen. Sur les principles variationnelles des milieux continus. *Bulletin de l'Académié de Belgique (Classe des Sciences)*, 35:841–846, 1949.
90. S. Dost and B. Tabarrok. Application of Hamilton principle to large deformation and flow problems. *Transactions on ASME, Journal of Applied Mechanics*, 2(2):285–290, 1979.
91. S. Drobot and A. Rybarski. A variational principle of hydrodynamics. *Archive of Rational Mechanics and Analysis*, 11:393–401, 1958.
92. V.T. Dubasov. Application of variational methods to solution of aerodynamic problems. *Trudi MATI*, 42:5–30, 1959.
93. A.M. Dykhne. Conductivity of two-phase two-dimensional system. *Journal of Experimental and Theoretical Physics*, 59(7):110–115, 1970.
94. E. Eckart. Variational principles of hydrodynamics. *Physics of Fluids*, 3:421–427, 1960.
95. I. Ekeland and R. Temam. *Convex analysis and variational problems*. North-Holland Pub. Co., Amsterdam, New York, 1976.
96. R.S. Ellis and J.S. Rosen. Asymptotic analysis of Gaussian integrals. I. Isolated minimum points. *Transactions of American Mathematical Society*, 273:447–481, 1982.
97. H. Ertel. Uber ein allgeneines Variationprinzip der Hydrodynamik. *Abhandlungen preussiche Akademie der Wissenschaften Physical Mathematics Klasse*, 7:S:30–41, 1939.
98. J.D. Eshelby. Energy relations and the energy-momentum tensor in continuum mechanics. In M.F. Kanninen, W.F. Adler, A.R. Rosenfield, and R.I. Jaffee, editors, *Inelastic behaviour of solids*, pp. 77–115. McGraw-Hill, New York, 1970.
99. L. Euler. *Dissertatio de principio minimae action is una cum examine objection Cl. Prof. Koenigii contra hoe principium factarum*. Berlin, 1753.
100. R. Feynman and A. Hibbs. *Quantum mechanics and path integrals*. McGraw-Hill, New York, 1965.
101. B. A. Fraijs de Vebeke. New variational principle for finite elastic displacements. *International Journal of Engineering Science*, 10:745–763, 1972.

102. M.I. Freidlin. Dirichlet problem for equation with periodic coefficients depending on small parameter. *Theory of Probability and its Applications*, 9(1):133–139, 1964.
103. M.I. Freidlin. *Functional integration and partial differential equations*, Princeton University Press, Princeton, N.J., 1985.
104. M.I. Freidlin and A.D. Wentzel. *Random perturbations of dynamical systems*. Springer-Verlag, Berlin, 1984.
105. K.O. Friedrichs. On the boundary-value problems of the theory of elasticity and Korn's inequality. *Annals of Mathematics*, 48(2):441–471, 1947.
106. S.L. Gavrilyuk. Acoustic properties of a multiphase model with micro-inertia. *European Journal of Mechanics B/Fluids*, 24:397–406, 2005.
107. S.L. Gavrilyuk and R. Saurel. Mathematical and numerical modeling to two-phase compressible flows with microinertia. *Journal of Computational Physics*, 175:326–360, 2002.
108. S.L. Gavrilyuk and V.M. Teshukov. Generalized vorticity and flow properties for bubbly liquid and dispersive shallow water equations. *Continuum Mechanics and Thermodynamics*, 13:365–382, 2001.
109. S.L. Gavrilyuk and V.M. Teshukov. Drag force acting on a bubble in a cloud of compressible spherical bubbles at large Reynolds number. *European Journal of Mechanics B/Fluids*, 24:468–477, 2005.
110. I.M. Gelfand and S.V. Fomin. *Calculus of variations*. Prentice-Hall, Englewood Cliffs, N.J., 1963.
111. V.L. German. Some theorems on anisotropic media. *Soviet Physics – Doklady*, 48(1), 1945.
112. J.H. Giese. Stream functions for threedimensional flows. *Journal of Mathematical Physics*, 30(I):1, 1951.
113. I.I. Gihman and A.V. Skorokhod. *The theory of stochastic processes*. Springer-Verlag, Berlin, 1980.
114. P. Glansdorff and I. Prigogine. *Thermodynamic theory of structure, stability and fluctuations*. Wiley-Interscience, London, New York, 1971.
115. A.L. Goldenveizer. *Theory of elastic thin shells*. Pergamon Press, New York, 1961.
116. H.H. Goldstine. *A history of the calculus of variations from the 17th through the 19th century*. Springer-Verlag, Berlin, 1980.
117. S.H. Gould. *Variational methods for eigenvalue problems; an introduction to the Weinstein method of intermediate problems*. Toronto University Press; Oxford University Press, Toronto London, 2nd edition, 1966.
118. M.A. Grinfeld. On thermodynamic stability of materials. *Doklady* Academii Nauk SSSR, 253(6):1349–1353, 1980.
119. M.A. Grinfeld. On the two types of heterogeneous phase-equilibria. *Doklady* Akademii Nauk SSSR, 258(3):54–57, 1981.
120. M.A. Grinfeld. Variational principles for steady flows of ideal fluid. *Doklady* Akademii Nauk SSSR, 262(1):54–57, 1982.
121. M.E. Gurtin. Variational principles for linear elastodynamics. *Archive for Rational Mechanics and Analysis*, 13:179–191, 1963.
122. M.E. Gurtin. Variational principles for linear initial-value problems. *Quarterly Applied Mathematics*, XXII(3):252–256, 1964.
123. W.R. Hamilton. Second essay on a general method in dynamics. *Philosophical Transactions of Royal Society*, (1):95–144, 1835.
124. G.H. Hardy, J.E. Littlewood, and G. Paolya. *Inequalities*. The University Press, Cambridge, 1934.
125. R. Hargreaves. A pressure-integral as kinetic potential. Philosophical Magazine, 1908.
126. Z. Hashin and S. Strikman. A variational approach to the theory of the behaviour of polycrystals. *Journal of the Mechanics and Physics of Solids*, 10:343–352, 1962.
127. Z. Hashin and S. Strikman. A variational approach to the theory of the elastic behaviour of multiphase materials. *Journal of the Mechanics and Physics of Solids*, 11:127–140, 1963.
128. J.W. Herivel. The derivation of equations of motion of an ideal fluid by Hamilton's principle. *Proceedings of the Cambridge Philosophical Society*, 51:344–349, 1955.

129. C. Hermann. Tensoren und Kristallsymmetric. *Zeitschrift fur Kristallographic*, Band 89 (H. 1):37–51, 1934.
130. H. Hertz. *The principles of mechanics, presented in a new form*. Dover Publications, New York, 1956.
131. R. Hill. New horizons in the mechanics of solids. *Journal of the Mechanics and Physics of Solids*, 5(1):66–74, 1956.
132. R. Hill. Elastic properties of reinforced solids: some theoretical principles. *Journal of the Mechanics and Physics of Solids*, 11(5):357–382, 1963.
133. I. Hlavacek. Variational principles for parabolic equations. *Aplikace Mathematics, Chekoslov. Akad. Ved.*, 14(4):278–293, 1969.
134. D.H. Hodges. *Nonlinear composite beam theory*. AIAA, New York, 2006.
135. E. Holder. Klassische und relativistische Gasdynamik Variations-problem. *Mathematical Nachrichten*, Band 4:366–371, 1950.
136. J. Holtsmark. Uber die Verbreitung von Spektrallinien. *Annalen der Physik*, Band 58: 577–630, 1919.
137. O. Horgan and J.K. Knowles. Eigenvalue problems associated with Korn inequalities. *Archive for Rational Mechanics and Analysis*, 40(5):130–141, 1971.
138. A.D. Ioffe and V.M. Tikhomirov. *Theory of extremal problems*. North-Holland Pub. Co., Amsterdam, New York, 1979.
139. H. Ito. Variational principle in hydrodynamics. *Progress in Theoretical Physics*, 9(2):117–131, 1953.
140. C.G.J. Jacobi. *Gesammelte Werke, Suppl.* G. Reimer, Berlin, 1884.
141. H. Jeffreys. What is Hamilton principle? *Quarterly Journal of Mechanics and Applied Mathematics*, 7:335–337, 1954.
142. V.V. Jikov, S.M. Kozlov, and O.A. Oleinik. *Homogenization of differential operators and integral functionals*. Springer-Verlag, Berlin, 1994.
143. F. John. Estimates for derivatives of the stresses in a thin shell and interior equations. *Communications of Pure and Applied Mathematics*, 12:235–267, 1965.
144. M. Kac. *Probability and Related Topics in Physical Sciences*. Interscience, New York, 1957.
145. J.A. Kamenjarzh. *Limit analysis of solids and structures*. CRC Press, Boca Raton, 1996.
146. J.B. Keller. A theorem on the conductivity of composite medium. *Journal of Mathematical Physics*, 5:548–549, 1964.
147. A. Khinchine. Korrelationstheorie der stationaren stochastischen Prozesse. *Mathematische Annalen*, 109(4):604–615, 1934.
148. G. Kirchhoff. *Vorlesungen uber mathematische Physik. Mechanik*. B.G. Teubner, Leipzig, 1876.
149. F. Klein. *Vorlesungen uber die Entwicklung der Mathematik*. Verlag von Julius Springer, Berlin, 1927.
150. W.T. Koiter. A consistent first approximation in the general theory of thin elastic shells. In *Proceedings of IUTAM symposium on theory of thin elastic shells*, pp. 12–33, 1960.
151. W.T. Koiter. On the mathematical foundations of shell theory. In *Proceedings of the international congress on mathematics Nice 1970*, volume 3, pp. 123–130. Gauthier-Villars, Paris, 1971.
152. W.T. Koiter and J.G. Simmonds. Foundation of shell theory. In *Theoretical and applied mechanics, Proceedings of the 13th international congress on theoretical and applied mechanics (Moscow, 1972)*, Springer, Berlin, 1973.
153. A. Korn. Uber einige Ungelichyngen, welche in der Theorie der elastischen und Schwingungen eine Rolle spiellen. *Bull. internal. cracovic akad. umiedet., Classe des sciences mathematiques et naturelles*, pp. 705–724, 1909.
154. S.M. Kozlov. Averaging of differential equations with almost periodic rapidly oscillating coefficients. *Soviet Mathematics – Doklady*, 236(5):1323–1326, 1977.
155. S.M. Kozlov. Averaging of random structures. *Soviet Mathematics – Doklady*, 241(5), 1978.
156. S.M. Kozlov. Conductivity of two-dimensional random media. *Russian Mathematical Survey*, 34(4):168–169, 1979.

157. V.V. Kozlov. Dynamics of the systems with non-integrable constraints. i. *Moscow University Mechanics Bulletin*, 37(3):92–100, 1982.
158. V.V. Kozlov. Dynamics of the systems with non-integrable constraints. ii. *Moscow University Mechanics Bulletin*, 37(4):70–76, 1982.
159. V.V. Kozlov. Dynamics of the systems with non-integrable constraints. iii. *Moscow University Mechanics Bulletin*, 37(3):102–111, 1983.
160. V.V. Kozlov. Dynamics of the systems with non-integrable constraints. iv. *Moscow University Mechanics Bulletin*, 37(5):76–83, 1987.
161. A.N. Kraiko. Variational principles for perfect gas flows with strong discontinuities expressed in Eulerian variables. *Journal of Applied Mathematics and Mechanics (PMM)*, 45(2):184–191, 1981.
162. A.L. Krylov. Justification of the Dirichlet principle for the first boundary value problem of nonlinear elasticity. *Soviet Mathematics – Doklady*, 146(1), 1962.
163. R. Kubo. *Journal of the Physical Society of Japan*, 12:570, 1957.
164. I.A. Kunin. *Elastic media with microstructure*, volume I,II. Springer-Verlag, Berlin, New York, 1982.
165. B.G. Kuznetsov. On the second Bateman variational principle. *Proceedings of Tomsk U.*, 144:117–124, 1959.
166. E.A. Kuznetsov and V.P. Ruban. Hamiltonian dynamics of vortex lines in hydrodynamic-type systems. *JETP Letters*, 67:1067–1081, 1998.
167. P. Ladeveze. On the validity of linear shell theories. In W.T. Koiter and G.K. Michaile, editors, *Theory of Shells*, pp. 369–343. North Holl. Publ., Amsterdam, 1980.
168. J.L. Lagrange. *Mecanique analytique*. Ve Courcier, Paris, 1811.
169. H. Lamb. *Hydrodynamics*. Dover Publications, New York, 6th edition, 1945.
170. C. Lanczos. *The variational principles of mechanics*. University of Toronto Press, Toronto, 1962.
171. L.D. Landau and E.M. Lifshits. *Mechanics*. Pergamon Press, Oxford, New York, 1976.
172. L.D. Landau and E.M. Lifshits. *Fluid mechanics*. Pergamon Press; Addison-Wesley Pub. Co., London Reading, Mass., 1987.
173. K.C. Le. *Vibrations of shells and rods*. Springer-Verlag, Berlin, New York, 1999.
174. K.C. Le and V.L. Berdichevsky. Energy distribution in a neutral gas of point vortices. *Journal of Statistical Physics*, 104:883–890, 2001.
175. C.M. Leech. Hamilton's principle applied to fluid mechanics. *The Quarterly Journal of Mechanics and Applied Mathematics*, XXX(Pt. I):107–130, 1977.
176. M Levinson. The complementary energy theorem in finite elasticity. *Journal of Applied Mechanics*, 32:826–828, 1965.
177. B.M. Levitan and V.V. Zhikov. *Almost periodic functions and differential equations*. Cambridge University Press, Cambridge, New York, 1982.
178. E.S. Levitin. On differential features of the optimum value of parametric problems in mathematical programming. *Soviet Mathematics – Doklady*, 15(2):603–607, 1974.
179. E.S. Levitin. On the local perturbation theory of a problem of mathematical programming in a Banach space. *Soviet Mathematics – Doklady*, 16(5):1354–1356, 1975.
180. E.S. Levitin. On the perturbation theory of non-smooth extremal problems with constraints. *Soviet Mathematics – Doklady*, 16(5):1389–1393, 1975.
181. E.S. Levitin and R.R. Tichatschke. *On smoothing of generalized max-functions via epsilon-regularization*. Trier : Univ., Mathematik/Informatik, 1996.
182. A.J. Lichtenberg and M.A. Liberman. *Regular and chaotic dynamics*. Springer-Verlag, Berlin, 1992.
183. L. Lichtenstein. *Grundlagen der Hydrodynamik*. Springer-Verlag, Berlin, 1929.
184. C.C. Lin. A new variational principle for isenergetic flow. *Quarterly Applied Mathematics*, 9(4):421–423, 1952.
185. C.C. Lin. Liquid helium. In *International school of physics course*, volume XXI, Academic Publisher, New York, 1963.
186. C.C. Lin and S.L Rubinov. On the flow behind curved shocks. *Journal of Mathematical Physics*, 27(2):105 – 129, 1949.

187. A.E.H. Love. *A treatise on the mathematical theory of elasticity*. Dover Publications, New York, 4th edition, 1944.
188. J.C. Luke. A variational principle for a fluid with a free surface. *Journal of Fluid Mechanics*, 27(2):375–397, 1967.
189. A.I. Lurie. *Theory of elasticity*. Springer-Verlag, Berlin, 2005.
190. K.A. Lurie. *Applied optimal control theory of distributed systems*. Plenum Press, New York, 1993.
191. K.A. Lurie and A.V. Cherkaev. G-closure of the set of anisotropic conductors in two dimensions. *Soviet Physics – Doklady*, 26(7):657–659, 1981.
192. M.V. Lurie. Application of variational principle for studying discontinuities in continuum media. *Journal of Applied Mathematics and Mechanics (PMM)*, 30(4):855–869, 1966.
193. M.V. Lurie. Application of variational principle for studying the propagation of discontinuities in continuum media. *Journal of Applied Mathematics and Mechanics (PMM)*, 33(4):586–592, 1969.
194. P.E. Lush and T.M Cherry. The variational method in hydrodynamics. *Quarterly Journal of Mechanics and Applied Mathematics*, 9(1):6–21, 1956.
195. E. Mach. *The science of mechanics*. Open Court Publishing Company, La Salle, I., London, 5th edition, 1942.
196. F. Magri. Variational formulation for every linear equation. *International Journal of Engineering Science*, 12:537–549, 1974.
197. Yu.I. Manin. Algebraic aspects of non-linear differential equations. *Journal of Soviet Mathematics*, 11:1–122, 1979.
198. A.R. Manwell. A variational principle for steady homoenergic compressible flow with finite shocks. *Wave Motion*, 2(1):83–90, 1980.
199. P. Marcellini and C. Sbordone. Sur quelques de G-coverergense et d'homogenneisation non lineaire. *Comptes Rendus de l'Academie des Sciences Paris*, 284:121–125, 1977.
200. V.A. Marchenko and E.I. Khruslov. *A boundary value problems for micro-inhomogeneous boundaries*. Nauk. Dumka, Kiev, 1974.
201. V.A. Marchenko, E.I.A. Khruslov, M. Goncharenko, and D. Shepelsky. *Homogenization of partial differential equations*, Birkhauster, Boston, 2006.
202. J. Marsden and A. Weinstein. Coadjoing orbits, vortices, and Clebsh variables for incompressible fluids. *Physica D*, 7:305–323, 1983.
203. E.F. Masur and C.H. Popelar. On the use of the complementary energy in the solution of buckling problems. *International Journal of Solids Structure*, 12(3):203–216, 1976.
204. P. Mauersberger. Erweitertes Hamiltonisches Prinzip in der Hydrodynamik. *Gerlands Betraege Geophysics*, 75(5):396–408, 1966.
205. P. Mauersberger. Variablentausch in Hamiltonischen Prinzip. *Monatsberiehte Deutsches Akademie der Wissenschaften zu Berlin*, 8(S.):289–298, 1966.
206. P. Mauersberger. Uber die Berucksichtigung freier Oberflachen in einer lokallen Fassung des Hamiltonischen Prinzip. *Monatsberiehte Deutsches Akademie der Wissenschaften zu Berlin*, 8(12):873–877, 1976.
207. A. Mayer. *Geschichte des Princips der Kleinsten Aktion*. Leipzig, 1877.
208. A.J. McConnell. *Application of tensor analysis*. Dover Publications, New York, 1957.
209. S.G. Mikhlin. *Variational methods in mathematical physics*. Macmillan, New York, 1964.
210. S.G. Mikhlin. *The problem of the minimum of a quadratic functional*. Holden-Day, San Francisco, 1965.
211. S.G. Mikhlin. *The numerical performance of variational methods*. Wolters-Noordhoff Publishing, Groningen, 1971.
212. S.G. Mikhlin. *Constants in some inequalities of analysis*. B.G. Teubner; Wiley, Stuttgart Chichester, New York, 1986.
213. L.M. Milne-Thomson. *Theoretical hydrodynamics*. Dover Publications, New York, 5th edition, 1996.
214. G.W. Milton. *The theory of composites*. Cambridge University Press, Cambridge, New York, 2002.

215. R.D. Mindlin. Second gradient of strain and surface tension in linear elaticity. *International Journal of Solids and Structures*, 1:417–438, 1965.
216. B.A. Misiura. Loss of accuracy of the classical theory of shells. *Soviet Physics – Doklady*, 27(5):423–425, 1982.
217. V.P. Mjasnikov and P.P. Mosolov. A proof of the Korn inequality. *Soviet Physics – Doklady*, 201(1), 1971.
218. H.K. Moffat. The degree of knottedness of tangled vortex lines, *Journal of Fluid Mechanics*, 35:117, 1969.
219. A.S. Monin and Yaglom A.M. *Statistical hydromechanics*. U.S. Dept. of Commerce, Washington, 1967.
220. D. Morgenstern. Herleitung der Plattentheorie aus der dreidimensionalen Elastizitats-theorie. *Archive for Rational Mechanics and Analysis*, 4(2):82–91, 1959.
221. P.P. Mosolov and V.P. Mjasnikov. *Variational methods in theory of visco-ideal plastic bodies*. MGU Publishing, Moscow, 1971.
222. P.P. Mosolov and V.P. Mjasnikov. *Mechanics of ideal plastic media*. Nauka, Moscow, 1981.
223. N.I. Muskhelishvili. *Some basic problems of the mathematical theory of elasticity; fundamental equations, plane theory of elasticity, torsion, and bending*. P. Noordhoff, Groningen, 1963.
224. P.M Naghdi. Foundations of elastic shell theory. *Progress in Solid Mechanics*, 4:1–90, 1963.
225. I.G. Napolitano. On the functional analysis derivation and generalization of hybrid variational method. *Meccanica*, 10(3):188–193, 1975.
226. I.G. Napolitano and C. Golia. Dual variational formulation of non-linear problem in fluid dynamics. In *23 Congr. naz. assoc. ital. mecc. teor: ed appl. (Cagliari, 1976), sez. 43, Bologna*, 1976.
227. S. Nemat-Nasser. General variational methods for elastic waves in composites. *Journal of Elasticity*, (2):73–90, 1972.
228. S. Nemat-Nasser. A note on complementary energy and Reissner principle in non-linear elasticity. *Iranian Journal of Science and Technology*, 6:95–101, 1977.
229. S. Nemat-Nasser, editor. Variational methods in the mechanics of solids. *Proceedings of IUTAM symposium on variational methods in the mechanics of solids*. Pergamon Press, New York, 1980.
230. R.I. Nigmatullin. *Foundations of mechanics of heterogeneous mixtures*. Nauka, Moscow, 1978.
231. F.I. Niordson. *A consistent refined shell theory*. 1980.
232. B. Noble and M.J. Sewell. On dual extremum principles in applied mathematics. *Journal of Institute of Mathematical Applications*, 9(2):123–193, 1972.
233. V.V. Novozhilov. *Thin shell theory*. P. Noordhoff, Groningen, 2d augm. and rev. edition, 1964.
234. J.F. Nye. Some geometrical relations in dislocated crystals. *Acta Metallurgica*, 1:153–162, 1953.
235. R.W. Ogden. A note on variational theorems in non-linear elastostatics. *Mathematical Proceedings of the Cambridge Philosophical Society*, 77:609–615, 1975.
236. I. Onsager. Reciprocal relations in irreversible processes. *Physical Review*, 31:405–416, 1931.
237. C. Oran. Complementary energy method for buckling. *Journal of Engineering Mechanics Division, ASCE*, 93(EMI):57–75, 1967.
238. G. Paolya and G. Szegio. *Isoperimetric inequalities in mathematical physics*, volume 27. Princeton University Press, Princeton, 1951.
239. I.E. Payne. Isoperimetric inequalities and their applications. *SIAM Review*, 9(3):453–488, 1967.
240. I.E. Payne and H.F. Weinberger. New bounds for solutions of second order elliptic partial differential equations. *Pakistan Journal of Mathematics*, 8:551–573, 1958.

241. L.E. Payne and H.F. Weinberger. On Korn inequality. *Archive for Rational Mechanics and Analysis*, 8(2):17–31, 1961.
242. P. Penfield. Hamilton's principle for fluids. *Journal of Physical Fluids*, 9(6):1184–1194, 1966.
243. A.G. Petrov. Reactions acting on a small body in two- dimensional vortex flow. *Soviet Physics – Doklady*, 23(1):18–19, 1978.
244. A.G. Petrov. Hamilton variational principle for motion of a contour of variable form in two-dimensional vortex flow. *Mechanics of Inhomogeneous Continua. Dynamics of continua*, (52):88–108, 1981.
245. W. Pietraszkiewicz. *Introduction to the non-linear theory of shells*. Ruhr-Universitat Bochum, 1977.
246. V.I. Piterbarg and V.R. Fatalov. Laplace method for probabilistic measures in Banach spaces. *Uspekhi Materiaux Nauk*, 50:57–150, 1995.
247. A. Poincare. Hertz's ideas in mechanics. In *addition to H. Hertz, Die Prizipien der Mechanik in neuem Zusammenhange dargestellt*. 1894.
248. L.S. Polak, editor. *Variational principles of mechanics*. Fizmatgiz, Moscow, 1959.
249. L.S. Pontrjagin. *The mathematical theory of optimal processes*. Wiley, New York, 1962.
250. W. Prager and J.L. Synge. Approximations in elasticity based on the concept of function space. *Quarterly Applied Mathematics*, 5(3):1–21, 1947.
251. M. Rasetti and T. Regge. Vortices in He II, current algebras and quantum nots. *Physica A*, 80:217, 1975.
252. J.W. Rayleigh. *Philosophical Magazine*, 26:776, 1913.
253. J.W.S. Rayleigh. *The theory of sound*. Dover, New York, 1945.
254. E. Reissner. The effect of transversal shear deformation on the bending of elastic plates. *Journal of Applied Mechanics*, 12(2):67–77, 1945.
255. D. Riabouchinsky. Sur un probleme de variation. *Comptes Rendus de l'Academie des Sciences Paris*, 185:84–87, 1927.
256. R. Rockafellar. *Convex analysis*. Princeton University Press, Princeton 1970.
257. O. Rodrigues. De la maniere d'employer le principe de a moindre action, pour obtenir les equations du mouvement rapportees aux variables independentes. *Correspondences sur l'Ecole Polytechnique*, 3(2):159–162, 1816.
258. D. Rogula. Variational principle for material coordinates as dependent variables. application in relativistic continuum mechanics. *Bulletin of the Academy of Polish Sciences; Services in Sciences and Technology*, 8(10):781–785, 1970.
259. P. Rosen. On variational principles for the solution of differential equations. *Journal of Chemical Physics*, 21:1220–1227, 1953.
260. V.P. Ruban. Variational principle for frozen-in vorticity interacting with sound waves. *Physical Review E*, 68(047):302, 2003.
261. V.V. Rumiantsev. On the Hamilton principle for nonholonomic systems. *Journal of Applied Mathematics and Mechanics*, 42(3):387–399, 1978.
262. M.Y. Ryazantseva. Flexural vibrations of symmetrical sandwich plates. *Mechanics of Solids*, 20(3):152–158, 1985.
263. M.Y. Ryazantseva. High-frequency vibrations of symmetrical sandwich plates. *Mechanics of Solids*, 24(5), 1989.
264. E. Sanchez-Palencia. Equation aux derivees partielles dans un type de millieux heterogenes. *Comptes Rendus de l'Academie Sciences Paris*, 272(Ser. A):1410–1413, 1971.
265. E. Sanchez-Palencia. Comportaments local et macroscopique d'un type de milieux physiques heterogenes. *International Journal of Engineering Science*, 12:331–351, 1974.
266. E. Sanchez-Palencia. *Non-homogeneous media and vibration theory*. Lecture Notes in Physics, 127, Springer-Verlag, N.Y., 1980.
267. J.L. Sanders. An improved first order approximation theory of thin shells. *NASA Report*, 24, 1959.
268. L. I. Sedov. *Introduction to the mechanics of a continuous medium*. Addison-Wesley Pub. Co., Reading, Mass., 1965.
269. L. I. Sedov. *A course in continuum mechanics*. Wolters-Noordhoff, Groningen, 1971.

270. L.I. Sedov. Some problems of designing new models of continuum medium. In *11 International Congress of Applied Mechanics (Munich, 1964)*, pp. 23–41. Springer-Verlag, Berlin, Heidelberg and New York, 1966.
271. R.I. Seliger and G.B. Whitham. Variational principles in continuum mechanics. *Proceedings of Royal Society*, 305(ser. A):1–25, 1968.
272. J. Serrin. *Mathematical principles of classical fluid mechanics*, volume VIII/I of *Handbuch der physik*. Berlin-Gottingen-Heidelberg, 1959.
273. M.J. Sewell. On reciprocal variational principle for perfect fluids. *Journal of Mathematics and Mechanics*, 12(4):495–504, 1963.
274. M.J. Sewell. On dual approximation principles and optimization in continuum mechanics. *Philosophical Transaction Royal Society of London, Series A*, 265(1162):319–351, 1969.
275. M. Shiffer. Sur les raports entre jes solutions des problems interieurs et celles des problemes exterieurs. *Comptes Rendus de l'Academie des Sciences Paris*, 244(22):2680–2683, 1957.
276. M. Shiffer. Sur la polarisation et la masse virtuelle. *Comptes Rendus de l'Academie des Sciences Paris*, 244(26):3118–3120, 1975.
277. M. Shiffman. On the existence of subsonic flows of a compressible fluid. *Journal of Rational Mechanics and Analysis*, 1(4):605–621, 1952.
278. A. Shnirelman. Lattice theory and the flows of an ideal fluid. *Russian Journal of Mathematical Physics*, 1:105–114, 1993.
279. B.A. Shoykhet. Existence theorems in linear shell theory. *Journal of Applied Mathematics and Mechanics (PMM)*, 38(3):527–553, 1974.
280. B.A. Shoykhet. An energy identity in physically non-linear elasticity theory and error estimates of the plate equations. *Journal of Applied Mathematics and Mechanics (PMM)*, 40(2):291-301, 1976.
281. B. Simon. *Functional integration and quantum physics*. Academic Press, New York, 1979.
282. A.V. Skorokhod. *Integration in Hilbert space*. Springer-Verlag, Berlin; New York, 1974.
283. D.A. Smith and C.V. Smith. When is Hamiltons principle an extremum principle. *AIAA Journal*, 12(11):1573–1576, 1974.
284. V.G. Soutyrine. Equations of the theory of springs. *Soviet Physics – Doklady*, 25(9):773–775, 1980.
285. J.J. Stephens. Alternate forms of the Herivel-Lin variational principle. *Physics of Fluids*, 10(1):76–77, 1967.
286. B.M. Strunin. Internal-stress distribution for a random positioning of dislocation. *Soviet Physics – Solid State*, 9:630, 1967.
287. H. Stumpf. Dual extremum principle and error bounds in the theory of plates with large deflection. *Archives of Mechanics and Stosowone*, 27(3):17–21, 1975.
288. H. Stumpf. Generating functionals and extremum principles in nonlinear elasticity with applications to nonlinear plate and shallow shell theory. In *Lecture notes in mathematics*, volume 503, pp. 500–510. Springer-Verlag, Berlin, 1976.
289. H. Stumpf. The principle of complementary energy in nonlinear plate theory. *Journal of Elasticity*, 6:95–104, 1976.
290. J.L. Synge. *The hypercycle in mathematical physics*. Cambridge University Press, 1957.
291. B. Tabarrok and J.Z. Johnston. Some variational principles for isentropic and isothermal gas flows. *Transaction on C.S.M.E.*, 3:187–192, 1975.
292. A. Taub. Hamilton's principle for perfect compressible fluids. In *Symposium in Applied Mathematics of American Mathematical Society*, volume 1, pp. 148–157, 1949.
293. W. Thomson and P.G. Tait. *Elements of natural philosophy*. London, 1867.
294. E. Tonti. Variational formulation of nonlinear differential equations. *Académie Royal de Belgique Bulletin de la classe des Sciences*, 55(I):137–165, 1969.
295. E. Tonti. On the variational formulation for linear initial value problems. *Annals of Mathematics in Pure Applied, Series 4*, 95:331–359, 1973.
296. S. Torquato. *Random heterogeneous materials: Microstructure and macroscopic properties*, volume 16. Springer-Verlag, New York, 2002.

297. L.M. Truskinovsky. Equilibrium phase interphases. *Soviet Physics – Doklady*, 27(7):551–553, 1982.
298. M.M. Vainberg. *Variational methods for the study of nonlinear operators*. Holden-Day, San Francisco, 1964.
299. M.M. Vainberg. *Variational method and method of monotone operators in the theory of nonlinear equations*. Wiley, New York, 1974.
300. A.M. Vaisman and M.A. Goldshtik. Deformation of a granular media. *Soviet Physics – Doklady*, 25(5):364–366, 1980.
301. R.S. Varahdan. *Large deviations and applications*. SIAM, New Delhi, 1984.
302. V.A. Vladimirov. On vibrodynamics of pendulum and submerged solid. *Journal of Mathematical Fluid Mechanics*, 7:397–412, 2005.
303. O.V. Voinov and A.G. Petrov. Lagrangian of a gas bubbles in nonuniform flow. *Soviet Physics – Doklady*, 18(6):372–374, 1973.
304. K.I. Voljak. Variational principle for compressible fluid. *Soviet Physics – Doklady*, 22(10):562–563, 1977.
305. V.V. Volovoi and D.H. Hodges. Theory of anisotropic thin-walled beams. *Journal of Applied Mechanics*, 67(3):453–459, 2000.
306. V.V. Volovoi and D.H. Hodges. Single- and multi-cell composite thin-walled beams. *AIAA Journal*, 40(5):960–965, 2002.
307. V.V. Volovoi, D.H. Hodges, V.L. Berdichevsky, and V.G. Soutyrine. Dynamic dispersion curves for non-homogeneous, anisotropic beams with cross sections of arbitrary geometry. *Journal of Sound and Vibration*, 215(5):1101–1120, 1998.
308. V.V. Volovoi, D.H. Hodges, V.L. Berdichevsky, and V.G. Soutyrine. Asymptotic theory for static behavior of elastic anisotropic I-beams. *International Journal of Solids and Structures*, 36:1017–1043, 1999.
309. I.I. Vorovich. Some results and problems in asymptotic theory of plate and shells. In *Proceedings of the First Allunion school on theory and numerical methods in shell and plate theory*, pp. 51–150. Tbilisi State University, Tbilisi, 1975.
310. C.T. Wang. A note on Bateman's variational principle for compressible fluid flow. *Quarterly Applied Mathematics*, 9:99–101, 1951.
311. C.T. Wang and G.V.R. Rao. A study of the non-linear characteristic of compressible flow equations by means of variational method. *Journal of Aerosol Science*, 17:343–348, 1950.
312. K. Washizu. *Variational methods in elasticity and plasticity*. Pergamon Press, Oxford, 1968.
313. K. Washizu. A note on the principle of stationary complementary energy in nonlinear elasticity. *Mechanics Today*, 5:509–522, 1980.
314. G. Wempner. Complementary theorems of solid mechanics. pp. 127–135. In [229], 1980.
315. D.R. Westbrook. A linear theory for anysotropic shells. In *3rd Canadian Congress on Applied Mechanics*, pp. 259–260, Calgary, 1971.
316. G. B. Whitham. *Linear and nonlinear waves*. Wiley, New York, 1974.
317. O.E. Widera. An anisotropic theory of plates. *Journal of Engineering Mathematics*, 3(3):7–21, 1969.
318. J.R. Willis. Bounds and self-consistent estimates for the overall properties of anisotropic composites. *Journal of the Mechanics and Physics of Solids*, 25:185–202, 1977.
319. A.M. Yaglom. Some classes of random fields in n-dimensional space similar to stationary random processes. *Theory of Probability and its Applications*, 2:292–337, 1957.
320. L.C. Young. *Lectures on the calculus of variations and optimal control theory*. Saunders, Philadelphia, 1969.
321. W.B. Yu, D.H. Hodges and V. Volovoi. Asymptotic construction of Reissner-like composite plate theory with accurate strain recovery. *International Journal of Solids and Structures*, 39(20):5185–5203, 2002.
322. W.B. Yu, D.H. Hodges and V. Volovoi. Asymptotic generalisation of Reissner-Mindlin theory: accurate three-dimensional recovery for composite shells. *Computer Methods in Applied Mechanics and Engineering*, 191(44):5087–5109, 2002.

323. W.B. Yu, D.H. Hodges and V. Volovoi. Asymptotically accurate 3-D recovery from Reissner-like composite plate finite elements, Computers & Structures, 87(7):439–454, 2003.
324. W.B. Yu and D.H. Hodges. Asymptotic approach for thermoelastic analysis of laminated composite plates. *Journal of Engineering Mechanics-ASCE*, 130(5):531–540, 2004.
325. V.I. Yudovich. Vibrodynamics and vibrogeometry of mechanical systems with constraints. *Uspekhi Mekhaniki*, N3:26–74, 2006.
326. V.E. Zakharov. Hamiltonian formalism for hydrodynamical models of plasma. *Journal of Experimental and Theoretical Physics*, 60(5):1714–1726, 1971.
327. V.E. Zakharov. The Hamiltonian formalism for waves in nonlinear media having dispersion. *Radiophysics and Quantum Electronics*, 17(4):326–343, 1974.
328. V.E. Zakharov and E.A. Kuznetsov. Hamiltonian formalism for nonlinear waves. *Physics-Uspekhi*, 40:1087–1116, 1997.
329. G. Zemplen. Kriterien fur die physikalische Bedutung des unstetigen Losungen der hydrodynamischen Bewegungsgleichungen. *Mathematical Annalen*, 61:437–449, 1905.
330. V.A. Zhelnorovich. Toward the theory of constructing the continuum mechanics models. *Soviet Physics – Doklady*, 176(2), 1967.
331. V.A. Zhelnorovich and L.I. Sedov. On the variational method of derivation of the equations of state for a material medium and a gravitational field. *Journal of Applied Mathematics and Mechanics (PMM)*, 42(5):827–838, 1978.
332. V.M. Zolotarev and B.M. Strunin. Internal-stress distribution for a random distribution of point defects. *Soviet Physics – Solid State*, 13, 1971.
333. L.M. Zubov. The stationary principle of complementary work in nonlinear theory of elasticity. *Journal of Applied Mathematics and Mechanics (PMM)*, 34(2):228–232, 1970.
334. L.M. Zubov. Variational principles of nonlinear theory of elasticity. *Journal of Applied Mathematics and Mechanics (PMM)*, 35(3):369–372, 1971.
335. L.M. Zubov. Variational principles of nonlinear elasticity theory, the case of composition small and finite deformations. *Journal of Applied Mathematics and Mechanics (PMM)*, 35(5):802–806, 1971.
336. L.M. Zubov. The representation of the displacement gradient of isotropic elastic body in terms of the Piola stress tensor. *Journal of Applied Mathematics and Mechanics (PMM)*, 40(6):1012–1019, 1976.

Index

W

Notation

x a point in three-dimensional space, x^i are its coordinates. Indices, i, j, k, l, m, run through values $1, 2, 3$.

In consideration of mathematical issues, x is a point in n-dimensional space, x^i are its coordinates, and small Latin indices, i, j, k, l, m, run through values $1, 2, \ldots, n$.

Usually, writing the arguments of a function, the indices are suppressed, and the notation, $f(x)$, is used for function $f(x^1, \ldots, x^n)$. The notation, $f(x^i)$, is used if it desirable to emphasize that f is a function of several arguments.

Summation is always conducted over repeated low and upper indices. Indices of vectors and tensors are written as low or upper indices depending on convenience and in accordance with the rule of summation over repeated low and upper index.

Indices, which do not have tensor nature are put usually in parentheses; for example, the boundary values of a function, u, is denoted by $u_{(b)}$.

t	time
R_n	n-dimensional space
R_3	three-dimensional Euclidean space
R_4	four-dimensional space-time
X^a	Lagrangian coordinates
a, b, c, d	small Latin indices run through values $1, 2, 3$ and correspond to projections on Lagrangian axes
$\alpha, \beta, \gamma, \delta$	small Greek indices run through values $1, 2$ and correspond to projections on a two-dimensional coordinate frame
g_{ij}, g^{ij}	components of the metric tensor in observer's frame
g_{ab}, g^{ab}	components of the metric tensor in Lagrangian frame
$\lVert a_{ij} \rVert$	matrix with the components, a_{ij}
$\lvert a_{ij} \rvert$	determinant of the matrix with the components a_{ij}
g	determinant of the matrix with the components g_{ij}
$\hat{g}$	determinant of the matrix with the components g_{ab}
$\hat{}$	this symbol marks quantities in Lagrangian coordinates in cases when an ambiguity appears without such a mark; it also marks a maximizer – the particular meaning is seen from the context.

$\circ$	this symbol marks quantities in the initial state
e_{ijk}	symbol Levi-Civita
ε_{ijk}	tensor Levi-Civita
δ_i^j	Kronecker's delta
x_a^i	distortion
X_i^a	inverse distortion
ε_{ab}	components of the strain tensor in Lagrangian coordinates
ε_{ij}	components of the strain tensor in Eulerian coordinates
ε	magnitude of deformation; small parameter
e_{ab}	components of the strain rate tensor in Lagrangian coordinates
e_{ij}	components of the strain rate tensor in Eulerian coordinates
σ^{ab}	components of the stress tensor in Lagrangian coordinates
σ^{ij}	components of the stress tensor in Eulerian coordinates
p_i^a	components of Piola-Kirchhoff's tensor
$\lvert x \rvert_{ab}$	modulus of distortion
α_a^i, α_j^i	orthogonal matrices
$a^{(-1)ij}$	the components of matrix inverse to the matrix $\lVert a_{ij} \rVert$
$T_{\underline{1}\underline{1}}$	physical (11)-component of the tensor T_{ij} (the corresponding indices are underlined)
$a_{(ij)}$	parentheses in indices mean symmetrization: $a_{(ij)} \equiv \frac{1}{2}\left(a_{ij} + a_{ji}\right)$
$a_{[ij]}$	brackets in indices mean antisymmetrization: $a_{[ij]} \equiv \frac{1}{2}\left(a_{ij} - a_{ji}\right)$
$u^{(b)}$	parenthesis for a single index are used to emphasize its non-tensor nature; e.g., $u^{(b)}$ usually denotes the boundary value of u
$b^\lambda_{(\alpha}\gamma_{\lambda\beta)}$	being combined with the contraction, the symmetrization does not act on the dummy index: $b^\lambda_{(\alpha}\gamma_{\lambda\beta)} \equiv \frac{1}{2}\left(b^\lambda_\alpha\gamma_{\lambda\beta} + b^\lambda_\beta\gamma_{\lambda\alpha}\right)$
$(i \to j)$	the expression in the previous parentheses with index i changed by j
$(i \leftrightarrow j)$	the expression in the previous parentheses with the substitution of indices: $i \to j,\ j \to i$
$\varkappa$	multi-index; it denotes a set of indices of various physical nature
$u^\varkappa$	field variables
u, x, X	sometimes we drop indices and write u instead of $u^\varkappa$, x instead of x^i, X instead of X^a.
U, F, K, S	densities of internal energy, free energy, kinetic energy and entropy per unit mass
$\mathcal{U}, \mathcal{F}, \mathcal{K}, \mathcal{S}$	total internal energy, free energy, kinetic energy and entropy of the body

ρ	mass density
$\partial_t \equiv \frac{\partial}{\partial t} \equiv (\cdot)_t = (\cdot)_{,t}$	time derivative at constant Eulerian coordinates, x
$\frac{d}{dt}$	time derivative at constant Lagrangian coordinates, X
$\partial_i \equiv \frac{\partial}{\partial x^i} = (\cdot)_{,i}$	partial space derivative in Eulerian variables
∇_i	covariant space derivatives in Eulerian coordinates
$\partial_a \equiv \frac{\partial}{\partial X^a} = (\cdot)_{,a}$	partial space derivatives in Lagrangian variables
∇_a	Lagrangian covariant space derivatives in the deformed state
$\mathring{\nabla}_a$	Lagrangian covariant space derivatives in the initial state
Δ	the Jacobian of transformation from Lagrangian to Eulerian coordinates
δ	variation at constant X
$\delta(x)$	δ-function; if x is a point in n-dimensional space, then $\delta(x)$ is the product of n one-dimensional δ-functions
$\theta(x)$	the step function: $\theta(x) = 0$ for $x < 0$, $\theta(x) = 1$ for $x \geq 0$
V	usually a region in three-dimensional space
∂V	boundary of region V
$\|V\|$	volume of region V
dV	volume element
Ω	usually a surface
$\|\Omega\|$	area of the surface Ω
Γ	usually a curve
$\|\Gamma\|$	length of curve Γ
$\partial V_f, \partial V_u$	parts of the boundary of elastic body in which one prescribes forces and displacements, respectively
$d^n x = dx^1 \ldots dx^n$	volume element in R_n
$d^3 x$	volume element in R_3 in Cartesian coordinates
dA	area element
I, J	usually functionals
I	usually action functional
$\check{I}$	minimum value of the functional
$\hat{I}$	maximum value of the functional
L	Lagrange function or Lagrangian
ϕ	symbol of empty set
$u \in (1.1)$	function u satisfies the constraint (1.1)
$*$	usually the symbol of Young-Fenchel transformation; complex conjugation in consideration of Fourier transformation
$\times$	symbol of Legendre transformation
$\equiv$	this sign is usually used for definitions

$[\varphi]$ difference of values of function φ on the two sides of the discontinuity surface

$[\varphi]_{t_0}^{t_1}$ difference of values of function φ at the instant t_1 and t_0

Breinigsville, PA USA
02 October 2009

225139BV00004B/3/P

9 783540 884668